The Finite Element Method
Fourth Edition

Volume 1

Basic Formulation and Linear Problems

THE FINITE ELEMENT METHOD

Fourth Edition

Volume 1

Basic Formulation and Linear Problems

O. C. Zienkiewicz, FRS

*Formerly Professor of Civil Engineering and
Head of the Department of Civil Engineering
University of Wales, Swansea*

and

R. L. Taylor

*Professor of Civil Engineering
University of California at Berkeley
Berkeley, California*

McGRAW-HILL BOOK COMPANY

London · New York · St Louis · San Francisco · Auckland
Bogotá · Guatemala · Hamburg · Lisbon · Madrid · Mexico
Montreal · New Delhi · Panama · Paris · San Juan
São Paulo · Singapore · Sydney · Tokyo · Toronto

Published by
McGRAW-HILL Book Company (UK) Limited
MAIDENHEAD · BERKSHIRE · ENGLAND

British Library Cataloguing in Publication Data

Zienkiewicz, O.C. (Olgierd Cecil)
  The finite element method.—4th ed.
  Vol. 1: Basic formulation and linear
  problems
  1. Engineering. Mathematics. Finite
  element methods
  I. Title II. Taylor, R.L.
  620'.001'515353
  ISBN 0-07-084174-8

Library of Congress Cataloging-in-Publication Data

Zienkiewicz, O. C.
  Finite element method.

  Bibliography: p.
  Includes index.
  Contents: v. 1. Basic formulation and linear
  problems.
  1. Structures, Theory of. 2. Continuum mechanics.
3. Finite element method. I. Taylor, R. L.
II. Title.
TA640.2.Z5 1988 620'.001'515353 88-862
ISBN 0-07-084174-8 (v. 1)

Cover illustration by courtesy of Travers Morgan Consulting Group

3 4 5   WL   9 10

Typeset by Santype International Limited, Salisbury, Wiltshire and printed
in Great Britain by Whitstable Litho Printers Limited, Whitstable, Kent

To
Our Families

Contents

Preface

It is just over twenty years since the *The Finite Element Method in Structural and Continuum Mechanics* was first published. This book, which was the first dealing with the finite element method, provided the base from which many further developments occurred. The expanding research and field of application of finite elements led to the second edition in 1971 and the third in 1977. The size of each of these volumes expanded geometrically (from 272 pages in 1967, 521 pages in 1971, to 787 pages in 1977). This was necessary to do justice to a rapidly expanding field of professional application and research. Even so, much filtering of the contents of the third edition was necessary to keep it within reasonable bounds.

As in essence the matters published in the third edition are still valid today and this forms a useful and widely used text and reference book, we have decided to publish an *expanded version* in two volumes. These will retain as far as possible the contents of the third edition and add or reinterpret matters which today have become of added importance.

The division of the contents between the two volumes follows the lines of instruction for which the book can serve either by self study, as we anticipate the book to be used widely by practising engineers, or in university courses for engineers and physicists. The first volume is thus devoted to the basic finite element approximation concepts and to simple linear, static, computations which even today provide the major part of the finite element usage.

We have relegated to the second volume all problems of dynamics, of non-linear solution techniques, and, indeed, the linear problems of plates and shells which introduce special difficulties and where optimal techniques are yet debated.

The contents of the first volume are slightly rearranged from those of the third edition with the first eight chapters dealing with problems of linear elasticity which for many are still the essence of the finite element method application. Chapter 9 generalizes the concepts and shows the position of the method in its application to the solution of all problems governed by appropriate differential equations for boundary value problems. In the early chapters little change from the third edition is made but addition of so-called hierarchic shape function forms (Chapter 7), of infinite elements (Chapter 8) and tensorial notation for those who prefer this form (Chapter 6), amplify the text.

Chapter 10 shows applications to various field problems and again is presented with minimal changes. However, Chapters 11 to 14 introduce new matter which today is the subject of increased activity.

In Chapter 11 we expand on the 'patch test' which is only briefly referred to in the previous editions. It is shown how this can be used rigorously in testing elements derived and, indeed, in their design.

Chapters 12 and 13 present the essence of 'mixed formulations' and show how such formulations can be used effectively in many problems. In particular, such problems as imposition of the incompressibility constraint vital to special structural and fluid mechanics problems are discussed in detail.

Chapter 14 introduces the reader to the new area of error estimation and adaptivity which was developed after the appearance of the third edition and which today is of vital importance in ensuring the reliability of computation.

The finite element method in its application is dependent on the skilful use of computers and efficient programming techniques. The concluding chapter of the book, Chapter 15, includes much of the experience of efficient programming assembled at the University of California at Berkeley, and the Institute of Numerical Methods in Engineering at the University of Wales, Swansea. The program system presented is not only one useful for instruction but can be used effectively for the solution of real problems on a variety of processes ranging from microcomputers to mainframes.†

Although much of the development presented in this book has been contributed by mathematicians we have tried to present matters in a language and manner accessible to practising engineers though endeavouring to preserve the rigour necessary.

OCZ and RLT

---

† Diskettes or tapes containing the complete code can be obtained from R. L. Taylor, Department of Civil Engineering, University of California, Berkeley CA 94720, USA.

Acknowledgements

To many friends in this field all over the world who, sharing the author's enthusiasm, have contributed through discussions and their own researches to many of the ideas here reported.

To my colleagues and research students at Swansea without whose effort this book could not have been written.

To the innumerable sponsoring agencies for supporting students and research.

Special thanks to Peter and Jackie Bettess for their scientific collaboration and for the work they put into making the subject index of this edition more useful to the reader.

Thanks are also due to many engineering industrial partners—be it civil, mechanical or aeronautical—who provided many of the interesting problems. Their need became our fun.

Finally, thanks are due to our wives for their help and forbearance with the numerous occasions when the task of writing, and indeed the involvement with finite elements, became a full-time preoccupation for us.

O. C. Zienkiewicz   R. L. Taylor

List of symbols

Below a list of principal symbols used in this book is presented for easy reference, although all are defined in the text as they occur. On many occasions, additional ones have to be used in a minor context and a *non-uniqueness* arises. It is hoped that appropriate test explanation will avoid confusion.

The symbols are listed roughly in the order of occurrence in chapter sequence.

Matrices and column vectors are denoted by bold symbols, e.g., $\mathbf{K}$ and $\mathbf{a}$ and $\mathbf{K}^T$ stands for transpose of $\mathbf{K}$. Dots are used to denote differentiation with respect to one variable, e.g., $\dfrac{d}{dt} \equiv \dot{\mathbf{a}}$, etc.

Chapter	*Symbol*	
1	$\mathbf{a}_i, \mathbf{a}$	nodal or global displacements; discrete problem parameters
	$\mathbf{q}_i^e$	nodal force at i due to element e
	$\mathbf{K}^e, \mathbf{K}$	stiffness matrix (element/global)
	$\mathbf{f}_{pi}^e$	nodal element force at i due to p, etc.
	$\mathbf{r}_i$	external concentrated nodal force
	σ	stress (vector)
	$\mathbf{L}, \mathbf{T}$	transformation matrices
	$\mathbf{b}$	alternative parameters
2, 3, 4, 5	$\mathbf{u}$	displacement vector (components u, v and w)
	$\hat{\mathbf{u}}$	approximation to $\mathbf{u}$
	ε	strain (vector)
	$\mathbf{S}$	strain operator
	$\mathbf{N}$	(displacement) shape function

$\mathbf{B} = \mathbf{L}\mathbf{N}$	strain shape function
$\mathbf{D}$	elasticity matrix
$\mathbf{b}$	body force (vector)
E	Young's modulus
v	Poisson's ratio
$\varepsilon_0, \boldsymbol{\sigma}_0$	initial strain or stress
$\delta\varepsilon; \delta\mathbf{u}$ etc.,	virtual (or variation of) ε; $\mathbf{k}$ etc.
$\mathbf{t}$	boundary traction
b_x, t_x, etc.,	x-components of body forces and tractions
$\varepsilon_x, \gamma_{xy}, \sigma_x, \tau_{xy}$	x-components of direct and shear strain or stress
U	strain energy
W	potential energy of loads
Π	total potential energy
$\mathbf{I}$	identity matrix
h	representative element dimension
ϕ	body force potential (or other scalar function)
$\boldsymbol{\phi}$	body force potential nodal values
$\mathbf{m}^{\mathrm{T}} = [1, 1, 0]$ or $[1, 1, 1, 0, 0, 0]$	
	matrix equivalent of Kronecker delta for two or three dimensional strain/stress vectors
$x, y, z, x', y', z', r, z, \theta$	
	Cartesian or cylindrical coordinates
θ	temperature change
α	coefficient of thermal expansion
p	internal pore pressure
$a_i\, b_i\, c_i$	constants in a polynomial expansion of function N_i

6	x_a	$a = 1, 2, 3$ cartesian coordinates subscript notation
	u_e	displacement parallel to x_a
	ε_{ab}	strain tensor component
	ω_{ab}	rotation tensor
	$\Lambda_{a'b}$	direction cosine between $x_{a'}$ and x_b
	σ_{ab}	stress tensor component
	$u_{a,b}$	derivative of u_a in direction x_b

7, 8	l_k^n	Lagrange polynomials
	$\xi, \eta, (\zeta)$	element, curvilinear, coordinates in two and three dimensions

	$L_1, L_2, (L_3)$	triangular (area) or tetrahedral (volume) coordinates
	$\mathbf{J}$	Jacobian matrix
	H_i, w_i	quadrature weights
9	Ω	any domain
	Γ	boundary of Ω
	$\mathbf{A(u)}, \mathbf{B(u)}$, etc.	operators defining governing differential equations and boundary conditions
	$\mathbf{u}, \boldsymbol{\phi}, \phi$	unknown function
	$\mathbf{v}$	'test' function
	$\mathbf{a}, \mathbf{b}$, etc.	nodal (or other) parameters defining the trial expansion $\mathbf{u} \simeq \mathbf{Na}$
	$\mathbf{w}_j$	'weight' function
	Π	a stationary functional
	$\mathbf{L}$	a linear differential operator
	$\mathbf{C(u)}$	constraint condition on $\mathbf{u}$
	λ	Lagrangian multiplier
	$\mathbf{n}^{\mathrm{T}} = [n_x, n_y, n_z]$	vector normal to boundary
	α	penalty number
	∇	gradient operator $= \left[\dfrac{\partial}{\partial x}, \dfrac{\partial}{\partial y}, \dfrac{\partial}{\partial z} \right]^{\mathrm{T}}$
	∇^{T}	divergence operator
10	$\mathbf{k}, k$	permeability matrix or coefficient
	$\mathbf{H}$	discretized problem matrix
	p	pressure
	ϕ	potential
	T	temperature
	$\mathbf{H}$	magnetic field strength
	$\mathbf{B}$	magnetic flux density
	$\mathbf{J}$	electric current
12	$\mathbf{A, B, C, Q, M}$	system matrices
	$\boldsymbol{\sigma}^*$	smoothed (projected) stress tensor
	ρ	convergence accelerator
	$\mathbf{K}, \mathbf{G}$	bulk and shear moduli
14	$\mathbf{e}$	local error
	$\mathbf{e}_\sigma$ or $\mathbf{e}_u$	local error in $\boldsymbol{\sigma}$ or $\mathbf{u}$
	$\|e\|; \|u\|$	energy norm of $\mathbf{e}$ or $\mathbf{u}$
	η	relative (percentage) error
	θ	effectivity index

$\hat{\sigma}$	stress; finite element approximation
σ^*	stress; smoothed or projected
$\mathbf{r}$	residual in domain
J	residual or discontinuity jump on interfaces
ζ	refinement indicator

1

Some preliminaries: the standard discrete system

1.1 Introduction

The limitations of the human mind are such that it cannot grasp the behaviour of its complex surroundings and creations in one operation. Thus the process of subdividing all systems into their individual components or 'elements', whose behaviour is readily understood, and then rebuilding the original system from such components to study its behaviour is a natural way in which the engineer, the scientist, or even the economist proceeds.

In many situations an adequate model is obtained using a finite number of well-defined components. Such problems we shall term *discrete*. In others the subdivision is continued indefinitely and the problem can only be defined using the mathematical fiction of an infinitesimal. This leads to differential equations or equivalent statements which imply an infinite number of elements. Such systems we shall term *continuous*.

With the advent of digital computers, *discrete* problems can generally be solved readily even if the number of elements is very large. As the capacity of all computers is finite, *continuous* problems can only be solved exactly by mathematical manipulation. Here, the available mathematical techniques usually limit the possibilities to oversimplified situations.

To overcome the intractability of the realistic type of continuum problem, various methods of *discretization* have from time to time been proposed both by engineers and mathematicians. All involve an *approximation* which, hopefully, is of such a kind that it approaches, as closely as desired, the true continuum solution as the number of discrete variables increases.

1

The discretization of continuous problems has been approached differently by mathematicians and engineers. The first have developed general techniques applicable directly to differential equations governing the problem, such as finite difference approximations,[1,2] various weighted residual procedures,[3,4] or approximate techniques of determining the stationarity of properly defined 'functionals'. The engineer, on the other hand, often approaches the problem more intuitively by creating an analogy between real discrete elements and finite portions of a continuum domain. For instance, in the field of solid mechanics McHenry,[5] Hrenikoff,[6] and Newmark[7] have, in the early forties, shown that reasonably good solutions to an elastic continuum problem can be obtained by substituting small portions of the continuum by an arrangement of simple elastic bars. Later, in the same context, Argyris[8] and Turner *et al.*[9] showed that a more direct, but no less intuitive, substitution of properties can be made much more directly by considering that small portions or 'elements' in a continuum behave in a simplified manner.

It is from the engineering 'direct analogy' view that the term 'finite element' has been born. Clough[10] appears to be the first to use this term, which implies in it a direct use of *standard methodology applicable to discrete systems*. Both conceptually and from the computational viewpoint, this is of the utmost importance. The first allows an improved understanding to be obtained; the second the use of a unified approach to the variety of problems and the development of standard computational procedures.

Since the early sixties much progress has been made, and today the purely mathematical and 'analogy' approaches are fully reconciled. It is the object of this text to present a view of the finite element method as *a general discretization procedure of continuum problems posed by mathematically defined statements*.

In the analysis of problems of a discrete nature, a standard methodology has been developed over the years. The civil engineer, dealing with structures, first calculates force–displacement relationships for each element of the structure and then proceeds to assemble the whole by following a well-defined procedure of establishing local equilibrium at each 'node' or connecting point of the structure. From such equations the solution of the unknown displacements becomes possible. Similarly, the electrical or hydraulic engineer, dealing with a network of electrical components (resistors, capacitances, etc.) or hydraulic conduits, first establishes a relationship between currents (flows) and potentials for individual elements and then proceeds to assemble the system by ensuring continuity of flows.

All such analyses follow a standard pattern which is universally adaptable to discrete systems. It is thus possible to define a *standard discrete system*, and this chapter will be primarily concerned with establishing the processes applicable to such systems. Much of what is presented here will be known to engineers, but some reiteration is at this stage advisable. As the treatment of elastic, solid structures has been the most developed area of activity this will be introduced first, followed by examples from other fields, before attempting a complete generalization.

The existence of a unified treatment of 'standard discrete problems' leads us to the first definition of the finite element process as a method of approximation to continuum problems such that

(a) the continuum is divided into a finite number of parts (elements), the behaviour of which is specified by a finite number of parameters, and

(b) the solution of the complete system as an assembly of its elements follows precisely the same rules as those applicable to *standard discrete problems*.

It will be found that most classical mathematical procedures of approximation fall into this category—as well as the various direct approximations used in engineering. It is thus difficult to determine the origins of the finite element method and the precise moment of its invention.

Table 1.1 shows the process of evolution which led to the present-day concepts of finite element analysis. Chapter 9 will give, in more detail, the mathematical basis which evolved from the classical landmarks.[11-20]

1.2 The structural element and system

To introduce the reader to the general concept of the discrete system we shall first consider a structural engineering example of linear elasticity.

Let Fig. 1.1 represent a two-dimensional structure assembled from individual components and interconnected at the nodes numbered 1 to *n*. The joints at the nodes, in this case, are pinned so that moments cannot be transmitted.

As a starting point it will be assumed that by separate calculation, or for that matter from the results of an experiment, the characteristics of each element are precisely known. Thus, if a typical element labelled (1) and associated with nodes 1, 2, 3 is examined, the forces acting at the nodes are uniquely defined by the displacements of these nodes, the distributed loading acting on the element (p), and its initial strain. The last may be due to temperature, shrinkage, or simply an initial 'lack of fit'. The forces and the corresponding displacements are defined by appropriate components (U, V and u, v) in a common coordinate system.

Listing the forces acting on all the nodes (three in the case illustrated)

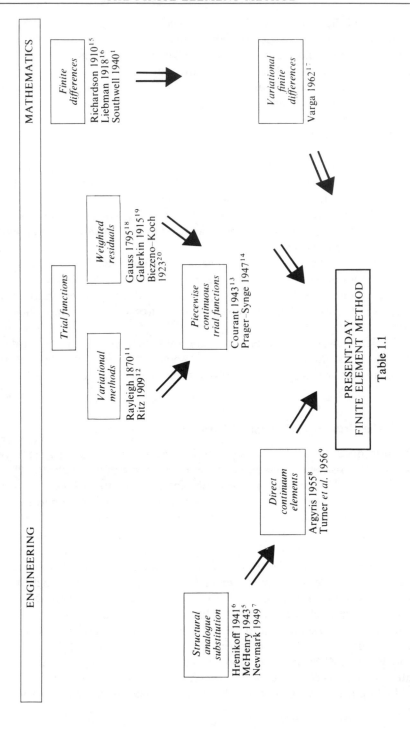

MATHEMATICS

Finite differences

Richardson 1910[15]
Liebman 1918[16]
Southwell 1940[1]

Variational finite differences

Varga 1962[17]

Weighted residuals

Gauss 1795[18]
Galerkin 1915[19]
Biezeno–Koch 1923[20]

Trial functions

Variational methods

Rayleigh 1870[11]
Ritz 1909[12]

Piecewise continuous trial functions

Courant 1943[13]
Prager–Synge 1947[14]

ENGINEERING

Direct continuum elements

Argyris 1955[8]
Turner *et al.* 1956[9]

Structural analogue substitution

Hrenikoff 1941[6]
McHenry 1943[5]
Newmark 1949[7]

PRESENT-DAY
FINITE ELEMENT METHOD

Table 1.1

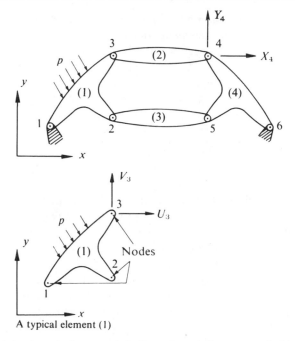

Fig. 1.1   A typical structure built up from interconnected elements

of the element (1) as a matrix† we have

$$\mathbf{q}^1 = \begin{Bmatrix} \mathbf{q}_1^1 \\ \mathbf{q}_2^1 \\ \mathbf{q}_3^1 \end{Bmatrix} \qquad \mathbf{q}_1^1 = \begin{Bmatrix} U_1 \\ V_1 \end{Bmatrix}, \qquad \text{etc.} \tag{1.1}$$

and for the corresponding nodal displacements

$$\mathbf{a}^1 = \begin{Bmatrix} \mathbf{a}_1^1 \\ \mathbf{a}_2^1 \\ \mathbf{a}_3^1 \end{Bmatrix} \qquad \mathbf{a}_1^1 = \begin{Bmatrix} u_1 \\ v_1 \end{Bmatrix}, \qquad \text{etc.} \tag{1.2}$$

Assuming linear elastic behaviour of the element, the characteristic relationship will always be of the form

$$\mathbf{q}^1 = \mathbf{K}^1 \mathbf{a}^1 + \mathbf{f}_p^1 + \mathbf{f}_{\varepsilon 0}^1 \tag{1.3}$$

† A limited knowledge of matrix algebra will be assumed throughout this book. This is necessary for reasonable conciseness and forms a convenient book-keeping form. For readers not familiar with the subject a brief appendix (Appendix 1) is included in which sufficient principles of matrix algebra are given to follow intelligently the development. Matrices (and vectors) will be distinguished by bold print throughout.

in which $\mathbf{f}_p^1$ represents the nodal forces required to balance any distributed loads acting on the element and $\mathbf{f}_{\varepsilon_0}^1$ the nodal forces required to balance any initial strains such as may be caused by temperature change if the nodes are not subject to any displacement. The first of the terms represents the forces induced by displacement of the nodes.

Similarly, the preliminary analysis or experiment will permit a unique definition of stresses or internal reactions at any specified point or points of the element in terms of the nodal displacements. Defining such stresses by a matrix $\boldsymbol{\sigma}^1$ a relationship of the form

$$\boldsymbol{\sigma}^1 = \mathbf{S}^1 \mathbf{a}^1 + \boldsymbol{\sigma}_p^1 + \boldsymbol{\sigma}_{\varepsilon_0}^1 \tag{1.4}$$

is obtained in which the last two terms are simply the stresses due to the distributed element loads or initial stresses respectively when no nodal displacement occurs.

The matrix $\mathbf{K}^e$ is known as the element stiffness matrix and the matrix $\mathbf{S}^e$ as the element stress matrix for an element (e).

Relationships Eqs (1.3) and (1.4) have been illustrated on an example of an element with three nodes and with the interconnection points capable of transmitting only two components of force. Clearly, the same arguments and definitions will apply generally. An element (2) of the hypothetical structure will possess only two points of interconnection; others may have quite a large number of such points. Similarly, if the joints were considered as rigid, three components of generalized force and of generalized displacement would have to be considered, the last corresponding to a moment and a rotation respectively. For a rigidly jointed, three-dimensional structure the number of individual nodal components would be six. Quite generally, therefore,

$$\mathbf{q}^e = \begin{Bmatrix} \mathbf{q}_1^e \\ \mathbf{q}_2^e \\ \vdots \\ \mathbf{q}_m^e \end{Bmatrix} \quad \text{and} \quad \mathbf{a}^e = \begin{Bmatrix} \mathbf{a}_1 \\ \mathbf{a}_2 \\ \vdots \\ \mathbf{a}_m \end{Bmatrix} \tag{1.5}$$

with each $\mathbf{q}_i$ and $\mathbf{a}_i$ possessing the same number of components or *degrees of freedom*. These quantities are conjugate to each other.

The stiffness matrices of the element will clearly always be square and of the form

$$\mathbf{K}^e = \begin{bmatrix} \mathbf{K}_{ii}^e & \mathbf{K}_{ij}^e & \cdots & \mathbf{K}_{im}^e \\ \vdots & \vdots & & \vdots \\ \mathbf{K}_{mi}^e & \cdots & \cdots & \mathbf{K}_{mm}^e \end{bmatrix} \tag{1.6}$$

in which $\mathbf{K}_{ii}^e$, etc., are submatrices which are again square and of the size

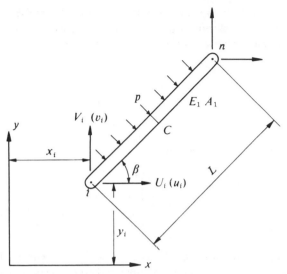

Fig. 1.2 A pin-ended bar

$l \times l$, where l is the number of force components to be considered at the nodes.

As an example, the reader can consider a pin-ended bar of a uniform section A and modulus E in a two-dimensional problem shown in Fig. 1.2. The bar is subject to a uniform lateral load p and a uniform thermal expansion strain

$$\varepsilon_0 = \alpha T$$

If the ends of the bar are defined by the coordinates x_i, y_i and x_n, y_n its length can be calculated as

$$L = \sqrt{[(x_n - x_i)^2 + (y_n - y_i)^2]}$$

and its inclination from the horizontal as

$$\beta = \tan^{-1} \frac{y_n - y_i}{x_n - x_i}$$

Only two components of force and displacement have to be considered at the nodes.

The nodal forces due to the lateral load are clearly

$$\mathbf{f}_p^e = \begin{Bmatrix} U_i \\ V_i \\ U_n \\ V_n \end{Bmatrix}_p = - \begin{Bmatrix} -\sin \beta \\ \cos \beta \\ -\sin \beta \\ \cos \beta \end{Bmatrix} \frac{pL}{2}$$

and represent the appropriate components of simple beam reactions, $pL/2$. Similarly, to restrain the thermal expansion ε_0 an axial force $(E\alpha TA)$ is needed, which gives the components

$$\mathbf{f}^e_{\varepsilon 0} = \begin{Bmatrix} U_i \\ V_i \\ U_n \\ V_n \end{Bmatrix}_{\varepsilon 0} = - \begin{Bmatrix} -\cos\beta \\ -\sin\beta \\ \cos\beta \\ \sin\beta \end{Bmatrix} (E\alpha TA)$$

Finally, the element displacements

$$\mathbf{a}^e = \begin{Bmatrix} u_i \\ v_i \\ u_n \\ v_n \end{Bmatrix}$$

will cause an elongation $(u_n - u_i)\cos\beta + (v_n - v_i)\sin\beta$. This, when multiplied by EA/L, gives the axial force whose components can again be found. Rearranging these in the standard form gives

$$\mathbf{K}^e\mathbf{a}^e = \begin{Bmatrix} U_i \\ V_i \\ U_n \\ V_n \end{Bmatrix}_\delta$$

$$= \frac{EA}{L} \begin{bmatrix} \cos^2\beta & \sin\beta\cos\beta & \vdots & -\cos^2\beta & -\sin\beta\cos\beta \\ \sin\beta\cos\beta & \sin^2\beta & \vdots & -\sin\beta\cos\beta & -\sin^2\beta \\ \cdots & \cdots & \vdots & \cdots & \cdots \\ -\cos^2\beta & -\sin\beta\cos\beta & \vdots & \cos^2\beta & \sin\beta\cos\beta \\ -\sin\beta\cos\beta & -\sin^2\beta & \vdots & \sin\beta\cos\beta & \sin^2\beta \end{bmatrix} \begin{Bmatrix} u_i \\ v_i \\ u_n \\ v_n \end{Bmatrix}$$

The components of the general Eq. (1.3) have thus been established for the elementary case discussed. It is again quite simple to find the stresses at any section of the element in the form of relation Eq. (1.4). For instance, if attention is focused on the mid-section C of the beam the extreme fibre stresses determined from the axial tension to the element and the bending moment can be shown to be

$$\boldsymbol{\sigma}^e_C = \begin{Bmatrix} \sigma_1 \\ \sigma_2 \end{Bmatrix}_C = \frac{E}{L} \begin{bmatrix} -\cos\beta, & -\sin\beta, & \cos\beta, & \sin\beta \\ -\cos\beta, & -\sin\beta, & \cos\beta, & \sin\beta \end{bmatrix} \mathbf{a}^e$$

$$+ \begin{Bmatrix} 1 \\ -1 \end{Bmatrix} \frac{pL^2}{8} \frac{d}{I} - \begin{Bmatrix} 1 \\ 1 \end{Bmatrix} E\alpha T$$

in which d is the half-depth of the section and I its second moment of area. All the terms of Eq. (1.4) can now be easily recognized.

For more complex elements more sophisticated procedures of analysis are required but the results are of the same form. The engineer will readily recognize that the so-called 'slope–deflection' relations used in analysis of rigid frames are only a special case of the general relations.

It may perhaps be remarked, in passing, that the complete stiffness matrix obtained for the simple element in tension turned out to be symmetric (as indeed was the case with some submatrices). This is by no means fortuitous but follows from the principle of energy conservation and from its corollary—the well-known Maxwell–Betti reciprocal theorem.

The element properties were assumed to follow a simple linear relationship. In principle, similar relationships could be established for non-linear materials, but discussion of such problems will be held over at this stage.

1.3 Assembly and analysis of a structure

Consider again the hypothetical structure of Fig. 1.1. To obtain a complete solution the two conditions of
(a) displacement compatibility and
(b) equilibrium
have to be satisfied throughout.

Any system of nodal displacements $\mathbf{a}$:

$$\mathbf{a} = \left\{ \begin{array}{c} \mathbf{a}_1 \\ \vdots \\ \mathbf{a}_n \end{array} \right\} \tag{1.7}$$

listed now for the whole structure in which all the elements participate, automatically satisfies the first condition.

As the conditions of overall equilibrium have already been satisfied *within* an element all that is necessary is to establish equilibrium conditions at the nodes of the structure. The resulting equations will contain the displacements as unknowns, and once these have been solved the structural problem is determined. The internal forces in elements, or the stresses, can easily be found by using the characteristics established *a priori* for each element by Eq. (1.4).

Consider the structure to be loaded by external forces $\mathbf{r}$:

$$\mathbf{r} = \left\{ \begin{array}{c} \mathbf{r}_1 \\ \vdots \\ \mathbf{r}_n \end{array} \right\} \tag{1.8}$$

applied at the nodes in addition to the distributed loads applied to the individual elements. Again, any one of the forces r_i must have the same number of components as that of the element reactions considered. In the example in question

$$r_i = \begin{Bmatrix} X_i \\ Y_i \end{Bmatrix} \tag{1.9}$$

as the joints were assumed pinned, but at this stage a generality with an arbitrary number of components will be assumed.

If now the equilibrium conditions of a typical node, i, are to be established, each component of r_i has, in turn, to be equated to the sum of the component forces contributed by the elements meeting at the node. Thus, considering *all* the force components we have

$$r_i = \sum_{e=1}^{m} q_i^e = q_i^1 + q_i^2 + \cdots \tag{1.10}$$

in which q_i^1 is the force contributed to node i by element 1, q_i^2 by element 2, etc. Clearly, only the elements which include point i will contribute non-zero forces, but for tidiness all the elements are included in the summation.

Substituting from the definition (1.3) the forces contributed to node i and noting that nodal variables a_i are common (thus omitting the superscript e), we have

$$r_i = \left(\sum_{e=1}^{m} K_{i1}^e \right) a_1 + \left(\sum_{e=1}^{m} K_{i2}^e \right) a_2 + \cdots + \sum_{e=1}^{m} f_i^e \tag{1.11}$$

where

$$f^e = f_p^e + f_{\varepsilon 0}^e$$

The summation again only concerns the elements which contribute to node i. If all such equations are assembled we have simply

$$Ka = r - f \tag{1.12}$$

in which the submatrices are

$$K_{ij} = \sum_{e=1}^{m} K_{ij}^e$$

$$f_i = \sum_{e=1}^{m} f_i^e \tag{1.13}$$

with summations including all elements. This simple rule for assembly is very convenient because as soon as a coefficient for a particular element is found it can be put immediately into the appropriate 'location' specified in the computer. *This general assembly process can be found to be the common and fundamental feature of all finite element calculations and should be well understood by the reader.*

If different types of structural elements are used and are to be coupled it must be remembered that the rules of matrix summation permit this to be done only if these are of identical size. The individual submatrices to be added have therefore to be built up of the same number of individual components of force or displacement. Thus, for example, if a member capable of transmitting moments to a node is to be coupled at that node to one which in fact is hinged, it is necessary to complete the stiffness matrix of the latter by insertion of appropriate (zero) coefficients in the rotation or moment positions.

1.4 The boundary conditions

The system of equations resulting from Eq. (1.12) can be solved once the prescribed support displacements have been substituted. In the example of Fig. 1.1, where both components of displacement of nodes 1 and 6 are zero, this will mean the substitution of

$$\mathbf{a}_1 = \mathbf{a}_6 = \begin{Bmatrix} 0 \\ 0 \end{Bmatrix}$$

which is equivalent to reducing the number of equilibrium equations (in this instance 12) by deleting the first and last pairs and thus reducing the total number of unknown displacement components to eight. It is, nevertheless, always convenient to assemble the equation according to relation Eq. (1.12) so as to include all the nodes.

Clearly, without substitution of a minimum number of prescribed displacements to prevent rigid body movements of the structure, it is impossible to solve this system, because the displacements cannot be uniquely determined by the forces in such a situation. This physically obvious fact will mathematically be interpreted in the matrix **K** being singular, i.e., not possessing an inverse. The prescription of appropriate displacements after the assembly stage will permit a unique solution to be obtained by deleting appropriate rows and columns of the various matrices.

If all the equations of a system are assembled, their form is

$$\mathbf{K}_{11}\mathbf{a}_1 + \mathbf{K}_{12}\mathbf{a}_2 + \cdots = \mathbf{r}_1 - \mathbf{f}_1$$
$$\mathbf{K}_{21}\mathbf{a}_1 + \mathbf{K}_{22}\mathbf{a}_2 + \cdots = \mathbf{r}_2 - \mathbf{f}_2 \qquad (1.14)$$

etc.

and it will be noted that if any displacement, such as $\mathbf{a}_1 = \bar{\mathbf{a}}_1$, is prescribed then the external 'force' $\mathbf{r}_1$ cannot be prescribed and remains unknown. The first equation could then be *deleted* and substitution of known values of $\mathbf{a}_1$ made in the remaining equations. This process is computationally cumbersome and the same objective is served by adding a large number, $\alpha\mathbf{I}$, to the coefficient $\mathbf{K}_{11}$ and replacing the right-hand side, $\mathbf{r}_1 - \mathbf{f}_1$, by $\bar{\mathbf{a}}_1\alpha$. If α is very much larger than other stiffness coefficients this alteration effectively replaces the first equation by the equation

$$\alpha\mathbf{a}_1 = \alpha\bar{\mathbf{a}}_1 \qquad (1.15)$$

that is, the required prescribed condition, but the whole system remains symmetric and minimal changes are necessary in the computation sequence. A similar procedure will apply to any other prescribed displacement. The above artifice has been introduced by Payne and Irons.[21] An alternative procedure avoiding the assembly of equations corresponding to nodes with prescribed boundary values will be presented in Chapter 15.

When all the boundary conditions are inserted the equations of the system can be solved for the unknown displacements and stresses, and internal forces in each element obtained.

1.5 Electrical and fluid networks

Identical principles of deriving element characteristics and of assembly will be found in many non-structural fields. Consider, for instance, an assembly of electrical resistances shown in Fig. 1.3.

If a typical resistance element, *ij*, is isolated from the system we can write by Ohm's law the relation between the currents *entering* the element at the ends and the end voltages as

$$J_i^e = \frac{1}{r^e}(V_i - V_j)$$

$$J_j^e = \frac{1}{r^e}(V_j - V_i)$$

or in matrix form

$$\begin{Bmatrix} J_i^e \\ J_j^e \end{Bmatrix} = \frac{1}{r^e}\begin{bmatrix} 1 & -1 \\ -1 & 1 \end{bmatrix}\begin{Bmatrix} V_i \\ V_j \end{Bmatrix}$$

which in our standard form is simply

$$\mathbf{J}^e = \mathbf{K}^e\mathbf{V}^e \qquad (1.16)$$

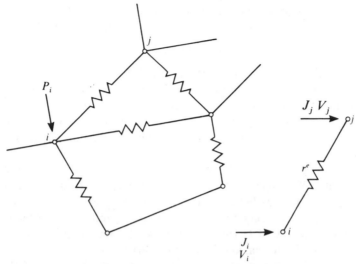

Fig. 1.3   A network of electrical resistances

This form clearly corresponds to the stiffness relationship (1.3); indeed if external current were supplied along the length of the element the element 'force' terms could also be found.

To assemble the whole network the continuity of potential at nodes is assumed and a current balance imposed there. If P_i now stands for an external input of current at node i we must have, with complete analogy to Eq. (1.11),

$$P_i = \sum_{j=1}^{j=m} \sum_{e=1}^{m} K_{ij}^e V_j \tag{1.17}$$

where the second summation is over all 'elements', and once again for all the nodes

$$\mathbf{P = KV} \tag{1.18}$$

in which

$$K_{ij} = \sum_{e=1}^{m} K_{ij}^e$$

Matrix notation in the above has been dropped as the quantities such as voltage and current and hence also the coefficients of the 'stiffness' matrix are scalars.

If the resistances were replaced by fluid-carrying pipes in which a laminar regime pertained, an identical formulation would once again result with V standing for the hydraulic head and J for flow.

For pipe networks usually encountered, however, the linear laws are in general not valid. Typically the flow–head relationship is of a form

$$J_i = c(V_i - V_j)^\gamma \qquad (1.19)$$

where the index γ lies between 0.5 and 0.7. Even now it would still be possible to write relationships in the form (1.16) noting, however, that the matrices $\mathbf{K}^e$ are no longer an array of constants but are known functions of $\mathbf{V}$. The final equations can once again be assembled but their form will be non-linear and in general iterative techniques of solution will be needed.

Finally it is perhaps of interest to mention the more general form of an electric network subject to an alternating current. It is customary to write the relationships between the current and voltage in *complex form* with the resistance being replaced by complex impedance. Once again the standard forms of (1.16) to (1.18) will be obtained but with each quantity divided into real and imaginary parts.

Identical solution procedures can be used if the equality of the real and imaginary quantities is considered at each stage. Indeed with modern digital computers it is possible to use standard programming making use of facilities available for dealing with complex numbers. Reference to some problems of this class will be made in a chapter dealing with vibration problems in the second volume.

1.6 The general pattern

To consolidate the concepts discussed in this chapter an example will be considered. This is shown in Fig. 1.4(a) where five discrete elements are interconnected. These may be of structural, electrical, or any other linear type. In the solution:

The first step is the determination of element properties from the geometric material and loading data. For each element the 'stiffness matrix' as well as the corresponding 'nodal loads' are found in the form of Eq. (1.3). Each element has its own identifying number and specified nodal connection. For example:

element	connection			
1	1	3	4	
2	1	4	2	
3	2	5		
4	3	6	7	4
5	4	7	8	5

Assuming that properties are found in global coordinates we can enter each 'stiffness' or 'force' component in its position of the global

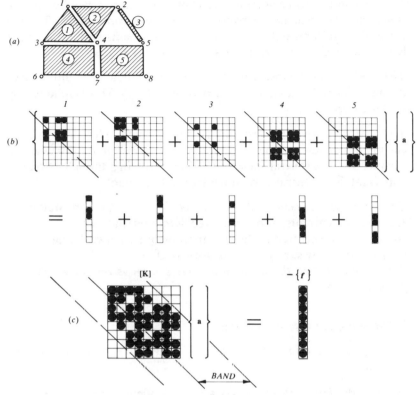

Fig. 1.4   The general pattern

matrix as shown in Fig. 1.4(*b*). Each shaded square represents a single coefficient or a submatrix of type $\mathbf{K}_{ij}$ if more than one quantity is being considered at the nodes. Here, for each element, its separate contribution is shown and the reader can verify the position of the coefficients. Note that various types of 'elements' considered here present no difficulty in specification. (All 'forces', including nodal ones, are here associated with elements for simplicity.)

The second step is the assembly of the final equations of the type given by Eq. (1.12). This is simply accomplished according to the rule of Eq. (1.13) by *simple addition* of all numbers in the appropriate space of the global matrix. The result is shown in Fig. 1.4(*c*) where the non-zero coefficients are indicated by shading.

As the matrices are symmetric only the half above the diagonal shown needs, in fact, to be found.

All the non-zero coefficients are confined within a *band* or *profile* which can be calculated *a priori* for the nodal connections. Thus in computer programs only the storage of the elements within the upper half of the profile is necessary, as shown in Fig. 1.4(*c*).

The third step is the insertion of prescribed boundary conditions into the final assembled matrix, as discussed in Sec. 1.3. This is followed by the final step.

The final step solves the resulting equation system. Here many different methods can be employed, some of which will be discussed in Chapter 15. The general subject of equation solving, though extremely important, is in general beyond the scope of this book.

The final step discussed above can be followed by substitution to obtain stresses, currents, or other desired *output* quantities.

All operations involved in the structural or other network analysis are thus of an extremely simple and repetitive kind.

We can now define *the standard discrete system* as one in which such conditions prevail.

1.7 The standard discrete system

In the *standard discrete system*, whether it is structural or any other kind, we find that:

1. A set of discrete parameters, say a_i, can be identified which describes simultaneously the behaviour of each element, e, and of the whole system. We shall call these the *system parameters*.
2. For each element a set of quantities q_i^e can be computed in terms of the system parameters a_i. The general function relationship can be non-linear

$$q_i^e = q_i^e(a) \tag{1.20}$$

but in many cases a linear form exists giving

$$q_i^e = K_{i1}^e a_1 + K_{i2}^e a_2 + \cdots + f_i^e \tag{1.21}$$

3. The *system equations* are obtained by a simple addition

$$r_i = \sum_{e=1}^{m} q_i^e \tag{1.22}$$

where r_i are system quantities (often prescribed as zero).

In the linear case this results in a system of equations

$$Ka + f = r \tag{1.23}$$

such that

$$\mathbf{K}_{ij} = \sum_{e=1}^{m} \mathbf{K}_{ij}^{e} \qquad \mathbf{f}_{i} = \sum_{e=1}^{m} \mathbf{f}_{i}^{e} \tag{1.24}$$

from which the solution for the system variables **a** can be found after imposing necessary boundary conditions.

The reader will observe that this definition includes the structural, hydraulic, and electrical examples already discussed. However, it is broader. In general neither linearity nor symmetry of matrices must exist—although in many problems this will arise naturally. Further, the narrowness of interconnections existing in usual elements is not essential.

While much further detail could be discussed (we refer the reader to specific books for more exhaustive studies in the structural context[22-24]), we feel that the general exposé given here should suffice for further study of this book.

Only one further matter relating to the change of discrete parameters need be mentioned here. The process of so-called transformation of coordinates is vital in many contexts and must be fully understood.

1.8 Transformation of coordinates

It is often convenient to establish the characteristics of an individual element in a coordinate system which is different from that in which the external forces and displacements of the assembled structure or system will be measured. A different coordinate system may, in fact, be used for every element, to ease the computation. It is a simple matter to transform the coordinates of the displacement and force components of Eq. (1.3) to any other coordinate system. Clearly, it is necessary to do so before an assembly of the structure can be attempted.

Let the local coordinate system in which the element properties have been evaluated be denoted by the prime suffix and the common coordinate system necessary for assembly be not annotated. The displacement components can be transformed by a suitable matrix of direction cosines **L** as

$$\mathbf{a}' = \mathbf{L}\mathbf{a} \tag{1.25}$$

As the corresponding force components must perform the same amount of work in either system†

$$\mathbf{q}^{T}\mathbf{a} = \mathbf{q}'^{T}\mathbf{a}' \tag{1.26}$$

† With ( )^{T} standing for transpose of the matrix.

On inserting (1.25) we have

$$\mathbf{q}^T\mathbf{a} = \mathbf{q'}^T\mathbf{La}$$

or

$$\mathbf{q} = \mathbf{L}^T\mathbf{q'} \tag{1.27}$$

The set of transformations given by (1.25) and (1.27) is called *contravariant*.

To transform 'stiffnesses' which may be available in local coordinates to global ones note that if we write

$$\mathbf{q'} = \mathbf{K'a'} \tag{1.28}$$

then by (1.27), (1.28), and (1.25)

$$\mathbf{q} = \mathbf{L}^T\mathbf{K'La}$$

or in global coordinates

$$\mathbf{K} = \mathbf{L}^T\mathbf{K'L} \tag{1.29}$$

The reader can verify the usefulness of the above transformations by re-working the sample example of the pin-ended bar, first establishing its stiffness in its length coordinates.

In many complex problems an external constraint of some kind may be imagined enforcing the requirement (1.25) with the number of degrees of freedom of $\mathbf{a}$ and $\mathbf{a'}$ being quite different. Even in such instances the relations (1.26) and (1.27) continue to be valid.

An alternative and more general argument can be applied to many other situations of discrete analysis. We wish to replace a set of parameters $\mathbf{a}$ in which the system equations have been written by another one related to it by a transformation matrix $\mathbf{T}$ as

$$\mathbf{a} = \mathbf{Tb} \tag{1.30}$$

In a linear case the system equations are of the form

$$\mathbf{Ka} = \mathbf{r} - \mathbf{f} \tag{1.31}$$

and on substitution we have

$$\mathbf{KTb} = \mathbf{r} - \mathbf{f} \tag{1.32}$$

The new system can be premultiplied simply by $\mathbf{T}^T$, yielding

$$(\mathbf{T}^T\mathbf{KT})\mathbf{b} = \mathbf{T}^T\mathbf{r} - \mathbf{T}^T\mathbf{f} \tag{1.33}$$

which will preserve the symmetry of equations if matrix $\mathbf{K}$ is symmetric. However, occasionally the matrix $\mathbf{T}$ is not square and expression (1.30) represents in fact *an approximation* in which a larger number of parameters **a** is *constrained*. Clearly the system of Eq. (1.32) gives more equations than necessary for a solution of the reduced set of parameters **b**, and the final expression (1.33) presents a reduced system which in some sense approximates to the original one.

We have thus introduced the basic idea of approximation, which will be the subject of subsequent chapters where infinite sets of quantities are reduced to finite sets.

References

1. R. V. SOUTHWELL, *Relaxation Methods in Theoretical Physics*, Clarendon Press, 1946.
2. D. N. DE G. ALLEN, *Relaxation Methods*, McGraw-Hill, 1955.
3. S. H. CRANDALL, *Engineering Analysis*, McGraw-Hill, 1956.
4. B. A. FINLAYSON, *The Method of Weighted Residuals and Variational Principles*, Academic Press, 1972.
5. D. MCHENRY, 'A lattice analogy for the solution of plane stress problems', *J. Inst. Civ. Eng.*, **21**, 59–82, 1943.
6. A. HRENIKOFF, 'Solution of problems in elasticity by the framework method', *J. Appl. Mech.*, **A8**, 169–75, 1941.
7. N. M. NEWMARK, 'Numerical methods of analysis in bars, plates and elastic bodies', in *Numerical Methods in Analysis in Engineering* (ed. L. E. Grinter), Macmillan, 1949.
8. J. H. ARGYRIS, *Energy Theorems and Structural Analysis*, Butterworth, 1960 (reprinted from *Aircraft Eng.*, 1954–5).
9. M. J. TURNER, R. W. CLOUGH, H. C. MARTIN, and L. J. TOPP, 'Stiffness and deflection analysis of complex structures', *J. Aero. Sci.*, **23**, 805–23, 1956.
10. R. W. CLOUGH, 'The finite element in plane stress analysis', *Proc. 2nd ASCE Conf. on Electronic Computation*, Pittsburgh, Pa., Sept. 1960.
11. LORD RAYLEIGH (J. W. STRUTT), 'On the theory of resonance', *Trans. Roy. Soc. (London)*, **A161**, 77–118, 1870.
12. W. RITZ, 'Über eine neue Methode zur Lösung gewissen Variations— Probleme der mathematischen Physik', *J. Reine Angew. Math.*, **135**, 1–61, 1909.
13. R. COURANT, 'Variational methods for the solution of problems of equilibrium and vibration', *Bull. Am. Math. Soc.*, **49**, 1–23, 1943.
14. W. PRAGER and J. L. SYNGE, 'Approximation in elasticity based on the concept of function space', *Q. J. Appl. Math.*, **5**, 241–69, 1947.
15. L. F. RICHARDSON, 'The approximate arithmetical solution by finite differences of physical problems', *Trans. Roy. Soc. (London)*, **A210**, 307–57, 1910.
16. H. LIEBMAN, 'Die angenäherte Ermittlung: harmonischen, functionen und konformer Abbildung', *Sitzber. Math. Physik Kl. Bayer Akad. Wiss. München*, **3**, 65–75, 1918.
17. R. S. VARGA, *Matrix Iterative Analysis*, Prentice-Hall, 1962.
18. C. F. GAUSS, See *Carl Friedrich Gauss Werks*, Vol. VII, Göttingen, 1871.
19. B. G. GALERKIN, 'Series solution of some problems of elastic equilibrium of rods and plates' (Russian), *Vestn. Inzh. Tech.*, **19**, 897–908, 1915.
20. C. B. BIEZENO and J. J. KOCH, 'Over een Nieuwe Methode ter Berekening van Vlokke Platen', *Ing. Grav.*, **38**, 25–36, 1923.

21. N. A. PAYNE and B. M. IRONS, Private communication, 1963.
22. R. K. LIVESLEY, *Matrix Methods in Structural Analysis*, 2nd ed., Pergamon Press, 1975.
23. J. S. PRZEMIENIECKI, *Theory of Matrix Structural Analysis*, McGraw-Hill, 1968.
24. H. C. MARTIN, *Introduction to Matrix Methods of Structural Analysis*, McGraw-Hill, 1966.

Finite elements of an elastic continuum—displacement approach

2.1 Introduction

The process of approximating the behaviour of a continuum by 'finite elements' which behave in a manner similar to the real, 'discrete', elements described in the previous chapter can be introduced through the medium of particular physical applications or as a general mathematical concept. We have chosen here to follow the first path, narrowing our view to a set of problems associated with structural mechanics which historically were the first to which the finite element method was applied. In Chapter 9 we shall generalize the concepts and show that the basic ideas are widely applicable.

In many phases of engineering the solution of stress and strain distributions in elastic continua is required. Special cases of such problems may range from two-dimensional plane stress or strain distributions, axisymmetric solids, plate bending, and shells, to fully three-dimensional solids. In all cases the number of interconnections between any 'finite element' isolated by some imaginary boundaries and the neighbouring elements is infinite. It is therefore difficult to see at first glance how such problems may be discretized in the same manner as was described in the preceding chapter for simpler structures. The difficulty can be overcome (and the approximation made) in the following manner:

1. The continuum is separated by imaginary lines or surfaces into a number of 'finite elements'.

2. The elements are assumed to be interconnected at a discrete number of nodal points situated on their boundaries. The displacements of these nodal points will be the basic unknown parameters of the problem, just as in the simple, discrete, structural analysis.
3. A set of functions is chosen to define uniquely the state of displacement within each 'finite element' in terms of its nodal displacements.
4. The displacement functions now define uniquely the state of strain within an element in terms of the nodal displacements. These strains, together with any initial strains and the constitutive properties of the material, will define the state of stress throughout the element and, hence, also on its boundaries.
5. A system of 'forces' concentrated at the nodes and equilibrating the boundary stresses and any distributed loads is determined, resulting in a stiffness relationship of the form of Eq. (1.3).

Once this stage has been reached the solution procedure can follow the standard discrete system pattern described earlier.

Clearly a series of approximations has been introduced. Firstly, it is not always easy to ensure that the chosen displacement functions will satisfy the requirement of displacement continuity between adjacent elements. Thus, the compatibility condition on such lines may be violated (though within each element it is obviously satisfied due to uniqueness of displacements implied in their continuous representation). Secondly, by concentrating the equivalent forces at the nodes, equilibrium conditions are satisfied in the overall sense only. Local violation of equilibrium conditions within each element and on its boundaries will usually arise.

The choice of element shape and of the form of the displacement function for specific cases leaves much choice to ingenuity and skill of the engineer, and obviously the degree of approximation which can be achieved will strongly depend on these factors.

The approach outlined here is known as the displacement formulation.[1,2]

So far, the process described is justified only intuitively, but what in fact has been suggested is equivalent to the minimization of the total potential energy of the system in terms of a prescribed displacement field. If this displacement field is defined in a suitable way, then convergence to the correct result must occur. The process is then equivalent to the well-known Ritz procedure. This equivalence will be proved in a later section of this chapter where also a discussion of the necessary convergence criteria will be made.

The recognition of the equivalence of the finite element method with a minimization process was late.[2,3] However, Courant in 1943[4]† and Prager and Synge[5] in 1947 proposed methods in essence identical.

This broader basis of the finite element method allows it to be extended to other continuum problems where a variational formulation is possible. Indeed, general procedures are now available for a finite element discretization of any problem defined by a properly constituted set of differential equations. Such generalizations will be discussed in Chapter 9, and throughout the book application to non-structural problems will be made. It will be found that the processes described in this chapter are essentially an application of trial-function and Galerkin-type approximations to a particular case of solid mechanics.

2.2 Direct formulation of finite element characteristics

The 'prescriptions' for deriving the characteristics of a 'finite element' of a continuum, which were outlined in general terms, will now be presented in more detailed mathematical form.

It is desirable to obtain results in a general form applicable to any situation, but to avoid introducing conceptual difficulties the general relations will be illustrated with a very simple example of plane stress analysis of a thin slice. In this a division of the region into triangular-shaped elements is used as shown in Fig. 2.1. Relationships of general validity will be underlined. Again, matrix notation will be implied.

2.2.1 *Displacement function.* A typical finite element, e, is defined by nodes, i, j, m, etc., and straight line boundaries. Let the displacements $\mathbf{u}$ at any point within the element be approximated as a column vector, $\hat{\mathbf{u}}$:

$$\mathbf{u} \approx \hat{\mathbf{u}} = \sum \mathbf{N}_i \mathbf{a}_i^e = [\mathbf{N}_i, \mathbf{N}_j, \ldots] \begin{Bmatrix} \mathbf{a}_i \\ \mathbf{a}_j \\ \vdots \end{Bmatrix}^e = \mathbf{N}\mathbf{a}^e \qquad (2.1)$$

in which the components of $\mathbf{N}$ are prescribed functions of position and $\mathbf{a}^e$ represents a listing of nodal displacements for a particular element.

In the case of plane stress, for instance,

$$\mathbf{u} = \begin{Bmatrix} u(x, y) \\ v(x, y) \end{Bmatrix}$$

† It appears that Courant had anticipated the essence of the finite element method in general, and of a triangular element in particular, as early as 1923 in a paper entitled 'On a convergence principle in calculus of variation', Kön. Gesellschaft der Wissenschaften zu Göttingen, Nachrichten, Berlin, 1923. He states: 'We imagine a mesh of triangles covering the domain . . . the convergence principles remain valid for each triangular domain.'

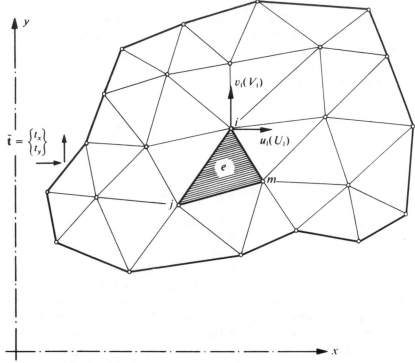

Fig. 2.1   A plane stress region divided into finite elements

represents horizontal and vertical movements of a typical point within the element and

$$\mathbf{a}_i = \begin{Bmatrix} u_i \\ v_i \end{Bmatrix}$$

the corresponding displacements of a node i.

The functions $\mathbf{N}_i$, $\mathbf{N}_j$, $\mathbf{N}_m$ have to be so chosen as to give appropriate nodal displacements when the coordinates of the corresponding nodes are inserted in Eq. (2.1). Clearly, in general,

$$\mathbf{N}_i(x_i, y_i) = \mathbf{I} \quad \text{(identity matrix)}$$

while

$$\mathbf{N}_i(x_j, y_j) = \mathbf{N}_i(x_m, y_m) = 0, \qquad \text{etc.}$$

which is simply satisfied by suitable linear functions of x and y.

If both the components of displacement are interpolated in an identical manner then we can write

$$\mathbf{N}_i = N_i \mathbf{I}$$

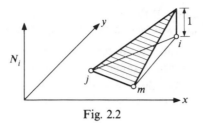

Fig. 2.2

and obtain N_i from Eq. (2.1) by noting that $N_i = 1$ at x_i and y_i but zero at other vertices.

The most obvious linear interpolation in the case of a triangle will yield the shape of N_i of the form shown in Fig. 2.2. Detailed expressions for such a linear interpolation are given in Chapter 3, but at this stage can be readily derived by the reader.

The functions **N** will be called *shape functions* and will be seen later to play a paramount role in finite element analysis.

2.2.2 *Strains*. With displacements known at all points within the element the 'strains' at any point can be determined. These will always result in a relationship that can be written in matrix notation as

$$\boxed{\boldsymbol{\varepsilon} = \mathbf{S}\mathbf{u}} \tag{2.2}$$

where $\mathbf{S}$ is a suitable linear operator. Using Eq. (2.1), the above equation can be approximated as

$$\boxed{\boldsymbol{\varepsilon} = \mathbf{B}\mathbf{a}} \tag{2.3}$$

with

$$\boxed{\mathbf{B} = \mathbf{S}\mathbf{N}} \tag{2.4}$$

For the plane stress case the relevant strains of interest are those occurring in the plane and are defined in terms of the displacements by well-known relations[6] which define the operator $\mathbf{S}$:

$$\boldsymbol{\varepsilon} = \left\{ \begin{array}{c} \varepsilon_x \\ \varepsilon_y \\ \gamma_{xy} \end{array} \right\} = \left\{ \begin{array}{c} \dfrac{\partial u}{\partial x} \\[2mm] \dfrac{\partial v}{\partial y} \\[2mm] \dfrac{\partial u}{\partial y} + \dfrac{\partial v}{\partial x} \end{array} \right\} = \left[\begin{array}{cc} \dfrac{\partial}{\partial x}, & 0 \\[2mm] 0, & \dfrac{\partial}{\partial y} \\[2mm] \dfrac{\partial}{\partial y}, & \dfrac{\partial}{\partial x} \end{array} \right] \left\{ \begin{array}{c} u \\ v \end{array} \right\}$$

With the shape functions $\mathbf{N}_i$, $\mathbf{N}_j$, and $\mathbf{N}_m$ already determined, the matrix

B will easily be obtained. If the linear form of these functions is adopted then, in fact, the strains will be constant throughout the element.

2.2.3 *Stresses.* In general, the material within the element boundaries may be subjected to initial strains such as may be due to temperature changes, shrinkage, crystal growth, and so on. If such strains are denoted by ε_0 then the stresses will be caused by the difference between the actual and initial strains.

In addition it is convenient to assume that at the outset of analysis the body is stressed by some known system of initial residual stresses σ_0 which, for instance, could be measured, but the prediction of which without the full knowledge of the material's history is impossible. These stresses can simply be added on to the general definition. Thus, assuming general linear elastic behaviour, the relationship between stresses and strains will be linear and of the form

$$\boxed{\sigma = D(\varepsilon - \varepsilon_0) + \sigma_0} \tag{2.5}$$

where **D** is an elasticity matrix containing the appropriate material properties.

Again, for the particular case of plane stress three components of stress corresponding to the strains already defined have to be considered. These are, in familiar notation

$$\sigma = \begin{Bmatrix} \sigma_x \\ \sigma_y \\ \tau_{xy} \end{Bmatrix}$$

and the **D** matrix may be simply obtained from the usual isotropic stress–strain relationship[6]

$$\varepsilon_x - (\varepsilon_x)_0 = \frac{1}{E}\sigma_x - \frac{v}{E}\sigma_y$$

$$\varepsilon_y - (\varepsilon_y)_0 = -\frac{v}{E}\sigma_x + \frac{1}{E}\sigma_y$$

$$\gamma_{xy} - (\gamma_{xy})_0 = \frac{2(1+v)}{E}\tau_{xy}$$

i.e., on solving,

$$D = \frac{E}{1-v^2} \begin{bmatrix} 1 & v & 0 \\ v & 1 & 0 \\ 0 & 0 & (1-v)/2 \end{bmatrix}$$

2.2.4 *Equivalent nodal forces.* Let

$$\mathbf{q}^e = \begin{Bmatrix} \mathbf{q}_i^e \\ \mathbf{q}_j^e \\ \vdots \end{Bmatrix}$$

define the nodal forces which are equivalent statically to the boundary stresses and distributed loads on the element. Each of the forces $\mathbf{q}_i^e$ must contain the same number of components as the corresponding nodal displacement $\mathbf{a}_i$ and be ordered in the appropriate, corresponding directions.

The distributed loads $\mathbf{b}$ are defined as those acting on a unit volume of material within the element with directions corresponding to those of the displacements $\mathbf{u}$ at that point.

In the particular case of plane stress the nodal forces are, for instance,

$$\mathbf{q}_i^e = \begin{Bmatrix} U_i \\ V_i \end{Bmatrix}$$

with components U and V corresponding to the directions of u and v displacements, and the distributed load is

$$\mathbf{b} = \begin{Bmatrix} b_x \\ b_y \end{Bmatrix}$$

in which b_x and b_y are the 'body force' components.

To make the nodal forces statically equivalent to the actual boundary stresses and distributed loads, the simplest procedure is to impose an arbitrary (virtual) nodal displacement and to equate the external and internal work done by the various forces and stresses during that displacement.

Let such a virtual displacement be $\delta \mathbf{a}^e$ at the nodes. This results, by Eqs (2.1) and (2.2), in displacements and strains within the element equal to

$$\delta \mathbf{u} = \mathbf{N} \, \delta \mathbf{a}^e \quad \text{and} \quad \delta \boldsymbol{\varepsilon} = \mathbf{B} \, \delta \mathbf{a}^e \tag{2.6}$$

respectively.

The work done by the nodal forces is equal to the sum of the products of the individual force components and corresponding displacements, i.e., in matrix language

$$\delta \mathbf{a}^{eT} \mathbf{q}^e \tag{2.7}$$

Similarly, the internal work per unit volume done by the stresses and distributed forces is

$$\delta \boldsymbol{\varepsilon}^T \boldsymbol{\sigma} - \delta \mathbf{u}^T \mathbf{b} \tag{2.8}$$

or†

$$\delta\mathbf{a}^{T}(\mathbf{B}^{T}\boldsymbol{\sigma} - \mathbf{N}^{T}\mathbf{b}) \tag{2.9}$$

Equating the external work with the total internal work obtained by integrating over the volume of the element, V^{e}, we have

$$\delta\mathbf{a}^{eT}\mathbf{q}^{e} = \delta\mathbf{a}^{eT}\left(\int_{V^{e}} \mathbf{B}^{T}\boldsymbol{\sigma} \, d(\text{vol}) - \int_{V^{e}} \mathbf{N}^{T}\mathbf{b} \, d(\text{vol})\right) \tag{2.10}$$

As this relation is valid for any value of the virtual displacement, the equality of the multipliers must exist. Thus

$$\mathbf{q}^{e} = \int_{V^{e}} \mathbf{B}^{T}\boldsymbol{\sigma} \, d(\text{vol}) - \int_{V^{e}} \mathbf{N}^{T}\mathbf{b} \, d(\text{vol}) \tag{2.11}$$

This statement is valid quite generally for any stress–strain relations. With the linear law of Eq. (2.5) we can write Eq. (2.11) as

$$\mathbf{q}^{e} = \mathbf{K}^{e}\mathbf{a}^{e} + \mathbf{f}^{e} \tag{2.12}$$

where

$$\mathbf{K}^{e} = \int_{V^{e}} \mathbf{B}^{T}\mathbf{D}\mathbf{B} \, d(\text{vol}) \tag{2.13a}$$

and

$$\mathbf{f}^{e} = -\int_{V^{e}} \mathbf{N}^{T}\mathbf{b} \, d(\text{vol}) - \int_{V^{e}} \mathbf{B}^{T}\mathbf{D}\boldsymbol{\varepsilon}_{0} \, d(\text{vol}) + \int_{V^{e}} \mathbf{B}^{T}\boldsymbol{\sigma}_{0} \, d(\text{vol}) \tag{2.13b}$$

In the last equation the three terms represent forces due to body forces, initial strain, and initial stress respectively. The relations have the characteristics of the discrete structural elements described in Chapter 1.

If the initial stress system is self-equilibrating, as must be the case with normal residual stresses, then the forces given by the initial stress term of Eq. (2.13b) are identically zero after assembly. Thus frequently evaluation of this force component is omitted. However, if for instance a machine part is manufactured out of a block in which residual stresses are present or if an excavation is made in rock where known tectonic

---

† Note that by rules of matrix algebra for transpose of products

$$(\mathbf{AB})^{T} = \mathbf{B}^{T}\mathbf{A}^{T}$$

stresses exist a removal of material will cause a force imbalance which results from the above term.

For the particular example of the plane stress triangular element these characteristics will be obtained by the appropriate substitution. It has already been noted that the **B** matrix in that example was not dependent on the coordinates; hence the integration will become particularly simple.

The interconnection and solution of the whole assembly of the elements follow the simple structural procedures outlined in Chapter 1. In general, external concentrated forces may exist at the nodes and the matrix

$$\mathbf{r} = \begin{Bmatrix} \mathbf{r}_1 \\ \mathbf{r}_2 \\ \vdots \\ \mathbf{r}_n \end{Bmatrix} \tag{2.14}$$

will be added to the consideration of equilibrium at the nodes.

A note should be added here concerning elements near the boundary. If, at the boundary, displacements are specified, no special problem arises as these can be satisfied by specifying some of the nodal parameters **a**. Consider, however, the boundary as subject to a distributed external loading, say $\bar{\mathbf{t}}$ per unit area. A loading term on the nodes of the element which has a boundary face A^e will now have to be added. By the virtual work consideration, this will simply result in

$$-\int_{A^e} \mathbf{N}^T \bar{\mathbf{t}} \, d(\text{area}) \tag{2.15}$$

with the integration taken over the boundary area of the element. It will be noted that $\bar{\mathbf{t}}$ must have the same number of components as **u** for the above expression to be valid.

Such a boundary element is shown again for the special case of plane stress in Fig. 2.1. An integration of this type is sometimes not carried out explicitly. Often by 'physical intuition' the analyst will consider the boundary loading to be represented simply by concentrated loads acting on the boundary nodes and calculate these by direct static procedures. In the particular case discussed the results will be identical.

Once the nodal displacements have been determined by solution of the overall 'structural' type equations, the stresses at any point of the element can be found from the relations in Eqs (2.3) and (2.5), giving

$$\boldsymbol{\sigma} = \mathbf{DB}\mathbf{a}^e - \mathbf{D}\boldsymbol{\varepsilon}_0 + \boldsymbol{\sigma}_0 \tag{2.16}$$

in which the typical terms of the relationship of Eq. (1.4) will be immediately recognized, the element stress matrix being

$$S^e = DB \qquad (2.17)$$

To this the stresses

$$\sigma_{\varepsilon_0} = -D\varepsilon_0 \quad \text{and} \quad \sigma_0 \qquad (2.18)$$

have to be added.

The absence of the term of stresses due to distributed loading σ_p^e needs a comment. It is due to the fact that the internal equilibrium within any element has not been considered, and only overall equilibrium conditions were established.

2.2.5 *Generalized nature of displacements, strains, and stresses.* The meaning of displacements, strains, and stresses in the illustrative case of plane stress was obvious. In many other applications, shown later in this book, this terminology may be applied to other, less obvious, quantities. For example, in considering plate elements the 'displacement' may be characterized by the lateral deflection and the slopes of the plate at a particular point. The 'strains' will then be defined as the curvatures of the middle surface and the 'stresses' as the corresponding internal bending moments.

All the expressions derived here are generally valid provided the sum product of displacement and corresponding load components represents truly the external work done, while that of the 'strain' and corresponding 'stress' components results in the total internal work.

2.3 Generalization to the whole region—internal nodal force concept abandoned

In the preceding section the virtual work principle was applied to a single element and the concept of equivalent nodal force was retained. The assembly principle thus followed the conventional, direct equilibrium, approach.

The idea of nodal forces contributed by elements replacing the continuous interaction is a conceptual difficulty although it has a considerable appeal to 'practical' engineers and does at times allow an interpretation which otherwise would not be obvious to the more rigorous mathematician. There is, however, no need to consider each element individually and the reasoning of the previous section may be applied directly to the whole continuum.

Equation (2.1) can be interpreted as applying to the whole structure, that is,

$$\mathbf{u} = \bar{\mathbf{N}}\mathbf{a} \tag{2.19}$$

in which **a** lists all the nodal points and

$$\bar{\mathbf{N}}_i = \mathbf{N}_i^e \tag{2.20}$$

when the point concerned is within a particular element e and i is a point associated with that element. If point i does not occur within the element

$$\bar{\mathbf{N}}_i = 0 \tag{2.21}$$

Matrix $\bar{\mathbf{B}}$ can be similarly defined and we shall drop the bar superscript considering simply that the shape functions, etc., are defined over the whole region V.

For any virtual displacement $\delta\mathbf{a}$ we can now write the sum of internal and external work for the whole region as

$$-\delta\mathbf{a}^\mathrm{T}\mathbf{r} = \int_V \delta\mathbf{u}^\mathrm{T}\mathbf{b}\ \mathrm{d}V + \int_A \delta\mathbf{u}^\mathrm{T}\bar{\mathbf{t}}\ \mathrm{d}A - \int_V \delta\boldsymbol{\varepsilon}^\mathrm{T}\boldsymbol{\sigma}\ \mathrm{d}V \tag{2.22}$$

In the above equation $\delta\mathbf{a}$, $\delta\mathbf{u}$, and $\delta\boldsymbol{\varepsilon}$ can be completely arbitrary, providing they stem from a continuous displacement assumption. If for convenience we assume they are simply variations linked by the relations (2.19) and (2.3) we obtain on substitution of the constitutive relation (2.5) a system of algebraic equations

$$\boxed{\mathbf{K}\mathbf{a} + \mathbf{f} = \mathbf{r}} \tag{2.23}$$

where

$$\boxed{\mathbf{K} = \int_V \mathbf{B}^\mathrm{T}\mathbf{D}\mathbf{B}\ \mathrm{d}V} \tag{2.24a}$$

and

$$\boxed{\mathbf{f} = -\int_V \mathbf{N}^\mathrm{T}\mathbf{b}\ \mathrm{d}V - \int_A \mathbf{N}^\mathrm{T}\bar{\mathbf{t}}\,\mathrm{d}A - \int_V \mathbf{B}^\mathrm{T}\mathbf{D}\boldsymbol{\varepsilon}_0\ \mathrm{d}V + \int_V \mathbf{B}^\mathrm{T}\boldsymbol{\sigma}_0\ \mathrm{d}V} \tag{2.24b}$$

The integrals are taken over the whole volume V and over the whole surface area A on which the tractions are given.

It is immediately obvious from the above that

$$\boxed{\mathbf{K}_{ij} = \sum \mathbf{K}_{ij}^e} \quad \boxed{\mathbf{f}_i = \sum \mathbf{f}_i^e} \tag{2.25}$$

by virtue of the property of definite integrals requiring that the total be the sum of the parts:

$$\int_V (\quad)\, dV = \sum \int_{V^e} (\quad)\, dV \tag{2.26}$$

The same is obviously true for the surface integrals in Eq. (2.25). We see thus that the 'secret' of the approximation possessing the required behaviour of a 'standard discrete system of Chapter 1' lies simply in the requirement of writing the approximation in an integral form.

The assembly rule as well as the whole derivation has been achieved without involving the concept of 'interelement forces'. In the remainder of this chapter the element superscript will be dropped unless specifically needed. Also no differentiation between element and system shape functions will be made.

However, an important point arises immediately. In considering the virtual work for the whole system [Eq. (2.22)] and equating this to the sum of the element contributions it is implicitly assumed that no discontinuity between adjacent elements develops. If such a discontinuity developed a contribution equal to the work done by the stresses in the separations would have to be added.

Put in other words, we require that the terms integrated in Eq. (2.26) be finite. These terms arise from the shape functions N_i used in defining the displacement $\mathbf{u}$ [by Eq. (2.19)] and its derivatives associated with the definition of strain [viz. Eq. (2.3)]. If, for instance, the 'strains' are defined by first derivatives of the functions N these must be continuous. In Fig. 2.3 we see how first derivatives of continuous functions may involve a 'jump' but are still finite, while second derivatives may become infinite. Such functions we call C_0 continuous.

In some problems the 'strain' in a generalized sense may be defined by second derivatives. In such cases we shall obviously require that both the function N and its slope (first derivative) be continuous. Such functions are more difficult to derive but we shall make use of them in plate and shell problems. The continuity involved now is called C_1 continuity.

2.4 Displacement approach as a minimization of total potential energy

The principle of virtual displacements used in the previous sections ensured satisfaction of equilibrium conditions within the limits prescribed by the assumed displacement pattern. Only if the virtual work equality for all, arbitrary, variations of displacement was ensured (prescribing only the boundary conditions) would the equilibrium be complete.

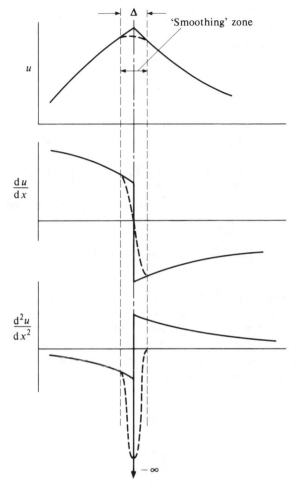

Fig. 2.3   Differentiation of function with slope discontinuity (C_0 continuous)

If the number of parameters of **a** which prescribes the displacement increases without limit then ever closer approximation of all equilibrium conditions can be ensured.

The virtual work principle as written in Eq. (2.22) can be restated in a different form if the virtual quantities δ**a**, δ**u**, and $\delta\varepsilon$ are considered as *variations* (or differentials) of the real quantities.

Thus, for instance, we can write

$$\delta\left(\mathbf{a}^{\mathrm{T}}\mathbf{r} + \int_{V} \mathbf{u}^{\mathrm{T}}\mathbf{b}\ \mathrm{d}V + \int_{A} \mathbf{u}^{\mathrm{T}}\bar{\mathbf{t}}\ \mathrm{d}A \right) = -\delta W \qquad (2.27)$$

for the first three terms of Eq. (2.22), where *W is the potential energy of the external loads.* The above is certainly true if **r**, **b**, and $\bar{\mathbf{t}}$ are conservative (or independent of displacement).

The last term of Eq. (2.22) can, for elastic materials, be written as

$$\delta U = \int_V \delta\boldsymbol{\varepsilon}^\mathsf{T}\boldsymbol{\sigma}\ \mathrm{d}V \tag{2.28}$$

where U is the 'strain energy' of the system. For the elastic, linear material described by Eq. (2.5) the reader can verify that

$$U = \frac{1}{2}\int_V \boldsymbol{\varepsilon}^\mathsf{T}\mathbf{D}\boldsymbol{\varepsilon}\ \mathrm{d}V - \int_V \boldsymbol{\varepsilon}^\mathsf{T}\mathbf{D}\boldsymbol{\varepsilon}_0\ \mathrm{d}V + \int_V \boldsymbol{\varepsilon}^\mathsf{T}\boldsymbol{\sigma}_0\ \mathrm{d}V \tag{2.29}$$

will, after differentiation, yield the correct expression providing **D** is a symmetric matrix. (This is indeed a necessary condition for single-valued U to exist.)

Thus instead of Eq. (2.22) we can write simply

$$\delta(U + W) = \delta(\Pi) = 0 \tag{2.30}$$

in which the quantity Π is called the *total potential energy.*

The above statement means that for equilibrium to be ensured the *total potential energy must be stationary* for variations of admissible displacements. The finite element equations derived in the previous section [Eqs (2.23) to (2.25)] are simply the statements of this variation with respect to displacements constrained to a finite number of parameters **a** and could be written as

$$\frac{\partial\Pi}{\partial\mathbf{a}} = \left\{ \begin{array}{c} \dfrac{\partial\Pi}{\partial\mathbf{a}_1} \\[2mm] \dfrac{\partial\Pi}{\partial\mathbf{a}_2} \\[1mm] \vdots \end{array} \right\} = 0 \tag{2.31}$$

It can be shown that in stable elastic situations the total potential energy is not only stationary but is a minimum.[7] *Thus the finite element process seeks such a minimum within the constraint of an assumed displacement pattern.*

The greater the degrees of freedom, the more closely will the solution approximate to the true one ensuring complete equilibrium, providing the true displacement can, in the limit, be approximated. The necessary convergence conditions for the finite element process could thus be derived. Discussion of these will, however, be deferred to a later section.

It is of interest to note that if true equilibrium requires an absolute minimum of the total potential energy, Π, an approximate finite element solution by displacement approach will always provide an approximate

Π greater than the correct one. *Thus a bound on the value of the total potential energy is always achieved.*

If the function Π could be specified, *a priori*, then the finite element equations could be derived directly by differentiation specified by Eq. (2.31).

The well-known Rayleigh[8]–Ritz[9] process of approximation frequently used in elastic analysis uses precisely this approach. The total potential energy expression is formulated and the displacement pattern is assumed to vary with a finite set of undetermined parameters. A set of simultaneous equations minimizing the total potential energy with respect to these parameters is set up. Thus the finite element process as described so far is identically the Rayleigh–Ritz procedure. The difference is only in the manner in which the displacements are prescribed. In the Ritz process traditionally used these are usually given by expressions valid throughout the whole region, thus leading to simultaneous equations in which no banding occurs and the coefficient matrix is full. In the finite element process this specification is usually piecewise, each nodal parameter influencing only adjacent elements, and thus a sparse and usually banded matrix of coefficients is found.

By its nature the conventional Ritz process is limited to relatively simple geometrical shapes of the total region while this limitation only occurs in finite element analysis in the element itself. Thus complex, realistic, configurations can be assembled from relatively simple element shapes.

A further difference in kind is in the usual association of the undetermined parameter with a particular nodal displacement. This allows a simple physical interpretation invaluable to an engineer. Doubtless much of the popularity of the finite element process is due to this fact.

2.5 Convergence criteria

The assumed shape functions limit the infinite degrees of freedom of the system, and the true minimum of the energy may never be reached, irrespective of the fineness of subdivision. To ensure convergence to the correct result certain simple requirements have to be satisfied. Obviously, for instance, the displacement function should be able to represent the true displacement distribution as closely as possible. It will be found that this is not so if the chosen functions are such that straining of an element is possible when this is subject to rigid body displacements. Thus, the first criterion that the displacement function must obey is as follows:

Criterion 1. The displacement function chosen should be such that it does not permit straining of an element to occur when the nodal displacements are caused by a rigid body displacement.

The self-evident condition can be violated easily if certain types of function are used; care must therefore be taken in the choice of displacement functions.

A second criterion stems from similar requirements. Clearly, as elements get smaller nearly constant strain conditions will prevail in them. If, in fact, constant strain conditions exist, it is most desirable for good accuracy that a finite size element is able to reproduce these exactly. It is possible to formulate functions that satisfy the first criterion but at the same time require a strain variation throughout the element when the nodal displacements are compatible with a constant strain solution. Such functions will, in general, not show a good convergence to an accurate solution and cannot, even in the limit, represent the true strain distribution. The second can therefore be formulated as follows:

> *Criterion* 2. The displacement function has to be of such a form that if nodal displacements are compatible with a constant strain condition such constant strain will in fact be obtained. (In this context again a generalized 'strain' definition is implied.)

It will be observed that Criterion 2 in fact incorporates the requirement of Criterion 1, as rigid body displacements are a particular case of constant strain—with a value of zero. This criterion was first stated by Bazeley *et al.*[10] in 1965. *Strictly, both criteria need only be satisfied in the limit as the size of the element tends to zero.* However, the imposition of these criteria on elements of finite size leads to improved accuracy, although in certain situations (such as illustrated by the axisymmetric analysis of Chapter 4) the imposition of the second one is not possible or essential.

Lastly, as already mentioned in Sec. 2.3, it is implicitly assumed in the derivation presented that no contribution to the virtual work arises at element interfaces. It therefore appears necessary that the following criterion be included:

> *Criterion* 3. The displacement functions should be so chosen that the strains at the interface between elements are finite (even though indeterminate).

This criterion implies a certain continuity of displacements between elements. In the case of strains being defined by first derivatives, as in the plane example quoted here, the displacements only have to be continuous. If, however, as in the plate and shell problems, the 'strains' are defined by second derivatives of deflections, first derivatives of these have also to be continuous.[2]

The above criteria are mathematically included in a statement of 'functional completeness' and the reader is referred for full mathematical discussion elsewhere.[11-16] The 'heuristic' proof of the convergence requirements given here is sufficient for practical purposes in all but the most pathological cases and we shall generalize all of the above criteria in Chapter 9.

2.6 Discretization error and convergence rate

In the foregoing sections we have assumed that the approximation to the displacement as represented by Eq. (2.1) will yield the exact solution in the limit as the size h of elements decrease. The arguments for this are simple: if the expansion is capable, in the limit, of reproducing exactly any displacement form conceivable in the continuum, then as the solution of each approximation is unique it must approach, in the limit of $h \to 0$, the unique exact solution. In some cases the exact solution is indeed obtained with a finite number of subdivisions (or even with one element only) if the *polynomial expansion is used in that element and if this can fit exactly the correct solution*. Thus, for instance, if the exact solution is of the form of a quadratic polynomial *and* the shape functions include all the polynomials of that order, the approximation will yield the exact answer.

The last argument helps in determining the order of convergence of the finite element procedure as the exact solution can always be expanded in the vicinity of any point (or node) i as a polynomial

$$\mathbf{u} = \mathbf{u}_i + \left(\frac{\partial \mathbf{u}}{\partial x}\right)_i (x - x_i) + \left(\frac{\partial \mathbf{u}_i}{\partial y}\right)_i (y - y_i) + \cdots \qquad (2.32)$$

If within an element of 'size' h a polynomial expansion of degree p is employed, this can fit locally the Taylor expansion up to that degree and, as x and y are of the order of magnitude h, the error in $\mathbf{u}$ will be of the order $O(h^{p+1})$. Thus, for instance, in the case of the plane elasticity problem discussed, we used a linear expansion and $p = 1$. We should therefore expect a *convergence* rate of order $O(h^2)$, i.e., the error in displacement being reduced to $\frac{1}{4}$ for a halving of the mesh spacing.

By a similar argument the strains (or stresses) which are given by mth derivatives of displacement should converge with an error of $O(h^{p+1-m})$, i.e., as $O(h)$ in the example quoted, where $m = 1$. The strain energy being given by the square of stresses will show an error of $O(h^{2(p+1-m)})$ or $O(h^2)$ in the plane stress example.

The arguments given here are perhaps a trifle 'heuristic' from a mathematical viewpoint—they are, however, true[16] and give correctly the orders of convergence, which can be expected to be achieved asymptotically as the element size tends to zero and if the exact solution does not contain singularities. Such singularities may result in infinite values of the coefficients in terms omitted in the Taylor expansion of Eq. (2.32) and invalidate the arguments. However, in many well-behaved problems the mere determination of the order of convergence often suffices to extrapolate the solution to the correct result. Thus, for instance, if the displacement converges at $O(h^2)$ and we have two approximate solutions u^1 and u^2 obtained with meshes of size h and $h/2$, we can write with u being the exact solution

$$\frac{u^1 - u}{u^2 - u} = \frac{O(h^2)}{O(h/2)^2} = 4 \tag{2.33}$$

From the above an (almost) exact solution u can be predicted. This type of extrapolation was first introduced by Richardson[17] and is of use if convergence is monotonic and nearly asymptotic.

We shall return to the important question of estimating errors due to the discretization process in Chapter 14 and will show that much more precise methods than those arising from convergence rate considerations are today possible. Indeed automatic mesh refinement processes are being introduced so that specified accuracy can be achieved.

The discretization error is not the only one possible in the finite element computation. In addition to obvious mistakes which can occur when using computers, errors due to *round-off* are always possible. With the computer operating on numbers rounded off to a finite number of digits, a reduction of accuracy occurs every time differences between 'like' numbers are being formed. In the process of equation solving many subtractions are necessary and accuracy decreases. Problems of matrix conditioning, etc., enter here and the user of the finite element method must at all times be aware of accuracy limitations which simply do not allow the exact solution ever to be obtained. Fortunately in many computations, by using modern machines which carry a large number of significant digits, these errors are often small!

The question of errors arising from the algebraic processes will be stressed in Chapter 15 dealing with computation procedures.

2.7 Displacement functions with discontinuity between elements—non-conforming elements and the patch test

In some cases considerable difficulty is experienced in finding displacement functions for an element which will automatically be continuous along the whole interface between adjacent elements.

As already pointed out, the discontinuity of displacement will cause infinite strains at the interfaces—a factor ignored in the formulation presented because the energy contribution is limited to the elements themselves.

However, if, in the limit, as the size of the subdivision decreases continuity is restored, then the formulation already obtained will still tend to the correct answer. This condition is always reached if

(a) a constant strain condition automatically ensures displacement continuity and

(b) the constant strain criterion of the previous section is satisfied.

To test that such continuity is achieved for any mesh configuration when using such *non-conforming* elements it is necessary to impose, on an arbitrary patch of elements, nodal displacements corresponding to any state of constant strain. *If nodal equilibrium is simultaneously achieved without the imposition of external, nodal, forces and if a state of constant stress is obtained, then clearly no external work has been lost through interelement discontinuity.*

Elements which pass such a *patch test* will converge, and indeed at times non-conforming elements will show a superior performance to conforming elements.

The patch test was first introduced by Irons[10] and has since been demonstrated to give a sufficient condition for convergence.[16,18,19] The concept of the patch test can be generalized to give information on the rate of convergence which can be expected from a given element.

On occasion we shall find that the use of 'non-conforming' elements leads to better results than with those satisfying *a priori* the continuity requirements. We shall return to this problem in detail in Chapter 11 where the test will be fully discussed.

2.8 Bound on strain energy in a displacement formulation

While the approximation obtained by the finite element displacement approach always overestimates the true value of Π, the total potential energy (the absolute minimum corresponding to the exact solution), this is not directly useful in practice. It is, however, possible to obtain a more useful limit in special cases.

Consider in particular the problem in which no 'initial' strains or initial stresses exist. Now by the principle of energy conservation the strain energy will be equal to the work done by the external loads which increase uniformly from zero.[20] This work done is equal to $-\frac{1}{2}W$ where W is the potential energy of the loads.

Thus

$$U + \tfrac{1}{2}W = 0 \tag{2.34}$$

or

$$\Pi = U + W = -U \tag{2.35}$$

whether an exact or approximate displacement field is assumed.

Thus in the above case the approximate solution always *underestimates* the value of U and a displacement solution is frequently referred to as the *lower bound solution*.

If only one external concentrated load R is present the strain energy bound immediately informs us that the deflection under this load has been underestimated (as $U = -\frac{1}{2}W = \frac{1}{2}\mathbf{r}^{\mathrm{T}}\mathbf{a}$). In more complex loading cases the usefulness of this bound is limited as neither deflections nor stresses, i.e., the quantities of real engineering interest, can be bounded.

It is important to remember that this bound on strain energy is only valid in the absence of any initial stresses or strains.

The expression for U in this case can be obtained from Eq. (2.29) as

$$U = \frac{1}{2} \int_V \boldsymbol{\varepsilon}^{\mathrm{T}} \mathbf{D} \boldsymbol{\varepsilon} \, \mathrm{d(vol)} \tag{2.36}$$

which becomes by Eq. (2.2) simply

$$U = \frac{1}{2}\mathbf{a}^{\mathrm{T}}\left[\int_V \mathbf{B}^{\mathrm{T}}\mathbf{D}\mathbf{B} \, \mathrm{d(vol)}\right]\mathbf{a} = \frac{1}{2}\mathbf{a}^{\mathrm{T}}\mathbf{K}\mathbf{a} \tag{2.37}$$

a 'quadratic' matrix form in which $\mathbf{K}$ is the 'stiffness' matrix previously discussed.

The above energy expression is always positive from physical considerations. It follows therefore that the matrix $\mathbf{K}$ occurring in all the finite element assemblies is not only symmetric but is 'positive definite' (a property defined in fact by the requirements that the quadratic form should always be greater than or equal to zero).

This feature is of importance when the numerical solution of the simultaneous equations involved is considered as simplifications arise in the case of 'symmetric positive definite' equations.

2.9 Direct minimization

The fact that the finite element approximation reduces to the problem of minimizing the total potential energy Π defined in terms of a finite number of nodal parameters led us to formulation of the simultaneous set of equations given symbolically by Eq. (2.31). This is the most usual and convenient approach, especially in linear solutions, but other search procedures, now well developed in the field of optimization, could be used to estimate the lowest value of Π. In this text we shall continue with the simultaneous equation process but the interested reader could well bear the alternative possibilities in mind.[21,22]

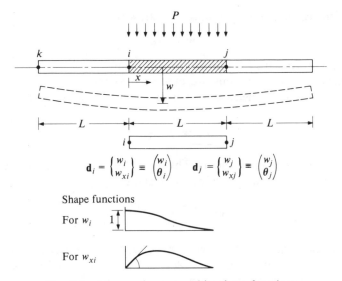

Fig. 2.4   A beam element and its shape functions

2.10 An example

The concepts discussed and the general formulation cited are a little abstract and readers may at this stage seek to test their grasp of the nature of the approximations derived. While detailed computations of a two-dimensional element system are performed using the computer, we can perform a simple hand calculation on a one-dimensional finite element of a beam. Indeed, this example will allow us to introduce the concept of generalized stresses and strains in a simple manner.

Consider the beam shown in Fig. 2.4. The generalized 'strain' here is the curvature. Thus we have

$$\varepsilon \equiv \kappa = -\frac{d^2w}{dx^2}$$

where w is the deflection, which is the basic unknown. The generalized stress (in the absence of shear deformation) will be the bending moment M, which is related to the 'strain' as

$$\sigma \equiv M = -EI\frac{d^2w}{dx^2}$$

Thus immediately we have, using the general notation of previous sections,

$$\mathbf{D} \equiv EI$$

If the displacement w is discretized we can write

$$w = \mathbf{Na}$$

for the whole system or, for an individual element, ij.

In this example the strains are expressed as the second derivatives of displacement and it is necessary to ensure that both w and its slope

$$w_x \equiv \frac{dw}{dx} = \theta$$

be continuous between elements. This is easily accomplished if the nodal parameters are taken as the values of w and the slope, w_x. Thus,

$$\mathbf{a}_i = \left\{ \begin{matrix} w \\ w_x \end{matrix} \right\}_i = \left\{ \begin{matrix} w_i \\ \theta_i \end{matrix} \right\}$$

The shape functions now will be derived. If we accept that in an element two nodes (i.e., four variables) define the deflected shape we can assume this to be given by a cubic

$$w = \alpha_1 + \alpha_2 x + \alpha_3 x^2 + \alpha_4 x^3$$

This will define the shape functions corresponding to w_i and w_{xi} by taking for each a cubic giving unity for the appropriate points ($x = 0; L$) and zero for other quantities, as shown in Fig. 2.4.

The expressions for the shape function can be written for the element shown as

$$\begin{aligned} \mathbf{N}_i &= \left[1 - 3(x/L)^2 + 2(x/L)^3, \quad L(x/L - 2(x/L)^2 + (x/L)^3 \right. \\ \mathbf{N}_j &= \left. \quad 3(x/L)^2 - 2(x/L)^3, \qquad L(-(x/L)^2 + (x/L)^3) \quad \right] \end{aligned}$$

Immediately we can write

$$\mathbf{B}_i = -\frac{d^2}{dx^2} \mathbf{N}_i = \frac{[6 - 12(x/L), (4 - 6(x/L))L]}{L^2}$$

$$\mathbf{B}_j = -\frac{d^2}{dx^2} \mathbf{N}_j = \frac{[-6 + 12(x/L), (2 - 6(x/L))L]}{L^2}$$

and the stiffness matrices for the element can be written as

$$\mathbf{K}_{ij}^e = \int_0^L \mathbf{B}_i^T EI \, \mathbf{B}_j \, dx$$

We shall leave the detailed calculation of this and the 'forces' corresponding to a uniformly distributed load p (assumed constant on ij and zero elsewhere) to the reader. It will be observed that the finally assembled equations for a node i are of the form linking three nodal displacements i, j, k. Explicitly these equations are for elements of equal length L.

$$EI\begin{bmatrix} -12/L^3, & -6/L^2 \\ 6/L^2, & 2/L \end{bmatrix}\begin{Bmatrix} w_k \\ \theta_k \end{Bmatrix} + EI\begin{bmatrix} 24/L^3, & 0 \\ 0, & 8/L \end{bmatrix}\begin{Bmatrix} w_i \\ \theta_i \end{Bmatrix}$$

$$+ EI\begin{bmatrix} -12/L^3, & +6/L^2 \\ -6/L^2, & 2/L \end{bmatrix}\begin{Bmatrix} w_j \\ \theta_j \end{Bmatrix} + \begin{Bmatrix} pL/2 \\ -pL^2/12 \end{Bmatrix} = 0$$

It is of interest to compare these with the *exact* form represented by the so-called 'slope–deflection' equations which can be found in standard texts on structural analysis.

Here it will be found that the finite element approximation has achieved the exact solution as the cubic polynomial was capable of representing it for a uniform load. For other distributed loads it is easy to show that the difference between the approximation and the exact equations decreases as the length of elements tends to zero.

2.11 Concluding remarks

The 'displacement' approach to analysis of elastic solids is still undoubtedly the most popular and easily understood procedure. In many of the following chapters we shall use the general formulae developed here in the context of linear elastic analysis (Chapters 3, 4, 5, and 8). These are also applicable in the context of non-linear analysis, the main variants being the definitions of the stresses, generalized strains, and other associated quantities. It is thus convenient to summarize the essential formulae, and this is done in Appendix 2.

In Chapter 9 we shall show that the procedures developed here are but a particular case of finite element discretization applied to the governing equilibrium equations written in terms of displacements.[23] Clearly, alternative starting points are possible. Some of these will be mentioned in Chapter 12.

References

1. R. W. CLOUGH, 'The finite element in plane stress analysis', *Proc. 2nd ASCE Conf. on Electronic Computation*, Pittsburgh, Pa., Sept. 1960.
2. R. W. CLOUGH, 'The finite element method in structural mechanics', Chapter 7 of *Stress Analysis* (eds O. C. Zienkiewicz and G. S. Holister), Wiley, 1965.
3. J. SZMELTER, 'The energy method of networks of arbitrary shape in problems of the theory of elasticity', *Proc. IUTAM Symposium on Non-Homogeneity in Elasticity and Plasticity* (ed. W. Olszak), Pergamon Press, 1959.
4. R. COURANT, 'Variational methods for the solution of problems of equilibrium and vibration', *Bull. Am. Math. Soc.*, **49**, 1–23, 1943.

5. W. PRAGER and J. L. SYNGE, 'Approximation in elasticity based on the concept of function space', *Quart. Appl. Math.*, **5**, 241–69, 1947.
6. S. TIMOSHENKO and J. N. GOODIER, *Theory of Elasticity*, 2nd ed., McGraw-Hill, 1951.
7. K. WASHIZU, *Variational Methods in Elasticity and Plasticity*, 2nd ed., Pergamon Press, 1975.
8. J. W. STRUTT (Lord Rayleigh), 'On the theory of resonance', *Trans. Roy. Soc. (London)*, **A161**, 77–118, 1870.
9. W. RITZ, 'Über eine neue Methode zur Lösung gewissen Variations—Probleme der mathematischen Physik', *J. Reine angew. Math.*, **135**, 1–61, 1909.
10. G. P. BAZELEY, Y. K. CHEUNG, B. M. IRONS, and O. C. ZIENKIEWICZ, 'Triangular elements in bending—conforming and non-conforming solutions', *Proc. Conf. Matrix Methods in Structural Mechanics*, Air Force Inst. Tech., Wright-Patterson AF Base, Ohio, 1965.
11. S. C. MIKHLIN, *The Problem of the Minimum of a Quadratic Functional*, Holden-Day, 1966.
12. M. W. JOHNSON and R. W. MCLAY, 'Convergence of the finite element method in the theory of elasticity', *J. Appl. Mech., Trans. Am. Soc. Mech. Eng.*, 274–8, 1968.
13. S. W. KEY, *A convergence investigation of the direct stiffness method*, PhD thesis, University of Washington, 1966.
14. T. H. H. PIAN and PING TONG, 'The convergence of finite element method in solving linear elastic problems', *Int. J. Solids Struct.*, **3**, 865–80, 1967.
15. E. R. DE ARRANTES OLIVEIRA, 'Theoretical foundations of the finite element method', *Int. J. Solids Struct.*, **4**, 929–52, 1968.
16. G. STRANG and G. J. FIX, *An Analysis of the Finite Element Method*, p. 106, Prentice-Hall, 1973.
17. L. F. RICHARDSON, 'The approximate arithmetical solution by finite differences of physical problems', *Trans. Roy. Soc. (London)*, **A210**, 307–57, 1910.
18. B. M. IRONS and A. RAZZAQUE, 'Experience with the patch test', in *Mathematical Foundations of the Finite Element Method* (ed. A. R. Aziz), pp. 557–87, Academic Press, 1972.
19. B. FRAEIJS DE VEUBEKE, 'Variational principles and the patch test', *Int. J. Num. Meth. Eng.*, **8**, 783–801, 1974.
20. B. FRAEIJS DE VEUBEKE, 'Displacement and equilibrium models in the finite element method', Chapter 9 of *Stress Analysis* (eds O. C. Zienkiewicz and G. S. Holister), Wiley, 1965.
21. R. L. FOX and E. L. STANTON, 'Developments in structural analysis by direct energy minimization', *JAIAA*, **6**, 1036–44, 1968.
22. F. K. BOGNER, R. H. MALLETT, M. D. MINICH, and L. A. SCHMIT, 'Development and evaluation of energy search methods in non-linear structural analysis', *Proc. Conf. Matrix Methods in Structural Mechanics*, Air Force Inst. Tech., Wright-Patterson AF Base, Ohio, 1965.
23. O. C. ZIENKIEWICZ and K. MORGAN, *Finite Elements and Approximation*, Wiley, 1983.

3

Plane stress and plane strain

3.1 Introduction

Two-dimensional elastic problems were the first successful examples of the application of the finite element method.[1,2] Indeed, we have already used this situation to illustrate the basis of the finite element formulation in Chapter 2 where the general relationships were derived. These basic relationships are given in Eqs (2.1) to (2.5) and (2.23) and (2.24), which for quick reference are summarized in Appendix 2.

In this chapter the particular relationships for the problem in hand will be derived in more detail, and illustrated by suitable practical examples, a procedure that will be followed throughout the remainder of the book.

Only the simplest, triangular, element will be discussed in detail but the basic approach is general. More elaborate elements to be discussed in later chapters would be introduced to the same problem in an identical manner.

The reader not familiar with the applicable basic definitions of elasticity is referred to elementary texts on the subject, in particular to the text by Timoshenko and Goodier,[3] whose notation will be widely used here.

In both problems of plane stress and plane strain the displacement field is uniquely given by the u and v displacements in directions of the cartesian, orthogonal x and y axes.

Again, in both, the only strains and stresses that have to be considered are the three components in the xy plane. In the case of *plane stress*, by definition, all other components of stress are zero and therefore give no contribution to internal work. In *plane strain* the stress in a direction perpendicular to the xy plane is not zero. However, by definition, the strain in that direction is zero, and therefore no contribution to internal work is made by this stress, which can in fact be explicitly evaluated from the three main stress components, if desired, at the end of all computation.

3.2 Element characteristics

3.2.1 *Displacement functions.* Figure 3.1 shows the typical triangular element considered, with nodes i, j, m numbered in an anticlockwise order.

The displacements of a node have two components

$$\mathbf{a}_i = \begin{pmatrix} u_i \\ v_i \end{pmatrix} \tag{3.1}$$

and the six components of element displacements are listed as a vector

$$\mathbf{a}^e = \begin{Bmatrix} \mathbf{a}_i \\ \mathbf{a}_j \\ \mathbf{a}_m \end{Bmatrix} \tag{3.2}$$

The displacements within an element have to be uniquely defined by these six values. The simplest representation is clearly given by two linear polynomials

$$u = \alpha_1 + \alpha_2 x + \alpha_3 y$$
$$v = \alpha_4 + \alpha_5 x + \alpha_6 y \tag{3.3}$$

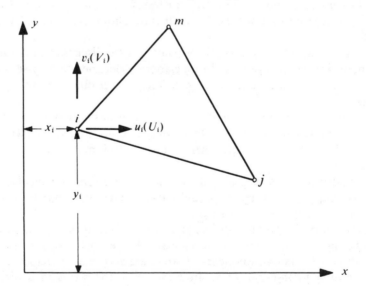

Fig. 3.1   An element of a continuum in plane stress or plane strain

The six constants α can be evaluated easily by solving the two sets of three simultaneous equations which will arise if the nodal coordinates are inserted and the displacements equated to the appropriate nodal displacements. Writing, for example,

$$u_i = \alpha_1 + \alpha_2 x_i + \alpha_3 y_i$$
$$u_j = \alpha_1 + \alpha_2 x_j + \alpha_3 y_j \qquad (3.4)$$
$$u_m = \alpha_1 + \alpha_2 x_m + \alpha_3 y_m$$

we can easily solve for α_1, α_2, and α_3 in terms of the nodal displacements u_i, u_j, u_m and obtain finally

$$u = \frac{1}{2\Delta} \left[(a_i + b_i x + c_i y)u_i + (a_j + b_j x + c_j y)u_j + (a_m + b_m x + c_m y)u_m \right]$$

$$(3.5a)$$

in which

$$a_i = x_j y_m - x_m y_j$$
$$b_i = y_j - y_m = y_{jm} \qquad (3.5b)$$
$$c_i = x_m - x_j = x_{mj}$$

with the other coefficients obtained by a cyclic permutation of subscripts in the order, i, j, m, and where†

$$2\Delta = \det \begin{vmatrix} 1 & x_i & y_i \\ 1 & x_j & y_j \\ 1 & x_m & y_m \end{vmatrix} = 2 \cdot \text{(area of triangle } ijm) \qquad (3.5c)$$

As the equations for the vertical displacement v are similar we also have

$$v = \frac{1}{2\Delta} \left[(a_i + b_i x + c_i y)v_i + (a_j + b_j x + c_j y)v_j + (a_m + b_m x + c_m y)v_m \right]$$

$$(3.6)$$

Though not strictly necessary at this stage we can represent the above relations, Eqs (3.5a) and (3.6), in the standard form of Eq. (2.1):

$$\mathbf{u} = \begin{Bmatrix} u \\ v \end{Bmatrix} = \mathbf{N}\mathbf{a}^e = [\mathbf{I}N_i, \mathbf{I}N_j, \mathbf{I}N_m]\mathbf{a}^e \qquad (3.7)$$

† *Note:* if coordinates are taken from the centroid of the element then

$$x_i + x_j + x_m = y_i + y_j + y_m = 0 \quad \text{and} \quad a_i = 2\Delta/3 = a_j = a_m$$

See also Appendix 4 for a summary of integrals for a triangle.

with **I** a two by two identity matrix, and

$$N_i = \frac{a_i + b_i x + c_i y}{2\Delta}, \qquad \text{etc.} \tag{3.8}$$

The chosen displacement function automatically guarantees continuity of displacements with adjacent elements because the displacements vary linearly along any side of the triangle and, with identical displacement imposed at the nodes, the same displacement will clearly exist all along an interface.

3.2.2 *Strain (total)*. The total strain at any point within the element can be defined by its three components which contribute to internal work. Thus

$$\varepsilon = \left\{ \begin{array}{c} \varepsilon_x \\ \varepsilon_y \\ \gamma_{xy} \end{array} \right\} = \left[\begin{array}{cc} \dfrac{\partial}{\partial x}, & 0 \\ 0, & \dfrac{\partial}{\partial y} \\ \dfrac{\partial}{\partial y}, & \dfrac{\partial}{\partial x} \end{array} \right] \left\{ \begin{array}{c} u \\ v \end{array} \right\} = \mathbf{S}\mathbf{u} \tag{3.9}$$

Substituting Eq. (3.7) we have

$$\varepsilon = \mathbf{B}\mathbf{a}^e = [\mathbf{B}_i, \mathbf{B}_j, \mathbf{B}_m] \left\{ \begin{array}{c} \mathbf{a}_i \\ \mathbf{a}_j \\ \mathbf{a}_m \end{array} \right\} \tag{3.10a}$$

with a typical matrix $\mathbf{B}_i$ given by

$$\mathbf{B}_i = \mathbf{S}N_i = \left[\begin{array}{cc} \dfrac{\partial N_i}{\partial x}, & 0 \\ 0, & \dfrac{\partial N_i}{\partial y} \\ \dfrac{\partial N_i}{\partial y}, & \dfrac{\partial N_i}{\partial x} \end{array} \right] = \frac{1}{2\Delta} \left[\begin{array}{cc} b_i, & 0 \\ 0, & c_i \\ c_i, & b_i \end{array} \right] \tag{3.10b}$$

This defines matrix **B** of Eq. (2.2) explicitly.

It will be noted that in this case the **B** matrix is independent of the position within the element, and hence the strains are constant throughout it. Obviously, the criterion of constant strain mentioned in Chapter 2 is satisfied by the shape functions.

3.2.3 *Initial strain (thermal strain)*. 'Initial' strains, i.e., strains which are independent of stress, may be due to many causes. Shrinkage, crystal growth, or, most frequently, temperature changes will, in general, result in an initial strain vector:

$$\varepsilon_0 = \begin{Bmatrix} \varepsilon_{x0} \\ \varepsilon_{y0} \\ \gamma_{xy0} \end{Bmatrix} \tag{3.11}$$

Although this initial strain may, in general, depend on the position within the element, it will usually be defined by average, constant, values. This is consistent with the constant strain conditions imposed by the prescribed displacement function.

Thus, for the case of *plane stress* in an isotropic material in an element subject to a temperature rise θ^e with a coefficient of thermal expansion α, we will have, for instance,

$$\varepsilon_0 = \begin{Bmatrix} \alpha\theta^e \\ \alpha\theta^e \\ 0 \end{Bmatrix} \tag{3.12}$$

as no shear strains are caused by a thermal dilatation.

In *plane strain* the situation is more complex. The presumption of plane strain implies that stresses perpendicular to the xy plane will develop due to thermal expansion even without the three main stress components, and hence the initial strain will be affected by the elastic constants.

It can be shown that in such a case

$$\varepsilon_0 = (1 + v) \begin{Bmatrix} \alpha\theta^e \\ \alpha\theta^e \\ 0 \end{Bmatrix} \tag{3.13}$$

where v is the Poisson's ratio.

Anisotropic materials present special problems, since the coefficients of thermal expansion may vary with direction. Let x' and y' in Fig. 3.2 show the principal directions of the material. The initial strain due to thermal expansion becomes, with reference to these coordinates for plane stress,

$$\varepsilon_0' = \begin{Bmatrix} \varepsilon_{x'0} \\ \varepsilon_{y'0} \\ \gamma_{x'y'0} \end{Bmatrix} = \begin{Bmatrix} \alpha_1\theta^e \\ \alpha_2\theta^e \\ 0 \end{Bmatrix} \tag{3.14}$$

where α_1 and α_2 are the expansion coefficients referred to the x' and y' axes respectively.

To obtain the strain components in the x, y system it is necessary to use an appropriate strain transformation matrix $\mathbf{T}$ giving

$$\varepsilon_0' = \mathbf{T}^T\varepsilon_0 \tag{3.15}$$

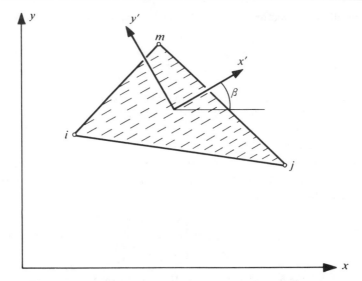

Fig. 3.2   An element of a stratified (transversely isotropic) material

With the β as defined in Fig. 3.2 it is easily verified that

$$
\mathbf{T} = \begin{bmatrix}
\cos^2 \beta & \sin^2 \beta & -2\sin\beta\cos\beta \\
\sin^2 \beta & \cos^2 \beta & 2\sin\beta\cos\beta \\
\sin\beta\cos\beta & -\sin\beta\cos\beta & \cos^2\beta - \sin^2\beta
\end{bmatrix}
$$

Thus, $\boldsymbol{\varepsilon}_0$ can be simply evaluated. It will be noted that no longer is the shear component of strain equal to zero in the xy coordinates.

3.2.4 *Elasticity matrix.* The matrix $\mathbf{D}$ of the relation Eq. (2.5)

$$
\boldsymbol{\sigma} = \begin{Bmatrix} \sigma_x \\ \sigma_y \\ \tau_{xy} \end{Bmatrix} = \mathbf{D} \left(\begin{Bmatrix} \varepsilon_x \\ \varepsilon_y \\ \gamma_{xy} \end{Bmatrix} - \boldsymbol{\varepsilon}_0 \right) \tag{3.16}
$$

can be explicitly stated for any material (excluding here $\boldsymbol{\sigma}_0$ which is simply additive).

Plane stress—isotropic material. For plane stress in an isotropic material we have, by definition,

$$
\varepsilon_x = \frac{\sigma_x}{E} - \frac{\nu\sigma_y}{E} + \varepsilon_{x0}
$$

$$
\varepsilon_y = -\frac{\nu\sigma_x}{E} + \frac{\sigma_y}{E} + \varepsilon_{y0} \tag{3.17}
$$

$$
\gamma_{xy} = \frac{2(1+\nu)\tau_{xy}}{E} + \varepsilon_{xy0}
$$

Solving the above for the stresses, we obtain matrix $\mathbf{D}$ as

$$\mathbf{D} = \frac{E}{1 - v^2} \begin{bmatrix} 1 & v & 0 \\ v & 1 & 0 \\ 0 & 0 & (1 - v)/2 \end{bmatrix} \tag{3.18}$$

in which E is the elastic modulus and v is Poisson's ratio.

Plane strain—isotropic material. In this case a normal stress σ_z exists in addition to the three other stress components. For the special case of isotropic thermal expansion we have

$$\varepsilon_x = \frac{\sigma_x}{E} - \frac{v\sigma_y}{E} - \frac{v\sigma_z}{E} + \alpha\theta^e$$

$$\varepsilon_y = -\frac{v\sigma_x}{E} + \frac{\sigma_y}{E} - \frac{v\sigma_z}{E} + \alpha\theta^e \tag{3.19}$$

$$\gamma_{xy} = \frac{2(1 + v)\tau_{xy}}{E}$$

but in addition

$$\varepsilon_z = 0 = -\frac{v\sigma_x}{E} - \frac{v\sigma_y}{E} + \frac{\sigma_z}{E} + \alpha\theta^e$$

On eliminating σ_z and solving for the three remaining stresses we obtain the previously quoted expression for the initial strain Eq. (3.13), and by comparison with Eq. (3.16), the matrix $\mathbf{D}$ is

$$\mathbf{D} = \frac{E(1 - v)}{(1 + v)(1 - 2v)} \begin{bmatrix} 1 & v/(1 - v) & 0 \\ v/(1 - v) & 1 & 0 \\ 0 & 0 & (1 - 2v)/[2(1 - v)] \end{bmatrix} \tag{3.20}$$

Anisotropic materials. For a completely anisotropic material, 21 independent elastic constants are necessary to define completely the three-dimensional stress–strain relationship.[4,5]

If two-dimensional analysis is to be applicable a symmetry of properties must exist, implying at most six independent constants in the $\mathbf{D}$ matrix. Thus, it is always possible to write

$$\mathbf{D} = \begin{bmatrix} d_{11} & d_{12} & d_{13} \\ & d_{22} & d_{23} \\ \text{symmetric} & & d_{33} \end{bmatrix} \tag{3.21}$$

to describe the most general two-dimensional behaviour. (The necessary symmetry of the $\mathbf{D}$ matrix follows from the general equivalent of the Maxwell–Betti reciprocal theorem and is a consequence of invariant energy irrespective of the path taken to reach a given strain state.)

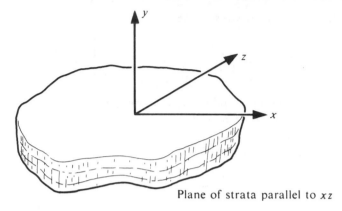

Plane of strata parallel to xz

Fig. 3.3   A stratified (transversely isotropic) material

A case of particular interest in practice is that of a 'stratified' or transversely isotropic material in which a rotational symmetry of properties exists within the plane of the strata. Such a material possesses only five independent elastic constants.

The general stress–strain relations give in this case, following the notation of Lekhnitskii[4] and taking now the y axis as perpendicular to the strata (neglecting initial strain) (Fig. 3.3),

$$\varepsilon_x = \frac{\sigma_x}{E_1} - \frac{v_2\sigma_y}{E_2} - \frac{v_1\sigma_z}{E_1}$$

$$\varepsilon_y = -\frac{v_2\sigma_x}{E_2} + \frac{\sigma_y}{E_2} - \frac{v_2\sigma_z}{E_2}$$

$$\varepsilon_z = -\frac{v_1\sigma_x}{E_1} - \frac{v_2\sigma_y}{E_2} + \frac{\sigma_z}{E_1}$$

$$\gamma_{xz} = \frac{2(1+v_1)}{E_1}\tau_{xz} \qquad (3.22)$$

$$\gamma_{xy} = \frac{1}{G_2}\tau_{xy}$$

$$\gamma_{yz} = \frac{1}{G_2}\tau_{yz}$$

in which the constants E_1, v_1 (G_1 is dependent) are associated with the behaviour in the plane of the strata and E_2, G_2, v_2 with a direction normal to these.

The $\mathbf{D}$ matrix in two dimensions becomes now, taking $E_1/E_2 = n$ and $G_2/E_2 = m$,

$$\mathbf{D} = \frac{E_2}{1 - nv_2^2} \begin{bmatrix} N & nv_2 & 0 \\ nv_2 & 1 & 0 \\ 0 & 0 & m(1 - nv_2^2) \end{bmatrix} \qquad (3.23)$$

for plane stress or

$$\mathbf{D} = \frac{E_2}{(1 + v_1)(1 - v_1 - 2nv_2^2)}$$

$$\times \begin{bmatrix} n(1 - nv_2^2) & nv_2(1 + v_1) & 0 \\ nv_2(1 + v_1) & (1 - v_1^2) & 0 \\ 0 & 0 & m(1 + v_1)(1 - v_1 - 2nv_2^2) \end{bmatrix} \qquad (3.24)$$

for plane strain.

When, as in Fig. 3.2, the direction of strata is inclined to the x axis then to obtain the $\mathbf{D}$ matrices in the universal coordinates a transformation is necessary. Taking $\mathbf{D}'$ as relating the stresses and strains in the inclined coordinate system (x', y') it is easy to show that

$$\mathbf{D} = \mathbf{T}\mathbf{D}'\mathbf{T}^\mathsf{T} \qquad (3.25)$$

where $\mathbf{T}$ is the same as given in Eq. (3.15).

If the stress systems $\boldsymbol{\sigma}'$ and $\boldsymbol{\sigma}$ correspond to $\boldsymbol{\varepsilon}'$ and $\boldsymbol{\varepsilon}$ respectively then by equality of work

$$\boldsymbol{\sigma}'^\mathsf{T}\boldsymbol{\varepsilon}' = \boldsymbol{\sigma}^\mathsf{T}\boldsymbol{\varepsilon}$$

or

$$\boldsymbol{\varepsilon}'^\mathsf{T}\mathbf{D}'\boldsymbol{\varepsilon}' = \boldsymbol{\varepsilon}^\mathsf{T}\mathbf{D}\boldsymbol{\varepsilon}$$

from which Eq. (3.25) follows on substitution of Eq. (3.15). (See also Chapter 1.)

3.2.5 *The stiffness matrix.* The stiffness matrix of the element ijm is defined from the general relationship Eq. (2.13) with the coefficients

$$\mathbf{K}_{ij}^e = \int \mathbf{B}_i^\mathsf{T}\mathbf{D}\mathbf{B}_j t \, \mathrm{d}x \, \mathrm{d}y \qquad (3.26)$$

where t is the thickness of the element and the integration is taken over the area of the triangle. If the thickness of the element is assumed to be constant, an assumption convergent to the truth as the size of elements decreases, then, as neither of the matrices contains x or y we have, simply,

$$\mathbf{K}_{ij}^e = \mathbf{B}_i^\mathsf{T}\mathbf{D}\mathbf{B}_j t\Delta \qquad (3.27)$$

where Δ is the area of the triangle [defined already by Eq. (3.5)]. This form is now sufficiently explicit for computation with the actual matrix operations being left to the computer.

3.2.6 *Nodal forces due to initial strain.* These are given directly by the expression Eq. (2.13) which, on performing the integration, becomes

$$(\mathbf{f}_i)^e_{\varepsilon_0} = -\mathbf{B}_i^{\mathsf{T}} \mathbf{D} \boldsymbol{\varepsilon}_0 \, t \Delta, \qquad \text{etc.} \tag{3.28}$$

These 'initial strain' forces are contributed to the nodes of an element in an unequal manner and require precise evaluation. Similar expressions are derived for initial stress forces.

3.2.7 *Distributed body forces.* In the general case of plane stress or strain each element of unit area in the xy plane is subject to forces

$$\mathbf{b} = \begin{Bmatrix} b_x \\ b_y \end{Bmatrix}$$

in the direction of the appropriate axes.

Again, by Eq. (2.13), the contribution of such forces to these at each node is given by

$$\mathbf{f}_i^e = -\int N_i \begin{Bmatrix} b_x \\ b_y \end{Bmatrix}$$

or by Eq. (3.7),

$$\mathbf{f}_i^e = -\begin{Bmatrix} b_x \\ b_y \end{Bmatrix} \int N_i \, \mathrm{d}x \, \mathrm{d}y, \qquad \text{etc.} \tag{3.29}$$

if the body forces b_x and b_y are constant. As N_i is not constant the integration has to be carried out explicitly. Some general integration formulae for a triangle are given in Appendix 3.

In this special case the calculation will be simplified if the origin of coordinates is taken at the centroid of the element. Now

$$\int x \, \mathrm{d}x \, \mathrm{d}y = \int y \, \mathrm{d}x \, \mathrm{d}y = 0$$

and on using Eq. (3.8)

$$\mathbf{f}_i^e = -\begin{Bmatrix} b_x \\ b_y \end{Bmatrix} \int \frac{a_i \, \mathrm{d}x \, \mathrm{d}y}{2\Delta} = -\begin{Bmatrix} b_x \\ b_y \end{Bmatrix} \frac{a_i}{2} = -\begin{Bmatrix} b_x \\ b_y \end{Bmatrix} \frac{\Delta}{3} \tag{3.30}$$

by relations noted on page 47.

Explicitly, for the whole element

$$\mathbf{f}^e = \left\{ \begin{array}{c} f_i^e \\ f_j^e \\ f_m^e \end{array} \right\} = - \left\{ \begin{array}{c} b_x \\ b_y \\ b_x \\ b_y \\ b_x \\ b_y \end{array} \right\} \frac{\Delta}{3} \tag{3.31}$$

which means simply that the total forces acting in the x and y directions due to the body forces are distributed to the nodes in three equal parts. This fact corresponds with physical intuition, and was often assumed implicitly.

3.2.8 *Body force potential.* In many cases the body forces are defined in terms of a body force potential ϕ as

$$b_x = - \frac{\partial \phi}{\partial x} \qquad b_y = - \frac{\partial \phi}{\partial y} \tag{3.32}$$

and this potential, rather than the values of b_x and b_y, is known throughout the region and is specified at nodal points. If ϕ^e lists the three values of the potential associated with the nodes of the element, i.e.,

$$\mathbf{\phi}^e = \left\{ \begin{array}{c} \phi_i \\ \phi_j \\ \phi_m \end{array} \right\} \tag{3.33}$$

and has to correspond with constant values of b_x and b_y, ϕ must vary linearly within the element. The 'shape function' of its variation will obviously be given by a procedure identical to that used in deriving Eqs (3.4) to (3.6), and yields

$$\phi = [N_i, N_j, N_m] \mathbf{\phi}^e \tag{3.34}$$

Thus,

$$b_x = - \frac{\partial \phi}{\partial x} = -[b_i, b_j, b_m] \frac{\mathbf{\phi}^e}{2\Delta}$$

and

$$b_y = - \frac{\partial \phi}{\partial y} = -[c_i, c_j, c_m] \frac{\mathbf{\phi}^e}{2\Delta} \tag{3.35}$$

The vector of nodal forces due to the body force potential will now replace Eq. (3.31) by

$$\mathbf{f}^e = \frac{1}{6} \begin{bmatrix} b_i, & b_j, & b_m \\ c_i, & c_j, & c_m \\ b_i, & b_j, & b_m \\ c_i, & c_j, & c_m \\ b_i, & b_j, & b_m \\ c_i, & c_j, & c_m \end{bmatrix} \boldsymbol{\phi}^e \qquad (3.36)$$

3.2.9 *Evaluation of stresses.* The formulae derived enable the full stiffness matrix of the structure to be assembled, and a solution for displacements to be obtained.

The stress matrix given in general terms in Eq. (2.16) is obtained by the appropriate substitutions for each element.

The stresses are, by the basic assumption, constant within the element. It is usual to assign these to the centroid of the element, and in most of the examples in this chapter this procedure is followed. An alternative consists of obtaining stress values at the nodes by averaging the values in the adjacent elements. Some 'weighting' procedures have been used in this context on an empirical basis but their advantage appears small.

It is usual to calculate the principal stresses and their directions for every element.

3.3 Examples—an assessment of accuracy

There is no doubt that the solution to plane elasticity problems as formulated in Sec. 3.2 is, in the limit of subdivision, an exact solution. Indeed at any stage of a finite subdivision it is an approximate solution as is, say, a Fourier series solution with a limited number of terms.

As already explained in Chapter 2 the total strain energy obtained during any stage of approximation will be below the true strain energy of the exact solution. In practice it will mean that the displacements, and hence also the stresses, will be underestimated by the approximation in its *general picture*. However, it must be emphasized that this is not necessarily true at every point of the continuum individually; hence the value of such a bound in practice is not great.

What is important for the engineer to know is the order of accuracy achievable in typical problems with a certain fineness of element subdivision. In any particular case the error can be assessed by comparison with known, exact, solutions or by a study of the convergence, using two or more stages of subdivision.

With the development of experience the engineer can assess *a priori* the order of approximation that will be involved in a specific problem tackled with a given element subdivision. Some of this experience will perhaps be conveyed by the examples considered in this book.

In the first place attention will be focused on some simple problems for which exact solutions are available.

3.3.1 *Uniform stress field.* If the exact solution is in fact that of a uniform stress field then, whatever the element subdivision, the finite element sol-

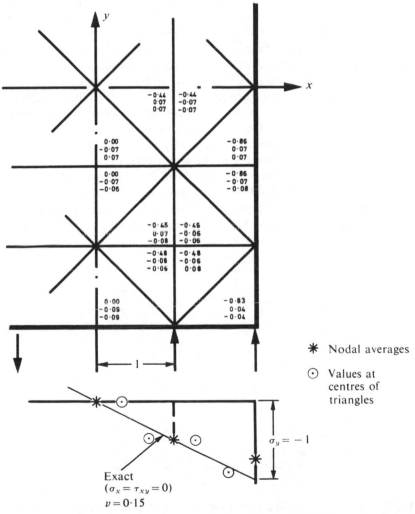

Fig. 3.4 Pure bending of a beam solved by a coarse subdivision into elements of triangular shape. (Values of σ_y, σ_x, and τ_{xy} listed in that order)

ution will coincide exactly with the exact one. This is an obvious corollary of the formulation; nevertheless it is useful as a first check of written computer programs.

3.3.2 *Linearly varying stress field.* Here, obviously, the basic assumption of constancy of stress within elements means that the solution will be approximate only. In Fig. 3.4 a simple example of a beam subject to constant bending moment is shown with a fairly coarse subdivision. It is readily seen that the axial (σ_y) stress given by the element 'straddles' the exact values and, in fact, if the constant stress values are associated with centroids of the elements and plotted, the best 'fit' line represents the exact stresses. (See Chapter 12 for optimal sampling points.)

The horizontal and shear stress components differ again from the exact values (which are simply zero). Again, however, it will be noted that they oscillate by equal, small amounts around the exact values.

At internal nodes, if the average of stresses of surrounding elements is taken it will be found that the exact stresses are very closely represented. The average at external faces is not, however, so good. The overall improvement in representing the stresses by nodal averages, as shown on Fig. 3.4, is often used in practice for improvement of the approximation.

A weighting of averages near the faces of the structure can further be used for refinement. Without being dogmatic on this point, it seems preferable, when accuracy demands this, simply to use a finer mesh subdivision. The matter of arriving logically at the subdivision with a specified accuracy will be discussed in Chapter 14.

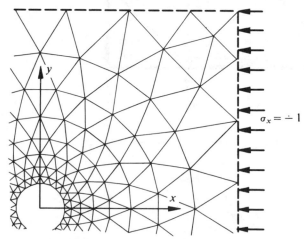

Fig. 3.5   A circular hole in a uniform stress field: (a) isotropic material; (b) stratified (orthotropic) material; $E_x = E_1 = 1$, $E_y = E_2 = 3$, $v_1 = 0.1$, $v_2 = 0$, $G_{xy} = 0.42$

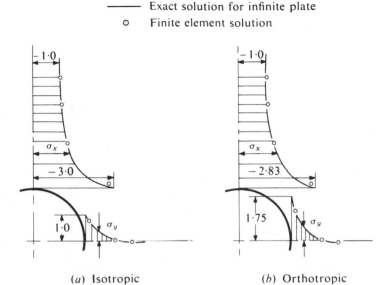

—————— Exact solution for infinite plate

o   Finite element solution

(a) Isotropic   (b) Orthotropic

Fig. 3.6 Comparison of theoretical and finite element results for cases (a) and
(b) of Fig. 3.5

3.3.3 *Stress concentration.* A more realistic test problem is shown in
Figs 3.5 and 3.6. Here the flow of stress around a circular hole in an
isotropic and in an anisotropic stratified material is considered when the
stress conditions are uniform.[6] A graded division into elements is used to
allow a more detailed study in the region where high stress gradients are
expected. The high degree of accuracy achievable can be assessed from
Fig. 3.6 where some of the results are compared against exact solutions.[3,7]

In later chapters we shall see that even more accurate answers can be
obtained with the use of more elaborate elements; however, the principles of the analysis remain identical.

3.4 Some practical applications

Obviously, the practical applications of the method are limitless, and the
finite element method has superseded experimental technique for plane
problems because of its high accuracy, low cost, and versatility. The ease
of treatment of material anisotropy, thermal stresses, or body force problems add to its advantages.

A few examples of actual applications to complex problems of engineering practice will now be given.

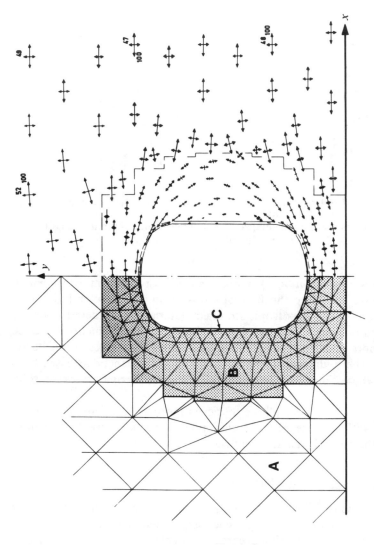

Restrained in y direction from movement

Fig. 3.7   A reinforced opening in a plate. Uniform stress field at a distance from opening $\sigma_x = 100$, $\sigma_y = 50$. Thickness of plate regions A, B, and C is in the ratio of 1 : 3 : 23

3.4.1 *Stress flow around a reinforced opening* (Fig. 3.7). In steel pressure vessels or aircraft structures, openings have to be introduced in the stressed skin. The penetrating duct itself provides some reinforcement round the edge and, in addition, the skin itself is increased in thickness to reduce the stresses due to the concentration effects.

Analysis of such problems treated as cases of plane stress presents no difficulties. The elements are so chosen as to follow the thickness variation, and appropriate values of this are assigned.

The narrow band of thick material near the edge can be represented either by special beam-type elements, or more easily in a standard program by very thin triangular elements of the usual type, to which appropriate thickness is assigned. The latter procedure was used in the problem shown in Fig. 3.7 which gives some of the resulting stresses near the opening itself. The fairly large extent of the region introduced in the analysis and the grading of the mesh should be noted.

3.4.2 *An anisotropic valley subject to tectonic stress*[6] (Fig. 3.8). A symmetrical valley subject to a uniform horizontal stress is considered. The material is stratified, and hence is 'transversely isotropic', and the direction of strata varies from point to point.

The stress plot shows the tensile region that develops. This phenomenon is of considerable interest to geologists and engineers concerned with rock mechanics.

3.4.3 *A dam subject to external and internal water pressures*[8,9] (Fig. 3.9). A buttress dam on a somewhat complex rock foundation is here analysed. The heterogeneous foundation region is subject to plane strain conditions while the dam itself is considered as a plate (plane stress) of variable thickness.

With external and gravity loading no special problems of analysis arise, though perhaps it should be mentioned that it was found worth while to 'automatize' the computation of gravity nodal loads.

When pore pressures are considered, the situation, however, requires perhaps some explanation.

It is well known that in a porous material the water pressure is transmitted to the structure as a *body force* of magnitude

$$b_x = -\frac{\partial p}{\partial x} \qquad b_y = -\frac{\partial p}{\partial y} \tag{3.37}$$

and that now the external pressure need not be considered.

The pore pressure p is, in fact, now a body force potential, as defined in Eq. (3.32). Figure 3.9 shows the element subdivision of the region and the outline of the dam. Figure 3.10(a) and (b) shows the stresses resulting

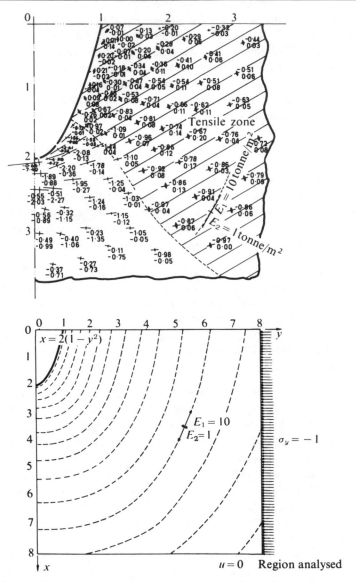

Fig. 3.8   A valley with curved strata subject to a horizontal tectonic stress (plane strain 170 nodes, 298 elements)

from gravity (applied to the dam only) and due to water pressure assumed to be acting as an external load or, alternatively, as an internal pore pressure. Both solutions indicate large tensile regions, but the increase of stresses due to the second assumption is important.

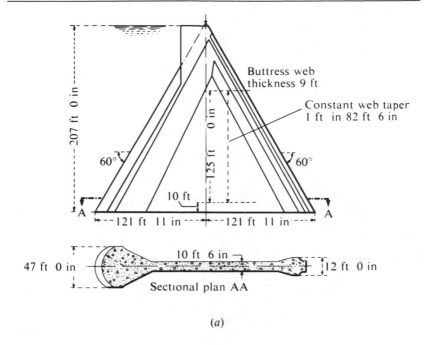

(a)

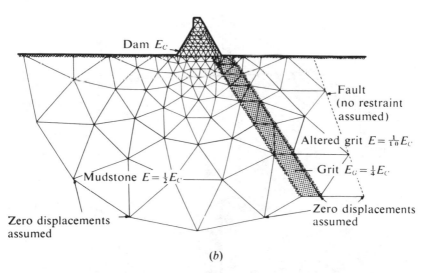

(b)

Fig. 3.9 Stress analysis of a buttress dam. Plane stress condition assumed in a dam and plane strain in foundation. (a) The buttress section analysed. (b) Extent of foundation considered and division into finite elements

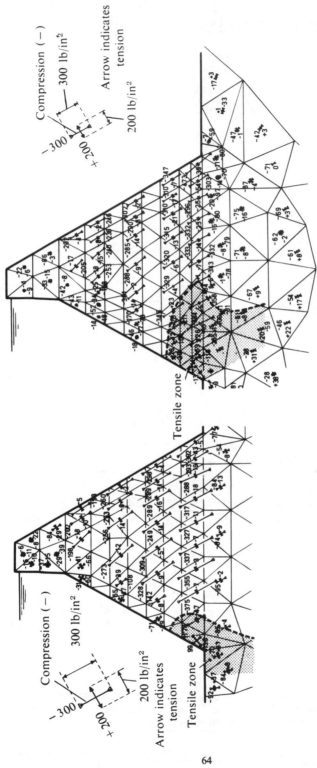

Below the foundation initial rock stresses should be superimposed

(a)

(b)

Fig. 3.10 Stress analysis of the buttress dam of Fig. 3.9. Principal stresses for gravity loads combined with water pressures, which are assumed to act (a) as external loads, (b) as body forces due to pore pressure

64

The stresses calculated here are the so-called 'effective' stresses. These represent the forces transmitted between the solid particles and are defined in terms of the *total* stresses σ and the pore pressures p by

$$\sigma' = \sigma + \mathbf{m}p \qquad \mathbf{m}^T = [1, 1, 0] \tag{3.38}$$

i.e., simply by removing the hydrostatic pressure component from the *total* stress.[10]

The effective stress is of particular importance in the mechanics of porous media such as occur in the study of soils, rocks, or concrete. The basic assumption in deriving the body forces of Eq. (3.37) is that only the effective stress is of any importance in deforming the solid phase. This leads immediately to another possibility of formulation.[11] If we examine the equilibrium conditions of Eq. (2.10) we note that this is written in terms of total stresses. Writing the constitutive relation, Eq. (2.5), in terms of effective stresses, i.e.,

$$\sigma' = \mathbf{D}'(\varepsilon - \varepsilon_0) + \sigma'_0 \tag{3.39}$$

and substituting into the equilibrium equation [Eq. (2.10)] we find that Eq. (2.12) is again obtained, with the stiffness matrix using the matrix $\mathbf{D}'$ and the force terms of Eq. (2.13b) being augmented by an additional force

$$-\int_{V^e} \mathbf{B}^T \mathbf{m}p \; d(\text{vol}) \tag{3.40}$$

or, if p is interpolated by shape functions N'_i, the force becomes

$$-\int_{V^e} \mathbf{B}^T \mathbf{m}\mathbf{N}' \; d(\text{vol})\bar{\mathbf{p}}^e \tag{3.41}$$

This alternative form of introducing the pore pressure effects allows a discontinuous interpolation of p to be used [as in Eq. (3.40) no derivatives occur] and this is now frequently used in practice.

3.4.4 *Cracking.* The tensile stresses in the previous example will doubtless cause the rock to crack. If a stable situation can develop when such a crack spreads then the dam can be considered safe.

Cracks can be introduced very simply into the analysis by assigning zero elasticity values to chosen elements. An analysis with a wide cracked wedge is shown in Fig. 3.11, where it can be seen that with the extent of the crack assumed no tension within the dam body develops.

A more elaborate procedure for following crack propagation and resulting stress redistribution can be developed.[12]

3.4.5 *Thermal stresses.* As an example of thermal stress computation the same dam is shown under simple temperature distribution assumptions. Results of this analysis are given in Fig. 3.12.

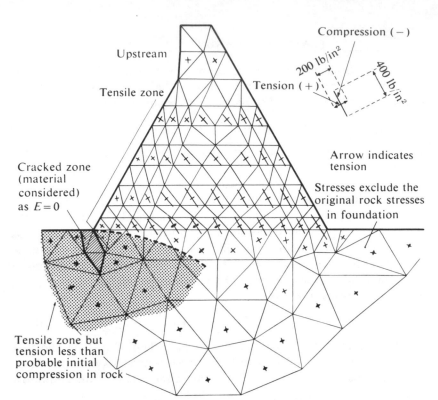

Fig. 3.11 Stresses in a buttress dam. An introduction of a 'crack' modifies stress distribution [same loading as Fig. 3.10(b)]

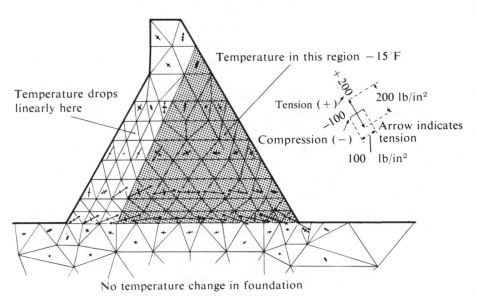

Fig. 3.12 Stress analysis of a buttress dam. Thermal stresses due to cooling of shaded area by 15°F ($E = 3 \times 10^6$ lb/in^2, $\alpha = 6 \times 10^{-6}/°$F)

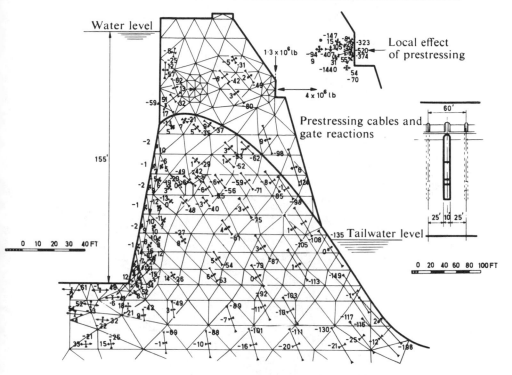

Fig. 3.13   A large barrage with piers and prestressing cables

3.4.6 *Gravity dams.* A buttress dam is a natural example for the application of finite element methods. Other types, such as gravity dams with or without piers and so on, can also be simply treated. Figure 3.13 shows an analysis of a large dam with piers and crest gates.

In this case an approximation of assuming a two-dimensional treatment in the vicinity of the abrupt change of section, i.e., where the piers join the main body of the dam, is clearly involved, but this leads to localized errors only.

It is important to note here how, in a single solution, the grading of element size is used to study concentration of stress at the cable anchorages, the general stress flow in the dam, and the foundation behaviour. The linear ratio of size of largest to smallest elements is of the order of 30 to 1 (the largest elements occurring in the foundation are not shown in the figure).

3.4.7 *Underground power station.* This last example, illustrated in Figs 3.14 and 3.15, shows an interesting large-scale application. Here principal stresses are plotted automatically. In this analysis many different components of σ_0, the initial stress, were used due to uncertainty of

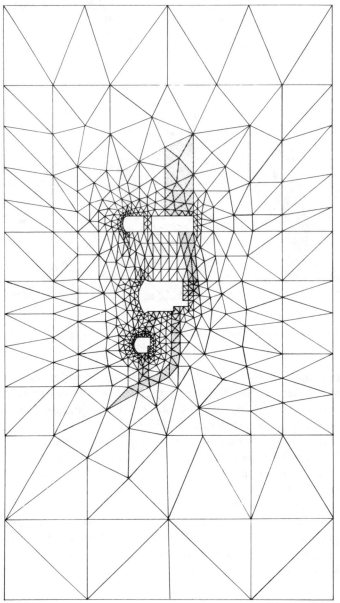

Fig. 3.14   An underground power station. Mesh used in analysis

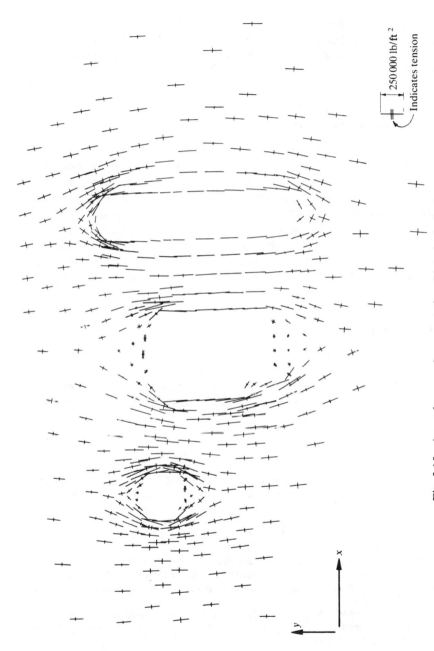

Fig. 3.15   An underground power station. Plot of principal stresses

knowledge about geological conditions. The rapid solution and plot of many results enabled the limits within which stresses vary to be found and an engineering decision arrived at. In this example, the exterior boundaries were taken far enough and 'fixed' ($u = v = 0$). However, a better treatment could be made using infinite elements as described in Sec. 8.13.

3.5 Special treatment of plane strain with an incompressible material

It will have been noted that the relationship Eq. (3.20) defining the elasticity **D** matrix for an isotropic material breaks down when Poisson's ratio reaches a value of 0.5 as the factor in the parentheses becomes infinite. A simple way of side-stepping the difficulty presented is to use values of Poisson's ratio approximating to 0.5 but not equal to it. Experience shows, however, that if this is done the solution deteriorates unless special formulations such as those discussed in Chapter 12 are used. An alternative procedure has been suggested by Herrmann.[13] This involves the use of a different variational formulation, and will be discussed in Chapters 12 and 13.

3.6 Concluding remark

In subsequent chapters, we shall introduce elements which give much greater accuracy for the same number of degrees of freedom in a particular problem. This has led to the belief that the simple triangle used here is completely superceded. In recent years, however, its very simplicity has led to its revival in practical use.

References

1. M. J. TURNER, R. W. CLOUGH, H. C. MARTIN, and L. J. TOPP, 'Stiffness and deflection analysis of complex structures', *J. Aero. Sci.*, **23**, 805–23, 1956.
2. R. W. CLOUGH, 'The finite element in plane stress analysis', *Proc. 2nd ASCE Conf. on Electronic Computation*, Pittsburgh, Pa., Sept. 1960.
3. S. TIMOSHENKO and J. N. GOODIER, *Theory of Elasticity*, 2nd ed., McGraw-Hill, 1951.
4. S. G. LEKHNITSKII, *Theory of Elasticity of an Anisotropic Elastic Body* (Translation from Russian by P. Fern), Holden Day, San Francisco, 1963.
5. R. F. S. HEARMON, *An Introduction to Applied Anisotropic Elasticity*, Oxford University Press, 1961.
6. O. C. ZIENKIEWICZ, Y. K. CHEUNG, and K. G. STAGG, 'Stresses in anisotropic media with particular reference to problems of rock mechanics', *J. Strain Analysis*, **1**, 172–82, 1966.
7. G. N. SAVIN, *Stress Concentration Around Holes* (Translation from Russian), Pergamon Press, 1961.

8. O. C. ZIENKIEWICZ and Y. K. CHEUNG, 'Buttress dams on complex rock foundations', *Water Power*, **16**, 193, 1964.
9. O. C. ZIENKIEWICZ and Y. K. CHEUNG, 'Stresses in buttress dams', *Water Power*, **17**, 69, 1965.
10. K. TERZHAGI, *Theoretical Soil Mechanics*, Wiley, 1943.
11. O. C. ZIENKIEWICZ, C. HUMPHESON, and R. W. LEWIS, 'A unified approach to soil mechanics problems, including plasticity and visco-plasticity', *Int. Symp. on Numerical Methods in Soil and Rock Mechanics*, Karlsruhe, 1975. See also Chapter 4 of *Finite Elements in Geomechanics* (ed. G. Gudehus), pp. 151–78, Wiley, 1977.
12. O. C. ZIENKIEWICZ, *Finite Element Method*, 3rd ed., McGraw-Hill, 1977.
13. L. R. HERRMANN, 'Elasticity equations for incompressible, or nearly incompressible materials by a variational theorem', *JAIAA*, **3**, 1896, 1965.

Axisymmetric stress analysis

4.1 Introduction

The problem of stress distribution in bodies of revolution (axisymmetric solids) under axisymmetric loading is of considerable practical interest. The mathematical problems presented are very similar to those of plane stress and plane strain as, once again, the situation is two dimensional.[1,2] By symmetry, the two components of displacements in any plane section of the body along its axis of symmetry define completely the state of strain and, therefore, the state of stress. Such a cross-section is shown in Fig. 4.1. If r and z denote respectively the radial and axial coordinates of a point, with u and v being the corresponding displacements, it can readily be seen that precisely the same displacement functions as those used in Chapter 3 can be used to define the displacements within the triangular element i, j, m shown.

The volume of material associated with an 'element' is now that of a body of revolution indicated on Fig. 4.1, and all integrations have to be referred to this.

The triangular element is again used mainly for illustrative purposes, the principles developed being completely general.

In plane stress or strain problems it was shown that internal work was associated with three strain components in the coordinate plane, the stress component normal to this plane not being involved due to zero values of either the stress or the strain.

In the axisymmetrical situation any radial displacement automatically induces a strain in the circumferential direction, and as the stresses in this direction are certainly non-zero, this fourth component of strain and of the associated stress has to be considered. Here lies the essential difference in the treatment of the axisymmetric situation.

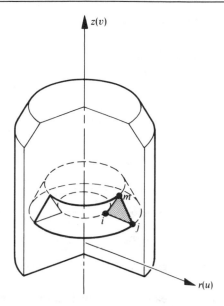

Fig. 4.1 Element of an axisymmetric solid

The reader will find the algebra involved in this chapter somewhat more tedious than that in the previous one but, essentially, identical operations are once again involved, following the general formulation of Chapter 2.

4.2 Element characteristics

4.2.1 *Displacement function.* Using the triangular shape of element (Fig. 4.1) with the nodes i, j, m numbered in the anticlockwise sense, we define the nodal displacement by its two components as

$$\mathbf{a}_i = \begin{Bmatrix} u_i \\ v_i \end{Bmatrix} \tag{4.1}$$

and the element displacements by the vector

$$a^e = \begin{Bmatrix} a_i \\ a_j \\ a_m \end{Bmatrix} \tag{4.2}$$

Obviously, as in Sec. 3.2.1, a linear polynomial can be used to define uniquely the displacements within the element. As the algebra involved is identical to that of Chapter 3 it will not be repeated here. The displacement field is now given again by Eq. (3.7):

$$\mathbf{u} = \begin{Bmatrix} u \\ v \end{Bmatrix} = [\mathbf{I}N_i, \ \mathbf{I}N_j, \ \mathbf{I}N_m]\mathbf{a}^e \tag{4.3}$$

with

$$N_i = \frac{a_i + b_i r + c_i z}{2\Delta}, \qquad \text{etc.}$$

and $\mathbf{I}$ a two-by-two identity matrix. In the above

$$a_i = r_j z_m - r_m z_j$$
$$b_i = z_j - z_m = z_{jm} \tag{4.4}$$
$$c_i = r_m - r_j = r_{mj}$$

etc., in cyclic order. Once again Δ is the area of the element triangle.

4.2.2 *Strain* (*total*). As already mentioned, four components of strain have now to be considered. These are, in fact, all the non-zero strain components possible in an axisymmetric deformation. Figure 4.2 illustrates and defines these strains and the associated stresses.

The strain vector defined below lists the strain components involved and defines them in terms of the displacements of a point. The expressions involved are almost self-evident and will not be derived here. The interested reader can consult a standard elasticity textbook[3] for the full

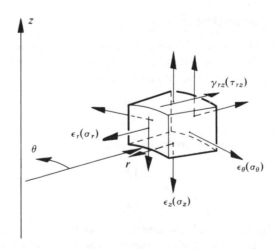

Fig. 4.2   Strains and stresses involved in the analysis of axisymmetric solids

derivation. We have thus

$$\varepsilon = \begin{Bmatrix} \varepsilon_z \\ \varepsilon_r \\ \varepsilon_\theta \\ \gamma_{rz} \end{Bmatrix} = \begin{Bmatrix} \dfrac{\partial v}{\partial z} \\ \dfrac{\partial u}{\partial r} \\ \dfrac{u}{r} \\ \dfrac{\partial u}{\partial z} + \dfrac{\partial v}{\partial r} \end{Bmatrix} = \mathbf{S}u \qquad (4.5)$$

Using the displacement functions defined by Eqs (4.3) and (4.4) we have

$$\varepsilon = \mathbf{B}a^e = [\mathbf{B}_i, \mathbf{B}_j, \mathbf{B}_m]a^e$$

in which

$$\mathbf{B}_i = \begin{bmatrix} 0 & , & \dfrac{\partial N_i}{\partial z} \\ \dfrac{\partial N_i}{\partial r} & , & 0 \\ \dfrac{1}{r} N_i & , & 0 \\ \dfrac{\partial N_i}{\partial z} & , & \dfrac{\partial N_i}{\partial r} \end{bmatrix} = \dfrac{1}{2\Delta} \begin{bmatrix} 0 & , & c_i \\ b_i & , & 0 \\ \dfrac{a_i}{r} + b_i + \dfrac{c_i z}{r} & , & 0 \\ c_i & , & b_i \end{bmatrix} , \quad \text{etc.} \quad (4.6)$$

With the **B** matrix now involving the coordinates r and z, the strains are no longer constant within an element as in the plane stress or strain case. This strain variation is due to the ε_θ term. If the imposed nodal displacements are such that u is proportional to r then indeed the strains will all be constant. As this is the only state of displacement coincident with a constant strain condition it is clear that the displacement function satisfies the basic criterion of Chapter 2.

4.2.3 *Initial strain (thermal strain)*. In general, four independent components of initial strain vector can be envisaged:

$$\varepsilon_0 = \begin{Bmatrix} \varepsilon_{z0} \\ \varepsilon_{r0} \\ \varepsilon_{\theta 0} \\ \gamma_{rz0} \end{Bmatrix} \qquad (4.7)$$

Although this can, in general, be variable within the element, it will be convenient to take the initial strain as constant there.

The most frequently encountered case of initial strain will be that due to a thermal expansion. For an isotropic material we shall have then

$$\varepsilon_0 = \left\{ \begin{array}{c} \alpha\theta^e \\ \alpha\theta^e \\ \alpha\theta^e \\ 0 \end{array} \right\} \tag{4.8}$$

where θ^e is the average temperature rise in an element and α is the coefficient of thermal expansion.

A general case of anisotropy need not be considered since axial symmetry would be impossible to achieve under such circumstances. A case of some interest in practice is that of a 'stratified' material, similar to the one discussed in Chapter 3, in which the plane of isotropy is normal to the axis of symmetry (Fig. 4.3). Here, two different expansion coefficients are possible: one in the axial direction α_z and another in the plane normal to it, α_r.

Now the initial thermal strain becomes

$$\varepsilon_0 = \left\{ \begin{array}{c} \alpha_z\,\theta^e \\ \alpha_r\,\theta^e \\ \alpha_r\,\theta^e \\ 0 \end{array} \right\} \tag{4.9}$$

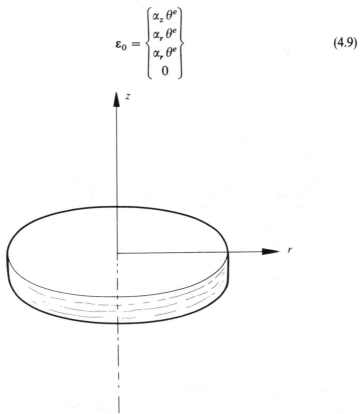

Fig. 4.3   Axisymmetrically stratified material

Practical cases of such 'stratified' anisotropy often arise in laminated or fibreglass construction of machine components.

4.2.4 *Elasticity matrix.* The elasticity matrix **D** linking the strains ε and the stresses σ in the standard form [Eq. (2.5)],

$$\sigma = \begin{Bmatrix} \sigma_z \\ \sigma_r \\ \sigma_\theta \\ \tau_{rz} \end{Bmatrix} = \mathbf{D}(\varepsilon - \varepsilon_0) + \sigma_0$$

needs now to be derived.

The anisotropic, 'stratified' material will be first considered, as the isotropic case can be simply presented as a special case.

Anisotropic, stratified, material (Fig. 4.3). With the z axis representing the normal to the planes of stratification we can rewrite Eqs (3.22) (again ignoring the initial strains and stresses for convenience) as

$$\varepsilon_z = \frac{\sigma_z}{E_2} - \frac{v_2 \sigma_r}{E_2} - \frac{v_2 \sigma_\theta}{E_2}$$

$$\varepsilon_r = -\frac{v_2 \sigma_z}{E_2} + \frac{\sigma_r}{E_1} - \frac{v_1 \sigma_\theta}{E_1}$$

$$\varepsilon_\theta = -\frac{v_2 \sigma_z}{E_2} - \frac{v_1 \sigma_r}{E_1} + \frac{\sigma_\theta}{E_1}$$

$$\gamma_{zr} = \frac{\tau_{zr}}{G_2}$$

(4.10)

Writing again

$$\frac{E_1}{E_2} = n \quad \text{and} \quad \frac{G_2}{E_2} = m$$

we have, on solving for the stresses, that

$$\mathbf{D} = \frac{E_2}{(1 + v_1)(1 - v_1 - 2nv_2^2)}$$

$$\times \begin{bmatrix} 1 - v_1^2 & nv_2(1 + v_1), & nv_2(1 + v_1), & 0 \\ & n(1 - nv_2^2), & (v_1 + nv_2^2)n, & 0 \\ & & n(1 - nv_2^2), & 0 \\ \text{symmetric} & , & m(1 + v_1) \times (1 - v_1 - 2nv_2^2) \end{bmatrix}$$

(4.11)

Isotropic material. For an isotropic material we can obtain the **D** matrix by taking

$$E_1 = E_2 = E \quad \text{or} \quad n = 1$$

and

$$v_1 = v_2 = v$$

and using the well-known relationship between isotropic elastic constants

$$\frac{G_2}{E_2} = \frac{G}{E} = m = \frac{1}{2(1 + v)}$$

Substituting in Eq. (4.11) we have now

$$\mathbf{D} = \frac{E(1 - v)}{(1 + v)(1 - 2v)} \begin{bmatrix} 1, & \dfrac{v}{1 - v}, & \dfrac{v}{1 - v}, & 0 \\ & 1 & , & \dfrac{v}{1 - v}, & 0 \\ & & 1 & , & 0 \\ & \text{symmetric} & & \dfrac{1 - 2v}{2(1 - v)} \end{bmatrix} \tag{4.12}$$

4.2.5 *The stiffness matrix.* The stiffness matrix of the element *ijm* can now be computed according to the general relationship Eq. (2.13). Remembering that the volume integral has to be taken over the whole ring of material we have

$$\mathbf{K}_{ij}^e = 2\pi \int \mathbf{B}_i^{\mathrm{T}} \mathbf{D} \mathbf{B}_j \, r \, dr \, dz \tag{4.13}$$

with **B** given by Eq. (4.6) and **D** by either Eq. (4.11) or Eq. (4.12), depending on the material.

The integration cannot now be performed as simply as was the case in the plane stress problem because the **B** matrix depends on the coordinates. Two possibilities exist: the first that of numerical integration and the second of an explicit multiplication and term-by-term integration.

The simplest approximate procedure is to evaluate $\bar{\mathbf{B}}$ for a centroidal point

$$\bar{r} = \frac{r_i + r_j + r_m}{3}$$

and

$$\bar{z} = \frac{z_i + z_j + z_m}{3}$$

In this case we have simply as a first approximation

$$\mathbf{K}_{ij}^e = 2\pi \bar{\mathbf{B}}_i^{\mathrm{T}} \mathbf{D} \bar{\mathbf{B}}_j \, \bar{r} \Delta \tag{4.14}$$

with Δ being the triangle area.

More elaborate numerical integration schemes could be used by evaluating the integrand at several points of the triangle. Such methods will be discussed in detail in Chapter 8. However, it can be shown that if the numerical integration is of such an order that the volume of the element is exactly determined by it, then in the limit of subdivision, the solution will converge to the exact answer.[4] The 'one point' integration suggested here is of such a type, as it is well known that the volume of a body of revolution is given exactly by the product of the area and the path swept around by its centroid. With the simple triangular element used here a fairly fine subdivision is in any case needed for accuracy and most practical programs use the simple approximation which, surprisingly perhaps, is in fact usually superior to the exact integration (see Chapter 11). One reason for this is the occurrence of logarithmic terms in the exact formulation. These involve ratios of the type r_i/r_m and, when the element is at a large distance from the axis, such terms tend to unity and evaluation of the logarithm is inaccurate.

4.2.6 *External nodal forces.* In the case of the two-dimensional problems of the previous chapter the question of assigning of the external loads was so obvious as not to need further comment. In the present case, however, it is important to realize that the nodal forces represent a combined effect of the force acting along the whole circumference of the circle forming the element 'node'. This point was already brought out in the integration of the expressions for the stiffness of an element, such integrations being conducted over the whole ring.

Thus, if $\bar{R}$ represents the radial component of force per unit length of the circumference of a node or a radius r, the external 'force' which will have to be introduced in the computation is

$$2\pi r \bar{R}$$

In the axial direction we shall, similarly, have

$$2\pi r \bar{Z}$$

to represent the combined effect of axial forces.

4.2.7 *Nodal forces due to initial strain.* Again, by Eq. (2.13),

$$\mathbf{f}^e = -2\pi \int \mathbf{B}^T \mathbf{D} \varepsilon_0 \, r \, dr \, dz \tag{4.15}$$

or noting that ε_0 is constant,

$$\mathbf{f}_i^e = -2\pi \left(\int \mathbf{B}_i^T r \, dr \, dz \right) \mathbf{D} \varepsilon_0 \tag{4.16}$$

The integration can be performed in a similar manner to that used in the determination of the stiffness.

It will be readily seen that, again, an approximate expression using a centroidal value is

$$f_i^e = -2\pi \bar{B}_i^T D \varepsilon_0 \, \bar{r} \Delta \qquad (4.17)$$

Initial stress forces are treated in an identical manner.

4.2.8 *Distributed body forces.* Distributed body forces, such as those due to gravity (if acting along the z axis), centrifugal force in rotating machine parts, or pore pressure, often occur in axisymmetric problems.

Let such forces be denoted by

$$\mathbf{b} = \begin{Bmatrix} b_r \\ b_z \end{Bmatrix} \qquad (4.18)$$

per unit volume of material in directions of r and z respectively. By the general Eq. (2.13) we have

$$\mathbf{f}_i^e = -2\pi \int \mathbf{I} N_i \begin{Bmatrix} b_r \\ b_z \end{Bmatrix} r \, dr \, dz \qquad (4.19)$$

Using a coordinate shift similar to that of Sec. 3.2.7 it is easy to show that the first approximation, if the body forces are constant, results in

$$\mathbf{f}_i^e = -2\pi \begin{Bmatrix} b_r \\ b_z \end{Bmatrix} \frac{\bar{r} \Delta}{3} \qquad (4.20)$$

Although this is not exact the error term will be found to decrease with reduction of element size and, as it is also self-balancing, it will not introduce inaccuracies. Indeed, as will be shown in Chapter 11, the convergence rate is maintained.

If the body forces are given by a potential similar to that defined in Sec. 3.2.8, i.e.,

$$b_r = -\frac{\partial \phi}{\partial r} \qquad b_z = -\frac{\partial \phi}{\partial z} \qquad (4.21)$$

and if this potential is defined linearly by its nodal values, an expression equivalent to Eq. (3.36) can again be used with the same degree of approximation.

In many problems the body forces vary proportionately to r. For example in rotating machinery we have centrifugal forces

$$b_r = \omega^2 \rho r \qquad (4.22)$$

where ω is the angular velocity and ρ the density of the material.

4.2.9 *Evaluation of stresses.* The stresses now vary throughout the element, as will be appreciated from Eqs (4.5) and (4.6). It is convenient now to evaluate the average stress at the centroid of the element. The stress matrix resulting from Eqs (4.6) and (2.3) gives there, as usual,

$$\bar{\sigma}^e = \mathbf{D}\bar{\mathbf{B}}\mathbf{a}^e - \mathbf{D}\varepsilon_0 + \sigma_0 \qquad (4.23)$$

It will be found that a certain amount of oscillation of stress values between elements occurs and better approximation can be achieved by averaging nodal stresses.

4.3 Some illustrative examples

Test problems such as those of a cylinder under constant axial or radial stress give, as indeed would be expected, solutions which correspond to exact ones. This is again an obvious corollary of the ability of the displacement function to reproduce constant strain conditions.

A problem for which an exact solution is available and in which almost linear stress gradients occur is that of a sphere subject to internal pressure. Figure 4.4(a) shows the centroidal stresses obtained using rather a coarse mesh, and the stress oscillation around the exact values should be noted. (This oscillation becomes even more pronounced at larger values of Poisson's ratio although the exact solution is independent of it.) In Fig. 4.4(b) the very much better approximation obtained by averaging the stresses at nodal points is shown, and in Fig. 4.4(c) a further improvement is given by element averaging. The close agreement with exact solution even for the very coarse subdivision used here shows the accuracy achievable. The displacements at nodes compared with the exact solution are given in Fig. 4.5.

In Fig. 4.6 thermal stresses in the same sphere are computed for a steady-state temperature variation shown. Again, excellent accuracy is demonstrated by comparison with the exact solution.

4.4 Practical applications

Two examples of practical applications of the programs available for axisymmetrical stress distribution are given here.

4.4.1 *A prestressed concrete reactor pressure vessel.* Figure 4.7 shows the stress distribution in a relatively simple prototype pressure vessel. Due to symmetry only one-half of the vessel is analysed, the results given here referring to the components of stress due to an internal pressure. Similar results due to the effect of prestressing cables are readily obtained by putting in the appropriate nodal loads due to these cables.

In Fig. 4.8 contours of equal major principal stresses caused by temperature are shown. The thermal state is due to a steady-state heat

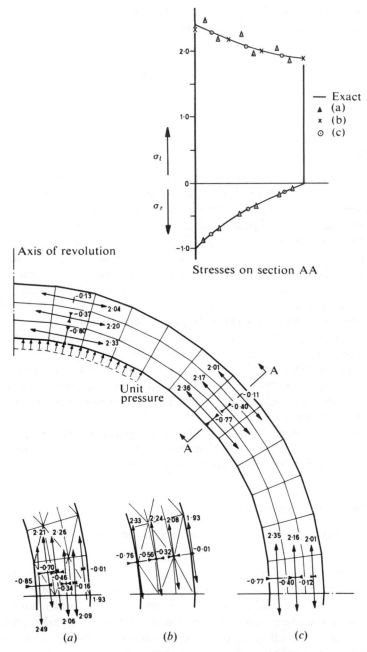

Fig. 4.4   Stresses in a sphere subject to an internal pressure (Poisson's ratio $v = 0.3$: (a) triangular mesh—centroidal values; (b) triangular mesh—nodal averages; (c) quadrilateral mesh obtained by averaging adjacent triangles

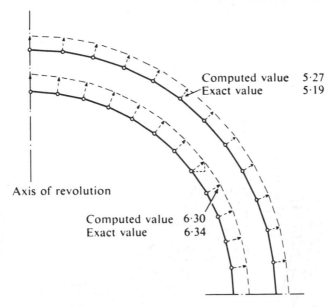

Fig. 4.5 Displacements of internal and external surfaces of sphere under loading of Fig. 4.4

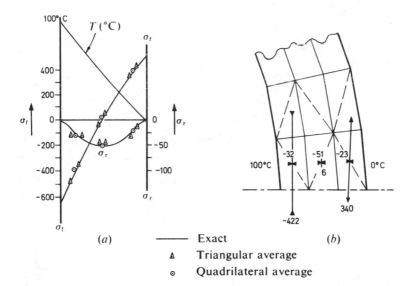

(a) ——— Exact (b)
▲ Triangular average
⊚ Quadrilateral average

Fig. 4.6 Sphere subject to steady-state heat flow (100°C internal temperature, 0°C external temperature): (a) temperature and stress variation on radial section; (b) 'quadrilateral' averages

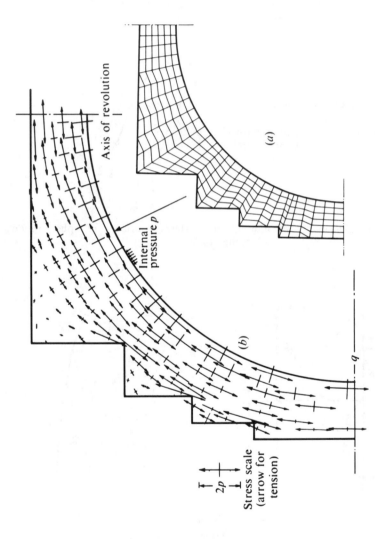

Fig. 4.7   A reactor pressure vessel. (*a*) 'Quadrilateral' mesh used in analysis; this was generated automatically by a computer. (*b*) Stresses due to a uniform internal pressure (automatic computer plot). Solution based on quadrilateral averages. (Poisson's ratio $v = 0.15$)

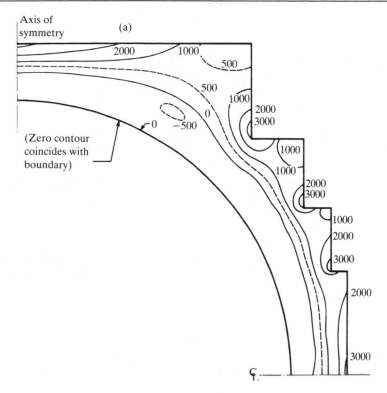

Axis of symmetry

(a)

(Zero contour coincides with boundary)

Fig. 4.8 A reactor pressure vessel. Thermal stresses due to steady-state heat conduction. Contours of major principal stress in pounds per square inch. (Interior temperature 400°C, exterior temperature 0°C, $\alpha = 5 \times 10^{-6}/°C$. $E = 2.58 \times 10^6$ lb/in, $v = 0.15$)

conduction and itself was found by the finite element method in a way described in Chapter 10.

4.4.2 *Foundation pile.* Figure 4.9 shows the stress distribution around a foundation pile penetrating two different strata. This non-homogeneous problem presents no difficulties and is treated by the standard program.

4.5 Non-symmetrical loading

The method described in the present chapter can be extended to deal with non-symmetrical loading. If the circumferential loading variation is expressed in circular harmonics then it is still possible to focus attention on one axial section although the degree of freedom is now increased to three.

Some details of this process are described in references 5 and 6.

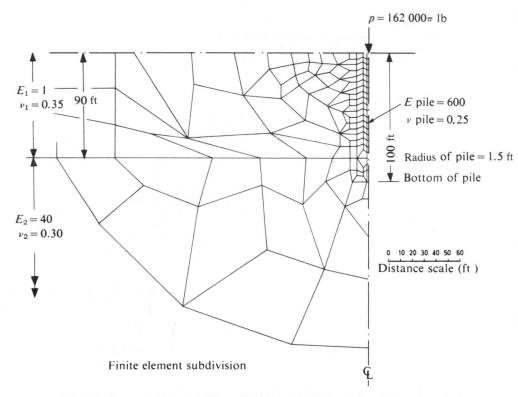

$p = 162\ 000\pi$ lb

$E_1 = 1$
$\nu_1 = 0.35$ 90 ft

E pile = 600
ν pile = 0.25

100 ft

Radius of pile = 1.5 ft

Bottom of pile

$E_2 = 40$
$\nu_2 = 0.30$

0 10 20 30 40 50 60
Distance scale (ft)

Finite element subdivision

$\underset{L}{\mathbb{C}}$

Fig. 4.9(a)   A pile in stratified soil. Irregular mesh and data for problem

4.6 Axisymmetry—plane strain and plane stress

In the previous chapter we have noted that plane stress and strain analysis was done in terms of three stress and strain components and, indeed, both cases would be generally incorporated in a single program with an indicator changing appropriate constants in the matrix **D**. Doing this loses track of the σ_z component in the plane strain case which has to be separately evaluated. Further, special expressions [viz. Eq. (3.13)] had to be used to introduce initial strains. This is inconvenient (especially when non-linear constitutive laws are used), and an alternative of writing the plane strain case in terms of four stress–strain components as a special case of axisymmetric analysis is highly recommended.

  If the axisymmetric strain definition of Eq. (4.5) is examined, we note that $r = \infty$ gives $\varepsilon_\theta \equiv 0$. Thus plane strain conditions are obtained. If we replace the coordinates

$$r \text{ and } z \qquad \text{by} \qquad x \text{ and } y$$

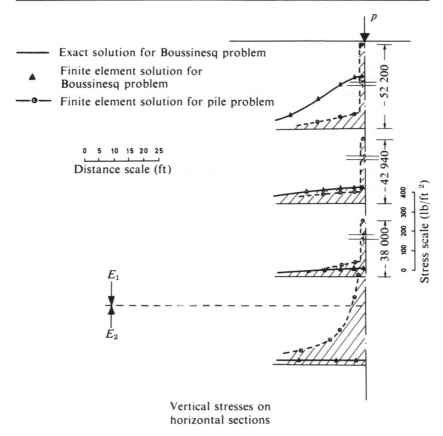

Fig. 4.9(b) A pile in stratified soil. Plot of vertical stresses on horizontal sections. Solution also plotted for Boussinesq problem obtained by making $E_1 = E_2 = E_{pile}$, and this is compared with exact values

and further change in the stiffness expressions the volume of integration

$$2\pi r \quad \text{to} \quad 1$$

the axisymmetric formulation becomes available from plane strain directly.

Plane stress conditions can similarly be incorporated, requiring in addition substitution of the axisymmetric **D** matrix by the Eqs (3.18) or (3.23) augmented by an appropriate zero row and column. Thus, at the cost of an additional storage of the fourth stress and strain component, all the cases discussed can be incorporated in a single format.

References

1. R. W. CLOUGH, Chapter 7 of *Stress Analysis* (eds O. C. Zienkiewicz and G. S. Holister), Wiley, 1965.
2. R. W. CLOUGH and Y. R. RASHID, 'Finite element analysis of axi-symmetric solids', *Proc. ASCE*, **91**, EM.1, 71, 1965.
3. S. TIMOSHENKO and J. N. GOODIER, *Theory of Elasticity*, 2nd ed., McGraw-Hill, 1951.
4. B. M. IRONS, 'Comment on "Stiffness matrices for section element" by I. R. Raju and A. K. Rao', *JAIAA*, **7**, 156–7, 1969.
5. E. L. WILSON, 'Structural analysis of axisymmetric solids', *JAIAA*, **3**, 2269–74, 1965.
6. O. C. ZIENKIEWICZ, *The Finite Element Method*, 3rd ed., McGraw-Hill, 1977.

5

Three-dimensional stress analysis

5.1 Introduction

It will have become obvious to the reader by this stage of the book that there is but one further step to apply the general finite element procedure to fully three-dimensional problems of stress analysis. Such problems embrace clearly all the practical cases, though for some, the various two-dimensional approximations give an adequate and more economical 'model'.

The simplest two-dimensional continuum element is a triangle. In three dimensions its equivalent is a tetrahedron, an element with four nodal corners—and this chapter will deal with the basic formulation of such an element. Immediately, a difficulty not encountered previously is presented. It is one of ordering of the nodal numbers and, in fact, of a suitable representation of a body divided into such elements.

The first suggestions for use of the simple tetrahedral element appear to be those of Gallagher et al.[1] and Melosh.[2] Argyris[3,4] elaborated further on the theme and Rashid and Rockenhauser[5] have shown that with the largest modern computers such a formulation can still be applied to realistic problems.

It is immediately obvious, however, that the number of simple tetrahedral elements which has to be used to achieve a given degree of accuracy has to be very large. This will result in very large numbers of simultaneous equations in practical problems, which may place a severe limitation on the use of the method in practice. Further, the band width of the resulting equation system becomes large, leading to big computer storage requirements.

To realize the order of magnitude of the problems presented let us assume that the accuracy of a triangle in two-dimensional analysis is comparable to that of a tetrahedron in three dimensions. If an adequate stress analysis of a square, two-dimensional region requires a mesh of some $20 \times 20 = 400$ nodes, the total number of simultaneous equations is around 800 given two displacement variables for node. (This is a fairly realistic figure.) The band width of the matrix involves 20 nodes (Chapter 15), i.e., some 40 variables.

An equivalent three-dimensional region is that of a cube with $20 \times 20 \times 20 = 8000$ nodes. The total number of simultaneous equations is now some 24 000 as three displacement variables have to be specified. Further, the band width now involves an interconnection of some $20 \times 20 = 400$ nodes or 1200 variables.

Given that with usual solution techniques the computation effort is roughly proportional to the number of equations and to the square of the band width, the magnitude of the problems can be appreciated. It is not surprising therefore that efforts to improve accuracy by use of complex elements with many degrees of freedom have been strongest in the area of three-dimensional analysis.[6–10] The development and practical application of such elements will be described in the following chapters. However, the presentation of this chapter gives all the necessary ingredients of formulation for three-dimensional elastic problems and so follows directly from the previous ones. Extension to more elaborate elements will be self-evident.

5.2 Tetrahedral element characteristics

5.2.1 *Displacement functions.* Figure 5.1 illustrates a tetrahedral element i, j, m, p in space defined by the x, y, and z coordinates.

The state of displacement of a point is defined by three displacement components, u, v, and w, in directions of the three coordinates x, y, and z. Thus

$$\mathbf{u} = \begin{Bmatrix} u \\ v \\ w \end{Bmatrix} \tag{5.1}$$

Just as in a plane triangle where a linear variation of a quantity was defined by its three nodal values, here a linear variation will be defined by the four nodal values. In analogy to Eq. (3.3) we can write, for instance,

$$u = \alpha_1 + \alpha_2 x + \alpha_3 y + \alpha_4 z \tag{5.2}$$

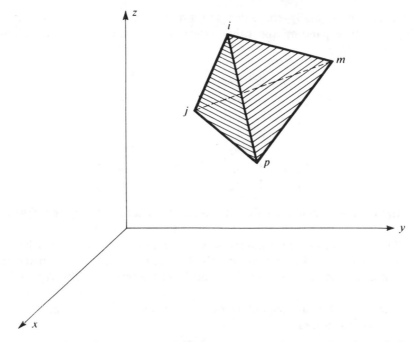

Fig. 5.1 A tetrahedral volume. (Always use a consistent order of numbering, e.g., starting with p count the other nodes in an anticlockwise order as viewed from p—*pijm* or *mipj*, etc.)

Equating the values of displacement at the nodes we have four equations of the type

$$u_i = \alpha_1 + \alpha_2 x_i + \alpha_3 y_i + \alpha_4 z_i, \qquad \text{etc.} \qquad (5.3)$$

from which α_1 to α_4 can be evaluated.

Again, it is possible to write this solution in a form similar to that of Eq. (3.5) by using a determinant form, i.e.,

$$u = \frac{1}{6V} [(a_i + b_i x + c_i y + d_i z)u_i + (a_j + b_j x + c_j y + d_j z)u_j$$

$$+ (a_m + b_m x + c_m y + d_m z)u_m + (a_p + b_p x + c_p y + d_p z)u_p] \qquad (5.4)$$

with

$$6V = \det \begin{vmatrix} 1 & x_i & y_i & z_i \\ 1 & x_j & y_j & z_j \\ 1 & x_m & y_m & z_m \\ 1 & x_p & y_p & z_p \end{vmatrix} \qquad (5.5a)$$

in which, incidentally, the value V represents the volume of the tetra-hedron. By expanding the other relevant determinants into their co-factors we have

$$
a_i = \det \begin{vmatrix} x_j & y_j & z_j \\ x_m & y_m & z_m \\ x_p & y_p & z_p \end{vmatrix} \qquad b_i = -\det \begin{vmatrix} 1 & y_j & z_j \\ 1 & y_m & z_m \\ 1 & y_p & z_p \end{vmatrix}
$$

$$
c_i = -\det \begin{vmatrix} x_j & 1 & z_j \\ x_m & 1 & z_m \\ x_p & 1 & z_p \end{vmatrix} \qquad d_i = -\det \begin{vmatrix} x_j & y_j & 1 \\ x_m & y_m & 1 \\ x_p & y_p & 1 \end{vmatrix}
$$
(5.5b)

with the other constants defined by cyclic interchange of the subscripts in the order p, i, j, m.

The ordering of nodal numbers p, i, j, m must follow a 'right-hand' rule obvious from Fig. 5.1. In this the first three nodes are numbered in an anticlockwise manner when viewed from the last one. (See Appendix 4.)

The element displacement is defined by the 12 displacement com-ponents of the nodes as

$$
\mathbf{a}^e = \begin{Bmatrix} \mathbf{a}_i \\ \mathbf{a}_j \\ \mathbf{a}_m \\ \mathbf{a}_p \end{Bmatrix}
$$
(5.6)

with

$$
\mathbf{a}_i = \begin{Bmatrix} u_i \\ v_i \\ w_i \end{Bmatrix} \qquad \text{etc.}
$$

We can write the displacements of an arbitrary point as

$$
\mathbf{u} = [\mathbf{I}N_i, \mathbf{I}N_j, \mathbf{I}N_m, \mathbf{I}N_p]\mathbf{a}^e
$$
(5.7)

with shape functions defined as

$$
N_i = \frac{a_i + b_i x + c_i y + d_i z}{6V}, \qquad \text{etc.}
$$
(5.8)

and $\mathbf{I}$ being a three by three identity matrix.

Once again the displacement functions used will obviously satisfy con-tinuity requirements on interfaces between various elements. This fact is a direct corollary of the linear nature of the variation of displacement.

5.2.2. *Strain matrix.* Six strain components are relevant in full three-dimensional analysis. The strain matrix can now be defined as

$$
\varepsilon = \begin{Bmatrix} \varepsilon_x \\ \varepsilon_y \\ \varepsilon_z \\ \gamma_{xy} \\ \gamma_{yz} \\ \gamma_{zx} \end{Bmatrix} = \begin{Bmatrix} \dfrac{\partial u}{\partial x} \\[6pt] \dfrac{\partial v}{\partial y} \\[6pt] \dfrac{\partial w}{\partial z} \\[6pt] \dfrac{\partial u}{\partial y} + \dfrac{\partial v}{\partial x} \\[6pt] \dfrac{\partial v}{\partial z} + \dfrac{\partial w}{\partial y} \\[6pt] \dfrac{\partial w}{\partial x} + \dfrac{\partial u}{\partial z} \end{Bmatrix} = \mathbf{S} u \qquad (5.9)
$$

following the standard notation of Timoshenko's elasticity text. Using Eqs (5.4) to (5.7) it is an easy matter to verify that

$$
\varepsilon = \mathbf{B} a^e = [\mathbf{B}_i, \mathbf{B}_j, \mathbf{B}_m, \mathbf{B}_p] a^e \qquad (5.10)
$$

in which

$$
\mathbf{B}_i = \begin{bmatrix} \dfrac{\partial N_i}{\partial x}, & 0, & 0 \\[6pt] 0, & \dfrac{\partial N_i}{\partial y}, & 0 \\[6pt] 0, & 0, & \dfrac{\partial N_i}{\partial z} \\[6pt] \dfrac{\partial N_i}{\partial y}, & \dfrac{\partial N_i}{\partial x}, & 0 \\[6pt] 0, & \dfrac{\partial N_i}{\partial z}, & \dfrac{\partial N_i}{\partial y} \\[6pt] \dfrac{\partial N_i}{\partial z}, & 0, & \dfrac{\partial N_i}{\partial x} \end{bmatrix} = \frac{1}{6V} \begin{bmatrix} b_i, & 0, & 0 \\ 0, & c_i, & 0 \\ 0, & 0, & d_i \\ c_i, & b_i, & 0 \\ 0, & d_i, & c_i \\ d_i, & 0, & b_i \end{bmatrix} \qquad (5.11)
$$

with other submatrices obtained in a similar manner simply by interchange of subscripts.

Initial strains, such as those due to thermal expansion, can be written in the usual way as a six-component vector which, for example, in an isotropic thermal expansion is simply

$$\boldsymbol{\varepsilon}_0 = \begin{Bmatrix} \alpha\theta^e \\ \alpha\theta^e \\ \alpha\theta^e \\ 0 \\ 0 \\ 0 \end{Bmatrix} \tag{5.12}$$

with α being the expansion coefficient and θ^e the average element temperature rise.

5.2.3 *Elasticity matrix.* With complete anisotropy the **D** matrix relating the six stress components to the strain components can contain 21 independent constants (see Sec. 3.2.4).

In general, thus,

$$\boldsymbol{\sigma} = \begin{Bmatrix} \sigma_x \\ \sigma_y \\ \sigma_z \\ \tau_{xy} \\ \tau_{yz} \\ \tau_{zx} \end{Bmatrix} = \mathbf{D}(\boldsymbol{\varepsilon} - \boldsymbol{\varepsilon}_0) + \boldsymbol{\sigma}_0 \tag{5.13}$$

Although no difficulty presents itself in computation when dealing with such materials, since the multiplication will never be carried out explicitly, it is convenient to recapitulate here the **D** matrix for an isotropic material. This, in terms of the usual elastic constants E (modulus) and v (Poisson's ratio), can be written as

$$\mathbf{D} = \frac{E(1-v)}{(1+v)(1-2v)}$$

$$\begin{bmatrix} 1, & v/(1-v), & v/(1-v), & 0 & , & 0 & , & 0 \\ & 1 & , & v/(1-v), & 0 & , & 0 & , & 0 \\ & & 1 & , & 0 & , & 0 & , & 0 \\ & \text{symmetric} & & \dfrac{1-2v}{2(1-v)}, & 0 & , & 0 \\ & & & & \dfrac{1-2v}{2(1-v)}, & 0 \\ & & & & & \dfrac{1-2v}{2(1-v)} \end{bmatrix} \tag{5.14}$$

5.2.4 *Stiffness, stress, and load matrices.* The stiffness matrix defined by the general relationship Eq. (2.10) can now be explicitly integrated since the strain and stress components are constant within the element.

The general *ij* submatrix of the stiffness matrix will be a three by three matrix defined as

$$\mathbf{K}_{ij}^e = \mathbf{B}_i^T \mathbf{D} \mathbf{B}_j \, V^e \tag{5.15}$$

where V^e represents the volume of the elementary tetrahedron.

The nodal forces due to the initial strain become, similarly to Eq. (3.28),

$$\mathbf{f}_i^e = -\mathbf{B}_i^T \mathbf{D} \boldsymbol{\varepsilon}_0 \, V^e \tag{5.16}$$

with a similar expression for forces due to initial stresses.

In fact, the similarity with the expressions and results of Chapter 3 is such that further explicit formulation is unnecessary. The reader will find no difficulty in repeating the various steps needed for the formulation of a computer program.

Distributed body forces can once again be expressed in terms of their b_x, b_y, and b_z components or in terms of the body force potential. Not surprisingly, it will once more be found that if the body forces are constant the nodal components of the total resultant are distributed in four equal parts [see Eq. (3.30)].

5.3 Composite elements with eight nodes

The division of a space volume into individual tetrahedra sometimes presents difficulties of visualization and could easily lead to errors in nodal numbering, etc., unless a fully automatic code is available. A more convenient subdivision of space is into eight-cornered brick elements. By sectioning a three-dimensional body parallel sections can be drawn and, each one being subdivided into quadrilaterals, a systematic way of element definition could be devised as in Fig. 5.2.

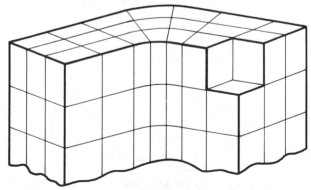

Fig. 5.2 A systematic way of dividing a three-dimensional object into 'brick'-type elements

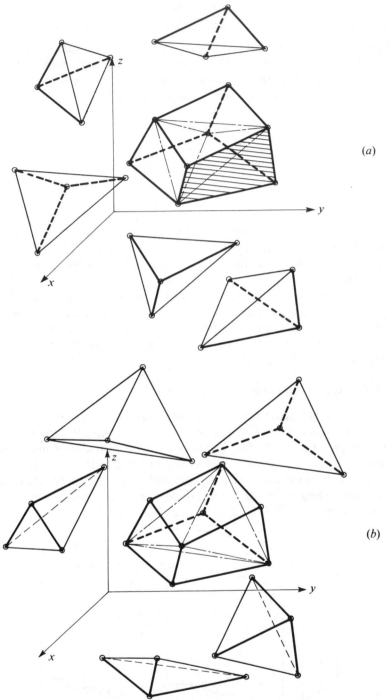

Fig. 5.3 Composite element with eight nodes and its subdivision into five tetra-hedra by alternatives (a) or (b)

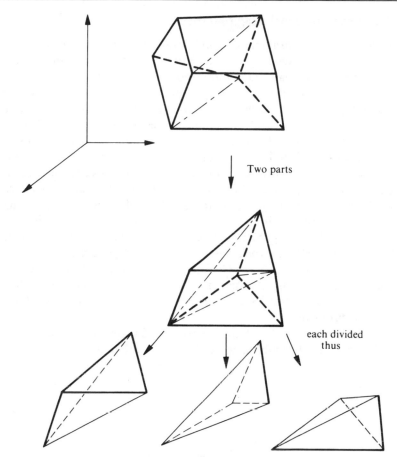

Fig. 5.4 A systematic way of splitting an eight-cornered brick into six tetrahedra

Such elements could be assembled automatically from several tetrahedra and the process of creating these tetrahedra left to a simple logical
program. For instance, Fig. 5.3 shows how a typical brick can be divided
into five tetrahedra in two (and only two) distinct ways. Indeed by
averaging the two types of subdivision a slight improvement of accuracy
can be obtained. Stresses could well be presented as averages for a whole
brick-like element or as final nodal averages.

In Fig. 5.4 an alternative subdivision of a brick into six tetrahedra is
shown. Here obviously the number of alternatives is very great.

In later chapters it will be seen how the basic bricks can be obtained
directly with more complex types of shape functions.

5.4 Examples and concluding remarks

A simple, illustrative example of application of simple, tetrahedral, elements is shown in Figs 5.5 and 5.6. Here the well-known Boussinesq problem of an elastic half-space with a point load is approximated to by analysing a cubic volume of space. Use of symmetry is made to reduce the size of the problem and the boundary displacements are prescribed in a manner shown in Fig. 5.5.[11] As zero displacements were prescribed at a finite distance below the load a correction obtained from the exact expression was applied before executing the plots shown in Fig. 5.6. Comparison of both stresses and displacement appears reasonable although it will be appreciated that the division is very coarse. However, even this trivial problem involved the solution of some 375 equations. More ambitious problems treated with simple tetrahedra are given in references 5 and 11. Figure 5.7, taken from the former, illustrates an analysis of a complex pressure vessel. Some 10 000 degrees of freedom are involved in this analysis. In Chapter 8 it will be seen how the use of complex elements permits a sufficiently accurate analysis to be performed with a much smaller total number of degrees of freedom for a very similar problem.

  Although we have in this chapter emphasized the easy visualization of a tetrahedral mesh through the use of brick-like subdivision, it is possible

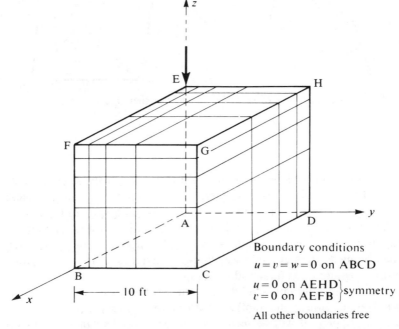

Boundary conditions

$u = v = w = 0$ on ABCD

$u = 0$ on AEHD
$v = 0$ on AEFB } symmetry

All other boundaries free

Fig. 5.5   The Boussinesq problem as one of three-dimensional stress analysis

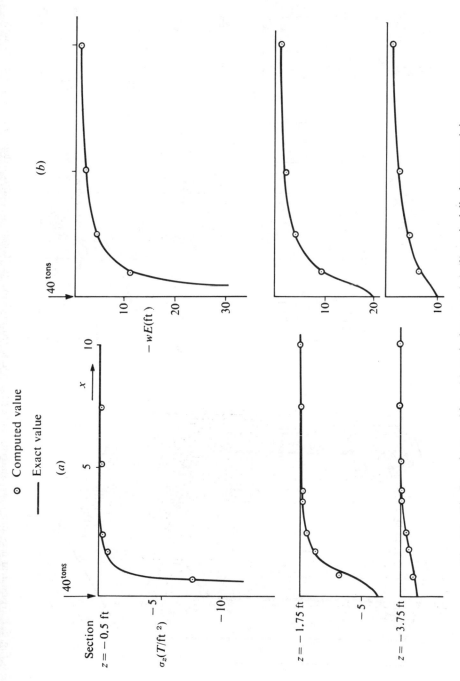

Fig. 5.6 The Boussineq problem: (a) vertical stresses (σ_z); (b) vertical displacements (w)

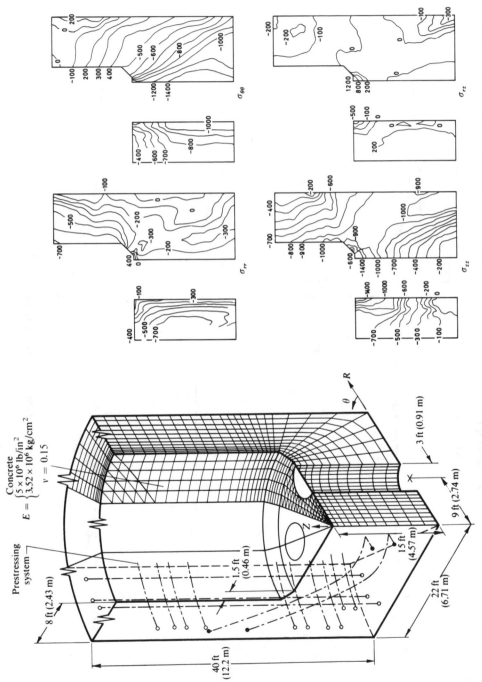

Fig. 5.7 A nuclear pressure vessel analysis using simple tetrahedral elements.[5] Geometry, subdivision, and some stress results

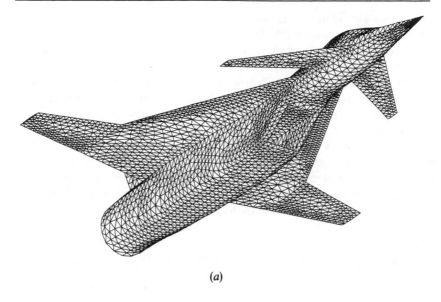

(a)

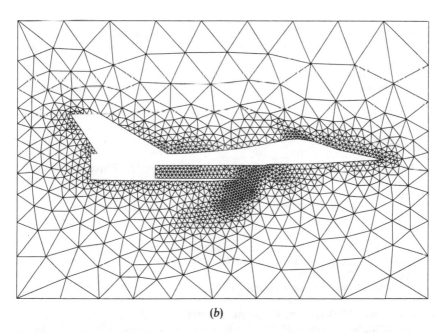

(b)

Fig. 5.8 An automatically generated mesh of tetrahedra for a specified mesh density in the exterior region on aircraft (a) and (b) an intersection of the mesh with the centreline plane

to generate automatically arbitrary tetrahedral meshes of great complexity with any prescribed mesh density distribution. The procedures follow the general pattern of automatic triangle generation[12] to which we shall refer in Chapter 14 when discussing efficient, adaptively constructed meshes, but, of course, the degree of complexity introduced is much greater in three dimensions. Some details of such a generator are described by Peraire et al.,[13] and Fig. 5.8 illustrates an intersection of such an automatically generated mesh with an outline of an aircraft. It is impractical to show the full plot of the mesh which contains over 30 000 nodes. The important point to note is that such meshes can be generated for any configuration which can be suitably described geometrically. Although this example concerns aerodynamics rather than elasticity, similar meshes can be generated in the latter context.

References

1. R. H. GALLAGHER, J. PADLOG, and P. P. BIJLAARD, 'Stress analysis of heated complex shapes', *ARS Journal*, 700–7, 1962.
2. R. J. MELOSH, 'Structural analysis of solids', *Proc. Am. Soc. Civ. Eng.*, **ST 4**, 205–23, Aug. 1963.
3. J. H. ARGYRIS, 'Matrix analysis of three-dimensional elastic media—small and large displacements', *JAIAA*, **3**, 45–51, Jan. 1965.
4. J. H. ARGYRIS, 'Three-dimensional anisotropic and inhomogeneous media—matrix analysis for small and large displacements', *Ingenieur Archiv.*, **34**, 33–55, 1965.
5. Y. R. RASHID and W. ROCKENHAUSER, 'Pressure vessel analysis by finite element techniques', *Proc. Conf. on Prestressed Concrete Pressure Vessels*, Inst. Civ. Eng., 1968.
6. J. H. ARGYRIS, 'Continua and discontinua', *Proc. Conf. Matrix Methods in Structural Mechanics*, Wright Patterson Air Force Base, Ohio, Oct. 1965.
7. B. M. IRONS, 'Engineering applications of numerical integration in stiffness methods', *JAIAA*, **4**, 2035–7, 1966.
8. J. G. ERGATOUDIS, B. M. IRONS, and O. C. ZIENKIEWICZ, 'Three dimensional analysis of arch dams and their foundations', *Proc. Symp. Arch Dams*, Inst. Civ. Eng., 1968.
9. J. H. ARGYRIS and J. C. REDSHAW, 'Three dimensional analysis of two arch dams by a finite element method', *Proc. Symp. Arch Dams*, Inst. Civ. Eng., 1968.
10. S. FJELD, 'Three dimensional theory of elastics', *Finite Element Methods in Stress Analysis* (eds I. Holand and K. Bell), Tech. Univ. of Norway, Tapir Press, Trondheim, 1969.
11. OLIVEIRA PEDRO, Thesis, Laboratorio Nacional-de Engenharia Civil, Lisbon, 1967.
12. J. PERAIRE, M. VAHDATI, K. MORGAN, and O. C. ZIENKIEWICZ, 'Adaptive remeshing for compressible flow computations', *J. Comp. Physics*, **72**, 449–466, 1987.
13. J. PERAIRE, J. PEIRO, L. FORMAGGIA, K. MORGAN, and O. C. ZIENKIEWICZ, 'Finite element Euler computations in three dimensions', *Int. J. Num. Meth. Eng.*, 1988 (to be published).

Tensor-indicial notation in the approximation of elasticity problems

6.1 Introduction

The matrix type of notation used in the preceding chapters for description of tensor quantities such as stresses and strains is compact and we believe easy to understand. However, in a computer program each quantity will still have to be identified by appropriate indices (viz. Chapter 15) and this conciseness does not always carry over its advantages. Further, many readers are accustomed to the use of indicial-tensor notation which is a standard tool of the 'elastician'. For this reason we shall recast the previous finite element formulation using such notation, which is used quite widely in the current finite element literature.

Some advantages of such reformulation vis-à-vis matrix form become apparent when evaluation of stiffness properties for isotropic media is considered. Here some multiplication operations previously necessary become redundant and complex programs can be written more economically.

If more complex elasticity problems involving large deformation have to be considered the use of tensorial notation is almost essential, to adopt, with relative ease, the results of the extensive literature dealing with such problems.

This chapter adds little new to the discretization ideas but repeats in a different language the results already presented.

6.2 Indicial notation

When describing cartesian coordinates or other *vector* quantities associated with these it is convenient to use Latin subscripts as follows:

$$x_a \qquad a = 1, 2, 3 \tag{6.1}$$

103

which describes (x_1, x_2, x_3) or (x, y, z) in previous notation. Similarly, for displacement we can write

$$u_a \qquad a = 1, 2, 3 \qquad (6.2)$$

This implies (u_1, u_2, u_3) or (u, v, w) previously used.

To avoid confusion with 'nodal' quantities to which previously we have also attached subscripts, we shall simply change their position to superscript. Thus

$$u_a^j \text{ has the same meaning as } \boldsymbol{u}_j \qquad (6.3)$$

used previously, etc.

In describing a stiffness coefficient two subscripts have been previously used but the submatrix $\mathbf{K}_{ij}$ implied 2×2 or 3×3 entries depending on whether two or three displacement components were involved. Now the scalar

$$K_{ab}^{ij} \qquad a, b = 1, 2, 3 \qquad (6.4)$$

defines completely the appropriate coefficient with term ab indicating the relative submatrix position (in this case for a three-dimensional displacement).

Note that for a symmetric matrix we have previously required that

$$\mathbf{K}_{ij} = \mathbf{K}_{ji}^{\mathrm{T}} \qquad (6.5\text{a})$$

In the indicial notation the same symmetry is implied if

$$K_{ab}^{ij} = K_{ba}^{ji} \qquad (6.5\text{b})$$

6.3 Derivatives and tensorial relations

In the indicial notation the derivative of any quantity with respect to a coordinate x_b is written compactly as

$$\frac{\partial}{\partial x_b} \equiv (\quad)_{,b} \qquad (6.6)$$

Thus we can write the 'gradient' of the displacement vector as

$$u_{a,b} \equiv \frac{\partial u_a}{\partial x_b} \qquad a, b = 1, 2, 3 \qquad (6.7)$$

and using the notion that (small) strains are the symmetric part of the displacement gradient,†

$$\varepsilon_{ab} = \tfrac{1}{2}(u_{a,\,b} + u_{b,\,a}) = \varepsilon_{ba} \qquad a, b = 1, 2, 3 \tag{6.8}$$

and (small) rotations are the skew or antisymmetric part of the displacement gradient:

$$\omega_{ab} = \tfrac{1}{2}(u_{a,\,b} - u_{b,\,a}) = -\omega_{ba} \qquad a, b = 1, 2, 3 \tag{6.9}$$

These expressions are analogous to those implied in Eq. (2.2). The elements of ε_{ab} and ω_{ab} may be represented by a matrix. For example,

$$\varepsilon_{ab} = \begin{bmatrix} \varepsilon_{11} & \varepsilon_{12} & \varepsilon_{13} \\ \varepsilon_{21} & \varepsilon_{22} & \varepsilon_{23} \\ \varepsilon_{31} & \varepsilon_{32} & \varepsilon_{33} \end{bmatrix} \qquad a, b = 1, 2, 3 \tag{6.10}$$

and from Eq. (6.8) we deduce that this matrix must be symmetric; i.e.,

$$\varepsilon_{12} = \varepsilon_{21}, \qquad \text{etc.} \tag{6.11}$$

We introduce here the coordinate transformation ideas by writing

$$x'_{a'} = \sum_{b=1}^{3} \Lambda_{a'b} x_b \qquad a', b = 1, 2, 3 \tag{6.12}$$

where x'_a denotes a new set of cartesian coordinates and

$$\Lambda_{a'b} = \cos (x'_{a'}, x_b) \tag{6.13}$$

define the direction cosines of the coordinate in a manner similar to that of Eq. (1.25).

The notation of a sum [e.g., see Eq. (6.12)] will occur often and we shall employ the convention that repeated indices within any term imply summation over the range of the index. Accordingly, instead of Eq. (6.12) we may write simply

$$x'_{a'} = \Lambda_{a'b} x_b \qquad a', b = 1, 2, 3 \tag{6.14}$$

to mean

$$x'_{a'} = \Lambda_{a'1} x_1 + \Lambda_{a'2} x_2 + \Lambda_{a'3} x_3 \qquad a' = 1, 2, 3 \tag{6.15}$$

In Eq. (6.14) a' is called a 'free index' whereas b is called a 'dummy index' since it may be replaced by any non-free index without changing the meaning of the term. Summation convention will be employed throughout the remainder of this chapter and the reader should ensure that this concept is fully understood before proceeding.

† Note that this definition is slightly different from that occurring in Chapters 2 to 5. Now $\varepsilon_{ij} = \tfrac{1}{2}\gamma_{ij}$ when $i \neq j$.

Using the notion of the direction cosines and the coordinate transformation of Eq. (6.14) we obtain similarly

$$u'_{a'} = \Lambda_{a'b} u_b \qquad a', b = 1, 2, 3 \tag{6.16}$$

After some differentiation using the definition (6.8) we obtain

$$\varepsilon'_{a'b'} = \Lambda_{a'c} \varepsilon_{cd} \Lambda_{b'd} \qquad a', b', c, d = 1, 2, 3 \tag{6.17}$$

Variables that transform according to Eq. (6.16) are called 'first-rank cartesian tensors' whereas quantities that transform according to Eq. (6.17) are called 'second-rank cartesian tensors'. Thus, the use of indicial notation in the context of rectangular cartesian coordinates will lead naturally to each structural mechanics variable being defined in terms of a cartesian tensor of an appropriate rank.

The elements of stress may be written as σ_{ab} for $a, b = 1, 2, 3$ and expanded in a matrix form similar to Eq. (6.10) for strains. Accordingly,

$$\sigma_{ab} = \begin{bmatrix} \sigma_{11} & \sigma_{12} & \sigma_{13} \\ \sigma_{21} & \sigma_{22} & \sigma_{23} \\ \sigma_{31} & \sigma_{32} & \sigma_{33} \end{bmatrix} \qquad a, b = 1, 2, 3 \tag{6.18}$$

The elements of stress also transform as a second-rank cartesian tensor. The symmetry of the stress tensor may be established by summing moments (angular momentum) about each of the coordinate axes to obtain

$$\sigma_{ab} = \sigma_{ba} \qquad a, b = 1, 2, 3 \tag{6.19}$$

Introducing a body force vector

$$b_a = (b_1, b_2, b_3) \qquad a = 1, 2, 3 \tag{6.20}$$

we may write the equilibrium equations (linear momentum balance) for an element of unit volume as

$$\sigma_{ba, b} + b_a = 0 \qquad a, b = 1, 2, 3 \tag{6.21}$$

where the repeated index again implies summation over the range of the index; i.e.,

$$\sigma_{ba, b} \equiv \sum_{b=1}^{3} \sigma_{ba, b}$$

$$= \sigma_{1a, 1} + \sigma_{2a, 2} + \sigma_{3a, 3} \tag{6.22}$$

Note that the free index a must appear in each term for the equation to be meaningful.

As a further example of the summation convention consider a term $\sigma_{ab} \varepsilon_{ab}$ which has units of work per unit volume. This term implies a double summation; hence summing first on a gives

$$\sigma_{ab}\varepsilon_{ab} = \sigma_{1b}\varepsilon_{1b} + \sigma_{2b}\varepsilon_{2b} + \sigma_{3b}\varepsilon_{3b} \tag{6.23}$$

and then summing on b gives

$$\sigma_{ab}\varepsilon_{ab} = \sigma_{11}\varepsilon_{11} + \sigma_{12}\varepsilon_{12} + \sigma_{13}\varepsilon_{13}$$

$$+ \sigma_{21}\varepsilon_{21} + \sigma_{11}\varepsilon_{11} + \sigma_{23}\varepsilon_{23}$$

$$+ \sigma_{31}\varepsilon_{31} + \sigma_{32}\varepsilon_{32} + \sigma_{33}\varepsilon_{33} \tag{6.24}$$

We may also use symmetry conditions on σ_{ab} and ε_{ab} to reduce the nine terms to six terms. Accordingly,

$$\sigma_{ab}\varepsilon_{ab} = \sigma_{11}\varepsilon_{11} + \sigma_{22}\varepsilon_{22} + \sigma_{33}\varepsilon_{33}$$

$$+ 2(\sigma_{12}\varepsilon_{12} + \sigma_{23}\varepsilon_{23} + \sigma_{31}\varepsilon_{31}) \tag{6.25}$$

Following a similar expansion we can show the meaning of the term

$$\sigma_{ab}\omega_{ab} \equiv 0 \qquad a, b = 1, 2, 3 \tag{6.26}$$

6.4 Elastic materials and finite element discretization

For an elastic material the most general linear relationship we may write to express the stress–strain characterization is

$$\sigma_{ab} = D_{abcd}(\varepsilon_{cd} - \varepsilon_{cd}^0) + \sigma_{ab}^0 \tag{6.27}$$

Equation (6.27) is the equivalent of Eq. (2.5) but now written in index notation. We note that the elastic moduli which appear in Eq. (6.27) involve four subscripts. By writing the constitutive equation with respect to the coordinates x'_a and using the transformation of stress and strain we may establish that the material moduli are elements of a 'fourth-rank' cartesian tensor. This transforms as

$$D'_{a'b'c'd'} = \Lambda_{a'e}\Lambda_{b'f}\Lambda_{c'g}\Lambda_{d'h}D_{efgh} \tag{6.28}$$

The elastic moduli for an isotropic linear elastic material may be written as

$$D_{abcd} = \delta_{ab}\delta_{cd}\lambda + (\delta_{ac}\delta_{bd} + \delta_{ad}\delta_{bc})\mu \tag{6.29}$$

where λ, μ are the Lamé constants and δ_{ab} is the Kronecker delta defined as

$$\delta_{ab} = \begin{cases} 1 & a = b \\ 0 & a \neq b \end{cases} \tag{6.30}$$

An isotropic, linear elastic material is thus characterized by two independent elastic constants. Instead of the Lamé constants we can use Young's modulus, E, and Poisson's ratio, v, to characterize the material. The Lamé constants may be deduced from Young's modulus and Poisson's ratio as

$$\mu = \frac{E}{2(1 + v)}$$

and   (6.31)

$$\lambda = \frac{vE}{(1 + v)(1 - 2v)}$$

If we now introduce the finite element displacement approximation given by Eq. (2.1) we may write, using indicial notation, for a single element

$$u_a \approx \hat{u}_a = N^i \bar{u}_a^i \qquad a = 1, 2, 3; i = 1, 2, \ldots, n \qquad (6.32)$$

where n is the total number of nodes on the element. The approximate strain in each element is given by the definition of Eq. (6.8) as

$$\hat{\varepsilon}_{ab} = \tfrac{1}{2}[N^i_{,b} \bar{u}_a^i + N^i_{,a} \bar{u}_b^i] \qquad a, b = 1, 2, 3; i = 1, 2, \ldots, n \quad (6.33)$$

The internal virtual work for an element is given by

$$\delta U^I = \int_{Ve} \delta \varepsilon_{ab} \, \sigma_{ab} \, dV \qquad (6.34)$$

Using Eqs (6.33) and (6.24) and noting symmetries in D_{abcd} we may write the internal virtual work for a linear elastic material as

$$\delta U^I = \delta \bar{u}_a^i \left(\int_{Ve} N^i_{,b} D_{abcd} N^j_{,d} \, dV \right) \bar{u}_c^j$$

$$- \delta \bar{u}_a^i \int_{Ve} N^i_{,b}(D_{abcd} \, \varepsilon_{cd}^0 - \sigma_{ab}^0) \, dV \qquad (6.35)$$

replacing the terms obtained in Chapter 2 in indicial notation.

The stiffness 'tensor' is now defined by

$$K_{ac}^{ij} = \int_{Ve} N^i_{,b} D_{abcd} N^j_{,d} \, dV \qquad (6.36)$$

When the elastic properties are constant over the element we may separate the integration from the material constants by defining

$$W_{bc}^{ij} = \int_{Ve} N^i_{,b} N^j_{,d} \, dV \qquad (6.37)$$

and then perform the summations with the material moduli

$$K_{ac}^{ij} = W_{bd}^{ij} D_{abcd}$$

In the case of isotropy a particularly simple result is obtained, giving

$$K_{ac}^{ij} = \lambda W_{ac}^{ij} + \mu[W_{ca}^{ij} + \delta_{ac}(W_{bb}^{ij})]$$

This allows the integration to be carried out using fewer arithmetic operations as compared with the use of matrix form.

The final equations of the system are written as

$$K_{ac}^{ij} u_c^j - f_a^i = 0 \tag{6.38}$$

and in this 'scalar' form every coefficient is simply identified. The reader can as a simple exercise complete the derivation of the force terms due to the initial strain ε_{ab}^0, body force b_a and external traction.

The tensorial notation is at times useful in clarifying individual terms, and this introduction could be helpful as a key to the reading of some current literature.

7

'Standard' and 'hierarchical' element shape functions: some general families of C_0 continuity

7.1 Introduction

In Chapters 3, 4, and 5 the reader was shown in some detail how linear elasticity problems could be formulated and solved using very simple finite element forms. Although the detailed algebra was concerned with shape functions which arose from triangular and tetrahedral shapes only it should by now be obvious that other element forms could equally well be used. Indeed, once the element and the corresponding shape functions are determined, subsequent operations follow a standard, well-defined path which could be entrusted to an algebraist not familiar with the physical aspects of the problem. It will be seen later that in fact it is possible to program a computer to deal with wide classes of problems by specifying the shape functions only. The choice of these is, however, a matter to which intelligence has to be applied and in which the human factor remains paramount. In this chapter some rules for generation of several families of one-, two-, and three-dimensional elements will be presented.

In the problems of elasticity illustrated in Chapters 3, 4, and 5 the displacement variable was a vector with two or three components and the shape functions were written in a matrix form. They were, however, derived for each component separately and in fact the matrix expressions in these were derived by multiplying a scalar function by an identity

matrix [e.g., Eqs (3.7), (4.3), and (5.7)]. We shall therefore concentrate in
this chapter on the scalar shape function forms, calling these simply N_i.

The shape functions used in the displacement formulation of elasticity
problems were such that to satisfy the convergence criteria of Chapter 2

(a) the continuity of the unknown only had to occur between elements
   (i.e., slope continuity is not required), or, in mathematical language,
   C_0 continuity was required;

(b) the function has to allow any arbitrary linear form to be taken so
   that the constant strain (constant first derivative) criterion could be
   observed.

The shape functions described in this chapter will require the satisfaction of these two criteria. They will thus be applicable to all the problems
of the preceding chapters and also to other problems which require these
conditions to be obeyed. For instance all problems of Chapter 10 can
use the forms here determined. Indeed they are applicable to any situation where the functional Π (see Chapter 9) is defined by derivatives of
the first order only.

The element families discussed will progressively have an increasing
number of degrees of freedom. The question may well be asked as to
whether any economic or other advantage is gained by increasing thus
the complexity of an element. The answer here is not an easy one
although it can be stated as a general rule that as the order of an element
increases so the total number of unknowns in a problem can be reduced
for a given accuracy of representation. Economic advantage requires,
however, a reduction of total computation and data preparation effort,
and this does not follow automatically for a reduced number of total
variables because, though equation-solving times may be reduced, the
time required for element formulation increases.

However, an overwhelming economic advantage in the case of three-
dimensional analysis has already been hinted at in Chapter 5 on three-
dimensional analysis.

The same kind of advantage arises on occasion in other problems but
in general the optimum element may have to be determined from case to
case.

In Sec. 2.6 of Chapter 2 we have shown that the order of error in the
approximation to the unknown function is $O(h^{p+1})$, where h is the
element 'size' and p is the complete polynomial present in the expansion.
Clearly, as the element shape functions increase in degree so will the
order of error increase, and convergence to the exact solution becomes
more rapid. While this says nothing about the magnitude of error at a
particular subdivision, it is clear that we should seek element shape functions with the highest complete polynomial for a given number of
degrees of freedom.

7.2 Standard and hierarchical concepts

The essence of the finite element method already stated in Chapter 2 is in approximating to the unknown (displacement) by an expansion given in Eq. (2.1). This for a scalar variable u can be written as

$$u \approx \hat{u} = \sum_{i=1}^{n} = N_i a_i = \mathbf{Na} \qquad (7.1)$$

where a_i are the unknown parameters to be determined.

We have explicitly chosen to identify such variables with the values of the unknown function at element nodes, thus making

$$u_i = a_i \qquad (7.2)$$

The shape functions so defined will be referred to as 'standard' ones and are the basis of most finite element programs. If polynomial expansions are used and the element satisfies Criterion 1 of Chapter 2 (which specifies that rigid body displacements cause no strain), it is clear that a constant value of a_i specified at all nodes must result in a constant value of $\hat{u}$:

$$\hat{u} = u_i \qquad (7.3)$$

By Eq. (7.1) it follows that

$$\sum_{i=1}^{n} N_i = 1 \qquad (7.4)$$

at all points of the domain. The first part of this chapter will deal with such *standard shape functions*.

A serious drawback exists, however, with 'standard' functions—as when element refinement is made totally new shape functions have to be generated and hence all calculations repeated. It would be of advantage to avoid this difficulty by considering the expression (7.1) as a *series* in which the shape function N_i does not depend on the number of nodes in the mesh n. This indeed is achieved with *hierarchic shape functions* to which the second part of this chapter is devoted.

The hierarchic concept is well illustrated by a one-dimensional (elastic bar) problem of Fig. 7.1. Here for simplicity elastic properties are taken as constant ($D = E$) and the body force b is assumed to vary in such a manner as to produce the exact solution shown on the figure (with zero displacements at both ends).

Two meshes are shown and a linear interpolation between nodal points assumed. For both standard and hierarchic form the coarse mesh gives

$$K_{11}^c a_1^c = f_1 \qquad (7.5)$$

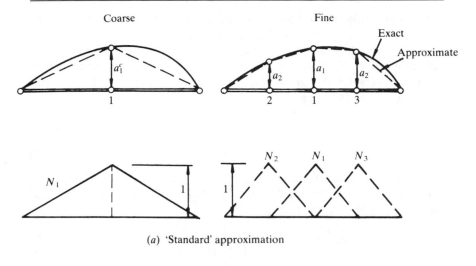

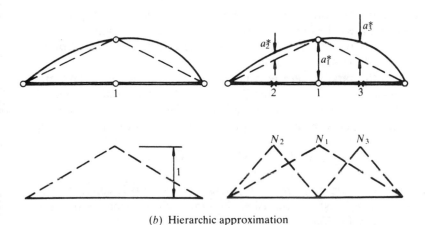

(b) Hierarchic approximation

Fig. 7.1 A one-dimensional problem of stretching of a uniform elastic bar by prescribed body forces

For a fine mesh two additional nodes are added and with the standard shape function the equations requiring solution are

$$\begin{bmatrix} K_{11}^F & K_{12}^F & 0 \\ K_{21}^F & K_{22}^F & K_{23}^F \\ 0 & K_{32}^F & K_{33}^F \end{bmatrix} \begin{Bmatrix} a_1 \\ a_2 \\ a_3 \end{Bmatrix} = \begin{Bmatrix} f_1 \\ f_2 \\ f_3 \end{Bmatrix} \tag{7.6}$$

In this form the zero matrices have been automatically inserted due to element interconnection which is here obvious, and we note that as no coefficients are the same, the new equations have to be resolved.

[Equation (2.13) shows how these coefficients are calculated and the reader is encouraged to work these out in detail.]

With 'hierarchic' form using shape functions shown, a similar form of equation arises and an identical approximation is achieved (being simply given by a series of straight segments). The *final* solution is identical but the meaning of the parameters $a_1 = a_1^*$ is now different, as shown in Fig. 7.1.

Quite generally,

$$K_{11}^F = K_{11}^c \tag{7.7}$$

as an identical shape function is used for the first variable. Further, in this particular case the off-diagonal coefficients are zero and the final equations become for the fine mesh

$$\begin{bmatrix} K_{11}^c & 0 & 0 \\ 0 & K_{22}^F & 0 \\ 0 & 0 & K_{33}^F \end{bmatrix} \begin{Bmatrix} a_1^* \\ a_2^* \\ a_3^* \end{Bmatrix} = \begin{Bmatrix} f_1 \\ f_2 \\ f_3 \end{Bmatrix} \tag{7.8}$$

The 'diagonality' feature is only true in the one-dimensional problem, but in general it will be found that the matrices obtained using hierarchic shape functions are nearly diagonal and hence imply better conditioning than those with standard shape functions.

Although the variables are now not subject to the obvious interpretation (as local displacement values), they can be easily transformed to those if desired. Though it is not usual to use hierarchic forms in linearly interpolated elements their derivation in polynomial form is simple and very advantageous.

Further advantages of this form of approximation are discussed in Chapter 14.

The reader should note that with hierarchic forms it is convenient to consider the finer mesh as still using the same, coarse, elements but now adding additional refining functions.

Hierarchic forms provide a link with other approximate series solutions. Many problems solved in classical literature by trigonometric, Fourier series, expansion are indeed particular examples of this approach.

Part 1 'Standard' shape functions

TWO-DIMENSIONAL ELEMENTS

7.3 Rectangular elements—some preliminary considerations

Conceptually (especially if the reader is conditioned by education to thinking in the cartesian coordinate system) the simplest element form of

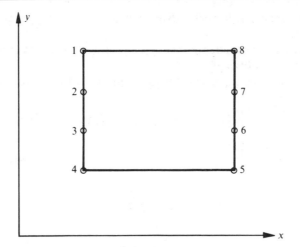

Fig. 7.2   A rectangular element

a two-dimensional kind is that of a rectangle with sides parallel to the x and y axes. Consider, for instance, a rectangle shown in Fig. 7.2 with nodal points numbered 1 to 8, located as shown, and at which the values of an unknown function u (here representing, for instance, one of the components of displacement) form the element parameters. How can suitable shape functions for this element be determined?

Let us first assume that u is expressed in a polynomial form in x and y. To ensure interelement continuity of u along the top and bottom sides the variation must be linear. Two points at which the function is common between elements lying above or below exist, and as two values determine uniquely a linear function, its identity all along these sides is ensured with that given by adjacent elements. Use of this fact was already made in specifying linear expansions for a triangle.

Similarly, if a cubic variation along the vertical sides is assumed, continuity will be preserved there as four values determine a unique cubic expansion. Conditions for satisfying the first criterion are now obtained.

To ensure the existence of constant values of the first derivative all that is necessary is that all the linear terms of the expansion be retained.

Finally, as eight points are to determine uniquely the variation of the function only eight coefficients of the expansion can be retained and thus we could write

$$u = \alpha_1 + \alpha_2 x + \alpha_3 y + \alpha_4 xy + \alpha_5 y^2 + \alpha_6 xy^2 + \alpha_7 y^3 + \alpha_8 xy^3 \quad (7.9)$$

The choice can in general be made unique by retaining the lowest possible expansion terms, though in this case apparently no such choice arises.† The reader will easily verify that all the requirements have now been satisfied.

Substituting coordinates of the various nodes a set of simultaneous equations will be obtained.

This can be written in exactly the same manner as was done for a triangle in Eq. (3.4) as

$$\begin{Bmatrix} u_1 \\ \vdots \\ u_8 \end{Bmatrix} = \begin{bmatrix} 1, \ x_1, \ y_1, \ x_1 y_1, \ y_1^2, \ x_1 y_1^2, \ y_1^3, \ x_1 y_1^3 \\ \cdot \ \ \cdot \ \ \cdot \ \ \cdot \ \ \cdot \ \ \cdot \ \ \cdot \ \ \cdot \ \ \cdot \ \ \cdot \\ \cdot \ \ \cdot \ \ \cdot \ \ \cdot \ \ \cdot \ \ \cdot \ \ \cdot \ \ \cdot \ \ \cdot \ \ \cdot \end{bmatrix} \begin{Bmatrix} \alpha_1 \\ \vdots \\ \alpha_8 \end{Bmatrix} \qquad (7.10)$$

or simply as

$$\mathbf{u}^e = \mathbf{C}\boldsymbol{\alpha} \qquad (7.11)$$

Formally,

$$\boldsymbol{\alpha} = \mathbf{C}^{-1}\mathbf{u}^e \qquad (7.12)$$

and we could write Eq. (7.9) as

$$u = \mathbf{P}\boldsymbol{\alpha} = \mathbf{P}\mathbf{C}^{-1}\mathbf{u}^e \qquad (7.13)$$

in which

$$\mathbf{P} = [1, \ x, \ y, \ xy, \ y^2, \ xy^2, \ y^3, \ xy^3] \qquad (7.14)$$

Thus the shape functions for the element defined by

$$u = \mathbf{N}\mathbf{u}^e = [N_1, N_2, \dots, N_8]\mathbf{u}^e \qquad (7.15)$$

can be found as

$$\mathbf{N} = \mathbf{P}\mathbf{C}^{-1} \qquad (7.16)$$

This process, frequently used in practice as it does not involve much ingenuity, has, however, some considerable disadvantages. Occasionally an inverse of $\mathbf{C}$ may not exist[1,2] and *always* considerable algebraic difficulty is experienced in obtaining an inverse in general terms suitable for all element geometries. It is therefore worth while to consider whether shape functions $N_i(x, y)$ can be written down directly. Before doing this some general properties of these functions have to be mentioned.

Inspection of the defining relation, Eq. (7.15), reveals immediately some important characteristics. Firstly, as this expression is valid for all components of $\mathbf{u}^e$,

$$N_i = 1$$

---

† Retention of a higher order term of expansion, ignoring one of lower order, will usually lead to a poorer approximation though still retaining convergence,[1] providing the linear terms are always included.

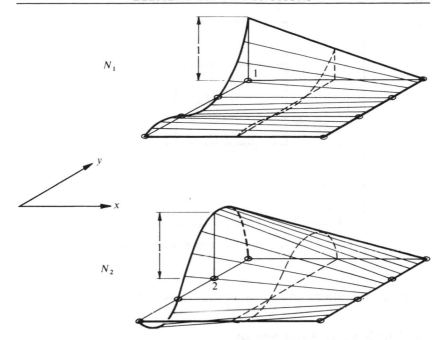

Fig. 7.3   Shape functions for elements of Fig. 7.2

at node i and is equal to zero at all other nodes. Further, the basic type of variation along boundaries defined for continuity purposes (e.g., linear in x and cubic in y in the above example) must be retained. The typical form of the shape functions for the elements considered is illustrated isometrically for two typical nodes in Fig. 7.3. It is clear that these could have been written down directly as a product of a suitable linear function in x with a cubic in y. The easy solution of this example is not always as obvious but given sufficient ingenuity, a direct derivation of shape function is always recommended.

It will be convenient to use normalized coordinates in further investigation. Such normalized coordinates are shown in Fig. 7.4 and are so chosen that on the faces of the rectangle their values are ± 1:

$$\xi = \frac{x - x_c}{a} \qquad d\xi = \frac{dx}{a}$$

$$\eta = \frac{y - y_c}{b} \qquad d\eta = \frac{dy}{b} \tag{7.17}$$

Once the shape functions are known in the normalized coordinates, translation into actual coordinates or transformation of the various

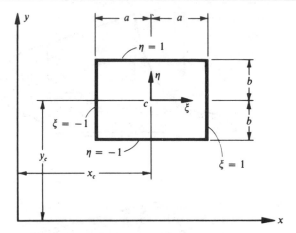

Fig. 7.4   Normalized coordinates for a rectangle

expressions occurring, for instance, in stiffness derivation is trivial.

7.4 Completeness of polynomials

The shape function derived in the previous section was of a rather special form [see Eq. (7.9)]. Only a linear variation with the coordinate x was permitted, while in y a full cubic was available. The complete polynomial contained in it was thus of order 1 and in general use a convergence order corresponding to a linear variation would occur despite an

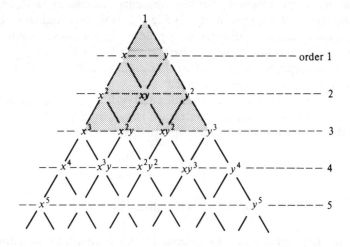

Fig. 7.5   The Pascal triangle. (Cubic expansion shaded—10 terms)

increase of the total number of variables. Only in situations where the linear variation in x corresponded closely to the exact solution would a higher order of convergence occur—and for this reason elements with such 'preferential' directions should be restricted to special use, e.g., in narrow beams or strips. In general, we shall seek element expansions which possess the highest order of a complete polynomial for a minimum of degrees of freedom. In this context it is useful to recall the Pascal triangle (Fig 7.5) from which the number of terms occurring in a polynomial in two variables x, y can be readily ascertained. For instance, first-order polynomials require three terms, second-order six terms, third-order ten terms, etc.

7.5 Rectangular elements—Lagrange family[3-6]

An easy and systematic method of generating shape functions of any order can be achieved by simple products of appropriate polynomials in the two coordinates. Consider an element shown in Fig. 7.6 in which a series of nodes, external and internal, is placed on a regular grid. It is required to determine a shape function for the point indicated by the heavy circle. Clearly the product of a fifth-order polynomial in ξ which has a value of unity at points of the second column of nodes and zero elsewhere and that of a fourth-order polynomial in η having unity on the coordinate corresponding to the top row of nodes and zero elsewhere satisfies all the interelement continuity conditions and gives unity at the nodal point concerned.

Polynomials in one coordinate having this property are known as Lagrange polynomials and can be written down directly as

$$l_k^n(\xi) = \frac{(\xi - \xi_0)(\xi - \xi_1) \cdots (\xi - \xi_{k-1})(\xi - \xi_{k+1}) \cdots (\xi - \xi_n)}{(\xi_k - \xi_0)(\xi_k - \xi_1) \cdots (\xi_k - \xi_{k-1})(\xi_k - \xi_{k+1}) \cdots (\xi_k - \xi_n)} \quad (7.18)$$

giving unity at ξ_k and passing through n points.

Thus in two dimensions, if we label the node by its column and node number, I, J, we have

$$N_i \equiv N_{IJ} = l_I^n(\xi)l_J^m(\eta) \quad (7.19)$$

where n and m stand for the number of subdivisions in each direction.

Figure 7.7 shows a few members of this unlimited family. Though it is easy to generate, the usefulness of this family is limited not only due to a large number of internal nodes present but also due to the poor curve-fitting properties of the higher order polynomials. It will be noticed that the expressions of shape function will contain some very high order terms while omitting some lower ones.

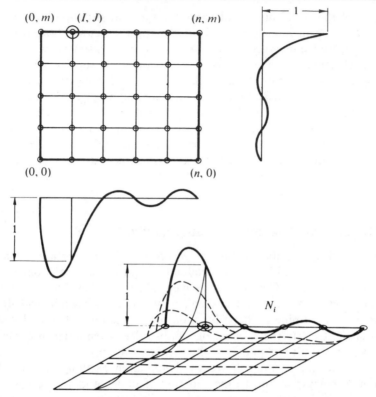

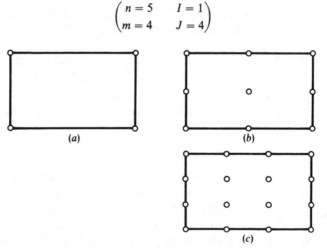

Fig. 7.6   A typical shape function for a lagrangian element

$$\begin{pmatrix} n = 5 & I = 1 \\ m = 4 & J = 4 \end{pmatrix}$$

(a)

(b)

(c)

Fig. 7.7   Three elements of the Lagrange family: (a) linear, (b) quadratic, and (c) cubic

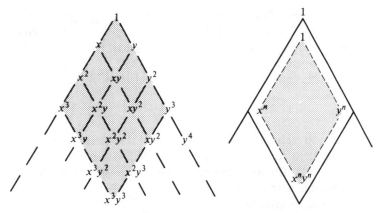

Fig. 7.8   Terms generated by a lagrangian expansion of order 3×3 (or $m \times n$). Complete polynomials of order 3 (or n)

Indeed, if we examine the polynomial terms present in a situation where $n = m$ we observe in Fig. 7.8, based on the Pascal triangle, that a very large number of *excessive* polynomial terms is present over those needed for a complete expansion.[7] However, when mapping of shape functions is considered (Chapter 8) some advantages occur for this family.

7.6 Rectangular elements—'serendipity' family[3,4]

It is usually most convenient to make the functions dependent on nodal values placed on the element boundary. Consider, for instance, the first

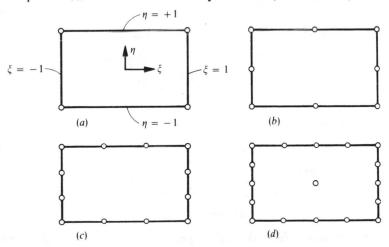

Fig. 7.9   Reactangles of boundary node (serendipity) family: (*a*) linear, (*b*) quadratic, (*c*) cubic, (*d*) quartic

three elements of Fig. 7.9. In each a progressively increasing and equal number of nodes is placed on the element boundary. The variation of function on the edges to ensure continuity is linear, parabolic, and cubic in increasing element development.

To achieve the shape function for the first element it is obvious that a product of the form

$$\tfrac{1}{4}(\xi + 1)(\eta + 1) \tag{7.20}$$

gives unity at top right corners where $\xi = \eta = 1$ and zero at all the other corners. Further, a linear variation of the shape function of all sides exists and hence continuity is satisfied. Indeed this element is identical to the lagrangian one with $n = 1$.

Introducing new variables

$$\xi_0 = \xi\xi_i \qquad \eta_0 = \eta\eta_i \tag{7.21}$$

the form

$$N_i = \tfrac{1}{4}(1 + \xi_0)(1 + \eta_0) \tag{7.22}$$

allows all shape functions to be written down in one expression.

As a linear combination of these shape functions yields any arbitrary linear variation of u, the second convergence criterion is satisfied.

The reader can verify that the following functions satisfy all the necessary criteria for quadratic and cubic members of the family.

'Quadratic' element
  Corner nodes:

$$N_i = \tfrac{1}{4}(1 + \xi_0)(1 + \eta_0)(\xi_0 + \eta_0 - 1) \tag{7.23}$$

  Mid-side nodes:

$$\xi_i = 0 \qquad N_i = \tfrac{1}{2}(1 - \xi^2)(1 + \eta_0)$$
$$\eta_i = 0 \qquad N_i = \tfrac{1}{2}(1 + \xi_0)(1 - \eta^2)$$

'Cubic' element
  Corner nodes:

$$N_i = \tfrac{1}{32}(1 + \xi_0)(1 + \eta_0)[-10 + 9(\xi^2 + \eta^2)] \tag{7.24}$$

  Mid-side nodes:

$$\xi_i = \pm 1 \qquad \text{and} \qquad \eta_i = \pm\tfrac{1}{3}$$
$$N_i = \tfrac{9}{32}(1 + \xi_0)(1 - \eta^2)(1 + 9\eta_0)$$

with the remaining mid-side node expression obtained by changing variables.

In the next, quartic, member[8] of this family a central node is added so that all terms of a complete fourth-order expansion would be available. This central node adds a shape function $(1 - \xi^2)(1 - \eta^2)$ which is zero on all outer boundaries.

The above functions have been originally derived by inspection, and progression to yet higher members is difficult and requires some ingenuity. It was therefore appropriate to name this family 'serendipity' after the famous princes of Serendip noted for their chance discoveries (Horace Walpole, 1754).

However, a quite systematic way of generating the 'serendipity' shape functions can be devised, which becomes apparent from Fig. 7.10 where the generation of a quadratic shape function is presented.[7,9]

As a starting point we observe that for *mid-side* nodes a lagrangian interpolation of a quadratic × linear type suffices to determine N_i at

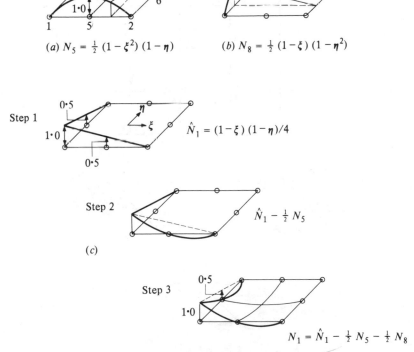

$$(a)\ N_5 = \tfrac{1}{2}(1 - \xi^2)(1 - \eta) \qquad (b)\ N_8 = \tfrac{1}{2}(1 - \xi)(1 - \eta^2)$$

Step 1 $\qquad \hat{N}_1 = (1 - \xi)(1 - \eta)/4$

Step 2 $\qquad \hat{N}_1 - \tfrac{1}{2} N_5$

(c)

Step 3

$$N_1 = \hat{N}_1 - \tfrac{1}{2} N_5 - \tfrac{1}{2} N_8$$

Fig. 7.10　Systematic generation of 'serendipity' shape functions

nodes 5 to 8. N_5 and N_8 are shown at Fig. 7.10(a) and (b). For a *corner* node, such as Fig. 7.10(c), we start with a bilinear $\hat{N}_1$ and note immediately that while $\hat{N}_1 = 1$ at node 1, it is not zero at nodes 5 or 8 (step 1). Successive subtraction of $\frac{1}{2}N_5$ (step 2) and $\frac{1}{2}N_8$ (step 3) ensures that a zero value is obtained at these nodes. The reader can verify that the expressions obtained coincide with those of Eqs (7.23) and (7.24).

Indeed, it should now be obvious that for all higher order elements the *mid-side* and *corner shape* functions can be generated by an identical process. For the former a simple multiplication of mth-order and first-order lagrangian interpolations suffices. For the latter a combination of bilinear corner functions, together with appropriate fractions of mid-side shape functions to ensure zero at appropriate nodes, is necessary.

Similarly, it is quite easy to generate shape functions for elements with different numbers of nodes along each side by a systematic algorithm. This may be very desirable if a transition between elements of different order is to be achieved, enabling a different order of accuracy in separate sections of a large problem to be studied. Figure 7.11 illustrates the

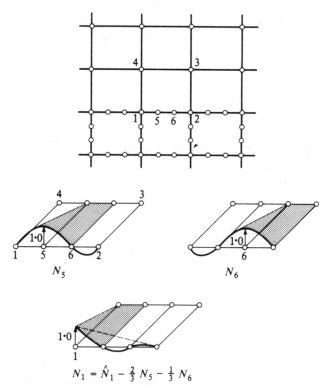

Fig. 7.11   Shape functions for a transition 'serendipity' element, cubic/linear

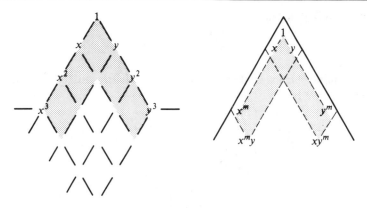

Fig. 7.12   Terms generated by edge shape functions in serendipity-type elements
(3 × 3 and $m \times m$)

necessary shape functions for a cubic/linear transition. Use of such special elements was first introduced in reference 9, but the simpler formulation used here is that of reference 7.

With the mode of generating shape functions for this class of elements available it is immediately obvious that fewer degrees of freedom are now necessary for a given complete polynomial expansion. Figure 7.12 shows this for a cubic element where only two surplus terms arise (as compared with six surplus terms in a lagrangian of the same order).

It is immediately evident, however, that the functions generated by nodes places only along the edges will not generate complete polynomials beyond the cubic. For higher order ones it is necessary to supplement the expansion by internal nodes (as was done in the quartic element of Fig. 7.9) or by the use of 'nodeless' variables (discussed in the next section) which contain the appropriate polynomial terms.

7.7 Elimination of internal variables before assembly—substructures

Internal nodes yield in the usual way the element properties (Chapter 2)

$$\frac{\partial \Pi^e}{\partial \mathbf{a}^e} = \mathbf{K}^e \mathbf{a}^e + \mathbf{f}^e \tag{7.25}$$

As $\mathbf{a}^e$ can be subdivided into parts which are common with other elements, $\bar{\mathbf{a}}^e$, and others which occur in the particular element only, $\bar{\bar{\mathbf{a}}}^e$, we can immediately write

$$\frac{\partial \Pi}{\partial \bar{\bar{\mathbf{a}}}^e} = \frac{\partial \Pi^e}{\partial \bar{\bar{\mathbf{a}}}^e} = 0$$

and eliminate $\bar{\bar{\mathbf{a}}}^e$ from further consideration. Writing Eq. (7.25) in a partitioned form we have

$$\frac{\partial \Pi^e}{\partial \mathbf{a}^e} = \begin{Bmatrix} \dfrac{\partial \Pi^e}{\partial \bar{\mathbf{a}}^e} \\ \dfrac{\partial \Pi^e}{\partial \bar{\bar{\mathbf{a}}}^e} \end{Bmatrix} = \begin{bmatrix} \bar{\mathbf{K}}^e & \hat{\mathbf{K}}^e \\ \hat{\mathbf{K}}^{e\mathrm{T}} & \bar{\bar{\mathbf{K}}}^e \end{bmatrix} \begin{Bmatrix} \bar{\mathbf{a}}^e \\ \bar{\bar{\mathbf{a}}}^e \end{Bmatrix} + \begin{Bmatrix} \bar{\mathbf{f}}^e \\ \bar{\bar{\mathbf{f}}}^e \end{Bmatrix}$$

$$= \begin{Bmatrix} \dfrac{\partial \Pi^e}{\partial \bar{\mathbf{a}}^e} \\ 0 \end{Bmatrix} \tag{7.26}$$

From the second set of equations given above we can write

$$\bar{\bar{\mathbf{a}}}^e = -(\bar{\bar{\mathbf{K}}}^e)^{-1}(\hat{\mathbf{K}}^{e\mathrm{T}}\bar{\mathbf{a}}^e + \bar{\bar{\mathbf{f}}}^e) \tag{7.27}$$

which on substitution yields

$$\frac{\partial \Pi^e}{\partial \bar{\mathbf{a}}^e} = \mathbf{K}^{*e}\bar{\mathbf{a}}^e + \mathbf{f}^{*e} \tag{7.28}$$

in which

$$\mathbf{K}^{*e} = \bar{\mathbf{K}}^e - \hat{\mathbf{K}}^e(\bar{\bar{\mathbf{K}}}^e)^{-1}\hat{\mathbf{K}}^{e\mathrm{T}}$$
$$\mathbf{f}^{*e} = \bar{\mathbf{f}}_e - \hat{\mathbf{K}}^e(\bar{\bar{\mathbf{K}}}^e)^{-1}\bar{\bar{\mathbf{f}}}^e \tag{7.29}$$

Assembly of the total region then follows, only considering the element boundary variables, thus giving a considerable saving in the equation-solving effort at the expense of a few additional manipulations carried out at the element stage.

Perhaps a structural interpretation of this elimination is desirable. What in fact is involved is the separation of a part of the structure from its surroundings and determination of its solution separately for any prescribed displacements at the interconnecting boundaries. $\mathbf{K}^{*e}$ is now simply the overall stiffness of the separated structure and $\mathbf{f}^{*e}$ the equivalent set of nodal forces.

If the triangulation of Fig. 7.13 is interpreted as an assembly of pin-jointed bars the reader will recognize immediately the well-known device of 'substructures' used frequently in structural engineering.

Such a substructure is in fact simply a complex element from which the internal degrees of freedom have been eliminated.

Immediately a new possibility for devising more elaborate, and presumably more accurate, elements is presented.

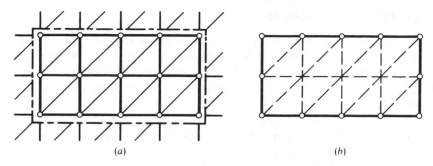

Fig. 7.13 Substructure of a complex element

Let Fig. 7.13(a) be interpreted as a continuum field subdivided into triangular elements. The substructure results in fact in one complex element shown in Fig. 7.13(b) with a number of boundary nodes.

The only difference from elements derived in previous sections is the fact that the unknown u is now not approximated internally by one set of smooth shape functions but by a series of piecewise approximations. This presumably results in a slightly poorer approximation but an economic advantage may arise if the total computation time for such an assembly is saved.

Substructuring is an important device in complex problems, particularly where a repetition of complicated components arises.

In simple, small-scale finite element analysis, much improved use of simple triangular elements was found by the use of simple subassemblies of the triangles (or indeed tetrahedra). For instance, a quadrilateral based on four triangles from which the central node is eliminated was found to give an economic advantage over a direct use of simple triangles (Fig. 7.14). This and other subassemblies based on triangles are discussed in detail by Doherty et al.[10]

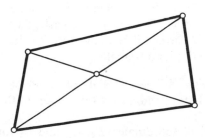

Fig. 7.14 A quadrilateral made up by four simple triangles

7.8 Triangular element family

The advantage of an arbitrary triangular shape in approximating to any boundary shape has been amply demonstrated in earlier chapters. Its apparent superiority here over the rectangular shapes needs no further discussion. The question of generating more elaborate elements needs to be further developed.

Consider a series of triangles generated on a pattern indicated in Fig. 7.15. The number of nodes in each member of the family is now such that a complete polynomial expansion, of the order needed for inter-element compatibility, is ensured. This follows by comparison with the Pascal triangle of Fig. 7.5 in which we see the number of nodes coinciding exactly with the number of polynomial terms required. This particular feature puts the triangle family in a special, privileged position, in which the inverse of the **C** matrices of Eq. (7.11) will always exist.[2]

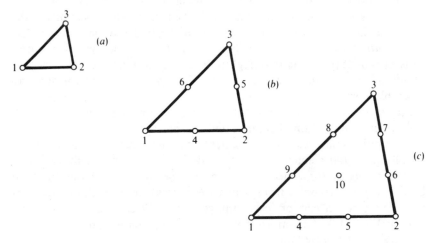

Fig. 7.15   Triangular element family: (a) linear, (b) quadratic, and (c) cubic

However, once again a direct generation of shape functions will be preferred—and indeed will be shown to be particularly easy.

Before proceeding further it is convenient to define a special set of normalized coordinates for a triangle.

7.8.1 *Area coordinates.* While cartesian directions parallel to the sides of a rectangle were a natural choice for that shape, in the triangle these are not convenient.

A convenient set of coordinates, L_1, L_2, and L_3 for a triangle 1, 2, 3 (Fig. 7.16), is defined by the following linear relation between these and

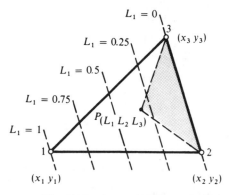

Fig. 7.16   Area coordinates

the cartesian system:

$$x = L_1 x_1 + L_2 x_2 + L_3 x_3$$
$$y = L_1 y_1 + L_2 y_2 + L_3 y_3 \qquad (7.30)$$
$$1 = L_1 + L_2 + L_3$$

To every set, L_1, L_2, L_3 (which are not independent, but are related by the third equation), corresponds a unique set of cartesian coordinates. At point 1, $L_1 = 1$ and $L_2 = L_3 = 0$, etc. A linear relation between the new and cartesian coordinates implies that contours of L_1 are equally placed straight lines parallel to side 2–3 on which $L_1 = 0$, etc.

Indeed it is easy to see that an alternative definition of the coordinate L_1 of a point P is by a ratio of the area of the shaded triangle to that of the total triangle:

$$L_1 = \frac{\text{area } P23}{\text{area } 123} \qquad (7.31)$$

Hence the name of area coordinates.

Solving Eq. (7.30) for x and y gives

$$L_1 = \frac{a_1 + b_1 x + c_1 y}{2\Delta}$$

$$L_2 = \frac{a_2 + b_2 x + c_2 y}{2\Delta} \qquad (7.32)$$

$$L_3 = \frac{a_3 + b_3 x + c_3 y}{2\Delta}$$

in which

$$\Delta = \tfrac{1}{2} \det \begin{vmatrix} 1 & x_1 & y_1 \\ 1 & x_2 & y_2 \\ 1 & x_3 & y_3 \end{vmatrix} = \text{area } 123 \qquad (7.33)$$

and

$$a_1 = x_2 y_3 - x_3 y_2$$

$$b_1 = y_2 - y_3$$

$$c_1 = x_3 - x_2$$

etc., with cyclic rotation of indices 1, 2, and 3.

The identity of expressions with those derived in Chapter 3 [Eqs (3.5b) and (3.5c)] is worth remarking upon.

7.8.2 *Shape functions.* For the first element of the series [Fig. 7.15(a)], the shape functions are simply the area coordinates. Thus

$$N_1 = L_1 \qquad N_2 = L_2 \qquad N_3 = L_3 \qquad (7.34)$$

This is obvious as each individually gives unity at one node, zero at others, and varies linearly everywhere.

To derive shape functions for other elements a simple recurrence relation can be derived.[2] However, it is very simple to write an arbitrary triangle of order M in a manner similar to that used for the lagrangian element of Sec. 7.5.

Denoting a typical node i by three numbers I, J, and K corresponding to the position of coordinates L_{1i}, L_{2i}, and L_{3i} we can write the shape function in terms of three lagrangian interpolations as [see Eq. (7.18)]

$$N_i = l_I^I(L_1) l_J^J(L_2) l_K^K(L_3) \qquad (7.35)$$

In the above l_I^I, etc., are given by expression (7.18), with L_1 taking the place of ξ, etc.

It is easy to verify that the above expression gives

$$N_i = 1 \qquad \text{at} \quad L_1 = L_{1I}, \quad L_2 = L_{2I}, \quad L_3 = L_{3I}$$

and zero at all other nodes.

The highest term occurring in the expansion is

$$L_1^I L_2^J L_3^K$$

and as

$$I + J + K \equiv M$$

for all points the polynomial is of order M.

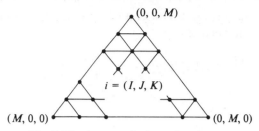

Fig. 7.17   A general triangular element

Expression (7.35) is valid for quite arbitrary distributions of nodes of the pattern given in Fig. 7.17 and simplifies if the spacing of the nodal lines is equal (i.e., $1/m$). The formula was first obtained by Argyris *et al.*[11] and formalized in a different manner by others.[7,12]

The reader can verify simply the shape functions for the second- and third-order elements as given below and indeed derive ones of any higher order easily.

Quadratic triangle [Fig. 7.15(*b*)]
   For corner nodes:

$$N_1 = (2L_1 - 1)L_1, \qquad \text{etc.}$$

Mid-side nodes:

$$N_4 = 4L_1L_2, \qquad \text{etc.} \qquad\qquad (7.36)$$

Cubic triangle [Fig. 7.15(*c*)]
   For corner nodes:

$$N_1 = \tfrac{1}{2}(3L_1 - 1)(3L_1 - 2)L_1, \qquad \text{etc.}$$

Mid-side nodes:

$$N_4 = \tfrac{9}{2}L_1L_2(3L_1 - 1), \qquad \text{etc.} \qquad\qquad (7.37)$$

and for the internal node:

$$N_{10} = 27L_1L_2L_3$$

The last shape again is a 'bubble' function giving zero contribution along boundaries—and this will be found to be useful in many other contexts (viz. mixed forms in Chapter 12).

The quadratic triangle was first derived by Veubeke[13] and used in the context of plane stress analysis by Argyris.[14]

When element matrices have to be evaluated it will follow that we are faced with integration of quantities defined in terms of area coordinates over the triangular region. It is useful to note in this context the following exact integration expression:

$$\iint_{\Delta} L_1^a L_2^b L_3^c \, dx \, dy = \frac{a! \, b! \, c!}{(a + b + c + 2)!} \, 2\Delta \tag{7.38}$$

ONE-DIMENSIONAL ELEMENTS

7.9 Linear elements

So far in this book the continuum was considered generally in two or three dimensions. 'One-dimensional' members, being of a kind for which exact solutions are generally available, were treated only as trivial examples in Chapter 2 and in Sec. 7.2. In many practical two- or three-dimensional problems such elements do in fact appear in conjunction with the more usual continuum elements—and a unified treatment is desirable. In the context of elastic analysis these elements may represent lines of reinforcement (plane and three-dimensional problems) or sheets of thin lining material in axisymmetric and three-dimensional bodies. In the context of field problems of the type to be discussed in Chapter 10 lines of drains in a porous medium of lesser conductivity can be envisaged.

Once the shape of such a function as displacement is chosen for an element of this kind its properties can be determined, noting, however, that such derived quantities as strain, etc., have to be considered only in one dimension.

Figure 7.18 shows such an element sandwiched between two adjacent quadratic-type elements. Clearly for continuity of the function a quadra-

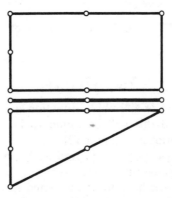

Fig. 7.18   A line element sandwiched between two-dimensional elements

tic variation of the unknown with the one variable ξ is all that is required. Thus the shape functions are given directly by the Lagrange polynomial as defined in Eq. (7.18).

THREE-DIMENSIONAL ELEMENTS

7.10 Rectangular prisms—'serendipity' family[4,9,15]

In a precisely analogous way to that given in previous sections equivalent elements of three-dimensional type can be described.

Now, for interelement continuity the simple rules given previously have to be modified. What is necessary to achieve is that along a whole face of an element the nodal values define a unique variation of the unknown function. With incomplete polynomials, this can be ensured only by inspection.

A family of elements shown in Fig. 7.19 is precisely equivalent to that of Fig. 7.9. Using now three normalized coordinates and otherwise following the terminology of Sec. 7.6 we have the following shape functions:

'Linear' element (8 nodes)

$$N_i = \tfrac{1}{8}(1 + \xi_0)(1 + \eta_0)(1 + \zeta_0) \tag{7.39}$$

'Quadratic' element (20 nodes)
   Corner nodes:

$$N_i = \tfrac{1}{8}(1 + \xi_0)(1 + \eta_0)(1 + \zeta_0)(\xi_0 + \eta_0 + \zeta_0 - 2) \tag{7.40}$$

   Typical mid-side node:

$$\xi_i = 0 \qquad \eta_i = \pm 1 \qquad \zeta_i = \pm 1$$

$$N_i = \tfrac{1}{4}(1 - \xi^2)(1 + \eta_0)(1 + \zeta_0)$$

'Cubic' elements (32 nodes)
   Corner node:

$$N_i = \tfrac{1}{64}(1 + \xi_0)(1 + \eta_0)(1 + \zeta_0)[9(\xi^2 + \eta^2 + \zeta^2) - 19] \tag{7.41}$$

   Typical mid-side node:

$$\xi_i = \pm \tfrac{1}{3} \qquad \eta_i = \pm 1 \qquad \zeta_i = \pm 1$$

$$N_i = \tfrac{9}{64}(1 - \xi^2)(1 + 9\xi_0)(1 + \eta_0)(1 + \zeta_0)$$

When $\zeta = 1 = \zeta_0$ the above expressions reduce to those of Eqs (7.22) to (7.24). Indeed such elements of the three-dimensional type can be

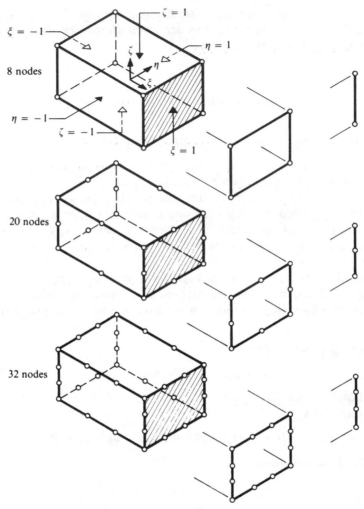

Fig. 7.19   Right prisms of boundary node (serendipity) family with corresponding sheet and line elements

joined in a compatible manner to sheet or line elements of the appropriate type as shown in Fig. 7.19.

Once again the procedure of generating the shape functions follows
that described in Figs 7.10 and 7.11 and once again elements with
varying degrees of freedom along the edges can be derived following the
same steps.

The equivalent of a Pascal triangle is now a tetrahedron and again we
can observe the small number of surplus degrees of freedom—a situation
even of greater importance than in two-dimensional analysis.

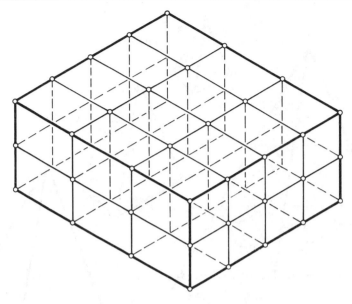

Fig. 7.20   Right prism of Lagrange family

7.11 Rectangular prisms—Lagrange family

Shape function for such elements, illustrated in Fig. 7.20, will be generated by a direct product of three Lagrange polynomials. Extending the notation of Eq. (7.19) we now have

$$N_i \equiv N_{IJK} = l_I^n l_J^m l_K^p \qquad (7.42)$$

for n, m, and p subdivisions along each side.

This element again is suggested by Ergatoudis[5] and elaborated upon by Argyris et al.[6] All the remarks about internal nodes and the limitation of the formulation made in Sec. 7.5 are applicable here and generally the practical application of such elements is inefficient.

7.12 Tetrahedral elements

The tetrahedral family shown in Fig. 7.21 not surprisingly exhibits properties similar to those of the triangle family.

Firstly, once again complete polynomials in three coordinates are achieved at each stage. Secondly, as faces are divided in a manner identical with that of the previous triangles, the same order of polynomial in two coordinates in the plane of the face is achieved and element compatibility ensured. No surplus terms in the polynomial occur.

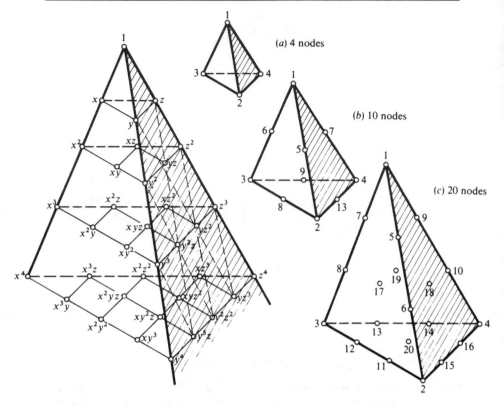

Fig. 7.21   The tetrahedron family: (a) linear, (b) quadratic, and (c) cubic

7.12.1 *Volume coordinates.* Once again special coordinates are introduced defined by (Fig. 7.22):

$$x = L_1 x_1 + L_2 x_2 + L_3 x_3 + L_4 x_4$$
$$y = L_1 y_1 + L_2 y_2 + L_3 y_3 + L_4 y_4 \qquad (7.43)$$
$$z = L_1 z_1 + L_2 z_2 + L_3 z_3 + L_4 z_4$$
$$1 = L_1 + L_2 + L_3 + L_4$$

The inversion of the above leads to expressions of type (7.32) and (7.33) with the constants which can be identified from Chapter 5 [Eqs (5.5)]. Again the physical nature of the coordinates can be identified as the ratio of volumes of tetrahedra based on an internal point P in the total volume, e.g., as shown in Fig. 7.22:

$$L_1 = \frac{\text{volume } P234}{\text{volume } 1234}, \qquad \text{etc.} \qquad (7.44)$$

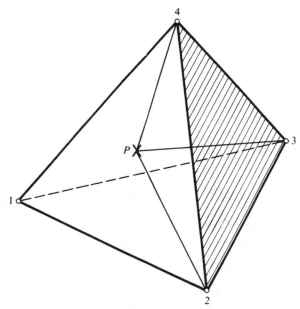

Fig. 7.22   Volume coordinates

7.12.2 *Shape function.* As the volume coordinates vary linearly with the cartesian ones from unity at one node to zero at the opposite face then shape functions for the linear element [Fig. 7.21(a)] are simply

$$N_1 = L_1 \qquad N_2 = L_2, \qquad \text{etc.} \tag{7.45}$$

Formulae for shape functions of higher order tetrahedra are derived in precisely the same manner as for the triangles by establishing appropriate Lagrange-type formulae similar to Eq. (7.35). Leaving this to the reader as a suitable exercise we quote the following:

'Quadratic' tetrahedron [Fig. 7.21(b)]
  For corner nodes:

$$N_1 = (2L_1 - 1)L_1, \qquad \text{etc.} \tag{7.46}$$

For mid-side nodes:

$$N_5 = 4L_1 L_2, \qquad \text{etc.}$$

'Cubic' tetrahedron
  Corner nodes:

$$N_1 = \tfrac{1}{2}(3L_1 - 1)(3L_1 - 2)L_1, \qquad \text{etc.} \tag{7.47}$$

Mid-side nodes:

$$N_5 = \tfrac{9}{2}L_1 L_2(3L_1 - 1), \qquad \text{etc.}$$

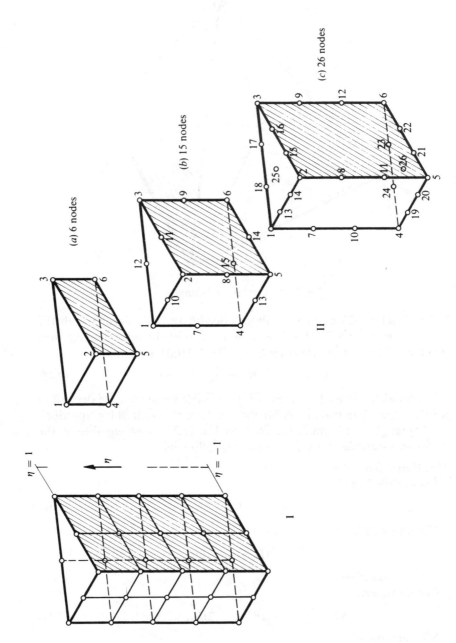

Fig. 7.23   Triangular prism elements (serendipity) family: (a) linear, (b) quadratic, and (c) cubic

Mid-face nodes:

$$N_{17} = 27L_1L_2L_3, \quad \text{etc.}$$

A useful integration formula again may be here quoted:

$$\iiint_{\text{vol}} L_1^a L_2^b L_3^c L_4^d \, \mathrm{d}x \, \mathrm{d}y \, \mathrm{d}z = \frac{a!\,b!\,c!\,d!}{(a + b + c + d + 3)!} \, 6V \qquad (7.48)$$

7.13 Other simple three-dimensional elements

The possibilities of simple shapes in three dimensions are greater, for obvious reasons, than in two dimensions. A quite useful series of elements can, for instance, be based on triangular prisms (Fig. 7.23). Here again variants of the product, Lagrange, approach or of the 'serendipity' type can be distinguished. The first element of both families is identical and indeed the shape functions for it are so obvious as not to need quoting.

For a 'quadratic' element illustrated in Fig. 7.23(b) the shape functions are

Corner nodes $L_1 = \xi_1 = 1$:

$$N_1 = \tfrac{1}{2}L_1(2L_1 - 1)(1 + \zeta) - \tfrac{1}{2}L_1(1 - \zeta^2) \qquad (7.49)$$

Mid-sides of triangles:

$$N_{10} = 2L_1L_2(1 + \zeta), \quad \text{etc.} \qquad (7.50)$$

Mid-sides of rectangle:

$$N_7 = L_1(1 - \zeta^2), \quad \text{etc.}$$

Such elements are not purely esoteric but have a practical application as 'fillers' in conjunction with 20-noded parallelepiped elements.

Part 2 Hierarchical shape functions

7.14 Hierarchic polynomials in one dimension

The general ideas of hierarchic approximation are introduced in Sec. 7.2 in the context of simple, linear, elements. The idea of generating higher order hierarchic forms is again simple. We shall start from a one-dimensional expansion as this has been shown to provide a basis for generation of two- and three-dimensional forms in the previous sections.

To generate a polynomial of order p along an element side we do not need to introduce nodes but can instead use parameters without an obvious physical meaning. As shown in Fig. 7.24, we could use here a linear expansion specified by 'standard' functions N_0 and N_1 and add to

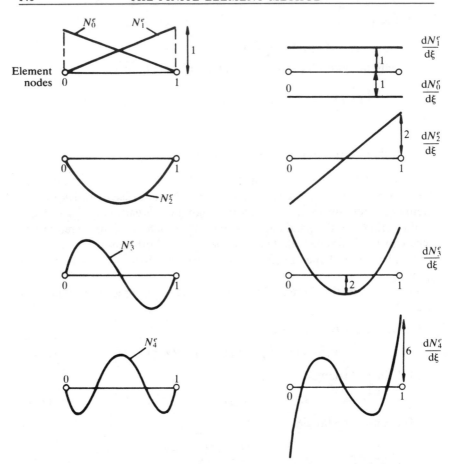

Fig. 7.24 Hierarchical element shape functions of nearly orthogonal form and their derivatives

this a series of polynomials always so designed as to have zero values at the ends of the range (i.e., points 0 and 1).

Thus for a quadratic approximation, we would write over the typical one-dimensional element, for instance,

$$\hat{u} = u_0 N_0 + u_1 N_1 + a_2 N_2 \qquad (7.51)$$

where

$$N_0 = -\frac{\xi - 1}{2} \qquad N_1 = \frac{\xi + 1}{2} \qquad N_2 = -(\xi - 1)(\xi + 1) \qquad (7.52)$$

using in the above the normalized x coordinate [viz. Eq. (7.17)].

We note that the parameter a_2 does in fact have a meaning in this case as it is the magnitude of the departure from linearity of the approximation $\hat{u}$ at the element centre, since N_2 has been chosen here to have the value of unity at that point.

In a similar manner, for a cubic element we simply have to add $a_3 N_3$ to the quadratic expansion of Eq. (7.51), where N_3 is any cubic of the form

$$N_3^e = \alpha_0 + \alpha_1 \xi + \alpha_2 \xi^2 + \alpha_3 \xi^3 \tag{7.53}$$

and which has zero values at $\xi = \pm 1$ (i.e., at nodes 0 and 1). Again an infinity of choices exists, and we could select a cubic of a simple form which has a zero value at the centre of the element and for which $dN_3/d\xi = 1$ at the same point. Immediately we can write

$$N_3^e = \xi(1 - \xi^2) \tag{7.54}$$

as the cubic function with the desired properties. Now the parameter a_3 denotes the departure of the slope at the centre of the element from that of the first approximation.

We note that we could proceed in a similar manner and define the fourth-order hierarchical element shape function as

$$N_4^e = \xi^2(1 - \xi^2) \tag{7.55}$$

but a physical identification of the parameter associated with this now becomes more difficult (even though it is not strictly necessary).

As we have already noted, the above set is not unique and many other possibilities exist. An alternative convenient form for the hierarchical functions is defined by

$$N_p^e(\xi) = \begin{cases} \dfrac{1}{p!}(\xi^p - 1) & p \text{ even} \\[2mm] \dfrac{1}{p!}(\xi^p - \xi) & p \text{ odd} \end{cases} \tag{7.56}$$

where $p(\geqslant 2)$ is the degree of the introduced polynomial.[16] This yields the set of shape functions:

$$N_2^e = \tfrac{1}{2}(\xi^2 - 1) \qquad N_3^e = \tfrac{1}{6}(\xi^3 - \xi)$$
$$N_4^e = \tfrac{1}{24}(\xi^4 - 1) \qquad N_5^e = \tfrac{1}{120}(\xi^5 - \xi) \tag{7.57}$$

etc.

We observe that all derivatives of N_p^e of second or higher order have the value zero at $\xi = 0$, apart from $d^p N_p^e/d\xi^p$, which equals unity at that point, and hence, when shape functions of the form given by Eq. (7.57) are used, we can identify the parameters in the approximation as

$$a_p^e = \left. \frac{d^p \hat{u}}{d\xi^p} \right|_{\xi=0} \qquad p \geqslant 2 \tag{7.58}$$

Such identification gives a general physical significance but is by no means necessary.

In two- and three-dimensional elements a simple identification of the hierarchic parameters on interfaces will automatically ensure C_0 continuity of the approximation.

As mentioned previously, an optimal form of hierarchical functions is one that results in a diagonal equation system. This can on occasion be achieved, or at least approximated, quite closely.

In the elasticity problems which we have discussed in the preceding chapters the element matrix $\mathbf{K}^e$ possesses terms of the form

$$K_{lm}^e = \int_{\Omega^e} k \frac{dN_l^e}{dx} \frac{dN_m^e}{dx} dx = \frac{2}{h} \int_{-1}^{1} k \frac{dN_l^e}{d\xi} \frac{dN_m^e}{d\xi} d\xi \qquad (7.59)$$

If shape function sets containing the appropriate polynomials can be found for which such integrals are zero for $l \neq m$, then orthogonality is achieved and the coupling between successive solutions disappears.

One set of polynomial functions which is known to possess this orthogonality property over the range $-1 \leqslant \xi \leqslant 1$ is the set of Legendre polynomials $P_p(\xi)$, and the shape functions could be defined in terms of integrals of these polynomials.[9] Here we define the Legendre polynomial of degree p by

$$P_p(\xi) = \frac{1}{(p-1)!} \frac{1}{2^{p-1}} \frac{d^p}{d\xi^p} [(\xi^2 - 1)^p] \qquad (7.60)$$

and integrating these polynomials to define

$$N_{p+1}^e = \int P_p(\xi) \, d\xi = \frac{1}{(p-1)! \, 2^{p-1}} \frac{d^{p-1}}{d\xi^{p-1}} [(\xi^2 - 1)^p] \qquad (7.61)$$

Evaluation for each p in turn gives

$$N_2^e = \xi^2 - 1 \qquad N_3^e = 2(\xi^3 - \xi) \qquad \text{etc.}$$

These differ from the element shape functions given by Eq. (7.57) only by a multiplying constant up to N_3^e, but for $p \geqslant 3$ the differences become significant. The reader can easily verify the orthogonality of the derivatives of these functions, which is useful in computation. A plot of these functions and their derivatives is given in Fig. 7.24.

7.15 Two- and three-dimensional, hierarchic, elements of the 'rectangle' or 'brick' type

In deriving 'standard' finite element approximations we have shown that all shape functions for the Lagrange family could be obtained by a

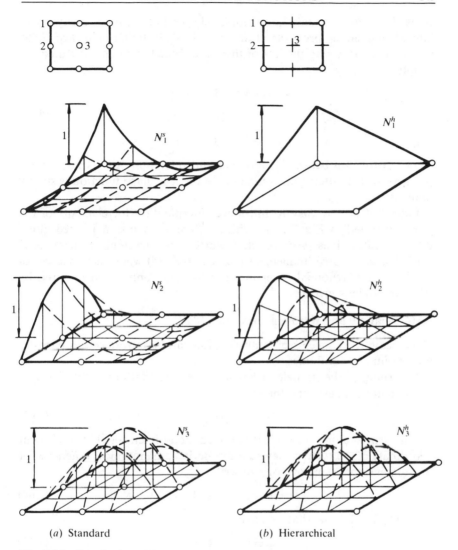

(a) Standard   (b) Hierarchical

Fig. 7.25   Standard and hierarchic shape functions corresponding to a lagrangian, quadratic element

simple multiplication of one-dimensional ones and those for serendipity
elements by a combination of such multiplications. The situation is even
simpler for hierarchic elements. Here *all* the shape functions can be
obtained by a simple multiplication process.

Thus, for instance, in Fig. 7.25 we show the shape functions for a
lagrangian nine-noded element and the corresponding hierarchical func-

tions. The latter not only have simpler shapes but are more easily calculated, being simple products of linear and quadratic terms of Eq. (7.56), (7.57), or (7.61). Using the last of these the three functions illustrated are simply

$$N_1 = -(\xi - 1)(\xi + 1)/4$$
$$N_2 = (\xi^2 - 1)(\xi - 1)/2 \tag{7.62}$$
$$N_3 = (\xi^2 - 1)(\xi^2 - 1)$$

The distinction between lagrangian and serendipity forms now disappears as for the latter in the present case the last shape function is simply omitted.

Indeed, it is now easy to introduce interpolation for elements of the type illustrated in Fig. 7.11 in which different expansion is used along different sides. This essential characteristic of hierarchical elements is exploited in adaptive refinement (viz. Chapter 14) where new degrees of freedom (or polynomial order increase) is made only when required by the magnitude of error.

7.16 Triangle and tetrahedron family[16,17]

Once again the concepts of multiplication can be introduced in terms of area (volume) coordinates.

Returning to the triangle of Fig. 7.16 we note that along the side 1–2, L_3 is identically zero, and therefore we have

$$(L_1 + L_2)_{1-2} = 1 \tag{7.63}$$

If ξ, measured along side 1–2, is the usual non-dimensional local element coordinate of the type we have used in deriving hierarchical functions for one-dimensional elements, we can write

$$L_1|_{1-2} = \tfrac{1}{2}(1 - \xi) \qquad L_2|_{1-2} = \tfrac{1}{2}(1 + \xi) \tag{7.64}$$

from which it follows that we have

$$\xi = (L_2 - L_1)_{1-2} \tag{7.65}$$

This suggests that we could generate hierarchical shape functions over the triangle by generalizing the one-dimensional shape function forms produced earlier. For example, using the expressions of Eq. (7.56), we associate with the side 1–2 the polynomial of degree $p(\geqslant 2)$ defined by

$$N^e_{p(1-2)} = \begin{cases} \dfrac{1}{p!} [(L_2 - L_1)^p - (L_1 + L_2)^p] & p \text{ even} \\[2mm] \dfrac{1}{p!} [(L_2 - L_1)^p - (L_2 - L_1)(L_1 + L_2)^{p-1}] & p \text{ odd} \end{cases} \tag{7.66}$$

It follows from Eq. (7.64) that these shape functions are zero at nodes 1 and 2. In addition, it can easily be shown that $N^e_{p(1-2)}$ will be zero all along the sides 3–1 and 3–2 of the triangle, and so C_0 continuity of the approximation $\hat{u}$ is assured.

It should be noted that in this case for $p \geqslant 3$ the number of hierarchical functions arising from the element sides in this manner is insufficient to define a complete polynomial of degree p, and internal hierarchical functions, which are identically zero on the boundaries, need to be introduced; for example, for $p = 3$ the function $L_1 L_2 L_3$ could be used, while for $p = 4$ the three additional functions $L_1^2 L_2 L_3$, $L_1 L_2^2 L_3$, $L_1 L_2 L_3^2$ could be adopted.

In Fig. 7.26 typical hierarchical linear, quadratic, and cubic trial functions for a triangular element are shown. Similar hierarchical shape functions could be generated from the alternative set of one-dimensional

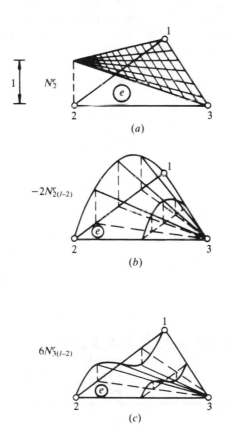

Fig. 7.26 Triangular elements and associated hierarchical shape functions of (a) linear, (b) quadratic, and (c) cubic form

shape functions defined in Eq. (7.61). Identical procedures are obvious in the context of tetrahedra.

7.17 Global and local finite element approximation

The very concept of hierarchic approximations (in which the shape functions are not affected by the refinement) means that it is possible to include in the expansion

$$u = \sum_{i=1}^{n} N_i a_i \tag{7.67}$$

functions N which are not local in nature. Such functions may, for instance, be the exact solutions of an analytical problem which in some way resembles the problem dealt with, but do not satisfy some boundary or inhomogeneity conditions. The 'finite element', local, expansions would here be a device for correcting this solution to satisfy the real conditions. Such use of global–local approximation has been suggested first by Mote[18] in a problem where the coefficients of this function were fixed. The example involved here is that of a rotating disc with cutouts (Fig. 7.27). The global, known, solution is the analytical one correspond-

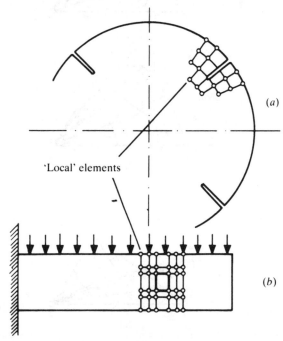

Fig. 7.27 Some possible uses of local–global (a) rotating slotted disc, (b) perforated beam

ing to a disc without cutout, and finite elements are added locally to modify the solution. Other examples of such 'fixed' solutions may well be those associated with point loads, where the use of the global approximation serves to eliminate the singularity modelled badly by the discretization.

In some problems the singularity itself is unknown and the appropriate function can be added with an unknown coefficient.

7.18 Improvement of conditioning with hierarchic forms

We have already mentioned that hierarchic element forms give a much improved equation conditioning due to their form which is more nearly diagonal. In Fig. 7.28 we show the 'condition number' (which is a

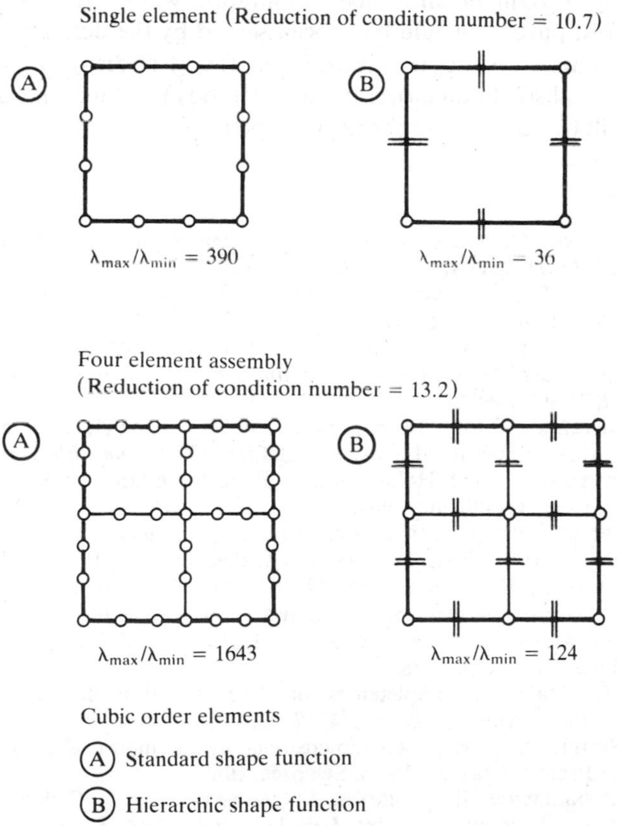

Single element (Reduction of condition number = 10.7)

$\lambda_{max}/\lambda_{min} = 390$ $\lambda_{max}/\lambda_{min} - 36$

Four element assembly
(Reduction of condition number = 13.2)

$\lambda_{max}/\lambda_{min} = 1643$ $\lambda_{max}/\lambda_{min} = 124$

Cubic order elements

(A) Standard shape function

(B) Hierarchic shape function

Fig. 7.28 Improvement of condition number (ratio of maximum to minimum eigenvalue of the stiffness matrix) by use of hierarchic form (elasticity isotropic $v = 0.15$)

measure of such diagonality and is defined in standard texts on linear algebra (see Appendix 1)) for a single cubic element and for an assembly of four cubic elements—using in their formulation standard and hierarchic forms. The improvement of the conditioning is a serious advantage of such forms and allows the use of iterative solution techniques to be easily adopted.[19]

7.19 Concluding remarks

An unlimited selection of element types has been presented here to the reader—and indeed equally unlimited alternative possibilities exist.[4,9] What of the use of such complex elements in practice? Putting aside the triangle and tetrahedron all the other elements are limited to situations where the real region is of a suitable shape which can be represented as an assembly of right prisms. Such a limitation would be so severe that little practical purpose would have been served by the derivation of such shape functions unless some way could be found of distorting these elements to fit realistic boundaries. In fact, methods for doing this are available and will be described in the next chapter.

References

1. P. C. DUNNE, 'Complete polynomial displacement fields for finite element methods', *Trans Roy. Aero. Soc.*, **72**, 245, 1968.
2. B. M. IRONS, J. G. ERGATOUDIS, and O. C. ZIENKIEWICZ, Comment on ref. 1, *Trans. Roy. Aero. Soc.*, **72**, 709–11, 1968.
3. J. G. ERGATOUDIS, B. M. IRONS, and O. C. ZIENKIEWICZ, 'Curved, isoparametric, quadrilateral elements for finite element analysis', *Int. J. Solids Struct.*, **4**, 31–42, 1968.
4. O. C. ZIENKIEWICZ et al., 'Iso-parametric and associated elements families for two and three dimensional analysis', Chapter 13 of *Finite Element Methods in Stress Analysis* (eds I. Holand and K. Bell), Tech. Univ. of Norway, Tapir Press, Norway, Trondheim, 1969.
5. J. G. ERGATOUDIS, *Quadrilateral elements in plane analysis: introduction to solid analysis*, M.Sc. thesis, University of Wales, Swansea, 1966.
6. J. H. ARGYRIS, K. E. BUCK, H. M. HILBER, G. MARECZEK, and D. W. SCHARPF, 'Some new elements for matrix displacement methods', *2nd Conf. on Matrix Methods in Struct. Mech.*, Air Force Inst. of Techn., Wright Patterson Base, Ohio, Oct. 1968.
7. R. L. TAYLOR, 'On completeness of shape functions for finite element analysis', *Int. J. Num. Meth. Eng.*, **4**, 17–22, 1972.
8. F. C. SCOTT, 'A quartic, two dimensional isoparametric element', Undergraduate Project, Univ. of Wales, Swansea, 1968.
9. O. C. ZIENKIEWICZ, B. M. IRONS, J. CAMPBELL, and F. C. SCOTT, 'Three dimensional stress analysis', *Int. Un. Th. Appl. Mech. Symposium on High Speed Computing in Elasticity*, Liége, 1970.
10. W. P. DOHERTY, E. L. WILSON, and R. L. TAYLOR, *Stress Analysis of Axisymmetric Solids Utilizing Higher-Order Quadrilateral Finite Elements*, Report

69–3, Structural Engineering Laboratory, Univ. of California, Berkeley, Jan. 1969.

11. J. H. ARGYRIS, I. FRIED, and D. W. SCHARPF, 'The TET 20 and the TEA 8 elements for the matrix displacement method', *Aero. J.*, **72**, 618–25, 1968.

12. P. SILVESTER, 'Higher order polynomial triangular finite elements for potential problems', *Int. J. Eng. Sci.*, **7**, 849–61, 1969.

13. B. FRAEIJS DE VEUBEKE, 'Displacement and equilibrium models in the finite element method", Chapter 9 of *Stress Analysis* (eds O. C. Zienkiewicz and G. S. Holister), Wiley, 1965.

14. J. H. ARGYRIS, 'Triangular elements with linearly varying strain for the matrix displacement method', *J. Roy. Aero. Soc. Tech. Note*, **69**, 711–13, Oct. 1965.

15. J. G. ERGATOUDIS, B. M. IRONS, and O. C. ZIENKIEWICZ, 'Three dimensional analysis of arch dams and their foundations', *Symposium on Arch Dams*, Inst. Civ. Eng., London, 1968.

16. A. G. PEANO, 'Hierarchics of conforming finite elements for elasticity and plate bending', *Comp. Math. and Applications*, **2**, 3–4, 1976.

17. J. P. DE S. R. GAGO, *A posteri error analysis and adaptivity for the finite element method*, Ph.D thesis, University of Wales, Swansea, 1982.

18. C. D. MOTE, 'Global-local finite element', *Int. J. Num. Meth. Eng.*, **3**, 565–74, 1971.

19. O. C. ZIENKIEWICZ, J. P. DE S. R. GAGO, and D. W. KELLY, 'The hierarchical concept in finite element analysis', *Computers and Structures*, **16**, 53–65, 1983.

Mapped elements and numerical integration— 'infinite' and 'singularity' elements

8.1 Introduction

In the previous chapter we have shown how some general families of finite elements can be obtained. A progressively increasing number of nodes and hence improved accuracy characterizes each new member of the family and presumably the number of such elements required to obtain an adequate solution decreases rapidly. To ensure that a small number of elements can represent a relatively complex form of the type that is liable to occur in real, rather than academic, problems, simple rectangles and triangles no longer suffice. This chapter is therefore concerned with the subject of distorting such simple forms into others of more arbitrary shape.

Elements of the basic one-, two-, or three-dimensional types will be 'mapped' into distorted forms in the manner indicated in Figs 8.1 and 8.2.

In these figures it is shown that the ξ, η, ζ, or $L_1 L_2 L_3 L_4$ coordinates can be distorted to a new, curvilinear set when plotted in a cartesian x, y, z space.

Not only can two-dimensional elements be distorted into others in two dimensions but the mapping of these can be taken into three dimensions as indicated by the flat sheet elements of Fig. 8.2 distorting into a three-dimensional space. This principle applies generally, providing some one-to-one correspondence between cartesian and curvilinear

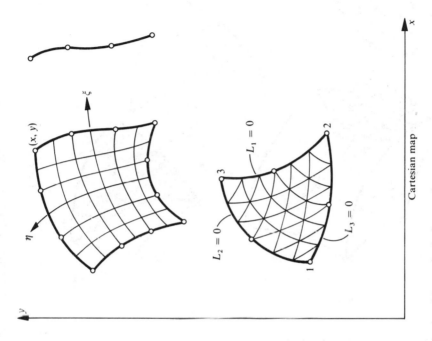

Fig. 8.1 Two-dimensional 'mapping' of some elements

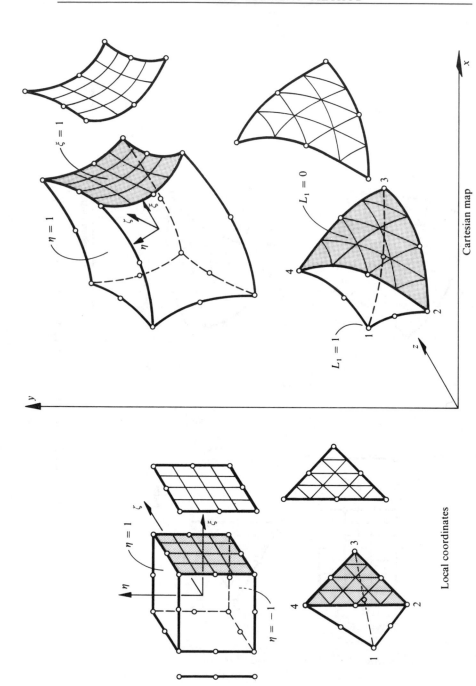

Fig. 8.2   Three-dimensional 'mapping' of some elements

coordinates can be established, i.e., once the mapping relations of the type

$$\begin{Bmatrix} x \\ y \\ z \end{Bmatrix} = f \begin{Bmatrix} \xi \\ \eta \\ \zeta \end{Bmatrix} \quad \text{or} \quad f \begin{Bmatrix} L_1 \\ L_2 \\ L_3 \\ L_4 \end{Bmatrix} \tag{8.1}$$

can be established.

Once such coordinate relationships are known, shape functions can be specified in local coordinates and by suitable transformations the element properties established in the global system.

In what follows we shall first discuss the so-called isoparametric form of relationship (8.1) which has found a great deal of practical application. Full details of this formulation will be given, including the establishment of element properties by numerical integration which are essential.

In the final section we shall show that many other coordinate transformations can be used effectively.

PARAMETRIC CURVILINEAR COORDINATES

8.2 Use of 'shape functions' in establishment of coordinate transformations

A most convenient method of establishing the coordinate transformations is to use the 'standard' type of shape functions we have already derived to represent the variation of the unknown function.

If we write, for instance, for each element

$$x = N_1' x_1 + N_2' x_2 + \cdots = \mathbf{N}' \begin{Bmatrix} x_1 \\ x_2 \\ \vdots \end{Bmatrix} = \mathbf{N}' \mathbf{x}$$

$$y = N_1' y_1 + N_2' y_2 + \cdots = \mathbf{N}' \begin{Bmatrix} y_1 \\ y_2 \\ \vdots \end{Bmatrix} = \mathbf{N}' \mathbf{y} \tag{8.2}$$

$$z = N_1' z_1 + N_2' z_2 + \cdots = \mathbf{N}' \begin{Bmatrix} z_1 \\ z_2 \\ \vdots \end{Bmatrix} = \mathbf{N}' \mathbf{z}$$

in which $\mathbf{N}'$ are standard shape functions given in terms of the local coordinates, then immediately a relationship of the required form is available. Further, the points with coordinates x_1, y_1, z_1, etc., will lie at appropriate points of the element boundary (as from the general definitions of the shape functions we know that these have a value of unity at the point in question and zero elsewhere). These points can establish nodes 'a priori'.

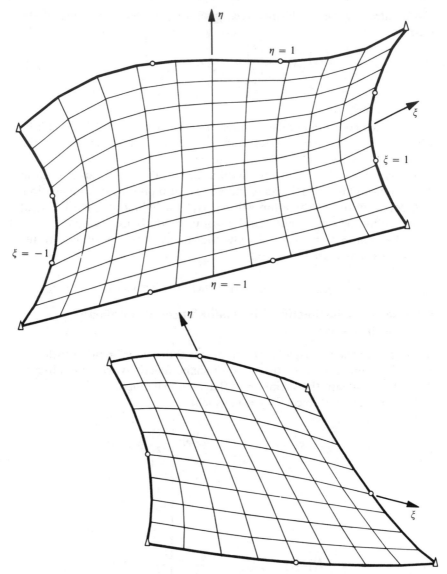

Fig. 8.3   Computer plots of curvilinear coordinates for cubic and parabolic elements (reasonable distortion)

To each set of local coordinates will correspond a set of global cartesian coordinates and in general only one such set. We shall see, however, that a non-uniqueness may arise, sometimes with violent distortion.

The concept of using such element shape functions for establishing curvilinear coordinates in the context of finite element analysis appears to have been first introduced by Taig.[1] In his first application basic linear quadrilateral relations were used. Irons[2,3] generalized the idea for other elements.

Quite independently the exercises of devising various practical methods of generating curved surfaces for purposes of engineering design led to the establishment of similar definitions by Coons[4] and Forrest,[5] and indeed today the subjects of surface definitions and analysis are drawing closer together due to this activity.

In Fig. 8.3 an actual distortion of elements based on the cubic and quadratic members of the 'serendipity' family is shown. It is seen here that a one-to-one relationship exists between the local (ξ, η) and global (x, y) coordinates. If the fixed points are such that a violent distortion occurs then a non-uniqueness may occur in the manner indicated for two situations in Fig. 8.4. Here at internal points of the distorted element two

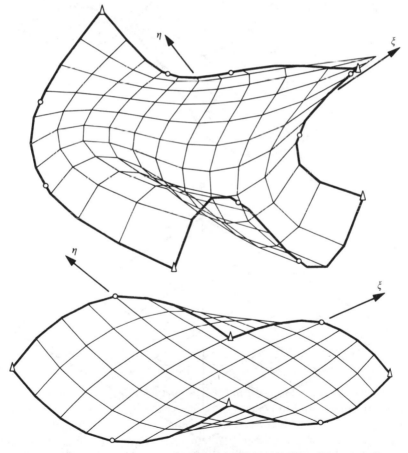

Fig. 8.4   Unreasonable element distortion leading to non-unique mapping and 'overspill'. Cubic and parabolic elements

sets of local coordinates are implied in addition to some internal points being mapped outside the element. Care must be taken in practice to avoid such gross distortion.

Figure 8.5 shows two examples of a two-dimensional (ξ, η) element mapped into a three-dimensional (x, y, z) space.

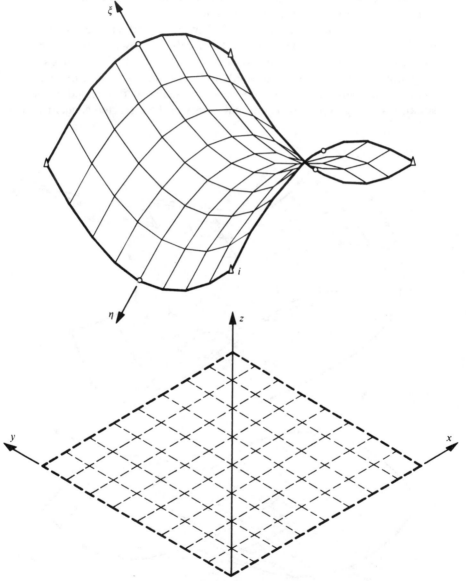

Fig. 8.5   Flat elements (of parabolic type) mapped into three dimensions

In this chapter we shall often refer to the basic element in undistorted, local, coordinates as a 'parent' element.

In Sec. 8.5 we shall define a quantity known as the jacobian determinant. The well-known condition for a *one-to-one* mapping (such as exists in Fig. 8.3 and does not in Fig. 8.4) is that the sign of this quantity should remain unchanged at all the points of the domain mapped.

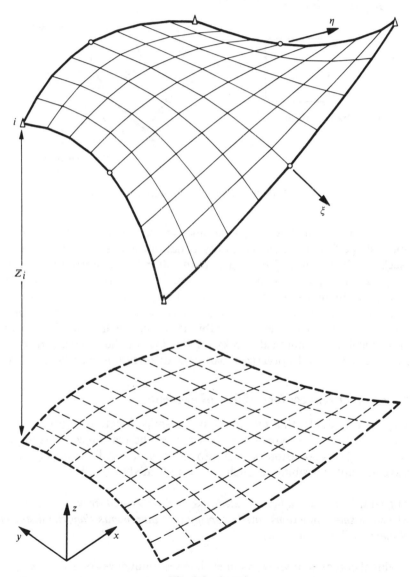

Fig. 8.5   (*cont.*)

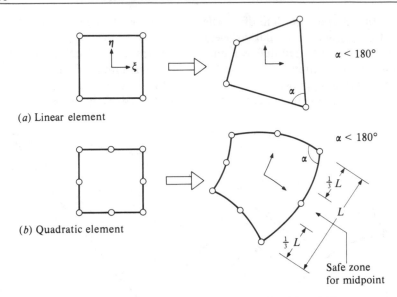

(a) Linear element

(b) Quadratic element

Fig. 8.6   Rules for uniqueness of mapping (a) and (b)

It can be shown that with a parametric transformation based on bi-linear shape functions, the necessary condition is that no internal angle [such as α in Fig. 8.6(a)] be greater than 180°.[6] In transformations based on parabolic-type 'serendipity' functions, it is necessary in addition to this requirement to ensure that the mid-side nodes are in the 'middle half' of the distance between adjacent corners[7] but a 'middle third' shown in Fig. 8.6 is safer. For cubic functions such general rules are impractical and numerical checks on the sign of the jacobian determinant are necessary. In practice a parabolic distortion is usually sufficient.

8.3 Geometrical conformability of elements

While it was shown that by the use of the shape function transformation each parent element maps uniquely a part of the real object, it is important that the subdivision of this into the new, curved, elements should leave no gaps. Possibility of such gaps is indicated in Fig. 8.7.

THEOREM 1. *If two adjacent elements are generated from 'parents' in which the shape functions satisfy continuity requirements then the distorted elements will be contiguous.*

This theorem is obvious, as in such cases uniqueness of any function u required by continuity is simply replaced by that of uniqueness of the x,

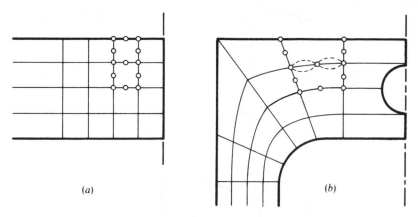

Fig. 8.7   Compatibility requirements in real subdivision of space

y, or z coordinate. As adjacent elements are given the same sets of coordinates at nodes, continuity is implied.

Nodes of the new distorted elements need not necessarily be placed only at points for which shape functions are specified. Other corresponding sets of nodes can be added on interfaces or boundaries.

8.4 Variation of the unknown function within distorted, curvilinear elements. Continuity requirements

With the shape of the element now defined by the shape functions $\mathbf{N'}$ the variation of the unknown, u, has to be specified before we can establish element properties. This is most conveniently given in terms of local, curvilinear coordinates by the usual expression

$$u = \mathbf{N}\mathbf{a}^e \tag{8.3}$$

where $\mathbf{a}^e$ lists the nodal values.

THEOREM 2. *If the shape functions $\mathbf{N}$ used in (8.3) are such that continuity of u is preserved in the parent coordinates then continuity requirements will be satisfied in distorted elements.*

The proof of this theorem follows the same lines as the previous section.

The nodal values may or may not be associated with the same nodes as used to specify the element geometry. For example, in Fig. 8.8 the points marked with a circle are used to define the element geometry. We

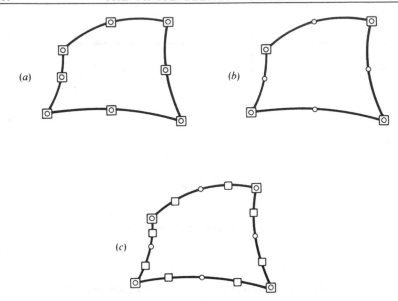

Fig. 8.8   Various element specifications: ○ point at which coordinate specified; □ points at which function parameter specified. (*a*) Isoparametric, (*b*) super-parametric, (*c*) subparametric

could use the values of the function defined at nodes marked with a square to define the variation of the unknown.

In Fig. 8.8(*a*) the same points define the geometry and the finite element analysis points. If then

$$\mathbf{N} = \mathbf{N}' \qquad\qquad (8.4)$$

i.e., the shape functions defining geometry and function are the same, the elements will be called *isoparametric*.

We could, however, use only the four corner points to define the variation of *u* [Fig. 8.8(*b*)]. Such an element we shall refer to as *super-parametric*, noting that the variation of geometry is more general than that of the actual unknown.

Similarly, if for instance we introduce more nodes to define *u* than are used to define geometry, *subparametric* elements will result [Fig. 8.8(*c*)]. Such elements will be found to be more often of use in practice.

While for mapping it is convenient to use 'standard' forms of shape functions the interpolation of the unknown can, of course, use hierarchic forms defined in the previous chapter. Once again the definitions of sub- and superparametric variations are applicable.

8.5 Evaluation of element matrices (transformation in ξ, η, ζ coordinates)

To perform finite element analysis the matrices defining element properties, e.g., stiffness, etc., have to be found. These will be of the form

$$\int_V \mathbf{G} \, dV \tag{8.5}$$

in which the matrix $\mathbf{G}$ depends on $\mathbf{N}$ or its derivatives with respect to *global coordinates*. As an example of this we have the stiffness matrix

$$\int_V \mathbf{B}^\mathrm{T}\mathbf{DB} \, dV \tag{8.6}$$

and associated load vectors

$$\int_V \mathbf{N}^\mathrm{T}\mathbf{b} \, dV \tag{8.7}$$

For each particular class of elastic problems the matrices of $\mathbf{B}$ are given explicitly by their components [see the general form of Eqs (3.10), (4.6), and (5.11)]. Quoting the first of these, Eq. (3.10), valid for plane problems we have

$$\mathbf{B}_i = \begin{bmatrix} \dfrac{\partial N_i}{\partial x}, & 0 \\[2mm] 0, & \dfrac{\partial N_i}{\partial y} \\[2mm] \dfrac{\partial N_i}{\partial y}, & \dfrac{\partial N_i}{\partial x} \end{bmatrix} \tag{8.8}$$

In the elasticity problems the matrix $\mathbf{G}$ is thus a function of the first derivatives of $\mathbf{N}$ and this situation will arise in many other classes of problem. In all C_0 continuity is needed and, as we have already noted, this is readily satisfied by the functions of Chapter 7, written now in terms of the curvilinear coordinates.

To evaluate such matrices we note that two transformations are necessary. In the first place, as N_i is defined in terms of local (curvilinear) coordinates, it is necessary to devise some means of expressing the global derivatives of the type occurring in Eq. (8.8) in terms of local derivatives.

In the second place the element of volume (or surface) over which the integration has to be carried out needs to be expressed in terms of the local coordinates with an appropriate change of limits of integration.

Consider, for instance, the set of local coordinates ξ, η, ζ and a corresponding set of global coordinates x, y, z. By the usual rules of partial differentiation we can write, for instance, the ξ derivative as

$$\frac{\partial N_i}{\partial \xi} = \frac{\partial N_i}{\partial x}\frac{\partial x}{\partial \xi} + \frac{\partial N_i}{\partial y}\frac{\partial y}{\partial \xi} + \frac{\partial N_i}{\partial z}\frac{\partial z}{\partial \xi} \tag{8.9}$$

Performing the same differentiation with respect to the other two coordinates and writing in matrix form we have

$$\begin{Bmatrix} \dfrac{\partial N_i}{\partial \xi} \\[2mm] \dfrac{\partial N_i}{\partial \eta} \\[2mm] \dfrac{\partial N_i}{\partial \zeta} \end{Bmatrix} = \begin{bmatrix} \dfrac{\partial x}{\partial \xi}, & \dfrac{\partial y}{\partial \xi}, & \dfrac{\partial z}{\partial \xi} \\[2mm] \dfrac{\partial x}{\partial \eta}, & \dfrac{\partial y}{\partial \eta}, & \dfrac{\partial z}{\partial \eta} \\[2mm] \dfrac{\partial x}{\partial \zeta}, & \dfrac{\partial y}{\partial \zeta}, & \dfrac{\partial z}{\partial \zeta} \end{bmatrix} \begin{Bmatrix} \dfrac{\partial N_i}{\partial x} \\[2mm] \dfrac{\partial N_i}{\partial y} \\[2mm] \dfrac{\partial N_i}{\partial z} \end{Bmatrix} = \mathbf{J} \begin{Bmatrix} \dfrac{\partial N_i}{\partial x} \\[2mm] \dfrac{\partial N_i}{\partial y} \\[2mm] \dfrac{\partial N_i}{\partial z} \end{Bmatrix} \tag{8.10}$$

In the above, the left-hand side can be evaluated as the functions N_i are specified in local coordinates. Further, as x, y, z are explicitly given by the relation defining the curvilinear coordinates [Eq. (8.2)], the matrix $\mathbf{J}$ can be found explicitly in terms of the local coordinates. This matrix is known as the *jacobian matrix*.

To find now the global derivatives we invert $\mathbf{J}$ and write

$$\begin{Bmatrix} \dfrac{\partial N_i}{\partial x} \\[2mm] \dfrac{\partial N_i}{\partial y} \\[2mm] \dfrac{\partial N_i}{\partial z} \end{Bmatrix} = \mathbf{J}^{-1} \begin{Bmatrix} \dfrac{\partial N_i}{\partial \xi} \\[2mm] \dfrac{\partial N_i}{\partial \eta} \\[2mm] \dfrac{\partial N_i}{\partial \zeta} \end{Bmatrix} \tag{8.11}$$

In terms of the shape function defining the coordinate transformation $\mathbf{N}'$ (which as we have seen are only identical with the shape functions $\mathbf{N}$ when isoparametric formulation is used) we have

$$\mathbf{J} = \begin{bmatrix} \sum \dfrac{\partial N_i'}{\partial \xi} x_i, & \sum \dfrac{\partial N_i'}{\partial \xi} y_i, & \sum \dfrac{\partial N_i'}{\partial \xi} z_i \\[3mm] \sum \dfrac{\partial N_i'}{\partial \eta} x_i, & \sum \dfrac{\partial N_i'}{\partial \eta} y_i, & \sum \dfrac{\partial N_i'}{\partial \eta} z_i \\[3mm] \sum \dfrac{\partial N_i'}{\partial \zeta} x_i, & \sum \dfrac{\partial N_i'}{\partial \zeta} y_i, & \sum \dfrac{\partial N_i'}{\partial \zeta} z_i \end{bmatrix} = \begin{bmatrix} \dfrac{\partial N_1'}{\partial \xi}, & \dfrac{\partial N_2'}{\partial \xi} & \cdots \\[3mm] \dfrac{\partial N_1'}{\partial \eta}, & \dfrac{\partial N_2'}{\partial \eta} & \cdots \\[3mm] \dfrac{\partial N_1'}{\partial \zeta}, & \dfrac{\partial N_2'}{\partial \zeta} & \cdots \end{bmatrix} \begin{bmatrix} x_1, & y_1, & z_1 \\[2mm] x_2, & y_2, & z_2 \\[2mm] \vdots & \vdots & \vdots \end{bmatrix} \tag{8.12}$$

To transform the variables and the region with respect to which the integration is made a standard process will be used which involves the determinant of $\mathbf{J}$. Thus, for instance, a volume element becomes

$$dx \, dy \, dz = \det \mathbf{J} \, d\xi \, d\eta \, d\zeta \qquad (8.13)$$

This type of transformation is valid irrespective of the number of coordinates used. For its justification the reader is referred to standard mathematical texts. A particularly lucid account of this is given by Murnaghan.[8]† (See also Appendix 5.)

Assuming that the inverse of $\mathbf{J}$ can be found we have now reduced the evaluation of the element properties to that of finding integrals of the form of Eq. (8.5).

More explicitly we can write this as

$$\int_{-1}^{1} \int_{-1}^{1} \int_{-1}^{1} \bar{\mathbf{G}}(\xi, \eta, \zeta) \, d\xi \, d\eta \, d\zeta \qquad (8.14)$$

if the curvilinear coordinates are of the normalized type based on the right prism. Indeed the integration *is carried out within such a prism* and not in the complicated distorted shape, thus accounting for the simple integration limits. One- and two-dimensional problems similarly will result in integrals with respect to one or two coordinates within simple limits.

While the limits of the integration are simple in the above case, unfortunately the explicit form of $\bar{\mathbf{G}}$ is not. Excepting the simplest elements, algebraic integration usually defies our mathematical skill, and numerical integration has to be resorted to. This, as will be seen from later sections, is not a severe penalty and has the advantage that algebraic errors are more easily avoided and that general programs, not tied to a particular element, can be written for various classes of problems. Indeed in such numerical calculations the inverses of $\mathbf{J}$ are never explicitly found.

8.5.1 *Surface integrals.* In elasticity and other applications, surface integrals frequently occur. Typical here are the expressions for evaluating the contributions of surface tractions [see Chapter 2, Eq. (2.24b)]:

$$\mathbf{f} = - \int_{A} \mathbf{N}^{\mathrm{T}} \mathbf{t} \, dA$$

---

† The determinant of the jacobian matrix is known in literature simply as 'the jacobian' and is often written as

$$\det \mathbf{J} \equiv \frac{\partial(x, y, z)}{\partial(\xi, \eta, \zeta)}$$

The element dA will generally lie on a surface where one of the coordinates (say ζ) is constant.

The most convenient process of dealing with the above is to consider dA as a vector oriented in the direction normal to the surface (see Appendix 5). For three-dimensional problems we form a vector product

$$
dA = \left\{ \begin{array}{c} \dfrac{\partial x}{\partial \xi} \\[2mm] \dfrac{\partial y}{\partial \xi} \\[2mm] \dfrac{\partial z}{\partial \xi} \end{array} \right\} \times \left\{ \begin{array}{c} \dfrac{\partial x}{\partial \eta} \\[2mm] \dfrac{\partial y}{\partial \eta} \\[2mm] \dfrac{\partial z}{\partial \eta} \end{array} \right\} \, d\xi \; d\eta
$$

and on substitution integrate within a domain $1 \leqslant \xi, \eta \leqslant 1$.

For two dimensions a line length dS arises and here the magnitude is simply

$$
dS = \left\{ \begin{array}{c} \dfrac{\partial x}{\partial \xi} \\[2mm] \dfrac{\partial y}{\partial \xi} \\[2mm] \dfrac{dz}{\partial \xi} \end{array} \right\} \, d\xi
$$

on constant η surfaces.

8.6 Element matrices. Area and volume coordinates

The general relationship [Eq. (8.2)] for coordinate mapping and indeed all the following theorems are equally valid for any set of local coordinates and could relate the local $L_1, L_2, \ldots$ coordinates used for triangles and tetrahedra in the previous chapter, to the global cartesian ones.

Indeed most of the discussion of the previous chapter is valid if we simply rename the local coordinates suitably. However, two important differences arise.

The first concerns the fact that the local coordinates are not independent and in fact number one more than the cartesian system. The matrix J would apparently therefore become rectangular and would not possess an inverse. The second is simply the difference of integration limits which have to correspond with a triangular or tetrahedral 'parent'.

The simplest, though perhaps not the most elegant, way out of the first difficulty is to consider the last variable as a dependent one. Thus, for example, we can introduce formally, in the case of the tetrahedra,

$$\xi = L_1$$
$$\eta = L_2$$
$$\zeta = L_3$$
$$1 - \xi - \eta - \zeta = L_4$$

(8.15)

(by definition of the previous chapter) and thus preserve without change Eq. (8.9) and all the equations up to Eq. (8.14).

As the functions N_i are given in fact in terms of L_1, L_2, etc., we must observe that

$$\frac{\partial N_i}{\partial \xi} = \frac{\partial N_i}{\partial L_1} \frac{\partial L_1}{\partial \xi} + \frac{\partial N_i}{\partial L_2} \frac{\partial L_2}{\partial \xi} + \frac{\partial N_i}{\partial L_3} \frac{\partial L_3}{\partial \xi} + \frac{\partial N_i}{\partial L_4} \frac{\partial L_4}{\partial \xi}$$

(8.16)

On using Eq. (8.15) this becomes simply

$$\frac{\partial N_i}{\partial \xi} = \frac{\partial N_i}{\partial L_1} - \frac{\partial N_i}{\partial L_4}$$

with the other derivatives obtainable by similar expressions.

The integration limits of Eq. (8.14) now change, however, to correspond with the tetrahedron limits. Typically

$$\int_0^1 \int_0^{1-\eta} \int_0^{1-\eta-\zeta} \bar{G}(\xi, \eta, \zeta) \, d\xi \, d\eta \, d\zeta$$

(8.17)

The same procedure clearly will apply in the case of triangular coordinates.

It must be noted that once again the expression $\bar{G}$ will necessitate numerical integration which, however, is carried out over the simple, undistorted, parent region whether this be triangular or tetrahedral.

Finally it should be remarked that any of the elements given in the previous chapter are capable of being mapped. In some, such as the triangular prism, both area and rectangular coordinates are used (Fig. 8.9).

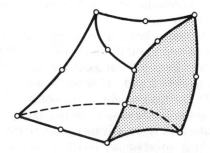

Fig. 8.9   A distorted triangular prism

The remarks regarding the dependence of coordinates apply once again with regard to the former but the processes of the present section should make procedures clear.

8.7 Convergence of elements in curvilinear coordinates

To consider the convergence aspects of the problem posed in curvilinear coordinates it is convenient to return to the starting point of the approximation where an energy functional Π, or an equivalent integral form (weak problem statement), was defined by volume integrals essentially similar to those of Eq. (8.5), in which the integrand was a function of u and its first derivatives.

Thus, for instance, the variational principles of the energy kind discussed in Chapter 2 (or others of Chapter 9) could be stated for a scalar function u as

$$\Pi = \int_\Omega F\left(u, \frac{\partial u}{\partial x}, \frac{\partial u}{\partial y}, x, y\right) d\Omega + \int_\Gamma E(u, \ldots) \, d\Gamma \qquad (8.18)$$

The coordinate transformation changes the derivatives of any function by the jacobian relation (8.11). Thus

$$\begin{Bmatrix} \dfrac{\partial u}{\partial x} \\ \dfrac{\partial u}{\partial y} \end{Bmatrix} = \mathbf{J}^{-1}(\xi, \eta) \begin{Bmatrix} \dfrac{\partial u}{\partial \xi} \\ \dfrac{\partial u}{\partial \eta} \end{Bmatrix} \qquad (8.19)$$

and the functional can be stated simply by a relationship of the form (8.18) with x, y, etc., replaced by ξ, η, etc., with the maximum order of differentiation unchanged.

It follows immediately that if the shape functions are so chosen in the curvilinear coordinate space as to observe the usual rules of convergence (continuity and presence of complete first-order polynomials), then convergence will occur. Further, all the arguments concerning the order of convergence with the element size h still hold, providing h is related to *the curvilinear coordinate system*.

Indeed, all that has been said above is applicable to problems involving higher derivatives and to most unique coordinate transformations. It should be noted that the patch test as conceived in the x, y, ... coordinate system (see Chapters 2 and 11) is no longer simply applicable and in principle should be applied with polynomial fields imposed in the curvilinear coordinates. In the case of isoparametric (or subparametric) elements the situation is more advantageous. Here a linear (constant derivative x, y) field is always reproduced by the curvilinear coordinate expansion, and thus the lowest order patch test will be passed in the standard manner on such elements.

The proof of this is simple. Consider a standard isoparametric expansion

$$u = \sum_{i=1}^{n} N_i a_i \equiv \mathbf{N a} \qquad \mathbf{N} = \mathbf{N}(\xi, \eta, \zeta) \tag{8.20}$$

with coordinates of nodes defining the transformation as

$$x = \sum N_i x_i \qquad y = \sum N_i y_i \qquad z = \sum N_i z_i \tag{8.21}$$

The question is under what circumstances it is possible for expression (8.20) to define a linear expansion in cartesian coordinates:

$$u = \alpha_1 + \alpha_2 x + \alpha_3 y + \alpha_4 z$$
$$\equiv \alpha_1 + \alpha_2 \sum N_i x_i + \alpha_3 \sum N_i y_i + \alpha_4 \sum N_i z_i \tag{8.22}$$

If we take

$$a_i = \alpha_1 + \alpha_2 x_i + \alpha_3 y_i + \alpha_4 z_i$$

and compare expression (8.20) with (8.22) we note that identity is obtained between these providing

$$\sum N_i = 1$$

As this is the usual requirement of standard element shape functions [viz. Eq. (7.4)] we can conclude that the following theorem is valid.

THEOREM 3. *The constant derivative condition will be satisfied for all isoparametric elements.*

As subparametric elements can always be expressed as specific cases of isoparametric transformation this theorem is obviously valid here also.

It is of interest to pursue the argument and to see under what circumstances higher polynomial expansions in cartesian coordinates can be achieved under various transformations. The simple linear case in which we 'guessed' the solution has now to be substituted by considering in detail the polynomial terms occurring in such expressions as (8.20) and (8.22) and establishing conditions for equating appropriate coefficients.

Consider a specific problem: the circumstances under which the bilinearly mapped quadrilateral of Fig. 8.10 can represent fully any quadratic cartesian expansion. We now have

$$x = \sum_1^4 N_i' x_i \qquad y = \sum_1^4 N_i' y_i \tag{8.23}$$

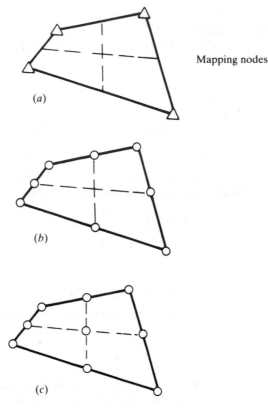

Mapping nodes

(a)

(b)

(c)

Fig. 8.10 Bilinear mapping of subparametric quadratic eight- and nine-noded element

and we wish to be able to reproduce

$$u = \alpha_1 + \alpha_2 x + \alpha_3 y + \alpha_4 x^2 + \alpha_5 xy + \alpha_6 y^2 \tag{8.24}$$

Noting that the bilinear form of N_i' contains such terms as 1, ξ, η and $\xi\eta$, the above can be written as

$$u = \beta_1 + \beta_2 \xi + \beta_3 \eta + \beta_4 \xi^2 + \beta_5 \xi\eta + \beta_6 \eta^2$$
$$+ \beta_7 \xi\eta^2 + \beta_8 \xi^2\eta + \beta_9 \xi^2\eta^2 \tag{8.25}$$

where β_1 to β_9 depend on the values of α_1 to α_6.

We shall now try to match the terms arising from quadratic expansions of the serendipity and lagrangian kinds shown in Fig. 8.10(b) and (c):

$$u = \sum_1^8 N_i a_i \tag{8.26a}$$

$$u = \sum_1^9 N_i a_i \qquad (8.26b)$$

where the appropriate terms are of a kind defined in the previous chapter.

For the eight-noded element (serendipity) [Fig. 8.10(b)] we can write (8.26a) directly using polynomial coefficients b_i, $i = 1$–8, in place of the nodal variables a_i (noting the terms occurring in the Pascal triangle) as

$$u = b_1 + b_2 \xi + b_3 \eta + b_4 \xi^2 + b_5 \xi\eta + b_6 \eta^2 + b_7 \xi\eta^2 + b_8 \xi^2\eta \qquad (8.27)$$

It is immediately evident that for arbitrary values of β_1 to β_9 it is impossible to match the coefficients b_1 to b_8 due to the absence of the term $\xi^2\eta^2$ in Eq. (8.27). [However if higher order (quartic, etc.) expansions of the serendipity kind were used such matching would evidently be possible and we could conclude that for linearly distorted elements the serendipity family of order four or greater will always represent quadratics.]

For the nine-noded, lagrangian, element [Fig. 8.10(c)] the expansion similar to (8.27) gives

$$u = b_1 + b_2 \xi + b_3 \eta + b_4 \xi^2 + \cdots + b_8 \xi^2\eta + b_9 \xi^2\eta^2 \qquad (8.28)$$

and the matching of the coefficients of Eqs (8.28) and (8.25) can be made directly.

We can conclude therefore that nine-noded elements represent better cartesian polynomials (when distorted linearly) and therefore are generally preferable in modelling smooth solutions.† An example of this is given in Fig. 8.11 where we consider the results of a finite element calculation with eight- and nine-noded elements respectively used to reproduce a simple beam solution in which we know that the exact answers are quadratic. With no distortion both elements give exact results but when distorted only the nine-noded element does so, with the eight-noded element giving quite wild stress fluctuation.

Similar arguments will lead to the conclusion that in three dimensions again only the lagrangian 27-noded element is capable of reproducing fully the quadratic in cartesian coordinates when bilinearly distorted.

The reader can extend simply the above arguments to higher order expansions and, for instance, to the modelling of cubic responses. Full discussion of such problems is given by Wachspress.[9]

---

† The authors are indebted to Prof. M. Crochet of Louvain University for pointing out this simple proof.

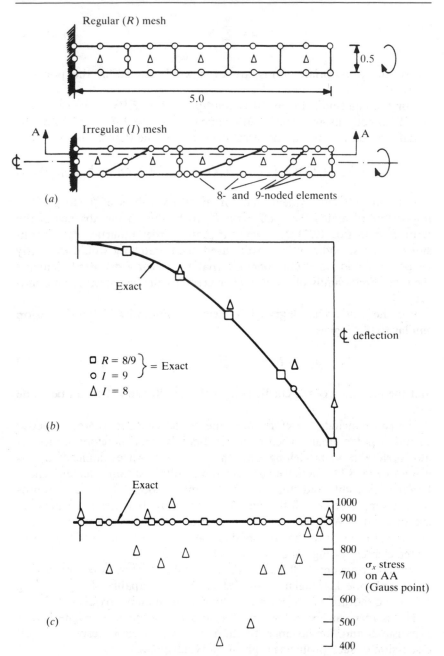

Fig. 8.11   Quadratic serendipity and Lagrange eight- and nine-noded elements in regular and distorted form. Elastic deflection of a beam under constant moment. Note poor results of eight-noded element

8.8 Numerical integration—one-dimensional

Already in Chapter 4 dealing with a relatively simple problem of axisymmetric stress distribution and simple triangular elements it was noted that exact integration of expressions for element matrices could be troublesome. Now for the more complex distorted elements numerical integration is essential.

Some principles of numerical integration will be summarized here together with tables of convenient numerical coefficients.

To find numerically the integral of a function of one variable we can proceed in one of several ways.[10]

8.8.1 *Newton–Cotes quadrature.*† In the most obvious procedure, points at which the function is to be found are determined *a priori*—usually at equal intervals—and a polynomial passed through the values of the function at these points and exactly integrated [Fig. 8.12(a)].

As '*n*' values of the function define a polynomial of degree $n - 1$, the errors will be of the order $O(h^n)$ where h is the element size. This leads to the well-known Newton–Cotes 'quadrature' formulae. The integrals can be written as

$$I = \int_{-1}^{1} f(\xi) \, d\xi = \sum_{1}^{n} H_i f(\xi_i) \tag{8.29}$$

for the range of integration between -1 and $+1$ [Fig. 8.12(a)]. For example, if $n = 2$, we have the well-known trapezoidal rule:

$$I = f(-1) + f(1) \tag{8.30}$$

for $n = 3$, the Simpson 'one-third' rule:

$$I = \tfrac{1}{3}[f(-1) + 4f(0) + f(1)] \tag{8.31}$$

and for $n = 4$:

$$I = \tfrac{1}{4}[f(-1) + 3f(-\tfrac{1}{3}) + 3f(\tfrac{1}{3}) + f(1)] \tag{8.32}$$

Formulae for n up to 21 are given by Kopal.[10]

8.2.2 *Gauss quadrature.* If in place of specifying the position of sampling points *a priori* we allow these to be located at points to be determined so as to aim for best accuracy, then for a given number of sampling points an increased accuracy can be obtained. Indeed, if we consider again that

$$I = \int_{-1}^{1} f(\xi) \, d\xi = \sum_{1}^{n} H_i f(\xi_i) \tag{8.33}$$

† 'Quadrature' is a term used alternatively to 'numerical integration'.

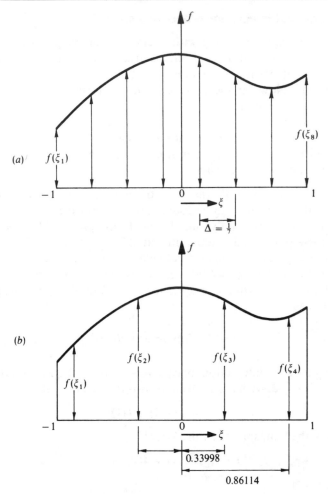

Fig. 8.12   (a) Newton–Cotes and (b) Gauss integrations. Each integrates exactly
a seventh-order polynomial [i.e., error $O(h^8)$]

and assume again a polynomial expression, it is easy to see that for n
sampling points we have $2n$ unknowns (H_i and ξ_i) and hence a polynomial of degree $2n - 1$ could be constructed and exactly integrated
[Fig. 8.12(b)]. The error thus is of order $O(h^{2n})$.

The simultaneous equations involved are difficult to solve, but some
mathematical manipulation will show that the solution can be obtained
explicitly in terms of Legendre polynomials. Thus this particular process
is frequently known as the Gauss–Legendre quadrature.

Table 8.1 shows the positions and weighting coefficients for gaussian
integration.

TABLE 8.1
ABSCISSAE AND WEIGHT COEFFICIENTS OF THE GAUSSIAN QUADRATURE FORMULA

$$\int_{-1}^{1} f(x) \, dx = \sum_{j=1}^{n} H_i \, f(a_j)$$

$\pm a$	H
$n = 1$	
0	2.000 000 000 000 000
$n = 2$	
0.577 350 269 189 626	1.000 000 000 000 000
$n = 3$	
0.774 596 669 241 483	0.555 555 555 555 556
0.000 000 000 000 000	0.888 888 888 888 889
$n = 4$	
0.861 136 311 594 953	0.347 854 845 137 454
0.339 981 043 584 856	0.652 145 154 862 546
$n = 5$	
0.906 179 845 938 664	0.236 926 885 056 189
0.538 469 310 105 683	0.478 628 670 499 366
0.000 000 000 000 000	0.568 888 888 888 889
$n = 6$	
0.932 469 514 203 152	0.171 324 492 379 170
0.661 209 386 466 265	0.360 761 573 048 139
0.238 619 186 083 197	0.467 913 934 572 691
$n = 7$	
0.949 107 912 342 759	0.129 484 966 168 870
0.741 531 185 599 394	0.279 705 391 489 277
0.405 845 151 377 397	0.381 830 050 505 119
0.000 000 000 000 000	0.417 959 183 673 469
$n = 8$	
0.960 289 856 497 536	0.101 228 536 290 376
0.796 666 477 413 627	0.222 381 034 453 374
0.525 532 409 916 329	0.313 706 645 877 887
0.183 434 642 495 650	0.362 683 783 378 362
$n = 9$	
0.968 160 239 507 626	0.081 274 388 361 574
0.836 031 107 326 636	0.180 648 160 694 857
0.613 371 432 700 590	0.260 610 696 402 935
0.324 253 423 403 809	0.312 347 077 040 003
0.000 000 000 000 000	0.330 239 355 001 260
$n = 10$	
0.973 906 528 517 172	0.066 671 344 308 688
0.865 063 366 688 985	0.149 451 349 150 581
0.679 409 568 299 024	0.219 086 362 515 982
0.433 395 394 129 247	0.269 266 719 309 996
0.148 874 338 981 631	0.295 524 224 714 753

For purposes of finite element analysis the complex calculations are involved in determining the values of f, the function to be integrated. Thus the Gauss processes, requiring the least number of such evaluations, are ideally suited and from now on will be used exclusively.

Other expressions for integration of functions of the type

$$I = \int_{-1}^{1} w(\xi) f(\xi) \, d\xi = \sum_{1}^{n} H_i f(\xi_i) \tag{8.34}$$

can be derived for prescribed forms of $w(\xi)$, again integrating up to a certain order of accuracy a polynomial expansion of $f(\xi)$.

8.9 Numerical integration—rectangular or right prism regions

The most obvious way of obtaining the integral

$$I = \int_{-1}^{1} \int_{-1}^{1} f(\xi, \eta) \, d\xi \, d\eta \tag{8.35}$$

is to first evaluate the inner integral keeping η constant, i.e.,

$$\int_{-1}^{1} f(\xi, \eta) \, d\xi = \sum_{j=1}^{n} H_j f(\xi_j, \eta) = \psi(\eta) \tag{8.36}$$

Evaluating the outer integral in a similar manner, we have

$$I = \int_{-1}^{1} \psi(\eta) \, d\eta = \sum_{i=1}^{n} H_j \psi(\eta_i)$$

$$= \sum_{i=1}^{n} H_i \sum_{j=1}^{n} H_j f(\xi, \eta_i)$$

$$= \sum_{i=1}^{n} \sum_{j=1}^{n} H_i H_j f(\xi_j, \eta_i) \tag{8.37}$$

For a right prism we have similarly

$$I = \int_{-1}^{1} \int_{-1}^{1} \int_{-1}^{1} f(\xi, \eta, \zeta) \, d\xi \, d\eta \, d\zeta$$

$$= \sum_{m=1}^{n} \sum_{j=1}^{n} \sum_{i=1}^{n} H_i H_j H_m f(\xi_i, \eta_j, \xi_m) \tag{8.38}$$

In the above, the number of integrating points in each direction was assumed to be the same. Clearly this is not necessary and on occasion it may be of advantage to use different numbers in each direction of integration.

It is of interest to note that in fact the double summation can be

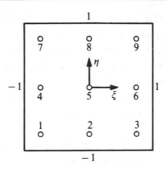

Fig. 8.13   Integrating points for $n = 3$ in a square region. (Exact for polynomial of fifth order in each direction)

readily interpreted as a single one over $(n \times n)$ points for a rectangle (or n^3 points for a cube). Thus in Fig. 8.13 we show the nine sampling points that result in exact integrals of order 5 in each direction.

However, we could approach the problem directly and require an exact integration of a fifth-order polynomial in two directions. At any sampling point two coordinates and a value of f have to be determined in a weighting formula of type

$$I = \int_{-1}^{1} \int_{-1}^{1} f(\xi, \eta) \, d\xi \, d\eta = \sum_{1}^{m} w_i f(\xi_i, \eta_i) \tag{8.39}$$

There it would appear that only seven points would suffice to obtain the same order of accuracy. Some such formulae for three-dimensional bricks have been derived by Irons[11] and used successfully.[12]

8.10 Numerical integration—triangular or tetrahedral regions

For a triangle, in terms of the area coordinates the integrals are of the form

$$I = \int_{0}^{1} \int_{0}^{1-L_1} f(L_1 L_2 L_3) \, dL_2 \, dL_1 \tag{8.40}$$

Once again we could use n Gauss points and arrive at a summation expression of the type used in the previous section. However, the limits of integration now involve the variable itself and it is convenient to use alternative sampling points for the second integration by use of a special Gauss expression for integrals of the type given by Eq. (8.34) in which w

is a linear function. These have been devised by Radau[13] and used successfully in the finite element context.[14] It is, however, much more desirable (and aesthetically pleasing) to use special formulae in which no bias is given to any of the natural coordinates L_i. Such formulae have been first derived by Hammer et al.[15] and Felippa[16] and a series of necessary sampling points and weights is given in Table 8.2.[17] (A more comprehensive list of higher order formulae derived by Cowper is given on p. 184 of reference 6.)

Similar extension for tetrahedra can obviously be made. Table 8.3 presents some such formulae based on reference 15.

TABLE 8.2

NUMERICAL INTEGRATION FORMULAE FOR TRIANGLES

Order	Figure	Error	Points	Triangular coordinates	Weights
Linear		$R = O(h^2)$	a	$\frac{1}{3}, \frac{1}{3}, \frac{1}{3}$	1
Quadratic		$R = O(h^3)$	a b c	$\frac{1}{2}, \frac{1}{2}, 0$ $0, \frac{1}{2}, \frac{1}{2}$ $\frac{1}{2}, 0, \frac{1}{2}$	$\frac{1}{3}$ $\frac{1}{3}$ $\frac{1}{3}$
Cubic		$R = O(h^4)$	a b c d	$\frac{1}{3}, \frac{1}{3}, \frac{1}{3}$ $0.6, 0.2, 0.2$ $0.2, 0.6, 0.2$ $0.2, 0.2, 0.6$	$-\frac{27}{48}$ $\frac{25}{48}$
Quintic		$R = O(h^6)$	a b c d e f g	$\frac{1}{3}, \frac{1}{3}, \frac{1}{3}$ $\alpha_1, \beta_1, \beta_1$ $\beta_1, \alpha_1, \beta_1$ $\beta_1, \beta_1, \alpha_1$ $\alpha_2, \beta_2, \beta_2$ $\beta_2, \alpha_2, \beta_2$ $\beta_2, \beta_2, \alpha_2$	0.225 000 000 0 0.132 394 152 7 0.125 939 180 5

with
$\alpha_1 = 0.059\,715\,871\,7$
$\beta_1 = 0.470\,142\,064\,1$
$\alpha_2 = 0.797\,426\,985\,3$
$\beta_2 = 0.101\,286\,507\,3$

TABLE 8.3

NUMERICAL INTEGRATION FORMULAE FOR TETRAHEDRA

No.	Order	Figure	Error	Points	Tetrahedral coordinates	Weights
1	Linear		$R = O(h^2)$	a	$\frac{1}{4}, \frac{1}{4}, \frac{1}{4}, \frac{1}{4}$	1
2	Quadratic		$R = O(h^3)$	a b c d	$\alpha, \beta, \beta, \beta$ $\beta, \alpha, \beta, \beta$ $\beta, \beta, \alpha, \beta$ $\beta, \beta, \beta, \alpha$ $\alpha = 0.585\,410\,20$ $\beta = 0.138\,196\,60$	$\frac{1}{4}$ $\frac{1}{4}$ $\frac{1}{4}$ $\frac{1}{4}$
3	Cubic		$R = O(h^4)$	a b c d e	$\frac{1}{4}, \frac{1}{4}, \frac{1}{4}, \frac{1}{4}$ $\frac{1}{2}, \frac{1}{6}, \frac{1}{6}, \frac{1}{6}$ $\frac{1}{6}, \frac{1}{2}, \frac{1}{6}, \frac{1}{6}$ $\frac{1}{6}, \frac{1}{6}, \frac{1}{2}, \frac{1}{6}$ $\frac{1}{6}, \frac{1}{6}, \frac{1}{6}, \frac{1}{2}$	$-\frac{4}{5}$ $\frac{9}{20}$ $\frac{9}{20}$ $\frac{9}{20}$ $\frac{9}{20}$

8.11 Required order of numerical integration

With numerical integration used to substitute the exact integration, an additional error is introduced into the calculation and the first impression is that this should be reduced as much as possible. Clearly the cost of numerical integration can be quite significant, and indeed in some early programs numerical formulation of element characteristics used a comparable amount of computer time to the subsequent solution of the equations. It is of interest, therefore, to determine (a) the minimum integration requirement permitting convergence and (b) the integration requirements necessary to preserve the rate of convergence which would result if exact integration were used.

It will be found later (Chapters 11 and 12) that it is in fact often a positive disadvantage to use higher orders of integration than those actually needed under (b) as, for very good reasons, a 'cancellation of errors' due to discretization and due to inexact integration occurs.

8.11.1 *Minimum order of integration for convergence.* In problems where the energy functional (or equivalent Galerkin integral statements) defines the approximation we have already stated that convergence can occur providing any arbitrary constant value of the mth derivatives can be reproduced. In the present case $m = 1$ and we thus require that in integrals of the form (8.5) a constant value of G be correctly integrated. Thus the volume of the element $\int_V dV$ needs to be evaluated correctly for convergence to occur. In curvilinear coordinates we thus can argue that $\int_V \det |J| \, d\zeta \, d\eta \, d\xi$ has to be evaluated exactly.[3,6]

However, even this condition is too demanding and we observe that, providing $\int_V d\zeta \, d\eta \, d\xi$ is correctly determined, convergence can occur. Thus any integration with order of error $O(h)$ suffices. We shall see that such a low integration order is often impractical, although we have in fact used this already in Chapter 4 for axisymmetric problems.

8.11.2 *Order of integration for no loss of convergence.* In a general problem we have already found that the finite element approximate evaluation of energy (and indeed all the other integrals in a Galerkin-type approximation, viz. Chapter 9) was exact to the order $2(p - m)$, where p was the degree of the complete polynomial present and m the order of differentials occurring in the appropriate expressions.

Providing the integration is exact to the order $2(p - m)$, or shows an error of $O(h^{2(p-m)+1})$, or less, then no loss of convergence order will occur.† If in curvilinear coordinates we take a curvilinear dimension h of an element, the same rule applies. For C_0 problems (i.e., $m = 1$) the integration formulae should be as follows:

$$p = 1, \quad \text{linear elements} \quad O(h)$$

$$p = 2, \quad \text{quadratic elements} \quad O(h^3)$$

$$p = 3, \quad \text{cubic elements} \quad O(h^5)$$

We shall make use of these results in practice, as will be seen later, but it should be noted that for a linear quadrilateral or triangle a single-point integration is adequate. For parabolic quadrilaterals (or bricks) 2×2 (or $2 \times 2 \times 2$), Gauss point integration is adequate and for parabolic triangles (or tetrahedra) three-point (and four-point) formulae of Tables 8.2 and 8.3 are needed.

The basic theorems of this section have been introduced and proved numerically in published work.[18–21]

8.11.3 *Matrix singularity due to numerical integration.* The final outcome of a finite element approximation in linear problems is an equation

---

† For an energy principle use of quadrature may result in loss of a bound for $\pi(\mathbf{a})$.

system
$$\mathbf{Ka} + \mathbf{f} = 0 \qquad\qquad (8.41)$$
in which the boundary conditions have been inserted and which should, on solution for the parameter **a**, give an approximate solution for the physical situation. If a solution is unique, as is the case with well-posed physical problems, the equation matrix **K** should be non-singular. We have *a priori* assumed that this was the case with exact integration and in general have not been disappointed. With numerical integration singularity may arise for low integration orders, and this may make such orders impractical. It is easy to show how, in some circumstances, a singularity of **K** must arise, but it is more difficult to prove that it will not. We shall, therefore, concentrate on the former case.

With numerical integration we replace the integrals by a weighted sum of independent linear relations between the nodal parameters **a**. These linear relations supply the only information from which the matrix **K** is constructed. *If the number of unknowns* **a** *exceeds the number of independent relations supplied at all the integrating points, then the matrix* **K** *must be singular.*

To illustrate this point we shall consider two-dimensional elasticity problems using linear and parabolic quadrilateral elements with one- and four-point quadratures respectively.

Here at each integrating point *three* independent 'strain relations' are used and the total number of independent relations equals 3 × (number of integration points). The number of unknowns **a** is simply 2 × (number of nodes) less restrained degrees of freedom.

In Fig. 8.14(*a*) and (*b*) we show a single element and an assembly of two elements supported by a minimum number of specified displacements eliminating rigid body motion. The simple calculation shows that only in the assembly of the quadratic elements is elimination of singularity possible, all the other cases remaining strictly singular.

In Fig. 8.14(*c*) a well-supported block of both kinds of elements is considered and here for both element types non-singular matrices may arise although local, near singularity may still lead to unsatisfactory results (viz. Chapter 11).

The reader may well consider the same assembly but supported again by the minimum restraint of three degrees of freedom. The assembly of linear elements with a single integrating point *will* be singular while the quadratic ones will, in fact, usually be well behaved.

For the reason just indicated, linear single-point integrated elements are used infrequently while the four-point quadrature is almost universal now for quadratic elements.

In Chapter 11 we shall return to the problem of convergence and will indicate dangers arising from local element singularities.

× Integrating point (3 independent relations)

○ Nodal point with 2 degrees of freedom

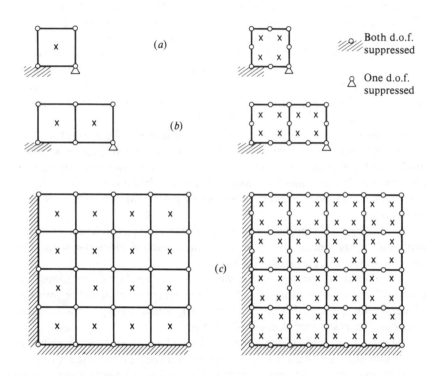

Fig. 8.14 Check on matrix singularity in two-dimensional elasticity problems
(*a*), (*b*), and (*c*)

However, it is of interest to mention that in Chapter 12 we shall in fact *seek* matrix singularity for special purposes by precisely the same arguments.

8.12 Generation of finite element meshes by mapping. Blending functions

It would have been observed that it is an easy matter to obtain a coarse subdivision of the analysis domain with a small number of isoparametric elements. If second- or third-degree elements are used, the fit of these to quite complex boundaries is reasonable, as shown in Fig. 8.15(*a*) where four parabolic elements specify a sectorial region. This number of elements would be too small for analysis purposes *but a simple subdivision into finer elements* can be done automatically by, say, assigning new positions of nodes of the central points of the curvilinear coordinates and deriving thus a larger number of similar elements, as shown in Fig. 8.15(*b*). Indeed, automatic subdivision could be carried out further to generate a field of triangular elements. The process thus allows, with a small number of original *input data*, to derive a finite element mesh of any refinement desirable. In reference 22 this type of mesh generation is developed for two- and three-dimensional solids and surfaces, and probably presents one of the most efficient means of subdivision.

The main drawback of the mapping and generation suggested is the fact that the originally circular boundaries in Fig. 8.15(*a*) are approximated by simple parabolae and a geometric error can be developed

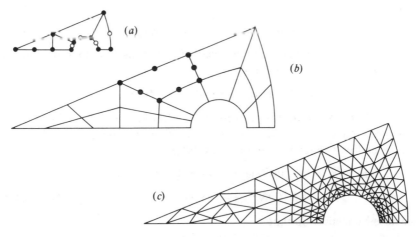

Fig. 8.15   Automatic mesh generation by parabolic isoparametric elements. (*a*) Specified mesh points. (*b*) Automatic subdivision into a small number of isoparametric elements. (*c*) Automatic subdivision into linear triangles

there. To overcome this difficulty another form of mapping, originally developed for representation of complex motor-car body shapes, can be adopted for this purpose.[23,24] In this mapping blending functions interpolate the unknown u in such a way as to satisfy *exactly* its variations along the edges of a square ξ, η domain. If the coordinates x and y are used in a parametric expression of the type given in Eq. (8.1), then any complex shape can be mapped by a single element. In reference 23 the region of Fig. 8.15 is in fact so mapped and a mesh subdivision obtained directly without any geometric error on the boundary.

The blending processes are of considerable importance and have been used to construct some interesting element families[25] (which in fact include the standard serendipity elements as a subclass). To explain the process we shall show how a function with prescribed variations along the boundaries can be interpolated.

Consider a region $-1 \leqslant \xi$, $\eta \leqslant 1$, shown in Fig. 8.16, on the edges of which an arbitrary function ϕ is specified [i.e., $\phi(-1, \eta)$, $\phi(1, \eta)$, $\phi(\xi, -1)$, $\phi(\xi, 1)$ are given]. The problem presented is that of interpolating a function $\phi(\xi, \eta)$ so that a smooth surface reproducing precisely the boundary values is obtained. Writing

$$N^1(\xi) = \frac{1 - \xi}{2} \qquad N^2(\xi) = \frac{1 + \xi}{2}$$

$$N^1(\eta) = \frac{1 - \eta}{2} \qquad N^2(\eta) = \frac{1 + \eta}{2}$$

$$(8.42)$$

for our usual, one-dimensional, linear interpolating functions, we note that

$$P_\eta \phi \equiv N^2(\eta)\phi(\xi, 1) + N^1(\eta)\phi(\xi, -1) \tag{8.43}$$

interpolates linearly between the specified functions in the η direction, as shown in Fig. 8.16(b). Similarly,

$$P_\xi \phi \equiv N^2(\xi)\phi(\eta, 1) + N^1(\xi)\phi(\eta, -1) \tag{8.44}$$

interpolates linearly in the ξ direction [Fig. 8.16(c)]. Constructing a third function which is a standard linear, lagrangian, interpolation of the kind we have already encountered [Fig. 8.16(d)], i.e.,

$$P_\xi P_\eta \phi = N^2(\xi)N^2(\eta)\phi(1, 1) + N^2(\xi)N^1(\eta)\phi(1, -1)$$
$$+ N^1(\xi)N^2\eta\phi(-1, 1) + N^1(\xi)N^1(\eta)\phi(-1, -1) \tag{8.45}$$

we note by inspection that

$$\phi = P_\eta \phi + P_\xi \phi - P_\xi P_\eta \phi \tag{8.46}$$

is a smooth surface interpolating exactly the boundary functions.

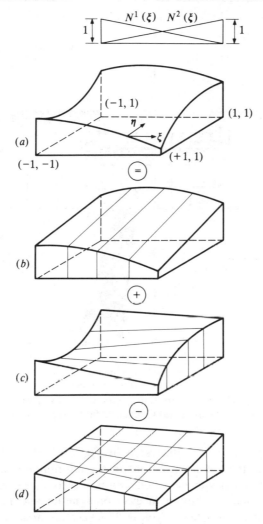

Fig. 8.16   Stages of construction of a blending interpolation (a), (b), (c), and (d)

Extension to functions with higher order blending is almost evident, and immediately the method of mapping the quadrilateral region $-1 \leqslant \xi, \eta \leqslant 1$ to any arbitrary shape is obvious.

8.13 Infinite domains and infinite elements

8.13.1 *Introduction.* In many problems of engineering and physics infinite or semi-infinite domains exist. A typical example from structural mechanics may, for instance, be that of three-dimensional (or

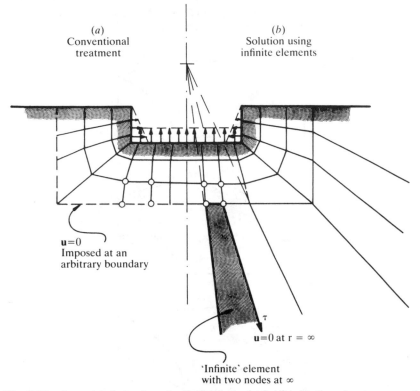

u=0
Imposed at an
arbitrary boundary

u=0 at r = ∞

'Infinite' element
with two nodes at ∞

Fig. 8.17 A semi-infinite domain. Deformations of a foundation due to removal of load following an excavation. (a) Conventional treatment and (b) use of infinite elements

axisymmetric) excavation, illustrated in Fig. 8.17. Here the problem is one of determining the deformations in a semi-infinite half-space due to the removal of loads with the specification of zero displacements at infinity. Similar problems abound in electromagnetics and fluid mechanics but the situation illustrated is typical. The question arises as to how such problems can be dealt with by a method of approximation in which elements of decreasing size are used in the modelling process. The first intuitive answer is the one illustrated in Fig. 8.17 where the infinite boundary condition is specified at a finite boundary placed at a *large distance* from the object. This, however, begs the question of what is a 'large distance' and obviously substantial errors may arise if this boundary is not placed far enough away. On the other hand, pushing this out excessively far necessitates the introduction of a large number of elements to model regions of relatively little interest to the analyst.

To overcome such, 'infinite', difficulties many methods have been proposed. In some a sequence of nesting grids is used and a recurrence relation derived.[26,27] In others a boundary-type exact solution is used and coupled to the finite element domain.[28,29] However, without doubt, the most effective and efficient treatment is the use of 'infinite elements'[30-33] pioneered originally by Bettess. In this process the conventional, finite, elements are coupled to elements of the type shown in Fig. 8.17(b) which model in a reasonable manner the material stretching to infinity.

The shape of such elements and their treatment is best accomplished by mapping[32,33] these onto a unit square (or a finite line in one dimension or cube in three dimensions). However, it is essential that the sequence of trial functions introduced in the mapped domain be such that it is complete and capable of modelling the true behaviour as the radial distance r increases. Here it would be advantageous if the mapped shape functions could approximate a sequence of the decaying form

$$\frac{C_1}{r} + \frac{C_2}{r^2} + \frac{C_3}{r^3} + \cdots \tag{8.47}$$

where C_i are arbitrary constants and r is the radial distance from the 'focus' of the problem.

In the next subsection we introduce a mapping function capable of doing just this.

8.13.2 *The mapping function.* Figure 8.18 illustrates the principles of generation of the derived mapping function.

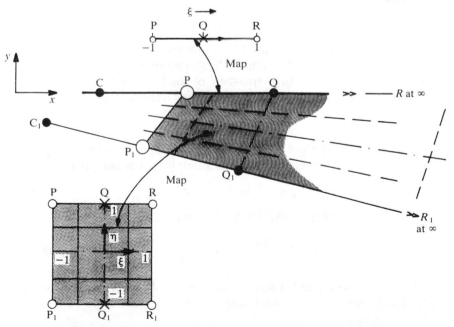

Fig. 8.18   Infinite line and element map. Linear η interpolation

We shall start with one-dimensional mapping along a line CPQ coinciding with the x direction. Consider the following function:

$$x = -\frac{\xi}{1-\xi} x_C + \left(1 + \frac{\xi}{1-\xi}\right) x_Q \qquad (8.48a)$$

and we immediately observe that

$$\xi = 0 \qquad \text{corresponds to } x = x_Q$$

$$\xi = 1 \qquad \text{corresponds to } x = \infty$$

$$\xi = -1 \quad \text{corresponds to } x = \frac{x_Q + x_C}{2} = x_P$$

where x_P is a point midway between Q and P.

Alternatively the mapping could be written directly in terms of the Q and P coordinates or by simple elimination of x_C. This gives, using our previous notation:

$$x = N_Q x_Q + N_P x_P$$

$$= \left(1 + \frac{2\xi}{1-\xi}\right) x_Q - \frac{2\xi}{1-\xi} x_P \qquad (8.48b)$$

Both forms give a mapping that is independent of the origin of the x coordinate as

$$N_Q + N_P = 1 = N_C + N_Q \qquad (8.49)$$

The significance of the point C is, however, of great importance. It represents the centre from which the 'disturbance' originates and, as we shall now show, allows the expansion of the form of Eq. (8.47) to be achieved on the assumption that r is measured from C. Thus

$$r = x - x_C \qquad (8.50)$$

If, for instance, the unknown function u is approximated by a polynomial function using, say, hierarchical shape functions and giving

$$u = \alpha_0 + \alpha_1 \xi + \alpha_2 \xi^2 + \alpha_3 \xi^3 + \cdots \qquad (8.51)$$

we can easily solve Eqs (8.48a) for ξ, obtaining

$$\xi = 1 - \frac{x_Q - x_C}{x - x_C} = 1 - \frac{x_Q - x_C}{r} \qquad (8.52)$$

Substitution into Eq. (8.51) shows that a series of the form given by Eq. (8.47) is obtained with the linear shape function in ξ corresponding to $1/r$ terms, quadratic to $1/r^2$, etc.

In one dimension the objectives specified have thus been achieved and the element will yield convergence as the degree of the polynomial expansion, p increases. Now a generalization to two or three dimensions is necessary. It is easy to see that this can be achieved by simple products of the one dimensional, infinite, mapping with a 'standard' type of shape function in η (and ζ) directions in the manner indicated in Fig. 8.18.

Firstly we generalize the interpolation of Eqs (8.48) for any straight line in the x, y, z space and write (for such a line as $C_1 P_1 Q_1$ in Fig. 8.18)

$$
x = -\frac{\xi}{1-\xi}\, x_{C_1} + \left(1 + \frac{\xi}{1+\xi}\right) x_{Q_1}
$$

$$
y = -\frac{\xi}{1-\xi}\, y_{C_1} + \left(1 + \frac{\xi}{1+\xi}\right) y_{Q_1} \qquad (8.53)
$$

$$
z = -\frac{\xi}{1-\xi}\, z_{C_1} + \left(1 + \frac{\xi}{1+\xi}\right) z_{Q_1} \qquad \text{(in three dimensions)}
$$

Secondly we complete the interpolation and map the whole $\xi\eta(\zeta)$ domain by adding a 'standard' interpolation in the $\eta(\zeta)$ directions. Thus for the linear interpolation shown we can write for element $PP_1 QQ_1 RR_1$ of Fig. 8.18,

$$
x = N_1(\eta)\left[-\frac{\xi}{1-\xi}\, x_C \left(1 + \frac{\xi}{1-\xi}\right) x_Q \right]
$$

$$
+ N_0(\eta)\left(-\frac{\xi}{1-\xi}\, x_{C_1} + \frac{\xi}{1-\xi}\, x_{Q_1} \right), \qquad \text{etc.} \qquad (8.54)
$$

with

$$
N_1(\eta) = \frac{1+\eta}{2} \qquad N_0(\eta) = \frac{1-\eta}{2}
$$

and map the points as shown.

In a similar manner we could use quadratic interpolations and map an element as shown in Fig. 8.19 by using quadratic functions in η.

Thus it is an easy matter to create infinite elements and join these to a standard element mesh as shown in Fig. 8.17(b). The reader will observe that in the generation of such element properties only the transformation jacobian matrix differs from standard forms, hence only this has to be altered in conventional programs.

The 'origin' or 'pole' of the coordinates C can be fixed arbitrarily for each radial line, as shown in Fig. 8.18. This will be done taking account of the knowledge of the physical solution expected.

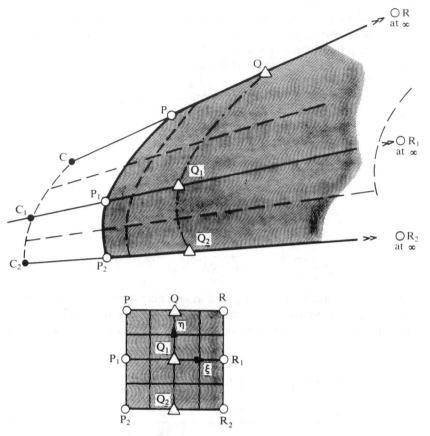

Fig. 8.19   Infinite element map. Quadratic η interpolation

In Fig. 8.20 we show a solution of the Boussinesq problem (a point load on an elastic half-space). Here results of using a fixed displacement or infinite elements are compared and the big changes in the solution noted. In this example the pole of each element was taken at the load point for obvious reasons.[33]

Figure 8.21 shows how similar infinite elements (of the linear kind) can give excellent results, even when combined with very few standard elements. In this example where a solution of the Laplace equation is used (viz. Chapter 10) for an irrotational fluid flow, the poles of the infinite elements are chosen at arbitrary points of the aerofoil centre-line.

In concluding this section it would be remarked that the use of infinite elements (as indeed of any other finite elements) must be tempered by background analytical knowledge and 'miracles' should not be expected.

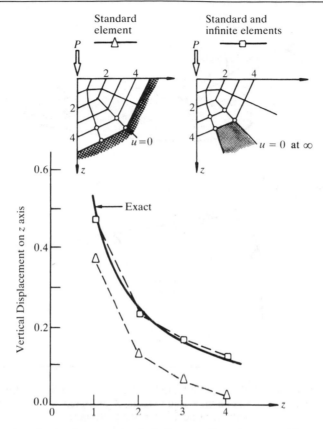

Fig. 8.20   A point load on an elastic half-space (Boussinesq problem). Standard linear elements and infinite line elements ($E = 1$, $v = 0.1$, $p = 1$)

Thus the user should not expect, for instance, such excellent results as those shown in Fig. 8.20 for a plane elasticity problem for the displacements. It is 'well known' that in this case the displacements under any load which is not self-equilibrated will be infinite and the numbers obtained from the computation will not be.

8.14 Singular elements by mapping for fracture mechanics, etc

In the study of fracture mechanics interest is often focused on the singularity point where such quantities as stress become (mathematically, but not physically) infinite. Near such singularities normal, polynomial-based, finite element approximations perform badly and attempts have

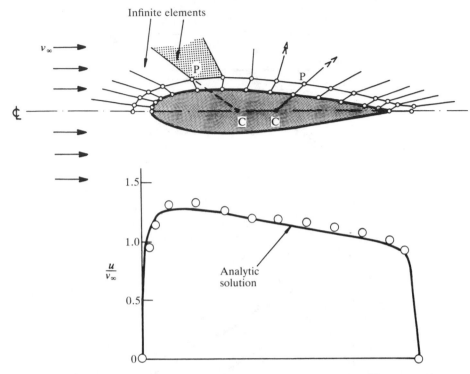

Fig. 8.21   Irrotational flow around NACA 0018 wing section.[29] (a) Mesh of
bilinear isoparametric and infinite elements. (b) Computed o and analytical —
results for velocity parallel to surface

frequently been made here to include special functions within an element
which can model the analytically known singular function. References 34
to 69 give an extensive literature survey of the problem and finite
element solution techniques. An alternative to the introduction of special
functions within an element—which frequently poses problems of enforcing continuity requirements with adjacent, standard, elements—lies in
the use of special mapping techniques.

An element of this kind, shown in Fig. 8.22(a), was introduced almost
simultaneously by Henshell and Shaw[65] and Barsoum[66,67] for quadrilaterals by a simple shift of the mid-side node in quadratic, isoparametric
elements to the quarter point.

It can now be shown (and we leave this exercise to the curious reader)
that along the element edges the derivatives $\partial u/\partial x$ (or strains) vary as
$1/\sqrt{r}$ where r is the distance from the corner node at which the singularity develops. Although good results are achievable with such elements
the singularity is, in fact, not well modelled on lines other than element

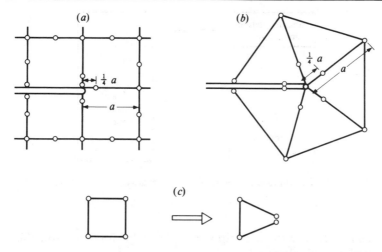

Fig. 8.22   Singular elements from degenerate isoparameters (a), (b), and (c)

edges. A development suggested by Hibbitt[68] achieves a better result by using triangular second-order elements for this purpose [Fig. 8.22(b)].

Indeed, the use of distorted or degenerate isoparametrics is not confined to elastic singularities. Rice[56] shows that in the case of plasticity a shear strain singularity of $1/r$ type develops and Levy et al.[49] use an isoparametric, linear quadrilateral to generate such a singularity by the simple device of coalescing two nodes but treating these displacements independently. A variant of this is developed by Rice and Tracey.[42]

The elements just described are evidently simple to implement without any changes in a standard finite element program.

8.15 A computational advantage of numerically integrated finite elements[70]

One considerable gain possible in numerically integrated finite elements is the versatility that can be achieved in a single computer program.

It will be observed that for a *given class of problems* the general matrices are always of the same form [see the example of Eq. (8.8)] in terms of the shape function and its derivatives.

To proceed to evaluation of the element properties it is necessary first to *specify the shape function* and its derivatives and, second, to *specify the order of integration*.

The computation of element properties is thus composed of three distinct parts as shown in Fig. 8.23. For a *given class of problems* it is only

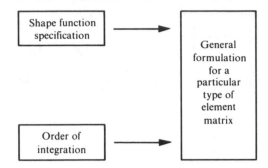

Fig. 8.23   Computation scheme for numerically integrated elements

necessary to change the prescription of the shape functions to achieve a variety of possible elements.

Conversely, the *same shape function* routines can be used in many different classes of problem, as is shown in Chapter 15.

Use of different elements, testing the efficiency of a new element in a given context, or extension of programs to deal with new situations can thus be readily achieved, and considerable algebra (with its inherent possibilities of mistakes) avoided.

The computer is thus placed in the position it deserves, i.e., of being the obedient slave capable of saving routine work.

The greatest practical advantage of the use of universal shape function routines is that they can be checked decisively for errors by a simple program with the patch test (viz. Chapter 11) playing the crucial role.

The incorporation of simple, exactly integrable, elements in such a system is, incidentally, not penalized as the time of exact and numerical integration in many cases is almost identical.

8.16 Some practical examples of two-dimensional stress analysis[71-77]

Some possibilities of two-dimensional analysis offered by curvilinear elements are illustrated in the following axisymmetric examples.

8.16.1 *Rotating disc* (Fig. 8.24). Here 18 elements only are needed to obtain an adequate solution. It is of interest to observe that all mid-side nodes of the cubic elements are generated within a program and need not be specified.

8.16.2 *Conical water tank* (Fig. 8.25). In this problem again cubic elements are used. It is worth noting that single-element thickness through-

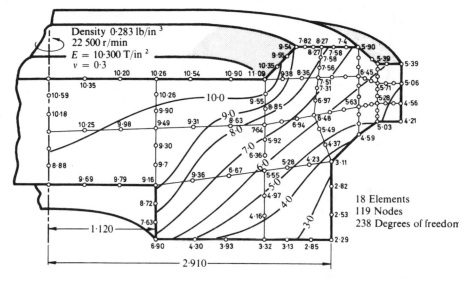

Fig. 8.24   A rotating disc—analysed with cubic elements

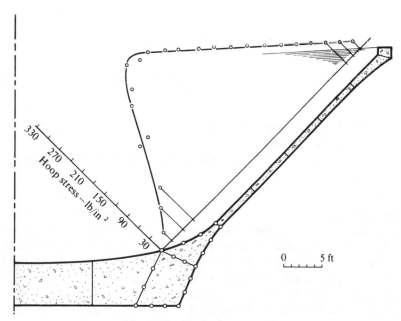

Fig. 8.25   Conical water tank

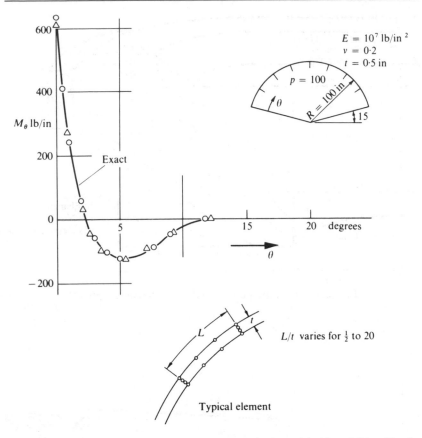

Fig. 8.26 *Encastré*, thin hemispherical shell. Solution with 15 and 24 cubic elements

out is adequate to represent the bending effects in both the thick and thin parts of the container. With simple triangular elements, as we have seen, several layers of elements would have been needed to give an adequate solution.

8.16.3 *A hemispherical dome* (Fig. 8.26). The possibilities of dealing with shells approached in the previous example are here further exploited to show how a limited number of elements can solve adequately a thin shell problem, with precisely the same program. This type of solution can be further improved upon from the economy viewpoint by making use of the well-known shell assumptions involving a linear variation of displacements across the thickness. Thus the number of degrees of freedom can be reduced. Methods of this kind will be dealt with in detail in the second volume of this text.

8.17 Three-dimensional stress analysis

In three-dimensional analysis, as was already hinted at in Chapter 5, the complex element presents a considerable economic advantage. Some typical examples are shown here in which the quadratic, serendipity-type formulation is used almost exclusively. In all problems integration using *three* Gauss points in each direction was used.

8.17.1 *Rotating sphere* (Fig. 8.27).[6] This example, in which the stresses due to centrifugal action are compared with exact values, is perhaps a test on the efficiency of highly distorted elements. Seven elements are used here and results show reasonable agreement with exact stresses.

8.17.2 *Arch dam in rigid valley.* This problem, perhaps a little unrealistic from the engineer's viewpoint, was the subject of a study carried out by a committee of the Institution of Civil Engineers and provided an excellent test for a convergence study of three-dimensional analysis. In Fig. 8.28 two subdivisions into quadratic and two into cubic elements are shown. In Fig 8.29 the convergence of displacements in the centre-line section is shown, indicating that quite remarkable accuracy can be achieved with even one element.

The comparison of stresses in Fig. 8.30 is again quite remarkable, though showing a greater 'oscillation' with coarse subdivision. The finest subdivision results can be taken as 'exact' from checks by models and alternative methods of analysis.

The above test problems illustrate the general applicability and accuracy. Two further illustrations typical of real situations are included.

8.17.3 *Pressure vessel* (Fig. 8.31): *an analysis of a biomechanic problem* (Fig. 8.32). Both show subdivisions sufficient to obtain a reasonable engineering accuracy. The pressure vessel, somewhat similar to the one indicated in Chapter 5, Fig. 5.7, shows the very considerable reduction of degrees of freedom possible with the use of more complex elements.

The example of Fig. 8.32 shows a perspective view of the elements used obtained directly from the analysis data on an automatic plotter. Such plots are not only helpful in visualization of the problem but also form an essential part of *data correctness checks* as any gross geometric error can be easily discovered. 'Connectivity' of all specified points is checked automatically.

The importance of avoiding data errors in complex three-dimensional problems should be obvious in view of their large usage of computer time. Such, and indeed other,[76] checking methods must form an essential part of any computation system.

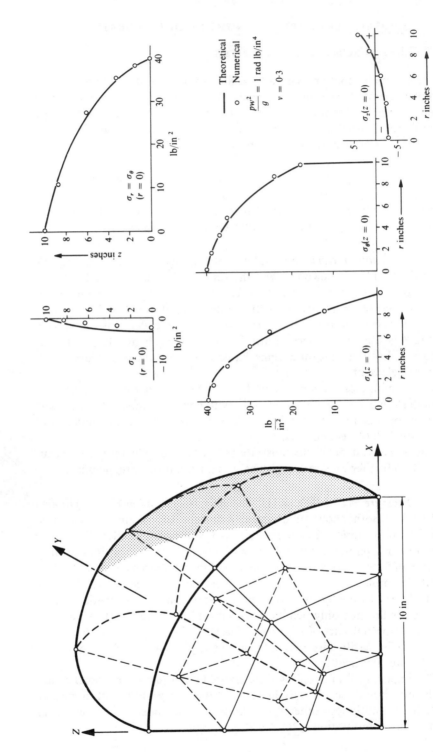

Fig. 8.27 A rotating sphere as a three-dimensional problem. Seven parabolic elements. Stresses along $z = 0$ and $r = 0$

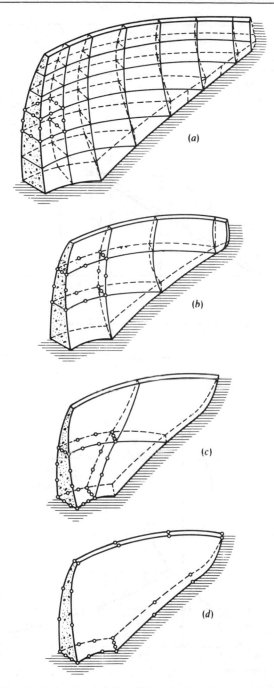

Fig. 8.28   Arch dam in a rigid valley—various element subdivisions

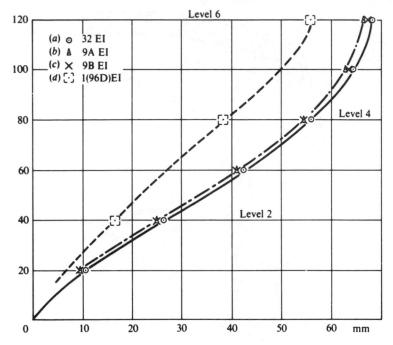

Fig. 8.29   Arch dam in a rigid valley—centre-line displacements

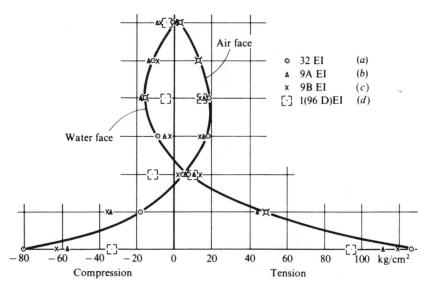

Fig. 8.30   Arch dam in a rigid valley—vertical stresses on centre-line

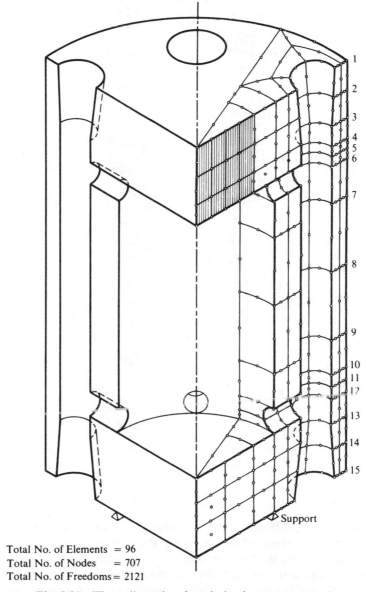

Total No. of Elements = 96
Total No. of Nodes   = 707
Total No. of Freedoms = 2121

Fig. 8.31   Three-dimensional analysis of a pressure vessel

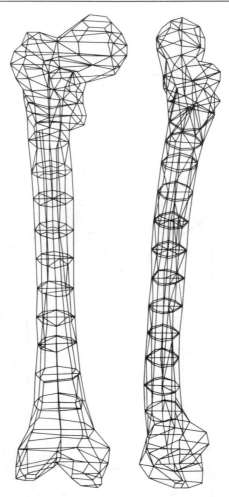

Fig. 8.32   A problem of biomechanics. Plot of linear element form only; curvature of elements omitted. Note degenerate element shapes

8.18 Symmetry and repeatability

In most of the problems shown, advantage of symmetry in loading and
geometry was taken when imposing the boundary conditions, thus
reducing the whole problem to manageable proportions. The use of symmetry conditions is so well known to the engineer and physicist that no
statement needs to be made about it explicitly. Less known, however,
appears to be the use of *repeatability*[78] when an identical structure (and)
loading is continuously repeated, as shown in Fig. 8.33 for an infinite
blade cascade. Here it is evident that each segment shown shaded

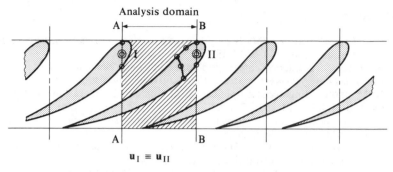

Fig. 8.33   Repeatability segments and analysis domain (shaded)

behaves identically to the next one, and thus such functions as velocities and displacements at corresponding points of AA and BB are simply identified, i.e.,

$$\mathbf{u}_I = \mathbf{u}_{II} .$$

This identification is made directly in a computer program.

Similar repeatability, in radial coordinates, occurs very frequently in problems involving turbine or pump impellers. Figure 8.34 shows a typical three-dimensional analysis of such a repeatable segment.

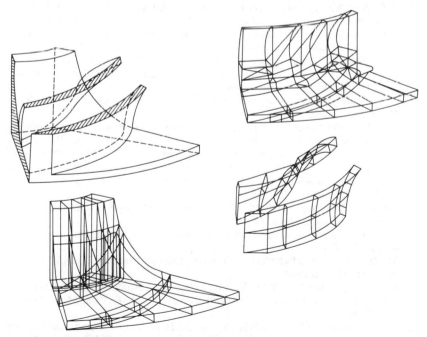

Fig. 8.34   Repeatable sector in analysis of an impeller

References

1. I. C. TAIG, *Structural analysis by the matrix displacement method*, Engl. Electric Aviation Report No. S017, 1961.
2. B. M. IRONS, 'Numerical integration applied to finite element methods', *Conf. Use of Digital Computers in Struct. Eng.*, Univ. of Newcastle, 1966.
3. B. M. IRONS, 'Engineering application of numerical integration in stiffness method', *JAIAA*, **14**, 2035–7, 1966.
4. S. A. COONS, *Surfaces for computer aided design of space form*, MIT Project MAC, MAC-TR-41, 1967.
5. A. R. FORREST, *Curves and Surfaces for Computer Aided Design*, Computer Aided Design Group, Cambridge, England, 1968.
6. G. STRANG and G. J. FIX, *An Analysis of the Finite Element Method*, pp. 156–63, Prentice-Hall, 1973.
7. W. B. JORDAN, *The plane isoparametric structural element*, General Elec. Co., Report KAPL-M-7112, Schenectady, New York, 1970.
8. F. D. MURNAGHAN, *Finite Deformation of an Elastic Solid*, Wiley, 1951.
9. E. L. WACHSPRESS, 'High order curved finite elements', *Int. J. Num. Meth. Eng.*, **17**, 735–45, 1981.
10. Z. KOPAL, *Numerical Analysis*, 2nd ed., Chapman & Hall, 1961.
11. B. M. IRONS, 'Quadrature rules for brick based finite elements', *Int. J. Num. Meth. Eng.*, **3**, 1971.
12. T. K. HELLEN, 'Effective quadrature rules for quadratic solid isoparametric finite elements', *Int. J. Num. Meth. Eng.*, **4**, 597–600, 1972.
13. RADAU, *Journ. de Math.*, **3**, 283, 1880.
14. R. G. ANDERSON, B. M. IRONS, and O. C. ZIENKIEWICZ, 'Vibration and stability of plates using finite elements', *Int. J. Solids Struct.*, **4**, 1031–55, 1968.
15. P. C. HAMMER, O. P. MARLOWE, and A. H. STROUD, 'Numerical integration over simplexes and cones', *Math. Tables Aids Comp.*, **10**, 130–7, 1956.
16. C. A. FELIPPA, *Refined finite element analysis of linear and non-linear two-dimensional structures*, Structures Materials Research Report No. 66–22, Oct. 1966, Univ. of California, Berkeley.
17. G. R. COWPER, 'Gaussian quadrature formulas for triangles', *Int. J. Num. Mech. Eng.*, **7**, 405–8, 1973.
18. G. J. FIX, 'On the effect of quadrature errors in the finite element method', *Advances in Computational Methods in Structural Mechanics and Design* (eds J. T. Oden, R. W. Clough, and Y. Yamamoto), pp. 55–68, University of Alabama Press, 1972. (See also *The Mathematical Foundations of the Finite Element Method with Applications to Differential Equations* (ed. A. K. Aziz), pp. 525–56, Academic Press, 1972.)
19. I. FRIED, 'Accuracy and condition of curved (isoparametric) finite elements', *J. Sound Vibration*, **31**, 345–55, 1973.
20. I. FRIED, 'Numerical integration in the finite element method', *Comp. Struc.*, **4**, 921–32, 1974.
21. M. ZLAMAL, 'Curved elements in the finite element method', *SIAM J. Num. Anal.*, **11**, 347–62, 1974.
22. O. C. ZIENKIEWICZ and D. V. PHILLIPS, 'An automatic mesh generation scheme for plane and curved element domains', *Int. J. Num. Meth. Eng.*, **3**, 519–28, 1971.
23. W. J. GORDON, 'Blending-function methods of bivariate and multivariate interpolation and approximation', *SIAM J. Num. Anal.*, **8**, 158–77, 1971.

24. W. J. GORDON and C. A. HALL, 'Construction of curvilinear co-ordinate systems and application to mesh generation', *Int. J. Num. Meth. Eng.*, **7**, 461–77, 1973.
25. W. J. GORDON and C. A. HALL, 'Transfinite element methods—blending-function interpolation over arbitrary curved element domains', *Numer. Math.*, **21**, 109–29, 1973.
26. R. W. THATCHER, 'On the finite element method for unbounded regions', *SIAM J. Numerical Analysis*, **15**, 3, pp. 466–76, June 1978.
27. P. SILVESTER, D. A. LOWTHER, C. J. CARPENTER and E. A. WYATT, 'Exterior finite elements for 2-dimensional field problems with open boundaries', *Proc. IEE*, **124**, No. 12, December 1977.
28. S. F. SHEN 'An aerodynamicist looks at the finite element method', in *Finite Elements in Fluids* (eds R. H. Gallagher et al.), Vol. 2, pp. 179–204, Wiley, 1975.
29. O. C. ZIENKIEWICZ, D. W. KELLY, and P. BETTESS, 'The coupling of the finite element and boundary solution procedures', *Int. J. Num. Meth. Eng.*, **11**, 355–75, 1977.
30. P. BETTESS, 'Infinite elements', *Int. J. Num. Meth. Eng.*, **11**, 53–64, 1977.
31. P. BETTESS and O. C. ZIENKIEWICZ, 'Diffraction and refraction of surface waves using finite and infinite elements', *Int. J. Num. Meth. Eng.*, **11**, 1271–90, 1977.
32. C. BEER and J. L. MEEK, 'Infinite domain elements', *Int. J. Num. Meth. Eng.*, **17**, 43–52, 1981.
33. O. C. ZIENKIEWICZ, C. EMSON and P. BETTESS, 'A novel boundary infinite element', *Int. J. Num. Meth. Eng.*, **19**, 393–404, 1983.
34. G. R. IRWIN, 'Fracture mechanics', in *Structural Mechanics, Proc. 1st Symp. on Naval Structural Mechanics* (eds J. N. Goodier and N. J. Hoff), pp. 557–94, Pergamon Press, 1960.
35. G. C. SIH (ed.), *Mechanics of Fracture—Vol. 1: Methods of Analysis and Solutions of Crack Problems*, Noordhoff, 1973.
36. Y. TADA, P. C. PARIS, and G. R. IRWIN, *The Stress Analysis of Cracks Handbook*, Del Research Corp., Hellertown, Penn., 1973.
37. J. F. KNOTT, *Fundamentals of Fracture Mechanics*, Butterworths, 1973.
38. R. H. GALLAGHER, 'Survey and evaluation of the finite element method in fracture mechanics analysis', in *Proc. 1st Int. Conf. on Structural Mechanics in Reactor Technology*, Vol. 6, Part L, pp. 637–53, Berlin, 1971.
39. N. LEVY, P. V. MARÇAL, and J. R. RICE, 'Progress in three-dimensional elastic–plastic stress analysis for fracture mechanics', *Nucl. Eng. Des.*, **17**, 64–75, 1971.
40. J. J. OGLESBY and O. LOMACKY, 'An evaluation of finite element methods for the computation of elastic stress intensity factors', *J. Eng. Ind.*, **95**, 177–83, 1973.
41. T. H. H. PIAN, 'Crack elements', in *Proc. World Congress on Finite Element Methods in Structural Mechanics*, Vol. 1, pp. F1–F39, Bournemouth, 1975.
42. J. R. RICE and D. M. TRACEY, 'Computational fracture mechanics', in *Numerical and Computer Methods in Structured Mechanics* (eds S. J. Fenves et al.), pp. 555–624, Academic Press, 1973.
43. E. F. RYBICKI and S. E. BENZLEY (eds), *Computational Fracture Mechanics*, ASME Special Publication, 1975.
44. A. A. GRIFFITHS, 'The phenomena of flow and rupture in solids', *Phil. Trans. Roy. Soc. (London)*, **A221**, 163–98, Oct. 1920.

45. B. Aamodt and P. G. Bergan, 'Propagation of elliptical surface cracks and nonlinear fracture mechanics by the finite element method', in *5th Conf. on Dimensioning and Strength Calculations*, Budapest, Oct. 1974.

46. P. G. Bergan and B. Aamodt, 'Finite element analysis of crack propagation in three-dimensional solids under cyclic loading', in *Proc. 2nd Int. Conf. on Structural Mechanics in Reactor Technology*, Vol. III, Part G-H, 1974.

47. J. L. Swedlow, 'Elasto-plastic cracked plates in plane strain', *Int. J. Fract. Mech.*, **5**, 33–44, March 1969.

48. T. Yokobori and A. Kamei, 'The size of the plastic zone at the tip of a crack in plane strain state by the finite element method', *Int. J. Fract. Mech.*, **9**, 98–100, 1973.

49. N. Levy, P. V. Marçal, W. J. Ostergren, and J. R. Rice, 'Small scale yielding near a crack in plane strain: a finite element analysis', *Int. J. Fract. Mech.*, **7**, 143–57, 1967.

50. J. R. Dixon and L. P. Pook, 'Stress intensity factors calculated generally by the finite element technique', *Nature*, **224**, 166, 1969.

51. J. R. Dixon and J. S. Strannigan, 'Determination of energy release rates and stress-intensity factors by the finite element method', *J. Strain Analysis*, **7**, 125–31, 1972.

52. V. B. Watwood, 'Finite element method for prediction of crack behavior', *Nucl. Eng. Des.*, **II** (No. 2), 323–32, March 1970.

53. D. F. Mowbray, 'A note on the finite element method in linear fracture mechanics', *Eng. Fract. Mech.*, **2**, 173–6, 1970.

54. D. M. Parks, 'A stiffness derivative finite element technique for determination of elastic crack tip stress intensity factors', *Int. J. Fract.*, **10**, 487–502, 1974.

55. T. K. Hellen, 'On the method of virtual crack extensions', *Int. J. Num. Meth. Eng.*, **9** (No. 1), 187–208, 1975.

56. J. R. Rice, 'A path-independent integral and the approximate analysis of strain concentration by notches and cracks', *J. Appl. Mech., Trans. Am. Soc. Mech. Eng.*, **35**, 379–86, 1968.

57. P. Tong and T. H. H. Pian, 'On the convergence of the finite element method for problems with singularity', *Int. J. Solids Struct.*, **9**, 313–21, 1972.

58. T. A. Cruse and W. Vanburen, 'Three dimensional elastic stress analysis of a fracture specimen with edge crack', *Int. J. Fract. Mech.*, **7**, 1–15, 1971.

59. E. Byskov, 'The calculation of stress intensity factors using the finite element method with cracked elements', *Int. J. Fract. Mech.*, **6**, 159–67, 1970.

60. P. F. Walsh, 'Numerical analysis in orthotropic linear fracture mechanics', *Inst. Eng. Australia, Civ. Eng., Trans.*, **15**, 115–19, 1973.

61. P. F. Walsh, 'The computation of stress intensity factors by a special finite element technique', *Int. J. Solids Struct.*, **7**, 1333–42, Oct. 1971.

62. A. K. Rao, I. S. Raju, and A. Murthy Krishna, 'A powerful hybrid method in finite element analysis', *Int. J. Num. Meth. Eng.*, **3**, 389–403, 1971.

63. W. S. Blackburn, 'Calculation of stress intensity factors at crack tips using special finite elements', in *The Mathematics of Finite Elements* (ed. J. R. Whiteman), pp. 327–36, Academic Press, 1973.

64. D. M. Tracey, 'Finite elements for determination of crack tip elastic stress intensity factors', *Eng. Fract. Mech.*, **3**, 255–65, 1971.

65. R. D. Henshell and K. G. Shaw, 'Crack tip elements are unnecessary', *Int. J. Num. Meth. Eng.*, **9**, 495–509, 1975.

66. R. S. Barsoum, 'On the use of isoparametric finite elements in linear fracture mechanics', *Int. J. Num. Meth. Eng.*, **10**, 25–38, 1976.

67. R. S. BARSOUM, 'Triangular quarter point elements as elastic and perfectly elastic crack tip elements', *Int. J. Num. Meth. Eng.*, **11**, 85–98, 1977.
68. H. D. HIBBITT, 'Some properties of singular isoparametric elements', *Int. J. Num. Meth. Eng.*, **11**, 180–4, 1977.
69. S. E. BENZLEY, 'Representation of singularities with isoparametric finite elements', *Int. J. Num. Meth. Eng.*, **8** (No. 3), 537–45, 1974.
70. B. M. IRONS, 'Economical computer techniques for numerically integrated finite elements', *Int. J. Num. Meth. Eng.*, **1**, 201–3, 1969.
71. O. C. ZIENKIEWICZ, B. M. IRONS, J. G. ERGATOUDIS, S. AHMAD, and F. C. SCOTT, 'Isoparametric and associated element families for two and three dimensional analysis', in *Proc. Course on Finite Element Methods in Stress Analysis* (eds I. Holand and K. Bell), Trondheim Tech. University, 1969.
72. B. M. IRONS and O. C. ZIENKIEWICZ, 'The isoparametric finite element system—a new concept in finite element analysis', *Proc. Conf. Recent Advances in Stress Analysis*, Royal Aero Soc., 1968.
73. J. G. ERGATOUDIS, B. M. IRONS, and O. C. ZIENKIEWICZ, 'Curved, iso-parametric, "quadrilateral" elements for finite element analysis', *Int. J. Solids Struct.*, **4**, 31–42, 1968.
74. J. G. ERGATOUDIS, *Isoparametric elements in two and three dimensional analysis*, Ph.D. thesis, University of Wales, Swansea, 1968.
75. J. G. ERGATOUDIS, B. M. IRONS, and O. C. ZIENKIEWICZ, 'Three dimensional analysis of arch dams and their foundations', *Symposium on Arch Dams*, Inst. Civ. Eng., London, 1968.
76. O. C. ZIENKIEWICZ, B. M. IRONS, J. CAMPBELL, and F. C. SCOTT, 'Three dimensional stress analysis', *Int. Un. Th. Appl. Mech. Symp. on High Speed Computing in Elasticity*, Liège, 1970.
77. O. C. ZIENKIEWICZ, 'Isoparametric and other numerically integrated elements', in *Numerical and Computer Methods in Structural Mechanics* (eds S. J. Fenves, N. Perrone, A. R. Robinson, and W. C. Schnobrich), pp. 13–41, Academic Press, 1973.
78. O. C. ZIENKIEWICZ and F. C. SCOTT, 'On the principle of repeatability and its application in analysis of turbine and pump impellers', *Int. J. Num. Meth. Eng.*, **9**, 445–52, 1972.

Generalization of the finite element concepts. Galerkin-weighted residual and variational approaches

9.1 Introduction

We have so far dealt with one possible approach to the approximate solution of the particular problem of linear elasticity. Many other continuum problems arise in engineering and physics and usually these problems are posed by appropriate differential equations and boundary conditions to be imposed on the unknown function or functions. It is the object of this chapter to show that all such problems can be dealt with by the finite element method.

Posing the problem to be solved in most general terms we find that we seek an unknown function **u** such that it satisfies a certain differential equation set

$$\mathbf{A(u)} = \begin{Bmatrix} A_1(\mathbf{u}) \\ A_2(\mathbf{u}) \\ \vdots \end{Bmatrix} = \mathbf{0} \tag{9.1}$$

in a 'domain' (volume, area, etc.) Ω (Fig. 9.1), together with certain boundary conditions

$$\mathbf{B(u)} = \begin{Bmatrix} B_1(\mathbf{u}) \\ B_2(\mathbf{u}) \\ \vdots \end{Bmatrix} = \mathbf{0} \tag{9.2}$$

on the boundaries Γ of the domain (Fig. 9.1).

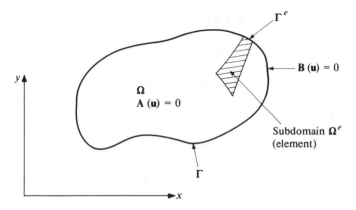

Fig. 9.1   Problem domain Ω and Γ boundary

The function sought may be a scalar quantity or may represent a vector of several variables. Similarly, the differential equation may be a single one or a set of simultaneous equations. It is for this reason that we have resorted in the above to matrix notation.

The finite element process, being one of approximation, will seek the solution in the approximation form

$$\mathbf{u} \approx \hat{\mathbf{u}} = \sum_{1}^{n} \mathbf{N}_i \mathbf{a}_i = \mathbf{N}\mathbf{a} \tag{9.3}$$

where $\mathbf{N}_i$ are shape functions prescribed in terms of independent variables (such as the coordinates x, y, etc.) and all or most of the parameters $\mathbf{a}_i$ are unknown.

We have seen that precisely the same form of approximation was used in the displacement approach to elasticity problems in the earlier chapters. We also noted there that (a) the shape functions were usually defined locally for elements or subdomains and (b) the properties of discrete systems were recovered if the approximating equations were cast in *an integral form* [viz. Eqs (2.22) to (2.26)].

With this object in mind we shall seek to cast the equation from which the unknown parameters $\mathbf{a}_i$ are to be obtained in an integral form

$$\int_{\Omega} \mathbf{G}_j(\hat{\mathbf{u}}) \, d\Omega + \int_{\Gamma} \mathbf{g}_j(\hat{\mathbf{u}}) \, d\Gamma = 0 \qquad j = 1 \text{ to } n \tag{9.4}$$

in which $\mathbf{G}_j$ and $\mathbf{g}_j$ prescribe known functions or operators.

These integral forms will permit the approximation to be obtained element by element and an assembly to be achieved by the use of procedures developed for *standard discrete systems* in Chapter 1, as, providing the functions $\mathbf{G}_j$ and $\mathbf{g}_j$ are integrable, we have

$$\int_{\Omega} \mathbf{G}_j \, d\Omega + \int_{\Gamma} \mathbf{g}_j \, d\Gamma = \sum_{e=1}^{m} \left(\int_{\Omega^e} \mathbf{G}_j \, d\Omega + \int_{\Gamma^e} \mathbf{g}_j \, d\Gamma \right) \qquad (9.5)$$

where Ω^e is the domain of each element and Γ^e its part of the boundary.

Two distinct procedures are available for obtaining the approximation in such integral forms. The first is the *method of weighted residuals* (known alternately as the Galerkin procedure), the second the determination of *variational functionals* for which stationarity is sought. We shall deal with both approaches in turn.

If the differential equations are linear, i.e., if we can write (9.1) and (9.2) as

$$\mathbf{A(u)} \equiv \mathbf{Lu} + \mathbf{p} = 0 \qquad \text{in } \Omega \qquad (9.6)$$

$$\mathbf{B(u)} \equiv \mathbf{Mu} + \mathbf{t} = 0 \qquad \text{on } \Gamma \qquad (9.7)$$

then the approximating equation system (9.4) will yield a set of linear equations of the form

$$\mathbf{Ka} + \mathbf{f} = 0 \qquad (9.8)$$

with

$$\mathbf{K}_{ij} = \sum_{e=1}^{m} \mathbf{K}_{ij}^e \qquad \mathbf{f}_i = \sum_{e=1}^{m} \mathbf{f}_i^e \qquad (9.9)$$

The reader not used to abstraction may well by now be confused in the meaning of the various terms. We shall introduce here some typical sets of differential equations for which we will seek solutions (and which may make the problems a little more definite).

Example 1. Steady-state heat conduction equations in a two-dimensional domain:

$$A(\phi) = \frac{\partial}{\partial x} \left(k \frac{\partial \phi}{\partial x} \right) + \frac{\partial}{\partial y} \left(k \frac{\partial \phi}{\partial y} \right) + Q = 0$$

$$B(\phi) = \phi - \bar{\phi} = 0 \qquad \text{on } \Gamma_\phi \qquad (9.10)$$

$$= k \frac{\partial \phi}{\partial n} - \bar{q} = 0 \qquad \text{on } \Gamma_q$$

where (with the n direction normal to Γ) $\phi \equiv \mathbf{u}$ indicates temperature, k is conductivity, and $\bar{\phi}$ and $\bar{q}$ are the prescribed values of temperature and heat flow on the boundaries.

In the above problem k and Q can be functions of position and, if the problem is non-linear, of ϕ or its derivatives.

Example 2. Steady-state heat conduction–convection equation in two dimensions:

$$A(\phi) = \frac{\partial}{\partial x}\left(k\,\frac{\partial \phi}{\partial y}\right) + \frac{\partial}{\partial y}\left(k\,\frac{\partial \phi}{\partial y}\right) + u\,\frac{\partial \phi}{\partial x} + v\,\frac{\partial \phi}{\partial y} + Q$$

$$= 0 \tag{9.11}$$

with boundary conditions as in the first example. Here u and v are known functions of position and represent velocities of the fluid through which heat transfer occurs.

Example 3. A system of three equations equivalent to the problem of Example 1:

$$\mathbf{A(u)} = \left\{ \begin{array}{c} \dfrac{\partial}{\partial x}(kq_x) + \dfrac{\partial}{\partial y}(kq_y) + Q \\[2mm] q_x - \dfrac{\partial \phi}{\partial x} \\[2mm] q_y - \dfrac{\partial \phi}{\partial y} \end{array} \right\} = 0 \tag{9.12}$$

in Ω and

$$\begin{aligned} \mathbf{B(u)} = \phi - \bar{\phi} &= 0 &&\text{on } \Gamma'_\phi \\ = q_n - \bar{q} &= 0 &&\text{on } \Gamma_q \end{aligned}$$

where q_n is the flux normal to the boundary.

Here the unknown function vector $\mathbf{u}$ corresponds to the set

$$\mathbf{u} = \left\{ \begin{array}{c} \phi \\ q_x \\ q_y \end{array} \right\}$$

This last example is typical of so-called *mixed formulation*. In such problems the number of dependent unknowns can always be reduced in the governing equations by suitable algebraic operation, still leaving a solvable problem [e.g., obtaining Eq. (9.10) from (9.12) by eliminating q_x and q_y].

If this can not be done [viz. Eq. (9.10)] we have an *irreducible formulation.*

Problems of mixed form present certain complexities in their solution which we shall discuss in Chapter 12.

In Chapter 10 we shall return to detailed examples of the above field and other examples will be introduced throughout the book. The three

sets of problems will, however, be useful in their full form or reduced to
one dimension (by suppressing the y variation) to illustrate the various
approaches used in this chapter.

9.2 Integral or 'weak' statements equivalent to the differential equations

As the set of differential equations [Eqs (9.1)] has to be zero at each
point of the domain Ω, it follows that

$$\int \mathbf{v}^T \mathbf{A}(\mathbf{u}) \, d\Omega \equiv \int [v_1 A_1(\mathbf{u}) + v_2 A_2(\mathbf{u}) + \cdots] \, d\Omega \equiv 0 \qquad (9.13)$$

where

$$\mathbf{v} = \begin{Bmatrix} v_1 \\ v_2 \\ \vdots \end{Bmatrix} \qquad (9.14)$$

is a set of arbitrary functions equal in number to the number of equations (or components of $\mathbf{u}$) involved.

The statement is, however, more powerful. *We can assert that if (9.13)
is satisfied for all $\mathbf{v}$ then the differential equations (9.1) must be satisfied at
all points of the domain.* The proof of the validity of this statement is
obvious if we consider the possibility that $\mathbf{A}(\mathbf{u}) \neq 0$ at any point or part
of the domain. Immediately, a function $\mathbf{v}$ can be found which makes the
integral of (9.13) non-zero, and hence the point is proved.

If the boundary conditions (9.12) are to be simultaneously satisfied,
then either we ensure such satisfaction by the choice of a function $\hat{\mathbf{u}}$ or
require that

$$\int_\Gamma \mathbf{v}^T \mathbf{B}(\mathbf{u}) \, d\Gamma \equiv \int_\Gamma [v_1 B_1(\mathbf{u}) + v_2 B_2(\mathbf{u}) + \cdots] \, d\Gamma = 0 \qquad (9.15)$$

for any set of functions $\mathbf{v}$.

Indeed, the integral statement that

$$\int_\Omega \mathbf{v}^T \mathbf{A}(\mathbf{u}) \, d\Omega + \int_\Gamma \bar{\mathbf{v}}^T \mathbf{B}(\mathbf{u}) \, d\Gamma = 0 \qquad (9.16)$$

is satisfied for all $\mathbf{v}$ and $\bar{\mathbf{v}}$ is equivalent to the satisfaction of the differential equations (9.1) and their boundary conditions (9.2).

In the above discussion it was implicity assumed that integrals such as
those in Eq. (9.16) are capable of being evaluated. This places certain
restrictions on the possible families to which the functions $\mathbf{v}$ or $\mathbf{u}$ must

belong. *In general we shall seek to avoid functions which result in any term in the integrals becoming infinite.*

Thus, in Eq. (9.16) we limit the choice of **v** and **v̄** to single, finite value functions without restricting the validity of previous statements.

What restrictions need to be placed on the functions $\mathbf{u}_1$, $\mathbf{u}_2$, ..., etc.? The answer depends obviously on the order of differentiation implied in the equations $\mathbf{A}(\mathbf{u})$ [or $\mathbf{B}(\mathbf{u})$]. Consider, for instance, a function **u** which is continuous but has a discontinuous slope in the x direction, as shown in Fig. 9.2. We imagine this discontinuity to be replaced by a continuous variable in a very small distance Δ and study the behaviour of the deriv-

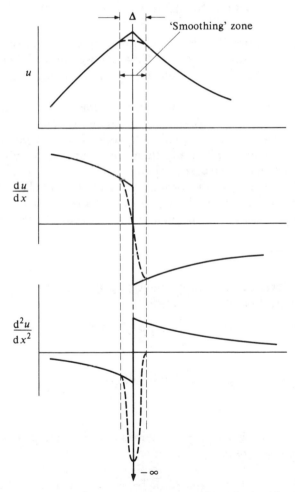

Fig. 9.2   Differentiation of function with slope discontinuity (C_0 continuous)

atives. It is easy to see that although the first derivative is not defined here, it can be integrated but the second derivative tends to infinity. The function illustrated would be a suitable choice for **u** if only first derivatives occurred in the differential equation. *Such a function is said to be C_0 continuous.*

In a similar way it is easy to see that if *n*th-order derivatives occur in any term of **A** or **B** then the function has to be such that its $n - 1$ derivatives are continuous (C_{n-1} continuity).

On many occasions it is possible to perform an integration by parts on Eq. (9.16) and replace it by an alternative statement of the form

$$\int_{\Omega} \mathbf{C(v)}^{\mathrm{T}} \mathbf{D(u)} \, d\Omega + \int_{\Gamma} \mathbf{E(\bar{v})}^{\mathrm{T}} \mathbf{F(u)} \, d\Gamma = 0 \tag{9.17}$$

In this the operators **C** to **F** contain lower order derivatives than those occurring in operators **A** and **B**. Now a lower order of continuity is required in the choice of the **u** function at a price of higher continuity for **v** and **v̄**.

The statement (9.17) is now more 'permissive' than the original problem posed by Eqs (9.1), (9.2), or (9.16) and is called a *weak form* of these equations. It is a somewhat surprising fact that often this weak form is more realistic physically than the original differential equation which implied an excessive 'smoothness' of the true solution.

Integral statements of the form of (9.16) and (9.17) will form the basis of finite element approximations, and we shall discuss them later in fuller detail. Before doing so we shall apply the new formulation to an example.

9.3 Weak form of the heat conduction equation—forced and natural boundary conditions

Consider now the integral form of Eq. (9.10). We can write the statement (9.16) as

$$\int_{\Omega} v \left[\frac{\partial}{\partial x} \left(k \frac{\partial \phi}{\partial x} \right) + \frac{\partial}{\partial y} \left(k \frac{\partial \phi}{\partial y} \right) + Q \right] dx \, dy + \int_{\Gamma_q} \bar{v} \left[k \frac{\partial \phi}{\partial n} - \bar{q} \right] d\Gamma = 0 \tag{9.18}$$

noting that v and $\bar{v}$ are scalar functions and presuming that one of the boundary conditions, i.e.,

$$\phi - \bar{\phi} = 0$$

is automatically satisfied by the choice of the functions ϕ.

Equation (9.18) can now be integrated by parts to obtain a weak form similar to Eq. (9.17). We shall make use here of general formulae for such integration (Green's formulae) which we derive in Appendix 6 and which will, on many occasions, be useful, i.e.

$$
\int_\Omega v \frac{\partial}{\partial x}\left(k\frac{\partial \phi}{\partial x}\right) dx\,dy \equiv -\int_\Omega \frac{\partial v}{\partial x}\left(k\frac{\partial \phi}{\partial x}\right) dx\,dy + \oint_\Gamma v\left(k\frac{\partial \phi}{\partial x}\right)n_x\,d\Gamma
$$

$$
\int_\Omega v \frac{\partial}{\partial y}\left(k\frac{\partial \phi}{\partial y}\right) dx\,dy \equiv -\int_\Omega \frac{\partial v}{\partial y}\left(k\frac{\partial \phi}{\partial y}\right) dx\,dy + \oint_\Gamma v\left(k\frac{\partial \phi}{\partial y}\right)n_y\,d\Gamma
$$

(9.19)

We have thus in place of Eq. (9.18)

$$
-\int_\Omega \left(\frac{\partial v}{\partial x}\,k\,\frac{\partial \phi}{\partial x} + \frac{\partial v}{\partial y}\,k\,\frac{\partial \phi}{\partial y} - Qv\right) dx\,dy + \oint_\Gamma vk\left(\frac{\partial \phi}{\partial x}\,n_x + \frac{\partial \phi}{\partial y}\,n_y\right) d\Gamma
$$

$$
+ \int_{\Gamma_q} \bar{v}\left[k\,\frac{\partial \phi}{\partial n} - \bar{q}\right] d\Gamma = 0 \qquad (9.20)
$$

Noting that the derivative along the normal is given as

$$
\frac{\partial \phi}{\partial n} \equiv \frac{\partial \phi}{\partial x}\,n_x + \frac{\partial \phi}{\partial y}\,n_y \qquad (9.21)
$$

and, further, making

$$
v = -\bar{v} \qquad (9.22)
$$

without loss of generality (as both functions are arbitrary), we can write Eq. (9.20) as

$$
\int_\Omega \mathbf{\nabla}^{\mathrm{T}} v k \mathbf{\nabla}\phi\,d\Omega - \int_\Omega vQ\,d\Omega - \int_{\Gamma_q} v\bar{q}\,d\Gamma - \int_{\Gamma_\phi} vk\,\frac{\partial \phi}{\partial n}\,d\Gamma = 0 \quad (9.23)
$$

where the operator $\mathbf{\nabla}$ is simply

$$
\mathbf{\nabla} = \left\{\begin{array}{c} \dfrac{\partial}{\partial x} \\[2mm] \dfrac{\partial}{\partial y} \end{array}\right\}
$$

We note
(a) that the variable ϕ has disappeared from the integrals taken along the boundary Γ_q and that the boundary condition

$$
B(\phi) = k\,\frac{\partial \phi}{\partial n} - \bar{q} = 0
$$

on that boundary is automatically satisfied—such a condition is known as a *natural boundary* condition—and

(b) that if the choice of ϕ is so restricted as to satisfy the *forced boundary conditions* $\phi - \bar{\phi} = 0$, we can omit the last term of Eq. (9.23) by restricting the choice of v to functions which give $v = 0$ on Γ_ϕ.

The form of Eq. (9.23) is the *weak form* of the heat conduction statement equivalent to Eq. (9.17). It admits discontinuous conductivity coefficients k and temperature ϕ which show discontinuous first derivatives—a real possibility not admitted in the differential form.

9.4 Approximation to integral formulations: the weighted residual–Galerkin method

If the unknown function **u** is approximated by the expansion (9.3), i.e.,

$$\mathbf{u} \approx \hat{\mathbf{u}} = \sum_1^r \mathbf{N}_i \mathbf{a}_i = \mathbf{N}\mathbf{a}$$

then it is clearly impossible to satisfy both the differential equation and the boundary conditions in a general case. The integral statements (9.16) or (9.17) allow an approximation to be made if, in place of *any function* **v**, we put a finite set of prescribed functions

$$\mathbf{v} = \mathbf{w}_j \qquad \bar{\mathbf{v}} = \bar{\mathbf{w}}_j \qquad j = 1 \text{ to } n \qquad (9.24)$$

where n is the number of unknown parameters $\mathbf{a}_i$ entering the problem $(n \leqslant r)$.

Equations (9.16) and (9.17) thus yield a set of ordinary equations from which parameters **a** can be determined, i.e., for Eq. (9.16) we have a set

$$\int_\Omega \mathbf{w}_j^\mathrm{T}\mathbf{A}(\mathbf{N}\mathbf{a}) \, d\Omega + \int_\Gamma \bar{\mathbf{w}}_j^\mathrm{T}\mathbf{B}(\mathbf{N}\mathbf{a}) \, d\Gamma = 0 \qquad j = 1 \text{ to } n \qquad (9.25)$$

or, from Eq. (9.17),

$$\int_\Omega \mathbf{C}(\mathbf{w}_j)^\mathrm{T}\mathbf{D}(\mathbf{N}\mathbf{a}) \, d\Omega + \int_\Gamma \mathbf{E}(\bar{\mathbf{w}}_j)^\mathrm{T}\mathbf{F}(\mathbf{N}\mathbf{a}) \, d\Gamma = 0 \qquad j = 1 \text{ to } n \quad (9.26)$$

If we note that $\mathbf{A}(\mathbf{N}\mathbf{a})$ represents the *residual or error* obtained by substitution of the approximation into the differential equation [and $\mathbf{B}(\mathbf{N}\mathbf{a})$, the residual of the boundary conditions], then Eq. (9.25) is a *weighted integral of such residuals*. The approximation thus may be called the *method of weighted residuals*.

In its classical sense it was first described by Crandall,[1] who points out the various forms used since the end of the last century. More recently a very full exposé of the method has been given by Finlayson.[2] Clearly, almost any set of independent functions $\mathbf{w}_j$ could be used for the purpose of weighting and, according to the choice of function, a different name can be attached to each process. Thus the various common choices are:

1. *Point collocation.*[3] $\mathbf{w}_j = \boldsymbol{\delta}_j$, where $\boldsymbol{\delta}_j$ is such that for $x \neq x_j$; $y \neq y_j$, $\mathbf{w}_j = 0$ but $\int_\Omega \mathbf{w}_j \, d\Omega = \mathbf{I}$ (unit matrix). This procedure is equivalent to simply making the residual zero at n points within the domain and integration is 'nominal' (incidentally although $\mathbf{w}_j$ defined here does not satisfy the integrability criteria of Sec. 9.2, it is nevertheless admissible in view of its properties).
2. *Subdomain collocation.*[4] $\mathbf{w}_j = \mathbf{I}$ in Ω_j and zero elsewhere. This essentially makes the integral of the error zero over the specified subdomain of the domain.
3. *The Galerkin method* (Bubnov–Galerkin).[5,6] $\mathbf{w}_j = \mathbf{N}_j$. Here simply the original shape (or basis) functions are used as weighting. This method, as we shall see, leads frequently (but by no means always) to symmetric matrices and for this and other reasons will be adopted in our finite element work almost exclusively.

The name of 'weighted residuals' is clearly much older than that of the 'finite element method'. The latter uses mainly locally based (element) functions in the expansion of Eq. (9.3) but the general procedures are identical. As the process leads always to equations which, being of integral form, can be obtained by summation of contributions from various subdomains, we choose to embrace all weighted residual approximations under the name of *generalized finite element method*. Frequently, simultaneous use of both local and 'global' trial functions will be found to be useful.

In mathematical literature the names of Petrov–Galerkin[6] are often associated with the use of weighting functions such that $\mathbf{w}_j \neq \mathbf{N}_j$. It is important to remark that the well-known *finite difference method* of approximation is a particular case of collocation with locally defined basis functions.

9.5 Examples

To illustrate the procedure of weighted residual approximation and its relation to the finite element process let us consider some specific examples.

Example 1. One-dimensional equation of heat conduction (Fig. 9.3). The problem here will be a one-dimensional representation of the heat con-

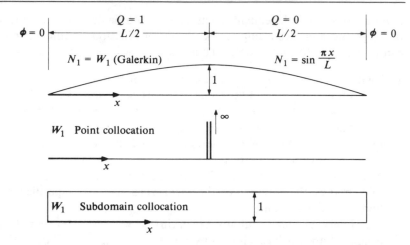

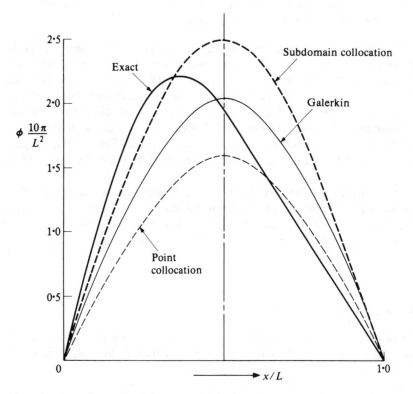

Fig. 9.3   One-dimensional heat conduction. (*a*) One-term solution using different weighting procedures. (*b*) Two-term solutions using different weighting procedures

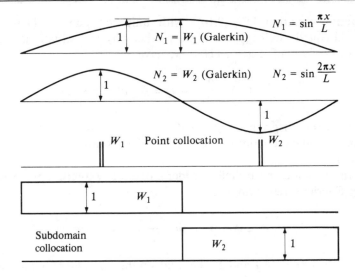

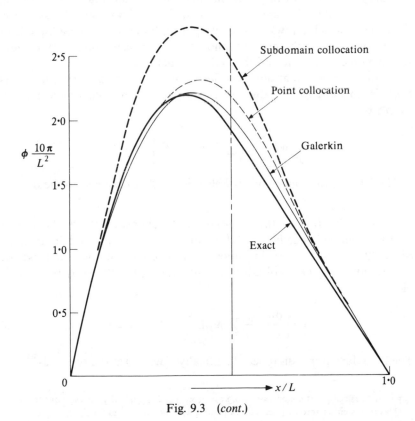

Fig. 9.3   (*cont.*)

duction equation [Eq. (9.10)] with unit conductivity. (This problem could equally well represent many other physical situations, e.g., deformation of a loaded string.) Here we have

$$A(\phi) = \frac{d^2\phi}{dx^2} + Q = 0 \quad (0 \leqslant x \leqslant L) \tag{9.27}$$

with $Q = Q(x)$ given by $Q = 1$ $(0 \leqslant x < L/2)$ and $Q = 0$ $(L/2 \leqslant x \leqslant L)$. The boundary conditions assumed will be simply $\phi = 0$ at $x = 0$ and $x = L$.

In the first case we shall consider a one- or two-term approximation of the Fourier series form, i.e.,

$$\phi \approx \hat{\phi} = \sum a_i \sin \frac{\pi x i}{L} \qquad N_i = \sin \frac{\pi x i}{L} \tag{9.28}$$

with $i = 1$ and $i = 1$ and 2. These satisfy the boundary conditions exactly and are continuous throughout the domain. We can thus use either Eq. (9.16) or Eq. (9.17) for approximation with equal validity. We shall use the former, which allows various weighting functions to be adopted. In Fig. 9.3 we present the problem and its solution using point collocation, subdomain collocation, and the Galerkin method.†

As the chosen expansion satisfies *a priori* the boundary conditions there is no need to introduce them into the formation, which is given simply by

$$\int_0^L w_j \left[\frac{d^2}{dx^2} \left(\sum N_i a_i \right) + Q \right] dx = 0 \tag{9.29}$$

The full working out of this problem is left as an exercise to the reader.

Of more interest to the standard finite element field is the use of piecewise defined (locally based) functions in place of the global functions of Eq. (9.28). Here, to avoid imposing slope continuity, we shall use the equivalent of Eq. (9.17) obtained by integrating Eq. (9.29) by parts. This yields

$$\int_0^L \left[\frac{dw_j}{dx} \frac{d}{dx} \sum N_i a_i - w_j Q \right] dx = 0 \tag{9.30}$$

The boundary terms disappear identically if $w_j = 0$ at the two ends.

---

† In the case of point collocation using $i = 1$, $x_i = L/2$, a difficulty arises about the value of Q (as this is either zero or one). The value of $\frac{1}{2}$ was therefore used for the example.

The above equations can be written as

$$\mathbf{Ka} + \mathbf{f} = 0 \tag{9.31}$$

where for each 'element' of length L^e,

$$
\begin{aligned}
K_{ji}^e &= \int_0^{L^e} \frac{dw_j}{dx} \frac{dN_i}{dx} \, dx \\[2mm]
f_j^e &= - \int_0^{L^e} w_j Q \, dx
\end{aligned}
\tag{9.32}
$$

with the usual rules of addition pertaining, i.e.,

$$K_{ji} = \int_0^L \frac{dw_j}{dx} \frac{dN_i}{dx} \, dx \qquad f_j^e = - \int_0^L w_j Q \, dx \tag{9.33}$$

In the computation we shall use the Galerkin procedure, i.e., $w_j = N_j$, and the reader will observe that the matrix $\mathbf{K}$ is then symmetric, i.e., $K_{ij} = K_{ji}$.

As the shape functions need only be of C_0 continuity, a piecewise linear approximation is conveniently used, as shown in Fig. 9.4. Considering a typical element ij shown, we can write (moving the origin of x to point i)

$$N_j = \frac{x}{L^e} \qquad N_i = \frac{L^e - x}{L^e} \tag{9.34}$$

giving, for a typical element,

$$K_{ij}^e = K_{ji}^e = -\frac{1}{L^e}$$

$$K_{ii} = K_{jj} = \frac{1}{L^e} \tag{9.35}$$

$$f_j^e = -QL^e/2 = f_i^e$$

Assembly of a typical equation at a node i is left to the reader, who is well advised to carry out the calculations leading to the results shown in Fig. 9.4 for a two- and four-element subdivision.

Some points of interest immediately arise if results of Figs 9.3 and 9.4 are compared. With smooth global shape functions the Galerkin method gives better overall results than those achieved for the same number of unknown parameters $\mathbf{a}$ with locally based functions. This we shall find to

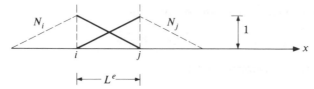

Locally based linear shape functions

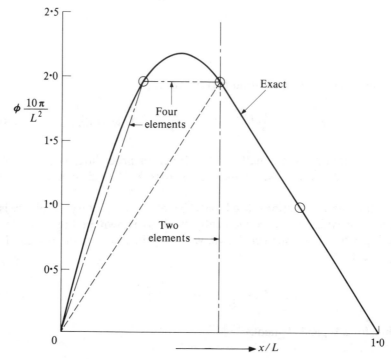

Fig. 9.4   Galerkin–finite element solution of problem of Fig. 9.3 using linear
locally based shaped functions

be the general case with higher order approximations, yielding a better accuracy. Further, it will be observed that the linear approximation has given the exact answers at the nodal points. This is a property of the particular equation being solved and unfortunately does not carry over to the general problems.[7] (See also Appendix 7.) Lastly, the reader will observe how easy it is to create equations with any degree of subdivision once the element properties [Eq. (9.35)] have been derived. This is not the case with global approximation where new integrations have to be carried out for each new parameter introduced. It is this repeatability feature that is one of the finite element advantages.

Example 2. Steady-state heat conduction–convection in two dimensions. The Galerkin formulation. We have already introduced the problem in Sec. 9.1 and defined it by Eq. (9.11) with appropriate boundary conditions. The equation differs only in the convective terms from that of simple heat conduction for which the weak form has already been obtained in Eq. (9.23). We can write the weighted residual equation immediately from this, substituting $v = w_j$ and adding the convective terms. Thus we have

$$\int_\Omega \nabla^T w_j \, k \nabla \hat{\phi} \, d\Omega - \int_\Omega w_j \left(u \frac{\partial \hat{\phi}}{\partial x} + v \frac{\partial \hat{\phi}}{\partial y} \right) d\Omega$$

$$- \int_\Omega w_j Q \, d\Omega - \int_{\Gamma_q} w_j \bar{q} \, d\Gamma = 0 \qquad (9.36)$$

with $\hat{\phi} = \sum N_i a_i$ being such that the prescribed values of $\hat{\phi}$ are given on the boundary $\Gamma = \phi$ and that $w_j = 0$ on that boundary.

Specializing to the Galerkin approximation, i.e., putting $w_j = N_j$, we have immediately a set of equations of the form

$$\mathbf{Ka} + \mathbf{f} = 0 \qquad (9.37)$$

with

$$K_{ji} = \int_\Omega \nabla^T N_j \, k \nabla N_i \, d\Omega - \int_\Omega \left(N_i u \frac{\partial N_i}{\partial x} + N_j v \frac{\partial N_i}{\partial y} \right) d\Omega$$

$$= \int_\Omega \left(\frac{\partial N_j}{\partial x} k \frac{\partial N_i}{\partial x} + \frac{\partial N_j}{\partial y} k \frac{\partial N_i}{\partial y} \right) d\Omega$$

$$- \int_\Omega \left(N_j u \frac{\partial N_i}{\partial x} + N_j v \frac{\partial N_i}{\partial y} \right) d\Omega \qquad (9.38a)$$

$$f_j = - \int_\Omega N_j Q \, d\Omega - \int_{\Gamma_q} N_j \bar{q} \, d\Gamma \qquad (9.38b)$$

Once again the components K_{ji} and f_i can be evaluated for a typical element or subdomain and systems equations built up by standard methods.

At this point it is important to mention that to satisfy the boundary conditions some of the parameters $\mathbf{a}_i$ have to be prescribed and approximation equations must be in number equal only to the unknown parameters. It is nevertheless convenient to form all equations for all parameters and prescribe the fixed values at the end using precisely the

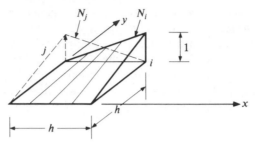

(a) Shape functions for a square C_0 element

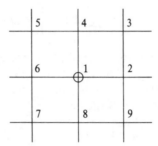

(b) Nodes 'connected' by equation for node 1

Fig. 9.5   A linear square element of C_0 continuity. (a) Shape functions for a square element. (b) 'Connected' equation for node 1

same techniques as we have described in Chapter 1 for the insertion of prescribed boundary conditions in standard discrete problems.

A further point concerning the coefficients of matrix **K** should here be noted. The first part, corresponding to the pure heat conduction equation, is symmetric ($K_{ij} = K_{ji}$) but the second is not and thus a system of non-symmetric equations needs to be solved. There is a basic reason for such non-symmetries which will be discussed in Sec. 9.11.

To make the problem concrete consider the domain Ω to be divided into regular square elements of side h (Fig. 9.5). To preserve C_0 continuity with nodes placed at corners, shape functions given as the product of the linear expansions can be written. For instance, for node i, as shown in Fig. 9.5,

$$N_i = \frac{x}{h}\frac{y}{h}$$

and for node j,

$$N_j = \frac{(h-x)}{h}\frac{y}{h}, \qquad \text{etc.}$$

With these shape functions the reader is invited to evaluate typical element contributions and to assemble the equations for point 1 of the mesh numbered as shown in Fig. 9.5. The result will be (if no boundary of type Γ_q is present and Q is assumed to be constant)

$$
\frac{8}{3} a_1 - \left(\frac{1}{3} - \frac{uh}{3k} - \frac{vh}{6k}\right)a_2 - \left(\frac{1}{3} - \frac{uh}{12k} - \frac{vh}{12k}\right)a_3 - \left(\frac{1}{3} - \frac{uh}{6k} - \frac{vh}{3k}\right)a_4
$$

$$
- \left(\frac{1}{3} + \frac{uh}{12k} - \frac{vh}{12k}\right)a_5 - \left(\frac{1}{3} + \frac{uh}{3k} - \frac{vh}{6k}\right)a_6 - \left(\frac{1}{3} + \frac{uh}{12k} + \frac{vh}{6k}\right)a_7
$$

$$
- \left(\frac{1}{3} - \frac{uh}{6k} + \frac{vh}{3k}\right)a_8 - \left(\frac{1}{3} + \frac{uh}{12k} + \frac{vh}{12k}\right)a_9 = 4h^2 Q \qquad (9.39)
$$

This equation is similar to those that would be obtained by using finite difference approximations to the same equations in a fairly standard manner.[8,9] In the example discussed some difficulties arise when the convective terms are large. In such cases the Galerkin weighting is not acceptable and other forms have to be used. This is discussed in detail in the second volume.

9.6 Virtual work as the 'weak form' of equilibrium equations for analysis of solids or fluids

In Chapter 2 we introduced the finite element by way of an application to the solid mechanics problem of elasticity. The integral statement necessary for formulation in terms of finite element approximation was supplied via the principle of *virtual work*, which was assumed to be so basic as not to merit proof. Indeed, to many this is so, and the virtual work principle is considered as a statement of mechanics more fundamental than the traditional equilibrium conditions of Newton's laws of motion. Others will argue with this view and will point out that all work statements are derived from the classical laws pertaining to the equilibrium of the particle. We shall therefore show in this section that the virtual work statement is simply a 'weak form' of equilibrium equations.

In a general three-dimensional continuum the equilibrium equations of an elementary volume can be written in terms of the components of the symmetric cartesian stress tensor as[10]

$$
\left\{
\begin{array}{l}
\dfrac{\partial \sigma_{xx}}{\partial x} + \dfrac{\partial \tau_{xy}}{\partial y} + \dfrac{\partial \tau_{xz}}{\partial z} \\[2mm]
\dfrac{\partial \sigma_{yy}}{\partial y} + \dfrac{\partial \tau_{yx}}{\partial x} + \dfrac{\partial \tau_{yz}}{\partial z} \\[2mm]
\dfrac{\partial \sigma_{zz}}{\partial z} + \dfrac{\partial \tau_{xz}}{\partial x} + \dfrac{\partial \tau_{yz}}{\partial y}
\end{array}
\right\}
+
\left\{
\begin{array}{l}
b_x \\[2mm]
b_y \\[2mm]
b_z
\end{array}
\right\}
= 0 \qquad (9.40)
$$

where $\mathbf{b}^T = [b_x,\ b_y,\ b_z]$ stands for the forces acting per unit volume (which may well include the acceleration effects).

In solid mechanics the six stress components will be some general functions of the components of the displacement

$$\mathbf{u}^T = [u,\ v,\ w] \tag{9.41}$$

and in fluid mechanics of the velocity vector $\mathbf{u}$, which has similar components. Thus Eq. (9.40) can be considered as a general equation of the form (9.1), i.e., $\mathbf{A}(\mathbf{u}) = 0$. To obtain a weak form we shall proceed as before, introducing an arbitrary weighting function vector $\delta\mathbf{u}$, defined as

$$\delta\mathbf{u}^T = [\delta u,\ \delta v,\ \delta w] \tag{9.42}$$

We can write now the integral statement of Eq. (9.13) as

$$\int_V \delta\mathbf{u}^T \mathbf{A}(\mathbf{u})\ dV = \int_V \left[\delta u\left(\frac{\partial\sigma_x}{\partial x} + \frac{\partial\tau_{xy}}{\partial y} + \frac{\partial\tau_{xz}}{\partial z} + b_x\right) \right.$$
$$\left. + \delta v(\cdots) + \delta w(\cdots) \right] dV \tag{9.43}$$

where V, the volume, is the problem domain.

Integrating each term by parts and rearranging we can write this as

$$-\int_V \left[\sigma_x \frac{\partial}{\partial x}(\delta u) + \tau_{xy}\left(\frac{\partial}{\partial x}(\delta u) + \frac{\partial}{\partial y}(\delta v)\right) + \cdots \right.$$
$$\left. - \delta u b_x - \delta v b_y - \delta w b_z \right] dV \tag{9.44}$$

$$+ \int_\Gamma [\delta u(\sigma_x n_x + \tau_{xy} n_y + \tau_{xz} n_z) + \delta v(\cdots) + \delta w(\cdots)]\ dA = 0$$

where Γ is the surface area of the solid (here again Green's formulae of Appendix 6 are used).

In the first set of bracketed terms we can recognize immediately the small strain operators acting on $\delta\mathbf{u}$—which can be termed a virtual displacement (or virtual velocity). We can therefore introduce virtual strain (or strain rate) defined as

$$\delta\boldsymbol{\varepsilon} = \begin{Bmatrix} \dfrac{\partial}{\partial x}(\delta u) \\[2mm] \dfrac{\partial}{\partial y}(\delta v) \\[2mm] \dfrac{\partial}{\partial z}(\delta w) \\[2mm] \vdots \end{Bmatrix} = \mathbf{S}\,\delta\mathbf{u} \tag{9.45}$$

where the strain operators are defined as in Chapter 2 [Eq. (2.4)].

Similarly, the terms in the second integral will be recognized as forces $\mathbf{t}$:

$$\mathbf{t} = [t_x, t_y, t_z] \tag{9.46}$$

acting per unit area of the surface A. Arranging the six stress components in a vector $\boldsymbol{\sigma}$ and similarly the six virtual strain (or rate of virtual strain) components in a vector $\delta\boldsymbol{\varepsilon}$, we can write Eq. (9.44) simply as

$$\int_V \delta\boldsymbol{\varepsilon}^T\boldsymbol{\sigma} \, dV - \int_V \delta\mathbf{u}^T\mathbf{b} \, dV - \int_\Gamma \delta\mathbf{u}^T\mathbf{t} \, d\Gamma = 0 \tag{9.47}$$

which is the virtual work statement used in Eqs (2.10) and (2.22) of Chapter 2.

We see from above that the virtual work statement is precisely the weak form of the equilibrium equations and is valid for non-linear as well as linear stress–strain (or stress–rate of strain) relations.

The finite element approximation which we have derived in Chapter 2 *is in fact a Galerkin formulation of the weighted residual process applied to the equilibrium equation.* Thus, if we take $\delta\mathbf{u}$ as the shape function

$$\delta\mathbf{u} = \mathbf{N} \tag{9.48}$$

where the displacement field is discretized, i.e.,

$$\mathbf{u} = \sum \mathbf{N}_i \mathbf{a}_i \tag{9.49}$$

together with the constitutive relation of Eq. (2.5), we shall determine once again all the basic expressions of Chapter 2 which are so essential to the solution of elasticity problems.

Similar expressions are vital to the formulation of equivalent fluid mechanics problems.

9.7 Partial discretization

In the approximation to the problem of solving the differential equation [Eqs (9.1)] by an expansion of the standard form of Eq. (9.3), we have assumed that the shape functions $\mathbf{N}$ included in them are *all* independent coordinates of the problem and that $\mathbf{a}$ was simply a set of constants. The final approximation equations were thus always of an algebraic form, from which a unique set of parameters could be determined.

In some problems it is convenient to proceed differently. Thus, for instance, if the independent variables are x, y, and z we could allow the parameters $\mathbf{a}$ to be functions of z and do the approximate expansion only in the domain of x, y, say $\bar{\Omega}$. Thus, in place of Eq. (9.3) we would have

$$\mathbf{u} = \mathbf{Na}$$

$$\mathbf{N} = \mathbf{N}(x, y) \tag{9.50}$$

$$\mathbf{a} = \mathbf{a}(z)$$

Clearly the derivatives of **a** with respect to z will remain in the final discretization and the result will be a set of ordinary differential equations with z as the independent variable. In linear problems such a set will have the appearance

$$\mathbf{Ka} + \mathbf{C\dot{a}} + \cdots + \mathbf{f} = 0 \tag{9.51}$$

where $\dot{\mathbf{a}} \equiv \left(\dfrac{\mathrm{d}}{\mathrm{d}z}\,\mathbf{a}\right)z$, etc.

Such partial discretization can obviously be used in different ways, but is particularly useful when the domain $\bar{\Omega}$ is not dependent on z, i.e., when the *problem is prismatic*. In such a case the coefficients of the ordinary differential Eq. (9.51) are independent of z and the solution of the system can frequently be carried out efficiently by standard analytical methods.

This type of partial discretization has been applied extensively by Kantorovitch[11] and is frequently known by his name. In the second volume we shall discuss such semi-analytical treatments in the context of prismatic solids where the final solution is obtained in terms of Fourier series. The most frequently encountered 'prismatic' problem is one involving the time variable, where the space domain $\bar{\Omega}$ is not subject to change. It is convenient by way of illustration to consider here the heat conduction in a two-dimensional equation in its transient state. This is obtained from Eq. (9.10) by addition of the heat storage term $c(\partial\phi/\partial t)$, where c is the specific heat. We now have a problem posed in a domain $\Omega(x, y, t)$ in which the following equation holds:

$$A(\phi) \equiv \frac{\partial}{\partial x}\left(k\,\frac{\partial\phi}{\partial x}\right) + \frac{\partial}{\partial y}\left(k\,\frac{\partial\phi}{\partial y}\right) + Q - c\,\frac{\partial\phi}{\partial t} = 0 \tag{9.52}$$

with boundary conditions identical to those of Eq. (9.10). Taking

$$\phi \approx \hat{\phi} = \sum N_i a_i \tag{9.53}$$

with $a_i = a_i(t)$ and $N_i = N_i(x, y)$ and using the Galerkin weighting procedure we follow precisely the steps outlined in Eqs (9.36) to (9.38) and arrive at a system of ordinary differential equations

$$\mathbf{Ka} + \mathbf{C}\,\frac{\mathrm{d}\mathbf{a}}{\mathrm{d}t} + \mathbf{f} = 0 \tag{9.54}$$

Here the expression for K_{ij} is identical with that of Eq. (9.38a) (convective terms neglected), f_i identical to Eq. (9.38b), and the reader can verify that the matrix $\mathbf{C}$ is defined by

$$C_{ij} = \int_\Omega N_i c N_j \, dx \, dy \qquad (9.55)$$

Once again the matrix $\mathbf{C}$ can be assembled from its element contribution. Various analytical and numerical procedures can be applied simply to the solution of such transient, ordinary, differential equations which, again, we shall discuss in the second volume. However, to illustrate the detail and the possible advantage of the process of partial discretization, we shall consider a very simple problem.

Example. Consider a square prism of size L in which the transient heat conduction equation [Eq. (9.52)] applies and assume that the rate of heat generation varies with time as

$$Q = Q_0 e^{-\alpha t} \qquad (9.56)$$

(this approximates a problem of heat development due to hydration of concrete). We assume that at $t = 0$, $\phi = 0$ throughout. Further, we shall take $\phi = 0$ on all boundaries throughout all times.

As a first approximation a shape function for a one-parameter solution is taken:

$$\phi = N_1 a_1$$

$$N_1 = \cos \frac{\pi x}{L} \cos \frac{\pi y}{L} \qquad (9.57)$$

with x and y measured from the centre (Fig. 9.6).
Evaluating the coefficients, we have

$$K_{11} = \int_{-L/2}^{L/2} \int_{-L/2}^{L/2} \left[k \left(\frac{\partial N_1}{\partial x} \right)^2 + k \left(\frac{\partial N_1}{\partial y} \right)^2 \right] dx \, dy = \frac{\pi^2 k}{2}$$

$$C_{11} = \int_{-L/2}^{L/2} \int_{-L/2}^{L/2} c N_1^2 \, dx \, dy = \frac{L^2 c}{4} \qquad (9.58)$$

$$f_1 = \int_{-L/2}^{L/2} \int_{-L/2}^{L/2} N_1 Q_0 e^{-\alpha t} \, dx \, dy = \frac{4 Q_0 L^2}{\pi^2} e^{-\alpha t}$$

This leads to an ordinary differential equation with one parameter a_1:

$$C_{11} \frac{da_1}{dt} + K_{11} a_1 + f_1 = 0 \qquad (9.59)$$

with $a_1 = 0$ when $t = 0$. The exact solution of this is easy to obtain, as is shown in Fig. 9.6 for specific values of the parameters α and $k/L^2 c$.

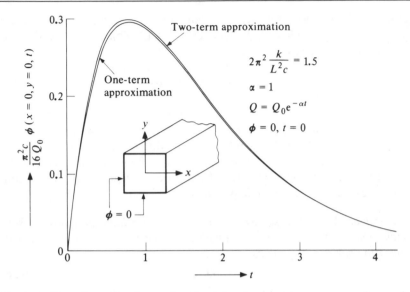

Fig. 9.6   Two-dimensional transient heat development in a square prism—plot of temperature at centre

On the same figure we show a two-parameter solution with

$$N_2 = \cos\frac{3\pi x}{L}\cos\frac{3\pi y}{L} \qquad (9.60)$$

which readers can pursue to test their grasp of the problem. The second component of the Fourier series is here omitted due to the required symmetry of solution.

The remarkable accuracy of the one-term approximation in this example should be noted.

9.8 Convergence

In the previous sections we have discussed how approximate solutions can be obtained by use of an expansion of the unknown function in terms of trial or shape functions. Further, we have stated the necessary conditions that such functions have to fulfil in order that the various integrals can be evaluated over the domain. Thus if various integrals contain only the values of N or its first derivatives then N has to be C_0 continuous. If second derivatives are involved, C_1 continuity is needed, etc. The problem to which we have not yet addressed ourselves consists of the questions of *just how good the approximation is* and *how it can be systematically improved to approach the exact answer.* The first question is more difficult to answer and presumes knowledge of the exact solution (see Chapter 14). The second is more rational and can be answered if we consider some systematic way in which the number of parameters **a** in the standard expansion of Eq. (9.3),

$$\hat{\mathbf{u}} = \sum_{1}^{r} \mathbf{N}_i \mathbf{a}_i$$

is presumed to increase.

In some of the examples we have assumed, in effect, a trigonometric Fourier-type series limited to a finite number of terms with a single form of trial function assumed over the whole domain. Here addition of new terms would be simply an extension of the number of the terms in the series included in the analysis, and as the Fourier series is known to be able to represent any desired function within any accuracy desired as the number of terms increases, we can talk about *convergence* of the approximation to the true solution as the number of terms increases.

In other examples of this chapter we have used locally based functions which are fundamental in the finite element analysis. Here we have tacitly assumed that *convergence occurs as the size of elements decreases and, hence, the number of* **a** *parameters specified at nodes increases.* It is with such convergence that we need to be concerned and we have already discussed this in the context of the analysis of elastic solids in Chapter 2 (Sec. 2.6).

Clearly we have now to determine

(*a*) that, as the number of elements increases, the unknown functions can be approximated as closely as required, and

(*b*) how the error decreases with the size, *h*, of the element subdivisions (*h* is here some typical dimension of an element).

The first problem is that of *completeness* of the expansion and we shall here assume that all trial functions are polynomials (or at least include certain terms of a polynomial expansion).

Clearly, as the approximation discussed here is to the weak, integral form typified by Eqs (9.13) or (9.17) it is necessary that every term occurring under the integral be in the limit capable of being approximated as nearly as possible and, in particular, giving a single constant value over an infinitesimal part of the domain Ω.

If a derivative of order m exists in any such terms, then it is obviously necessary for the local polynomial to be at least of the order m so that, in the limit, such a constant value be obtained.

We will thus state that a necessary condition for the expansion to be convergent is the *criterion of completeness*: that a constant value of the *m*th derivative be attainable in the element domain (if *m*th derivatives occur in the integral form) when the size of any element tends to zero.

This criterion is automatically ensured if the polynomials used in the shape function N are complete to the *m*th order. This criterion is also equivalent to the one of constant strain postulated in Chapter 2 (Sec. 2.5). This, however, has to be satisfied only in the limit $h \to 0$.

If the actual order of a complete polynomial used in the finite element expansion is $p \geqslant m$, then *the order of convergence* can be ascertained by seeing how closely such a polynomial can follow the local Taylor expansion of the unknown **u**. Clearly the order of error will be simply $O(h^{p+1})$ since only terms of order p can be rendered correctly.

The knowledge of the order of convergence helps in ascertaining how good the approximation is if studies on several decreasing mesh sizes are conducted though in Chapter 14 we shall see this asymptotic convergence rate is seldom reached if singularities occur in the problem. Once again we have reestablished some of the conditions discussed in Chapter 2.

We shall not discuss, at this stage, approximations which do not satisfy the postulated continuity requirements except to remark that once again, in many cases, convergence and indeed improved results can be obtained (see Chapter 11).

In the above we have referred to the convergence of a given element type as the size of this is reduced. This is sometimes referred to as *h convergence*.

On the other hand, it is possible to consider a subdivision into elements of a given size and to obtain convergence to the exact solution by increasing the polynomial order p of the element. This is referred to as *p convergence*, which obviously is assured.

In general p convergence is more rapid per degree of freedom introduced. We shall discuss both types further in Chapter 14.

VARIATIONAL PRINCIPLES

9.9 What are 'variational principles'?

What are variational principles and how can they be useful in the approximation to continuum problems? It is to these questions that the following sections are addressed.

First a definition: a 'variational principle' specifies a scalar quantity (functional) Π, which is defined by an integral form

$$\Pi = \int_{\Omega} F\left(\mathbf{u}, \frac{\partial}{\partial x}\mathbf{u}, \ldots\right) d\Omega + \int_{\Gamma} E\left(\mathbf{u}, \frac{\partial}{\partial x}\mathbf{u}, \ldots\right) d\Gamma \qquad (9.61)$$

in which $\mathbf{u}$ is the unknown function and F and E are specified operators. The solution to the continuum problem is a function $\mathbf{u}$ which makes Π stationary with respect to small changes $\delta\mathbf{u}$. Thus, for a solution to the continuum problem, the 'variation' is

$$\delta\Pi = 0 \tag{9.62}$$

If a 'variational principle' can be found, then immediately means are established for obtaining approximate solutions in the standard, integral form suitable for finite element analysis.

Assuming a trial function expansion in the usual form [Eq. (9.3)]

$$\mathbf{u} \approx \hat{\mathbf{u}} = \sum_1^n \mathbf{N}_i \mathbf{a}_i$$

we can insert this into Eq. (9.61) and write

$$\delta\Pi = \frac{\partial\Pi}{\partial\mathbf{a}_1}\,\delta\mathbf{a}_1 + \frac{\partial\Pi}{\partial\mathbf{a}_2}\,\delta\mathbf{a}_2 + \cdots = \frac{\partial\Pi}{\partial\mathbf{a}_n}\,\delta\mathbf{a}_n = 0 \tag{9.63}$$

This being true for any variations $\delta\mathbf{a}$ yields a set of equations

$$\frac{\partial\Pi}{\partial\mathbf{a}} = \left\{ \begin{array}{c} \dfrac{\partial\Pi}{\partial\mathbf{a}_1} \\ \vdots \\ \dfrac{\partial\Pi}{\partial\mathbf{a}_n} \end{array} \right\} = 0 \tag{9.64}$$

from which parameters $\mathbf{a}_i$ are found. The equations are of an integral form necessary for the finite element approximation as the original specification of Π was given in terms of domain and boundary integrals.

The process of finding stationarity with respect to trial function parameters $\mathbf{a}$ is an old one and is associated with the names of Rayleigh[12] and Ritz.[13] It has become extremely important in finite element analysis which, to many investigators, is typified as a 'variational process'.

If the functional Π is 'quadratic', i.e., if the function $\mathbf{u}$ and its derivatives occur in powers not exceeding 2, then Eq. (9.64) reduces to a standard linear form similar to Eq. (9.8), i.e.,

$$\frac{\partial\Pi}{\partial\mathbf{a}} \equiv \mathbf{K}\mathbf{a} + \mathbf{f} = 0 \tag{9.65}$$

It is easy to show that the matrix $\mathbf{K}$ will now always be symmetric. To do this let us consider a variation of the vector $\partial\Pi/\partial\mathbf{a}$ generally. This we can write as

$$\delta\left(\frac{\partial \Pi}{\delta \mathbf{a}}\right) = \left[\begin{array}{c} \frac{\partial}{\partial \mathbf{a}_1}\left(\frac{\partial \Pi}{\partial \mathbf{a}_1}\right)\delta \mathbf{a}_1, \frac{\partial}{\partial \mathbf{a}_2}\left(\frac{\partial \Pi}{\partial \mathbf{a}_1}\right)\delta \mathbf{a}_2, \cdots \\ \vdots \end{array}\right] \equiv \mathbf{K}_T\,\delta \mathbf{a} \qquad (9.66)$$

in which $\mathbf{K}_T$ is generally known as the tangent matrix of significance in non-linear analysis. Now it is easy to see that

$$\mathbf{K}_{Tij} = \frac{\partial^2 \Pi}{\partial \mathbf{a}_i\,\partial \mathbf{a}_j} = \mathbf{K}_{Tji}^T \qquad (9.67)$$

Hence $\mathbf{K}_T$ is symmetric.

For a quadratic functional we have, from Eq. (9.65),

$$\delta\left(\frac{\partial \Pi}{\partial \mathbf{a}}\right) = \mathbf{K}\,\delta \mathbf{a} \qquad \text{or} \qquad \mathbf{K} = \mathbf{K}^T \qquad (9.68)$$

and hence symmetry must exist.

The fact that *symmetric matrices will arise whenever a variational principle exists is one of the most important merits of variational approaches for discretization.* However, symmetric forms will frequently arise directly from the Galerkin process. In such cases we simply conclude that the variational principle exists but we shall not need to use it directly.

How then do 'variational principles' arise and is it always possible to construct these for continuous problems?

To answer the first part of the question we note that frequently the physical aspects of the problem can be stated directly in a variational principle form. Such theorems as minimization of total potential energy to achieve equilibrium in mechanical systems, least energy dissipation principles in viscous flow, etc., may be known to the reader and are considered by many as the basis of formulation. We have already referred to the first of these in Sec. 2.4 of Chapter 2.

Variational principles of this kind are 'natural' ones but unfortunately they do not exist for all continuum problems for which well-defined differential equations may be formulated.

However, there is another category of variational principles which we may call 'contrived'. Such contrived principles can always be constructed for any differentially specified problems either by extending the number of unknown functions **u** by additional variables known as Lagrange multipliers, or by procedures imposing a higher degree of continuity requirements such as least square problems. In subsequent sections we shall discuss, respectively, such 'natural' and 'contrived' variational principles.

Before proceeding further it is worth noting that, in addition to symmetry arising in equations derived by variational means, sometimes further motivation arises. When 'natural' variational principles exist the quantity Π may be of specific interest itself. If this arises a variational approach possesses the merit of easy evaluation of this functional.

The reader will observe that if the functional is 'quadratic' and yields Eq. (9.65), then we can write the approximate 'functional' Π simply as

$$\Pi = \tfrac{1}{2}\mathbf{a}^T\mathbf{K}\mathbf{a} + \mathbf{a}^T\mathbf{f} \tag{9.69}$$

That this is true the reader can observe by simple differentiation.†

9.10 'Natural' variational principles and their relation to governing differential equations

9.10.1 *Euler equations.* If we consider the definitions of Eqs (9.61) and (9.62) we observe that for stationarity we can write, after performing some differentiations,

$$\delta\Pi = \int_\Omega \delta\mathbf{u}^T\mathbf{A}(\mathbf{u})\ d\Omega + \int_\Gamma \delta\mathbf{u}^T\mathbf{B}(\mathbf{u})\ d\Gamma = 0 \tag{9.70}$$

As the above has to be true for any variations $\delta\mathbf{u}$, we must have

$$\mathbf{A}(\mathbf{u}) = 0 \quad \text{in } \Omega$$

and $\hspace{9cm}$ (9.71)

$$\mathbf{B}(\mathbf{u}) = 0 \quad \text{on } \Gamma$$

If **A** corresponds precisely to the differential equations governing the problem and **B** to its boundary conditions, then the variational principle is a *natural* one. Equations (9.70) and (9.71) are known as the Euler differential equations corresponding to the variational principle requiring the stationarity of Π. It is easy to show that for any variational principle a corresponding set of Euler equations can be established. The reverse is unfortunately not true, i.e., only certain forms of differential equations

---

† Observe that

$$\delta\Pi = \tfrac{1}{2}\delta(\mathbf{a}^T)\mathbf{K}\mathbf{a} + \tfrac{1}{2}\mathbf{a}^T\mathbf{K}\ \delta\mathbf{a} + \delta\mathbf{a}^T\mathbf{f}$$

As **K** is symmetric,

$$\delta\mathbf{a}^T\mathbf{K}\mathbf{a} \equiv \mathbf{a}^T\mathbf{K}\ \delta\mathbf{a}$$

Hence

$$\delta\Pi = \delta\mathbf{a}^T(\mathbf{K}\mathbf{a} + \mathbf{f}) = 0$$

which is true for all $\delta\mathbf{a}^T$ and hence

$$\mathbf{K}\mathbf{a} + \mathbf{f} = 0$$

are Euler equations of a variational functional. In the next section we shall consider the conditions necessary for the existence of variational principles and give a prescription for the establishment of Π from a set of suitable linear differential equations. In this section we shall continue to assume that the form of the variational principle is known.

To illustrate the process let us consider now a specific example. Suppose we specify the problem by requiring the stationarity of a functional

$$\Pi = \int_{\Omega} \left[\frac{1}{2} k \left(\frac{\partial \phi}{\partial x} \right)^2 + \frac{1}{2} k \left(\frac{\partial \phi}{\partial y} \right)^2 - Q\phi \right] d\Omega - \int_{\Gamma_q} \bar{q}\phi \, d\Gamma \quad (9.72)$$

in which k and Q depend only on position and $\delta\phi$ such that $\delta\phi = 0$ on Γ_ϕ where Γ_ϕ and Γ_q are bounding the domain Ω.

We now perform the variation. This can be written following rules of differentiation as

$$\delta\Pi = \int_{\Omega} \left[k \frac{\partial \phi}{\partial x} \delta\left(\frac{\partial \phi}{\partial x} \right) + k \frac{\partial \phi}{\partial y} \delta\left(\frac{\partial \phi}{\partial y} \right) - Q \, \delta\phi \right] d\Omega - \int_{\Gamma_q} (\bar{q} \, \delta\phi) \, d\Gamma$$

$$(9.73)$$

As

$$\delta\left(\frac{\partial \phi}{\partial x} \right) = \frac{\partial}{\partial x} (\delta\phi) \quad (9.74)$$

we can integrate by parts (as in Sec. 9.3) and, noting that $\delta\phi = 0$ on Γ_ϕ, obtain

$$\delta\Pi = -\int_{\Omega} \delta\phi \left[\frac{\partial}{\partial x} \left(k \frac{\partial \phi}{\partial x} \right) + \frac{\partial}{\partial y} \left(k \frac{\partial \phi}{\partial y} \right) + Q \right] d\Omega$$

$$+ \int_{\Gamma_q} \delta\phi \left(k \frac{\partial \phi}{\partial n} - \bar{q} \right) d\Gamma = 0 \quad (9.75a)$$

This is of the form of Eq. (9.70) and we immediately observe that the Euler equations are

$$A(\phi) = \frac{\partial}{\partial x} \left(k \frac{\partial \phi}{\partial y} \right) + \frac{\partial}{\partial y} \left(k \frac{\partial \phi}{\partial y} \right) + Q \qquad \text{in } \Omega$$

$$(9.75b)$$

$$B(\phi) = k \frac{\partial \phi}{\partial n} - \bar{q} = 0 \qquad \qquad \text{on } \Gamma_q$$

If ϕ is so prescribed that $\phi = \bar{\phi}$ on Γ_ϕ and $\delta\phi = 0$ on that boundary, then the problem is precisely the one we have already discussed in Sec. 9.3 and the functional (9.72) specifies the *two-dimensional heat conduction problem* in an alternative way.

In this case we have 'guessed' the functional but the reader will observe that the variation operation could have been carried out for any functional specified and corresponding *Euler* equations could have been established.

Let us continue the problem to obtain an approximate solution of the linear heat conduction problem. Taking, as usual,

$$\phi \approx \hat{\phi} = \sum N_i a_i = \mathbf{N a} \tag{9.76}$$

we substitute this approximation into the expression for the functional Π [Eq. (9.72)] and obtain

$$\Pi = \int_\Omega \frac{1}{2} k \left(\sum \frac{\partial N_i}{\partial x} a_i \right)^2 d\Omega + \int_\Omega \frac{1}{2} k \left(\sum \frac{\partial N_i}{\partial y} a_i \right)^2 d\Omega$$
$$- \int_\Omega Q \sum N_i a_i \, d\Omega - \int_{\Gamma_q} \bar{q} \sum N_i a_i \, d\Gamma \tag{9.77}$$

On differentiation with respect to a typical parameter a_j we have

$$\frac{\partial \Pi}{\partial a_j} = \int_\Omega k \left(\sum \frac{\partial N_i}{\partial x} a_i \right) \frac{\partial N_j}{\partial x} \, d\Omega + \int k \left(\sum \frac{\partial N_i}{\partial y} a_i \right) \frac{\partial N_j}{\partial y} \, d\Omega$$
$$- \int_\Omega Q N_j \, d\Omega - \int_{\Gamma_q} \bar{q} N_j \, d\Gamma \tag{9.78}$$

and a system of equations for solution of the problem is

$$\mathbf{K a} + \mathbf{f} = 0 \tag{9.79}$$

with

$$K_{ij} = K_{ji} = \int_\Omega k \frac{\partial N_i}{\partial x} \frac{\partial N_j}{\partial x} \, d\Omega + \int_\Omega k \frac{\partial N_i}{\partial y} \frac{\partial N_j}{\partial y} \, d\Omega$$
$$f_j = - \int_\Omega N_j Q \, d\Omega - \int_{\Gamma_q} N_j \bar{q} \, d\Gamma \tag{9.80}$$

The reader will observe that the approximation equations are here identical with those obtained in Sec. 9.5 for the same problem using the Galerkin process. No special advantage accrues to the variational formulation here—and indeed we can predict now that Galerkin and variational procedures must give the same answer for cases where natural variational principles exist.

9.10.2 *Relation of the Galerkin method to approximation via variational principles.* In the preceding example we have observed that the approximation obtained by the use of a natural variational principle and by the use of the Galerkin weighting process proved identical. That this is the case follows directly from Eq. (9.70), in which the variation was derived in terms of the original differential equations and the associated boundary conditions.

If we consider the usual trial function expansion [Eq. (9.3)]

$$\mathbf{u} \approx \hat{\mathbf{u}} = \mathbf{Na}$$

we can write the variation of this approximation as

$$\delta\hat{\mathbf{u}} = \mathbf{N}\,\delta\mathbf{a} \tag{9.81}$$

and inserting the above into (9.70) yields

$$\delta\Pi = \delta\mathbf{a}^T \int_\Omega \mathbf{N}^T\mathbf{A}(\mathbf{Na})\,d\Omega + \delta\mathbf{a}^T \int_\Gamma \mathbf{N}^T\mathbf{B}(\mathbf{Na})\,d\Gamma \tag{9.82}$$

The above form, being true of all $\delta\mathbf{a}$, requires that the expression under the integrals should be zero. The reader will immediately recognize this as simply the Galerkin form of the weighted residual statement discussed earlier [Eq. (9.25)], and identity is hereby proved.

We need to underline, however, that this is only true if the Euler equations of the variational principle coincide with the governing equations of the original problems. The Galerkin process thus retains its greater range of applicability.

At this stage another point must be made, however. If we consider a *system* of governing equations [Eq. (9.1)]

$$\mathbf{A}(\mathbf{u}) = \begin{Bmatrix} A_1(\mathbf{u}) \\ A_2(\mathbf{u}) \\ \vdots \end{Bmatrix} = \mathbf{0}$$

with $\hat{\mathbf{u}} = \mathbf{Na}$, the Galerkin weighted residual equation becomes (disregarding the boundary conditions)

$$\int_\Omega \mathbf{N}^T\mathbf{A}(\hat{\mathbf{u}})\,d\Omega = 0 \tag{9.83}$$

This form is not unique as the equation systems $\mathbf{A}$ can be ordered in a number of ways. Only one such ordering will correspond precisely with the Euler equations of a variational principle (if this exists) and the reader can verify that for an equation system weighted in the Galerkin manner at best only one arrangement of the vector $\mathbf{A}$ results in a symmetric set of equations.

As an example, consider, for instance, the one-dimensional heat conduction problem (Example 1, Sec. 9.5) redefined as an equation system with two unknowns, ϕ being the temperature and q the heat flow. Disregarding at this stage the boundary conditions we can write these equations as

$$\mathbf{A(u)} = \left\{ \begin{array}{c} q - \dfrac{d\phi}{dx} \\[2mm] \dfrac{dq}{dx} \end{array} \right\} + \left\{ \begin{array}{c} 0 \\ Q \end{array} \right\} = 0 \qquad (9.84)$$

or as a linear equation system,

$$\mathbf{A(u)} \equiv \mathbf{Lu} + \mathbf{b} = 0$$

in which

$$\mathbf{L} \equiv \begin{bmatrix} 1, & -\dfrac{d}{dx} \\[3mm] \dfrac{d}{dx}, & 0 \end{bmatrix} \qquad \mathbf{b} = \left\{ \begin{array}{c} 0 \\ Q \end{array} \right\} \qquad \mathbf{u} = \left\{ \begin{array}{c} q \\ \phi \end{array} \right\} \qquad (9.85)$$

Writing the trial function in which a different interpolation is used for each function

$$\mathbf{u} = \sum \mathbf{N}_i \mathbf{a}_i \qquad \mathbf{N}_i = \begin{bmatrix} N_i^1 & 0 \\ 0 & N_i^2 \end{bmatrix}$$

and applying the Galerkin process, we arrive at a usual linear equation system with

$$\mathbf{K}_{ij} = \int_\Omega \mathbf{N}_i^T \mathbf{L} \mathbf{N}_j \, dx = \int_\Omega \begin{bmatrix} N_i^1 N_j^1, & -N_i^1 \dfrac{d}{dx} N_j^2 \\[3mm] N_i^2 \dfrac{d}{dx} N_j^1, & 0 \end{bmatrix} dx \qquad (9.86)$$

This form yields a symmetric equation† system after integration by parts, and

$$K_{ij} = K_{ji} \qquad (9.87)$$

† As

$$\int N_i^1 \frac{d}{dx} N_j^2 \, dx \equiv - \int N_j^2 \frac{d}{dx} N_i^1 \, dx + \text{boundary terms}$$

If the order of equations were simply reversed, i.e., using

$$\mathbf{A(u)} = \begin{bmatrix} \dfrac{\mathrm{d}q}{\mathrm{d}x} \\[2ex] q - \dfrac{\mathrm{d}\phi}{\mathrm{d}x} \end{bmatrix} + \begin{Bmatrix} Q \\ 0 \end{Bmatrix} = 0 \tag{9.88}$$

application of the Galerkin process would now lead to non-symmetric equations quite different from those arising using the variational principle. The second type of Galerkin approximation would clearly be less desirable due to loss of symmetry in the final equations. It is easy to show that the first system corresponds precisely to Euler equations of a variational functional.

9.11 Establishment of natural variational principles for linear, self-adjoint differential equations

9.11.1 *General theorems.* General rules for deriving natural variational principles from non-linear differential equations are complicated and even the tests necessary to establish the existence of such variational principles are not simple. Much mathematical work has been done, however, in this context by Veinberg,[14] Tonti,[15] Oden,[16] and others.

For linear differential equations the situation is much simpler and a thorough study is available in the works of Mikhlin,[17,18] and in this section a brief presentation of such rules is given.

We shall consider here only the establishment of variational principles for a linear system of equations with *forced* boundary conditions, implying only variation of functions which yield $\delta \mathbf{u} = 0$ on their boundaries. The extension to include natural boundary conditions is simple and will be omitted.

Writing a linear system of differential equations as

$$\mathbf{A(u)} \equiv \mathbf{Lu} + \mathbf{b} = 0 \tag{9.89}$$

in which $\mathbf{L}$ is a linear, differential, operator it can be shown that natural variational principles require that the operator $\mathbf{L}$ be such that

$$\int_{\Omega} \boldsymbol{\psi}^{\mathrm{T}} \mathbf{L} \boldsymbol{\gamma} \, \mathrm{d}\Omega = \int \boldsymbol{\gamma}^{\mathrm{T}} \mathbf{L} \boldsymbol{\psi} \, \mathrm{d}\Omega + \text{b.t.} \tag{9.90}$$

for any two function sets $\boldsymbol{\psi}$ and $\boldsymbol{\gamma}$. In the above, 'b.t.' stands for boundary terms which we disregard in the present context. The property required in the above operator is called one of *self-adjointness* or *symmetry*.

If the operator $\mathbf{L}$ is self-adjoint, the variational principle can be written immediately as

$$\Pi = \int_\Omega [\tfrac{1}{2}\mathbf{u}^T\mathbf{L}\mathbf{u} + \mathbf{u}^T\mathbf{b}] \, d\Omega + \text{b.t.} \tag{9.91}$$

To prove the veracity of the last statement a variation needs to be considered. We thus write

$$\delta\Pi = \int_\Omega [\tfrac{1}{2}\delta\mathbf{u}^T\mathbf{L}\mathbf{u} + \tfrac{1}{2}\mathbf{u}^T\delta(\mathbf{L}\mathbf{u}) + \delta\mathbf{u}^T\mathbf{b}] \, d\Omega + \text{b.t.} \tag{9.92}$$

Noting that for any linear operator

$$\delta(\mathbf{L}\mathbf{u}) \equiv \mathbf{L} \, \delta\mathbf{u} \tag{9.93}$$

and that $\mathbf{u}$ and $\delta\mathbf{u}$ can be treated as any two independent functions, by identity (9.90) we can write Eq. (9.92) as

$$\delta\Pi = \int_\Omega \delta\mathbf{u}^T[\mathbf{L}\mathbf{u} + \mathbf{b}] \, d\Omega + \text{b.t.} \tag{9.94}$$

We observe immediately that the term in the brackets, i.e., the Euler equation of the functional, is identical with the original equation postulated, and therefore the variational principle is verified.

The above gives a very simple test and a prescription for the establishment of natural variational principles for differential equations of the problem.

Consider, for instance, two examples.

Example 1. This is a problem governed by the differential equation similar to the heat conduction equation, e.g.,

$$\nabla^2\phi + c\phi + Q = 0 \tag{9.95}$$

with c and Q being dependent on position only.

The above can be written in the general form of Eq. (9.89), with

$$\mathbf{L} \equiv \left[\frac{\partial^2}{\partial x^2}, \frac{\partial^2}{\partial y^2}, c \right] \qquad \mathbf{b} \equiv Q \tag{9.96}$$

Verifying that self-adjointness applies (which we leave to the reader as an exercise), we immediately have a variational principle

$$\Pi = \int_\Omega \left[\frac{1}{2} \, \phi \left(\frac{\partial^2\phi}{\partial x^2} + \frac{\partial^2\phi}{\partial y^2} + c\phi \right) + Q\phi \right] dx \, dy \tag{9.97}$$

with ϕ satisfying the forced boundary condition, i.e., $\phi = \bar{\phi}$ on Γ_ϕ. Integration by parts of the first two terms results in

$$\Pi = -\int_\Omega \left[\frac{1}{2}\left(\frac{\partial\phi}{\partial x}\right)^2 + \frac{1}{2}\left(\frac{\partial\phi}{\partial y}\right)^2 - \frac{1}{2}c\phi^2 - Q\phi \right] dx \, dy \qquad (9.98)$$

on noting that boundary terms with prescribed ϕ do not alter the principle.

Example 2. This problem concerns the equation system discussed in the previous section [Eqs (9.84) and (9.85)]. Again self-adjointness of the operator can be tested—and found satisfied. We now write the functional as

$$\Pi = \int_\Omega \left(\frac{1}{2}\begin{Bmatrix} q \\ \phi \end{Bmatrix}^{\mathrm{T}} \begin{bmatrix} 1, & -\frac{d}{dx} \\ \frac{d}{dx}, & 0 \end{bmatrix} \begin{Bmatrix} q \\ \phi \end{Bmatrix} + \begin{Bmatrix} q \\ \phi \end{Bmatrix}^{\mathrm{T}} \begin{Bmatrix} 0 \\ q \end{Bmatrix} \right) dx$$

$$= \int_\Omega \left(q^2 - q\frac{d\phi}{\partial x} + \phi\frac{dq}{dx} + \phi q \right) dx \qquad (9.99)$$

The verification of the correctness of the above, by executing a variation, is left to the reader.

These two examples illustrate the simplicity of application of the general expressions. The reader will observe that self-adjointness of the operator will generally exist if even orders of differentiation are present. For odd orders the self-adjointness is only possible if the operator is a 'skew'-symmetric matrix such as occurs in the second example.

9.11.2 *Adjustment for self-adjointness.* On occasion a linear operator which is not self-adjoint can be adjusted so that self-adjointness is achieved without altering the basic equation. Consider, for instance, a problem governed by the following differential equation of a standard linear form:

$$\frac{d^2\phi}{dx^2} + \alpha\frac{d\phi}{dx} + \beta\phi + q = 0 \qquad (9.100)$$

In this equation α and β are some functions of x. It is easy to see that the operator $\mathbf{L}$ is now a scalar:

$$L \equiv \frac{d^2}{dx^2} + \alpha\frac{d}{dx} + \beta \qquad (9.101)$$

and is not self-adjoint.

Let p be some, as yet undetermined, function of x. We shall show that it is possible to convert Eq. (9.100) to a self-adjoint form by multiplying it by this function. The new operator becomes

$$\bar{L} = pL \tag{9.102}$$

To test for symmetry with any two functions ψ and γ we write

$$\int_\Omega \psi(pL\gamma)\,dx = \int_\Omega \left(\psi p\,\frac{d^2\gamma}{dx^2} + \psi p\alpha\,\frac{d\gamma}{dx} + \psi p\beta\gamma\right)dx \tag{9.103}$$

On integration of the first term, by parts, we have (b.t. denoting boundary terms)

$$\int_\Omega \left(-\frac{d(\psi p)}{dx}\,\frac{d\gamma}{dx} + \psi p\alpha\,\frac{d\gamma}{dx} + \beta\psi p\gamma\,dx\right)dx + \text{b.t.}$$

$$= \int \left[-\frac{d\psi}{dx}\,p\,\frac{d\gamma}{dx} + \psi\,\frac{d\gamma}{dx}\left(p\alpha - \frac{dp}{dx}\right) + \psi p\beta\gamma\right]dx + \text{b.t.} \tag{9.104}$$

Symmetry (and therefore self-adjointness) is now achieved in the first and last terms. The middle term will only be symmetric if it disappears, i.e., if

$$p\alpha - \frac{dp}{dx} = 0 \tag{9.105}$$

or

$$\frac{dp}{p} = \alpha\,dx$$

$$p = e^{\int \alpha\,dx} \tag{9.106}$$

By using this value of p the operator is made self-adjoint and a variational principle for the problem of Eq. (9.100) is easily found.

Procedure of this kind has been used by Guymon et al.[19] to derive variational principles for a convective diffusion equation which is not self-adjoint. (We have noted such lack of symmetry in the equation in Example 2, Sec. 9.5.) We discuss this problem further in Appendix 7.

A similar method for creating variational functionals can be extended to a special case of non-linearity of Eq. (9.89) when

$$\mathbf{b} = \mathbf{b}(\mathbf{u}, x, \ldots) \tag{9.107}$$

If Eq. (9.92) is inspected we note that we could write

$$\delta(\mathbf{u}^T\mathbf{b}) = \delta(\mathbf{g}) \tag{9.108}$$

if

$$\mathbf{g} = \int \mathbf{b}^{\mathrm{T}}\, \mathrm{d}\mathbf{u}$$

This integration is generally quite easy to accomplish.

9.12 Maximum, minimum, or a saddle point?

In discussing variational principles so far we have assumed simply that at the solution point $\delta\Pi = 0$, or that the functional is stationary. It is often desirable to know whether Π is at a maximum, minimum, or simply at a 'saddle point'. If a maximum or a minimum is involved, then the approximation will always be 'bounded', i.e., will provide approximate values of Π which are either smaller or larger than the correct ones. This in itself may be of practical significance.

When, in elementary calculus, we consider a stationary point of a function Π of one variable a, we investigate the rate of change of $\mathrm{d}\Pi$ with $\mathrm{d}a$ and write

$$\mathrm{d}(\mathrm{d}\Pi) = \mathrm{d}\left(\frac{\partial\Pi}{\partial a}\,\mathrm{d}a\right) = \frac{\partial^2\Pi}{\partial a^2}\,(\mathrm{d}a)^2 \tag{9.109}$$

The sign of the second derivative determines whether Π is a minimum, maximum, or simply stationary (saddle point), as shown in Fig. 9.7. By analogy in the calculus of variation we shall consider changes of $\delta\Pi$.

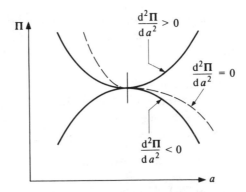

Fig. 9.7   Maximum, minimum, and a 'saddle' point for a functional Π of one variable

Noting the general form of this quantity given by Eq. (9.63) and the notion of the second derivative of Eq. (9.66) we can write, in terms of discrete parameters,

$$\delta(\delta\Pi) \equiv \delta\left(\frac{\partial\Pi}{\partial\mathbf{a}}\right)^{\mathrm{T}} \delta\mathbf{a}$$

$$= \delta\mathbf{a}^{\mathrm{T}} \, \delta\left(\frac{\partial\Pi}{\partial\mathbf{a}}\right)$$

$$= \delta\mathbf{a}^{\mathrm{T}}\mathbf{K}_{\mathrm{T}} \, \delta\mathbf{a} \tag{9.110}$$

If, in the above, $\delta(\delta\Pi)$ is always negative then Π is obviously reaching a maximum, if it is always positive then Π is a minimum, but if the sign is indeterminate this shows only the existence of a saddle point.

As $\delta\mathbf{a}$ is an arbitrary vector this statement is equivalent in requiring the matrix $\mathbf{K}_{\mathrm{T}}$ to be negative definite for a maximum *or* positive definite for a minimum. The form of the matrix $\mathbf{K}_{\mathrm{T}}$ (or in linear problems of $\mathbf{K}$ which is identical to it) is thus of great importance in variational problems.

9.13 Constrained variational principles. Lagrange multipliers and adjoint functions

9.13.1 *Lagrangian multipliers.* Consider the problem of making a functional Π stationary, subject to the unknown $\mathbf{u}$ obeying some set of additional differential relationships

$$\mathbf{C}(\mathbf{u}) = 0 \quad \text{in } \Omega \tag{9.111}$$

We can introduce this constraint by forming another functional

$$\bar{\Pi}(\mathbf{u}, \lambda) = \Pi(\mathbf{u}) + \int_{\Omega} \lambda^{\mathrm{T}}\mathbf{C}(\mathbf{u}) \, d\Omega \tag{9.112}$$

in which λ is some set of functions of the independent coordinates in domain Ω known as the *lagrangian multipliers*. The variation of the new functional is now

$$\delta\bar{\Pi} = \delta\Pi + \int_{\Omega} \delta\lambda^{\mathrm{T}}\mathbf{C}(\mathbf{u}) \, d\Omega + \int_{\Omega} \lambda^{\mathrm{T}} \, \delta\mathbf{C}(\mathbf{u}) \, d\Omega \tag{9.113}$$

and this is zero providing $\mathbf{C}(\mathbf{u}) = 0$ (and hence $\delta\mathbf{C} = 0$) and, simultaneously,

$$\delta\Pi = 0 \tag{9.114}$$

In a similar way, constraints can be introduced at some points or over boundaries of the domain. For instance, if we require that $\mathbf{u}$ obey

$$\mathbf{E}(\mathbf{u}) = 0 \qquad \text{on } \Gamma \tag{9.115}$$

we would add to the original functional the term

$$\int_\Gamma \lambda^T \mathbf{E}(\mathbf{u}) \, d\Gamma \tag{9.116}$$

with λ now being an unknown function defined only on Γ. Alternatively, if the constraint $\mathbf{C}$ is applicable only at one or more points of the system, then the simple addition of $\lambda^T \mathbf{C}(\mathbf{u})$ at these points to the general functional Π will introduce a discrete number of constraints.

It appears, therefore, possible to always introduce additional functions λ and modify a functional to include any prescribed constraints. In the 'discretization' process we shall now have to use trial functions to describe both $\mathbf{u}$ and λ.

Writing, for instance,

$$\hat{\mathbf{u}} = \sum \mathbf{N}_i \mathbf{a}_i = \mathbf{N}\mathbf{a} \qquad \hat{\lambda} = \sum \bar{\mathbf{N}}_i \mathbf{b}_i = \bar{\mathbf{N}}\mathbf{b} \tag{9.117}$$

we shall obtain a set of equations

$$\frac{\partial \Pi}{\partial \mathbf{c}} = \left\{ \begin{array}{c} \dfrac{\partial \Pi}{\partial \mathbf{a}} \\[2mm] \dfrac{\partial \Pi}{\partial \mathbf{b}} \end{array} \right\} = 0 \qquad \mathbf{c} = \left\{ \begin{array}{c} \mathbf{a} \\ \mathbf{b} \end{array} \right\} \tag{9.118}$$

from which both the sets of parameters $\mathbf{a}$ and $\mathbf{b}$ can be obtained. It is somewhat paradoxical that the 'constrained' problem has resulted in a larger number of unknown parameters than the original one and, indeed, complicated the solution. We shall, nevertheless, find practical use for lagrangian multipliers in formulating some physical variational principles, and will make use of these in a more general context in Chapter 12.

The point about increasing the number of parameters to introduce a constraint may perhaps be best illustrated in a simple algebraic situation in which we require a stationary value of a quadratic function of two variables a_1 and a_2:

$$\Pi = 2a_1^2 - 2a_1 a_2 + a_2^2 + 18a_1 + 6a_2 \tag{9.119}$$

subject to a constraint

$$a_1 - a_2 = 0 \tag{9.120}$$

The obvious way to proceed would be to insert directly the equality 'constraint' and obtain

$$\Pi = a_1^2 + 24a_1 \tag{9.121}$$

and write, for stationarity,

$$\frac{\partial \Pi}{\partial a_1} = 0 = 2a_1 + 24 \qquad a_1 = a_2 = -12 \tag{9.122}$$

Introducing a lagrangian multiplier λ we can alternatively find the stationarity of

$$\bar{\Pi} = 2a_1^2 - 2a_1 a_2 + a_2^2 + 18a_1 + 6a_2 + \lambda(a_1 - a_2) \tag{9.123}$$

and write *three* simultaneous equations

$$\frac{\partial \bar{\Pi}}{\partial a_1} = 0 \qquad \frac{\partial \bar{\Pi}}{\partial a_2} = 0 \qquad \frac{\partial \bar{\Pi}}{\partial \lambda} = 0 \tag{9.124}$$

The solution of the above system yields again the correct answer

$$a_1 = a_2 = -12 \qquad \lambda = 6$$

but at considerably more effort. Unfortunately, in most continuum problems the direct elimination of constraints cannot be so simply accomplished.†

Before proceeding further it is of interest to investigate the form of equations resulting from the modified functional Π of Eq. (9.112). If the original functional Π gave as its Euler equations a system

$$\mathbf{A}(\mathbf{u}) = 0 \tag{9.125}$$

then we have

$$\delta \bar{\Pi} = \int_{\Omega} \delta \mathbf{u}^{\mathrm{T}} \mathbf{A}(\mathbf{u}) \, \mathrm{d}\Omega + \int_{\Omega} \delta \lambda^{\mathrm{T}} \mathbf{C}(\mathbf{u}) \, \mathrm{d}\Omega + \int_{\Omega} \lambda^{\mathrm{T}} \, \delta \mathbf{C} \, \mathrm{d}\Omega \tag{9.126}$$

Substituting the trial functions (9.117) we can write, if the constraints are a linear set of equations,

$$\mathbf{C}(\mathbf{u}) = \mathbf{L}_1 \mathbf{u} + \mathbf{C}_1$$

$$\delta \bar{\Pi} = \delta \mathbf{a}^{\mathrm{T}} \int_{\Omega} \mathbf{N}^{\mathrm{T}} \mathbf{A}(\hat{\mathbf{u}}) \, \mathrm{d}\Omega + \delta \mathbf{b}^{\mathrm{T}} \int_{\Omega} \bar{\mathbf{N}}^{\mathrm{T}} (\mathbf{L}_1 \hat{\mathbf{u}} + \mathbf{C}_1) \, \mathrm{d}\Omega \tag{9.127}$$

$$+ \delta \mathbf{a}^{\mathrm{T}} \int_{\Omega} (\mathbf{L}_1 \mathbf{N})^{\mathrm{T}} \hat{\lambda} \, \mathrm{d}\Omega = 0$$

---

† In the finite element context, Szabo and Kassos[20] use such direct elimination; however, this involves considerable algebraic manipulation.

As this has to be true for all variations $\delta\mathbf{a}$ and $\delta\mathbf{b}$, we have a system of equations

$$\int_\Omega \mathbf{N}^T\mathbf{A}(\hat{\mathbf{u}}) \, d\Omega + \int_\Omega (\mathbf{L}_1\mathbf{N})^T\hat{\boldsymbol{\lambda}} \, d\Omega = 0$$

$$\int_\Omega \bar{\mathbf{N}}^T(\mathbf{L}_1\hat{\mathbf{u}} + \mathbf{C}_1) \, d\Omega = 0 \tag{9.128}$$

For linear equations $\mathbf{A}$, the first term of the first equation, is precisely the ordinary, unconstrained, variational approximation

$$\mathbf{Ka} + \mathbf{f} \tag{9.129}$$

and inserting again the trial functions (9.117) we can write the approximated Eq. (9.128) as a linear system:

$$\mathbf{K}_c\,\mathbf{c} = \begin{bmatrix} \mathbf{K} & , & \mathbf{K}_{ab} \\ \mathbf{K}_{ab}^T, & \mathbf{0} \end{bmatrix}\begin{Bmatrix} \mathbf{a} \\ \mathbf{b} \end{Bmatrix} + \begin{Bmatrix} \mathbf{f} \\ \mathbf{0} \end{Bmatrix} = 0 \tag{9.130}$$

with

$$\mathbf{K}_{ab}^T = \int_\Omega \bar{\mathbf{N}}^T\mathbf{L}_1\mathbf{N} \, d\Omega \tag{9.131}$$

Clearly the equation system is symmetric but now possesses zeros on the diagonal, and therefore the variational principle Π is merely stationary. Further, computational difficulties may be encountered unless the solution process allows for zero diagonal terms.

9.13.2 Identification of lagrangian multipliers. Forced boundary conditions and modified variational principles.

Although the lagrangian multipliers were introduced as a mathematical fiction necessary for the enforcement of certain external constraints required to satisfy the original variational principle, we shall find that in most physical situations they can be identified with certain physical quantities of importance to the original mathematical model. Such an identification will follow immediately from the definition of the variational principle established in Eq. (9.112) and through the second of the Euler equations corresponding to it. The variation $\delta\bar{\Pi}$, written in Eq. (9.113), supplies through its first two terms the original Euler equation of the problem corresponding to the functional Π and the constraint equation. The last term can always be rewritten as

$$\int_\Omega \boldsymbol{\lambda}^T \, \delta\mathbf{C}(\mathbf{u}) \, d\Omega \equiv \int_\Omega \delta\mathbf{u}^T\mathbf{R}(\boldsymbol{\lambda}, \mathbf{u}) \, d\Omega + \text{b.t.} \tag{9.132}$$

imposing the requirements that

$$\mathbf{R}(\lambda, \mathbf{u}) = 0 \tag{9.133}$$

This supplies the identification of λ.

In the literature of variational calculation such identification arises frequently and the reader is referred to the excellent text by Washizu[21] for numerous examples.

Here we shall introduce this identification by means of the example considered in Sec. 9.10.1. As we have noted, the variational principle of Eq. (9.72) established the governing equation and the natural boundary conditions of the heat conduction problem providing the forced boundary condition

$$\mathbf{C}(\phi) = \phi - \bar{\phi} = 0 \tag{9.134}$$

was satisfied on Γ_ϕ in the choice of the trial function for ϕ.

The above, forced boundary, condition can, however, be considered as a constraint on the original problem. We can write the constrained variational principle as

$$\bar{\Pi} = \Pi + \int_{\Gamma_\phi} \lambda(\phi - \bar{\phi}) \, d\Gamma \tag{9.135}$$

where Π is given by Eq. (9.72).

Performing the variation we have

$$\delta\bar{\Pi} = \delta\Pi + \int_{\Gamma_\phi} \delta\lambda(\phi - \bar{\phi}) \, d\Gamma + \int_{\Gamma_\phi} \delta\phi\lambda \, d\Gamma \tag{9.136}$$

$\delta\Pi$ is now given by the expression (9.75a) augmented by an integral

$$\int_{\Gamma_\phi} \delta\phi k \frac{\partial \phi}{\partial n} \, d\Gamma \tag{9.137}$$

which previously was disregarded (as we had assumed that $\delta\phi = 0$ on Γ_ϕ). In addition to the conditions of Eq. (9.75b), we now require that

$$\int_{\Gamma_\phi} \delta\lambda(\phi - \bar{\phi}) \, d\Gamma + \int_{\Gamma_\phi} \delta\phi\left(\lambda + k\frac{\partial \phi}{\partial n}\right) d\Gamma = 0 \tag{9.138}$$

which must be true for all variations $\delta\lambda$ and $\delta\phi$. The first simply reiterates the constraint

$$\phi - \bar{\phi} = 0 \qquad \text{on } \Gamma_\phi \tag{9.139}$$

The second *defines* λ as

$$\lambda = -k \frac{\partial \phi}{\partial n} \tag{9.140}$$

Noting that $k(\partial\phi/\partial n)$ is equal to the flux $-q$ on the boundary Γ_ϕ, a physical identification of the multiplier has been achieved.

The identification of the lagrangian variable leads to the possible establishment of a modified variational principle in which λ is replaced by the identification.

We could thus write a new principle for the above example:

$$\bar{\Pi} = \Pi - \int_{\Gamma_\phi} k \frac{\partial\phi}{\partial n} (\phi - \bar{\phi}) \, d\Gamma \qquad (9.141)$$

in which once again Π is given by the expression (9.72) but ϕ is not constrained to satisfy any boundary conditions. Use of such modified variational principles can be made to restore interelement continuity and appears to have been introduced for that purpose first by Kikuchi and Ando.[22] In general these present interesting new procedures for establishing useful variational principles.

A further extension of such principles has been made use of by Chen and Mei[23] and Zienkiewicz et al.[24] Washizu[21] discusses many such applications in the context of structural mechanics. The reader can verify that the variational principle expressed in Eq. (9.141) leads to an automatic satisfaction of all the necessary boundary conditions in the example considered.

The use of modified variational principles restores the problem to the original number of unknown functions or parameters and is thus computationally advantageous.

9.13.3 *A general variational principle: adjoint functions and operators.* The lagrangian multiplier procedure leads to an obvious procedure of 'creating' a variational principle for any set of equations even if the operators are not self-adjoint:

$$\mathbf{A}(\mathbf{u}) = 0 \qquad (9.142)$$

Treating all the above equations as a set of constraints we can obtain such a general variational functional simply by putting $\Pi = 0$ in Eq. (9.112) and writing

$$\bar{\Pi} = \int_\Omega \lambda^T \mathbf{A}(\mathbf{u}) \, d\Omega \qquad (9.143)$$

now requiring stationarity for all variations of $\delta\lambda$ and $\delta\mathbf{u}$. The new variational principle has, however, been introduced at the expense of doubling the number of variables in the discretized situation. Treating the case of linear equations only, i.e.,

$$\mathbf{A}(\mathbf{u}) = \mathbf{L}\mathbf{u} + \mathbf{g} = 0 \qquad (9.144)$$

and discretizing we note, going through the steps involved in Eqs (9.126) to (9.130), that the final system of equations now takes the form

$$\begin{bmatrix} 0 & K_{ab} \\ K_{ab}^T & 0 \end{bmatrix} \begin{Bmatrix} a \\ b \end{Bmatrix} + \begin{Bmatrix} 0 \\ f \end{Bmatrix} = 0 \qquad (9.145)$$

with

$$K_{ab}^T = \int_\Omega \bar{N} L^T N \ d\Omega$$

$$f = \int_\Omega \bar{N}^T g \ d\Omega \qquad (9.146)$$

The equations are completely decoupled and the second set can be solved independently for all the parameters **a** describing the unknowns in which we were originally interested without consideration of parameters **b**. It will be observed that this second set of equations is *identical with an, apparently arbitrary, weighted residual process.* We have thus completed the full circle and obtained the weighted residual forms of Sec. 9.4 from a general variational principle.

The function λ which appears in the variational principle of Eq. (9.143) is known as the *adjoint function to* **u**.

By performing a variation on Eq. (9.143) it is easy to show that the Euler equations of the principle are such that

$$A(u) = 0 \qquad (9.147)$$

and

$$A^*(u) = 0 \qquad (9.148)$$

where the operator A^* is such that

$$\int \lambda^T \ \delta(Au) \ d\Omega = \int \delta u^T A^*(\lambda) \ d\Omega \qquad (9.149)$$

The operator A^* is known as the adjoint operator and will exist only in linear problems (see Appendix 7).

For the full significance of the adjoint operator the reader is advised to consult mathematical texts.[25]

9.14 Constrained variational principles. Penalty functions and the least square method

9.14.1 *Penalty functions.* In the previous section we have seen how the process of introducing lagrange multipliers allows constrained variational principles to be obtained at the expense of increasing the total number of unknowns. Further, we have shown that even in linear problems the algebraic equations which have to be solved are now complicated by having zero diagonal terms. In this section we shall consider an alternative procedure of introducing constraints which does not possess these drawbacks.

Considering once again the problem of obtaining stationarity of Π with a set of constraint equations $\mathbf{C}(\mathbf{u}) = 0$ in domain Ω, we note that the product

$$\mathbf{C}^{\mathrm{T}}\mathbf{C} = C_1^2 + C_2^2 + \cdots \tag{9.150}$$

where

$$\mathbf{C}^{\mathrm{T}} = [C_1, C_2, \ldots]$$

must always be a quantity which is positive or zero. Clearly, the latter value is found when the constraints are satisfied and clearly the variation

$$\delta(\mathbf{C}^{\mathrm{T}}\mathbf{C}) = 0 \tag{9.151}$$

as the product reaches that minimum.

We can now immediately write a new functional

$$\bar{\bar{\Pi}} = \Pi + \alpha \int \mathbf{C}^{\mathrm{T}}(\mathbf{u})\mathbf{C}(\mathbf{u}) \, \mathrm{d}\Omega \tag{9.152}$$

in which α is a 'penalty number' and then require the stationarity for the constrained solution. If Π itself is a minimum of the solution then α should be a positive number. The solution obtained by the stationarity of the functional $\bar{\bar{\Pi}}$ will satisfy the constraints only approximately. The larger the value of α the better will be the constraints achieved. Further, it seems obvious that the process is best suited to cases where Π is a minimum (or maximum) principle—but success can be obtained even with purely saddle point problems. The process is equally applicable to constraints applied on boundaries or simple discrete constraints. In this latter case integration is dropped.

To clarify ideas let us once again consider the algebraic problem of Sec. 9.13, in which the stationarity of a functional given by Eq. (9.119) was sought subject to a constraint. With the penalty function approach we now could seek the minimum of a functional

$$\bar{\bar{\Pi}} = 2a_1^2 - 2a_1a_2 + a_2^2 + 18a_1 + 6a_2 + \alpha(a_1 - a_2)^2 \tag{9.153}$$

with respect to the variation of both parameters a_1 and a_2. Writing the two simultaneous equations

$$\frac{\partial \bar{\bar{\Pi}}}{\partial a_1} = 0 \qquad \frac{\partial \bar{\bar{\Pi}}}{\partial a_2} = 0 \tag{9.154}$$

we find that as α is increased we approach the correct solution. In Table 9.1 the results are set out demonstrating the convergence.

TABLE 9.1

$\alpha =$	1	2	6	10	100
$a_1 =$	-12.00	-12.00	-12.00	-12.00	-12.00
$a_2 =$	-13.50	-13.00	-12.43	-12.78	-12.03

The reader will observe that in a problem formulated in the above manner the constraint introduces no additional unknown parameters—but neither does it decrease their original number. The process will always result in strongly positive definite matrices if the original variational principle is one of a minimum.

In practical application the method of penalty functions has proved quite effective,[26] and indeed is often introduced intuitively. One such 'intuitive' application was already made when we enforced the value of boundary parameters in the manner indicated in Chapter 1, Sec. 1.4.

In the example presented there (and frequently practised in the real assembly of discretized finite element equations), the forced boundary conditions are not introduced *a priori* and the problem gives, on assembly, a singular system of equations

$$\mathbf{Ka} + \mathbf{f} = 0 \tag{9.155}$$

which can be obtained from a functional (providing $\mathbf{K}$ is symmetric)

$$\Pi = \tfrac{1}{2}\mathbf{a}^T\mathbf{Ka} + \mathbf{a}^T\mathbf{f} \tag{9.156}$$

Introducing a prescribed value of a_1, i.e., writing

$$a_1 - \bar{a}_1 = 0 \tag{9.157}$$

the functional can be modified to

$$\bar{\Pi} = \Pi + \alpha(a_1 - \bar{a}_1)^2 \tag{9.158}$$

yielding

$$\bar{K}_{11} = K_{11} + 2\alpha \qquad \bar{f}_1 = f_1 - 2\alpha\bar{a}_1 \tag{9.159}$$

and giving no change in any of the other matrix coefficients. This is precisely the procedure adopted in Chapter 1 (page 12) for modifying the equations, to introduce prescribed values of a_1 (2α here replacing α, the 'large number' of Sec. 1.4). Many applications of such a 'discrete' kind are discussed by Campbell.[27]

In the second example we shall consider the problem of beam deflection discussed in Chapter 2 (Sec. 2.10). This problem can be stated as the minimization of total potential energy given by

$$\Pi = \frac{1}{2}\int_0^L EI\left(\frac{d^2w}{dx^2}\right)^2 dx - \int_0^L wq\,dx \tag{9.160}$$

As the above formulation requires w to be modelled with C_1 continuity it is of interest to investigate a possibility of a reformulation imposing only C_0 continuity. Such an alternative form would be to require the minimization of

$$\Pi = \int_0^L \frac{1}{2} EI\left(\frac{d\theta}{dx}\right)^2 dx - \int_0^L wq\, dx \qquad (9.161)$$

subject to the constraint

$$C \equiv \frac{dw}{dx} - \theta = 0 \qquad (9.162)$$

θ here is obviously the slope approximation and Π is now a function of two variables, θ and w, which can be interpolated with C_0 continuity.

A modified variational principle using the penalty function can now be introduced:

$$\bar{\Pi} = \Pi + \alpha \int_0^L \left(\frac{dw}{dx} - \theta\right)^2 \qquad (9.163)$$

where α is a large number.

The structural engineer will immediately recognize that the physical meaning of α is that of shear rigidity, i.e.,

$$\alpha = \frac{1}{2} GA \qquad (9.164)$$

and that the formulation presented is simply that for a beam in which the slopes and rotations of sections vary independently and the additional term stands for the strain energy absorbed in shear.

The thick shell and plate elements to be discussed in the second volume are but an extension of the process given here.

It is easy to show in another context[26,28] that the use of a high Poisson's ratio ($v \to 0.5$) for the study of incompressible solids or fluids is in fact equivalent to the introduction of a penalty term to suppress any compressibility allowed by an arbitrary displacement variation.

The use of the penalty function in the finite element context presents certain difficulties.

Firstly, the constrained functional of Eq. (9.152) leads to equations of the form

$$(\mathbf{K}_1 + \alpha \mathbf{K}_2)\mathbf{a} + \mathbf{f} = 0 \qquad (9.165)$$

where $\mathbf{K}_1$ derives from the original functions and $\mathbf{K}_2$ from the constraints. As α increases the above equation degenerates:

$$\mathbf{K}_2 \mathbf{a} = -\mathbf{f}/\alpha \to 0$$

and $\mathbf{a} = 0$ unless the matrix $\mathbf{K}_2$ is singular. This singularity does not always arise and we shall discuss means of its introduction in Chapter 12.

Secondly, with large but finite values of α numerical difficulties will be encountered. Noting that discretization errors can be of comparable magnitude to those due to not *satisfying* the constraint, we can make

$$\alpha = \text{constant } (1/h)^n$$

ensuring a limiting convergence to the correct answer. Fried[29,30] discusses this problem in detail.

A more general discussion of the whole topic is given in reference 31 and in Chapter 12 where the relationship between lagrangian constraints and penalty forms is made clear.

9.14.2 *Least square approximation.* In Sec. 9.13.3 we have shown how a constrained variational principle procedure could be used to construct a general variational principle if the constraints become simply the governing equations of the problem

$$\mathbf{C}(\mathbf{u}) = \mathbf{A}(\mathbf{u}) \tag{9.166}$$

Obviously the same procedure can be used in the context of the penalty function approach by setting $\Pi = 0$ in Eq. (9.152). We can thus write a 'variational principle'

$$\bar{\bar{\Pi}} = \int_\Omega (A_1^2 + A_2 + \cdots)\, d\Omega = \int_\Omega \mathbf{A}^{\mathrm{T}}(\mathbf{u})\mathbf{A}(\mathbf{u})\, d\Omega \tag{9.167}$$

for any set of differential equations. In the above equation the boundary conditions are assumed to be satisfied by $\mathbf{u}$ (forced boundary condition) and the parameter α is dropped as it becomes simply a multiplier.

Clearly, the above statement is simply a requirement that the sum of the squares of the residuals of the differential equations should be a minimum at the correct solution. This minimum is obviously zero at that point, and the process is simply the well-known *least square method* of approximation.

It is equally obvious that we could obtain the correct solution by minimizing any functional of the form

$$\bar{\bar{\Pi}} = \int_\Omega (p_1 A_1^2 + p_2 A_2^2 + \cdots)\, d\Omega = \int_\Omega \mathbf{A}^{\mathrm{T}}(\mathbf{u})\mathbf{p}\mathbf{A}(\mathbf{u})\, d\Omega \tag{9.168}$$

in which p_1, p_2, ..., etc., are positive valued functions or constants and $\mathbf{p}$ is a diagonal matrix:

$$\mathbf{p} = \begin{bmatrix} p_1 & & & 0 \\ & p_2 & & \\ & & p_3 & \\ 0 & & & \ddots \end{bmatrix} \tag{9.169}$$

The above alternative form is sometimes convenient as it puts different importance to the satisfaction of individual components of the equation and allows additional freedom in the choice of the approximate solution. Once again this weighting function could be so chosen as to ensure a constant ratio of terms contributed by various elements—although this has not yet been put into practice.

Least square methods of the kind shown above are a very powerful alternative procedure for obtaining integral forms from which an approximate solution can be started, and have been used with considerable success.[32,33] As the least square variational principles can be written for *any* set of differential equations without introducing additional variables, we may well enquire what is the difference between these and the *natural variational principles* discussed previously. On performing a variation in a specific case the reader will find that the Euler equations which are obtained no longer give the original differential equations but give higher order derivatives of these. This introduces a possibility of spurious solutions if incorrect boundary conditions are used. Further, higher order continuity of trial function is now generally needed. This may be a serious drawback but frequently can be by-passed by stating the problem originally as a set of lower order equations.

We shall now consider the general form of discretized equations resulting from the least square approximation for linear equation sets (again neglecting boundary conditions which are enforced). Thus, if we take

$$\mathbf{A(u)} = \mathbf{Lu} + \mathbf{b} \tag{9.170}$$

and take the usual trial function approximation

$$\hat{\mathbf{u}} = \mathbf{Na} \tag{9.171}$$

we can write, substituting into (9.168),

$$\bar{\Pi} = \int_\Omega [(\mathbf{LN})\mathbf{a} + \mathbf{b}]^T \mathbf{p}[(\mathbf{LN})\mathbf{a} + \mathbf{b}] \, d\Omega \tag{9.172}$$

and

$$\delta\bar{\Pi} = \int_\Omega \delta\mathbf{a}^T(\mathbf{LN})^T\mathbf{p}[(\mathbf{LN})\mathbf{a} + \mathbf{b}] \, d\Omega + \int_\Omega [(\mathbf{LN})\mathbf{a} + \mathbf{b}]^T\mathbf{p}(\mathbf{LN}) \, \delta\mathbf{a} \, d\Omega \tag{9.173}$$

or, as $\mathbf{p}$ is symmetric,

$$\delta\bar{\bar{\Pi}} = \delta\mathbf{a}^T\left\{\left[2\int_\Omega (\mathbf{LN})^T\mathbf{p}(\mathbf{LN})\ d\Omega\right]\mathbf{a} + \int_\Omega (\mathbf{LN})^T\mathbf{pb}\ d\Omega\right\} \quad (9.174)$$

This yields immediately the approximation equation in the usual form:

$$\mathbf{Ka} + \mathbf{f} = 0 \quad (9.175)$$

and the reader can observe that the matrix $\mathbf{K}$ is symmetric and positive definite.

To illustrate an actual example, consider a problem governed by Eq. (9.95) for which we have already obtained a *natural* variational principle [Eq. (9.98)] in which only first derivatives were involved requiring C_0 continuity for $\mathbf{u}$. Now, if we use the operator $\mathbf{L}$ and term $\mathbf{b}$ defined by Eq. (9.96), we have a set of approximating equations with

$$K_{ij} = 2\int_\Omega (\nabla^2 N_i + cN_i)(\nabla^2 N_j + cN_j)\ dx\ dy$$

$$f_i = \int_\Omega (\nabla^2 N_i + cN_i)Q\ dx\ dy \quad (9.176)$$

The reader will observe that now a C_1 continuity is needed for the trial functions $\mathbf{N}$.

An alternative avoiding this difficulty is to write Eq. (9.95) as a first-order system. This can be written as

$$\mathbf{A}(\mathbf{u}) = \left\{\begin{array}{c} \dfrac{\partial\phi_x}{\partial x} + \dfrac{\partial\phi_y}{\partial y} + c\phi + Q \\[2mm] \dfrac{\partial\phi}{\partial x} - \phi_x \\[2mm] \dfrac{\partial\phi}{\partial y} - \phi_y \end{array}\right\} = 0 \quad (9.177)$$

or, introducing a vector $\mathbf{u}$,

$$\mathbf{u}^T = [\phi, \phi_x, \phi_y] = (\mathbf{Na})^T \quad (9.178)$$

as the unknown we can write the standard linear form (9.170) as

$$\mathbf{Lu} + \mathbf{b} = 0$$

where

$$
\mathbf{L} = \begin{bmatrix} c, & \dfrac{\partial}{\partial x}, & \dfrac{\partial}{\partial y} \\[2mm] \dfrac{\partial}{\partial x}, & -1, & 0 \\[2mm] \dfrac{\partial}{\partial y}, & 0, & -1 \end{bmatrix} \qquad \mathbf{b} = \begin{Bmatrix} Q \\ 0 \\ 0 \end{Bmatrix} \tag{9.179}
$$

The reader can now perform the substitution into Eq. (9.174) to obtain the approximation equations in a form requiring only C_0 continuity—introduced, however, at the expense of additional variables. Use of such forms has been made extensively in the finite element context.[32,33]

9.15 Concluding remarks—finite differences and boundary methods

This very extensive chapter presents the general possibilities of using the finite element processes in almost any mathematical or mathematically modelled physical problem. The essential approximation processes have been given in as simple a form as possible, at the same time presenting a fully comprehensive picture which should allow the reader to understand much of the literature and indeed to experiment with new permutations. In the chapters that follow we shall apply to various physical problems only a limited selection of the methods to which allusion has been made. In some we shall show, however, that certain extensions of the process are possible (Chapter 12) and in another (Chapter 11) how a violation of some of the rules here expounded can be accomplished with benefit.

The numerous approximation procedures discussed fall into several categories. To remind the reader of these, we present in Table 9.2 a comprehensive catalogue of the methods used here and in Chapter 2. The only aspect of the finite element process mentioned in that table that has not been discussed here is that of a *direct physical method*. In such models an 'atomic' rather than continuum concept is the starting point. While much interest exists in the possibilities offered by such models, their discussion is outside the scope of this book.

In all the continuum processes discussed the first step is always the choice of suitable shape or trial functions. A few simple forms of such functions have been introduced as the need demanded it but any of the more elaborate forms discussed in elasticity examples are applicable. Indeed, the reader who has mastered the essence of the present chapter will have little difficulty in applying the finite element method to any

TABLE 9.2
FINITE ELEMENT APPROXIMATION

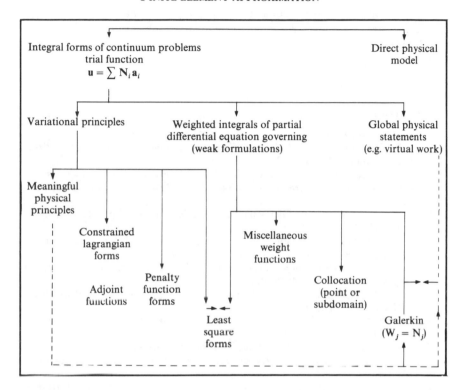

suitably defined physical problem. For further reading references 34 to 38 could be consulted.

The methods listed do not include specifically two well-known techniques, i.e., *finite difference* methods and *boundary solution* methods (sometimes known as boundary elements). In the general sense these belong under the category of the *generalized finite element* method here discussed.[34]

1. Finite difference procedures always represent an approximation based on local, discontinuous, shape functions with collocation weighting applied (although the actual derivation of the approximation algorithm is based on Taylor expansions and is generally simpler).

As Galerkin or variational approaches give, in the energy sense, the best approximation, this subset has only the merit of computational simplicity and often an accuracy loss.

2. Boundary solution methods choose the trial functions such that the governing equation is automatically satisfied. Starting thus from the general approximation equation (9.25), we note that only boundary terms are retained. We shall return to such approximations in Chapter 13.

References

1. S. H. CRANDALL, *Engineering Analysis*, McGraw-Hill, 1956.
2. B. A. FINLAYSON, *The Method of Weighted Residuals and Variational Principles*, Academic Press, 1972.
3. R. A. FRAZER, W. P. JONES, and S. W. SKEN, *Approximations to functions and to the solutions of differential equations*, Aero. Research Committee Report 1799, 1937.
4. C. B. BIEZENO and R. GRAMMEL, *Technische Dynamik*, p. 142, Springer-Verlag, 1933.
5. B. G. GALERKIN, 'Series solution of some problems of elastic equilibrium of rods and plates' (Russian), *Vestn. Inzh. Tech.*, **19**, 897–908, 1915.
6. Also attributed to Bubnov, 1913: see S. C. MIKHLIN, *Variational Methods in Mathematical Physics*, Macmillan, 1964.
7. P. TONG, 'Exact solution of certain problems by the finite element method', *J. AIAA*, **7**, 179–180, 1969.
8. R. V. SOUTHWELL, *Relaxation Methods in Theoretical Physics*, Clarendon Press, 1946.
9. R. S. VARGA, *Matrix Iterative Analysis*, Prentice-Hall, 1962.
10. S. TIMOSHENKO and J. N. GOODIER, *Theory of Elasticity*, 2nd ed., McGraw-Hill, 1951.
11. L. V. KANTOROVITCH and V. I. KRYLOV, *Approximate Methods of Higher Analysis*, Wiley (International), 1958.
12. J. W. STRUTT (Lord Rayleigh), 'On the theory of resonance', *Trans. Roy. Soc. (London)*, **A161**, 77–118, 1870.
13. W. RITZ, 'Über eine neue Methode zur Lösung gewissen Variations—Probleme der mathematischen Physik', *J. Reine angew. Math.*, **135**, 1–61, 1909.
14. M. M. VEINBERG, *Variational Methods for the Study of Nonlinear Operators*, Holden-Day, 1964.
15. E. TONTI, 'Variational formulation of non-linear differential equations', *Bull. Acad. Roy. Belg. (Classe Sci.)*, **55**, 137–65 and 262–78, 1969.
16. J. T. ODEN, 'A general theory of finite elements—I: Topological considerations', pp. 205–21, and 'II: Applications', pp. 247–60, *Int. J. Num. Meth. Eng.*, **1**, 1969.
17. S. C. MIKHLIN, *Variational Methods in Mathematical Physics*, Macmillan, 1964.
18. S. C. MIKHLIN, *The Problems of the Minimum of a Quadratic Functional*, Holden-Day, 1965.
19. G. L. GUYMON, V. H. SCOTT, and L. R. HERRMANN, 'A general numerical solution of the two-dimensional differential–convection equation by the finite element method', *Water Res.*, **6**, 1611–15, 1970.

20. B. A. SZABO and T. KASSOS, 'Linear equation constraints in finite element approximations', *Int. J. Num. Meth. Eng.*, **9**, 563–80, 1975.
21. K. WASHIZU, *Variational Methods in Elasticity and Plasticity*, 2nd ed., Pergamon Press, 1975.
22. F. KIKUCHI and Y. ANDO, 'A new variational functional for the finite element method and its application to plate and shell problems', *Nucl. Eng. Des.*, **21**, 95–113, 1972.
23. H. S. CHEN and C. C. MEI, *Oscillations and water forces in an offshore harbour*, Ralph M. Parsons Laboratory for Water Resources and Hydrodynamics, Report 190, Cambridge, Mass., 1974.
24. O. C. ZIENKIEWICZ, D. W. KELLY, and P. BETTESS, 'The coupling of the finite element method and boundary solution procedures', *Int. J. Num. Meth. Eng.*, **11**, 355–75, 1977.
25. I. STAKGOLD, *Boundary Value Problems of Mathematical Physics*, Macmillan, 1967.
26. O. C. ZIENKIEWICZ, 'Constrained variational principles and penalty function methods in the finite element analysis', *Lecture Notes in Mathematics*, No. 363, pp. 207–214, Springer-Verlag, 1974.
27. J. CAMPBELL, *A finite element system for analysis and design*, Ph.D. thesis, Swansea, 1974.
28. D. J. NAYLOR, 'Stresses in nearly incompressible materials for finite elements with application to the calculation of excess pore pressures', *Int. J. Num. Meth. Eng.*, **8**, 443–60, 1974.
29. I. FRIED, 'Finite element analysis of incompressible materials by residual energy balancing', *Int. J. Solids Struct.*, **10**, 993–1002, 1974.
30. I. FRIED, 'Shear in C^0 and C^1 bending finite elements', *Int. J. Solids Struct.*, **9**, 449–60, 1973.
31. O. C. ZIENKIEWICZ and E. HINTON, 'Reduced integration, function smoothing and non-conformity in finite element analysis', *J. Franklin Inst.*, **302**, 443–61, 1976.
32. P. P. LYNN and S. K. ARYA, 'Finite elements formulation by the weighted discrete least squares method', *Int. J. Num. Meth. Eng.*, **8**, 71–90, 1974.
33. O. C. ZIENKIEWICZ, D. R. J. OWEN, and K. N. LEE, 'Least square finite element for elasto-static problems—use of reduced integration', *Int. J. Num. Meth. Eng.*, **8**, 341–58, 1974.
34. O. C. ZIENKIEWICZ and K. MORGAN, *Finite Elements and Approximation*, Wiley, 1983.
35. E. B. BECKER, G. F. CAREY, and J. T. ODEN, *Finite Elements; An Introduction*, Vol. 1, Prentice Hall, 1981.
36. I. FRIED, *Numerical Solution of Differential Equations*, Academic Press, New York, 1979.
37. A. J. DAVIES, *The Finite Element Method*, Clarendon, Oxford, 1980.
38. C. A. T. FLETCHER, *Computational Galerkin Methods*, Springer Verlag, 1984.

10

Steady-state field problems—heat conduction, electric and magnetic potential, fluid flow, etc.

10.1 Introduction

While, in detail, most of the previous chapters dealt with problems of an elastic continuum the general procedures can be applied to a variety of physical problems. Indeed, some such possibilities have been indicated in Chapter 9 and here more detailed attention will be given to a particular but wide class of such situations.

Primarily we shall deal with situations governed by the general 'quasi-harmonic' equation, the particular cases of which are the well-known Laplace and Poisson equations.[1-6] The range of physical problems falling into this category is large. To list but a few frequently encountered in engineering practice we have:

Heat conduction
Seepage through porous media
Irrotational flow of ideal fluids
Distribution of electrical (or magnetic) potential
Torsion of prismatic shafts
Bending of prismatic beams, etc.
Lubrication of pad bearings

The formulation developed in this chapter is applicable equally to all, and hence little reference will be made to the actual physical quantities. Isotropic or anisotropic regions can be treated with equal ease.

Two-dimensional problems are discussed in the first part of the chapter. A generalization to three dimensions follows. It will be observed that the same, C_0, 'shape functions' as those used previously in two- or three-dimensional formulation of elasticity problems will again be encountered. The main difference will be that now only one unknown scalar quantity (the unknown function) is associated with each point in space. Previously, several unknown quantities, represented by the displacement vector, were sought.

In Chapter 9 we indicated both the 'weak form' and a variational principle applicable to the Poisson and Laplace equations (see Secs 9.3 and 9.10.1). In the following sections we shall generalize these approaches to a general, quasi-harmonic equation and indicate the ranges of applicability of a *single, unified, approach* by which one computer program can solve a large variety of physical problems.

10.2 The general quasi-harmonic equation

10.2.1 *The general statement.* In many physical situations we are concerned with the *diffusion* or flow of some quantity such as heat, mass, or a chemical, etc. In such problems the rate of transfer per unit area, **q**, can be written in terms of its cartesian components as

$$\mathbf{q}^T = [q_x, q_y, q_z] \tag{10.1}$$

If the rate at which the relevant quantity is generated (or removed) per unit volume is Q, then for steady-state flow the balance or continuity requirement gives

$$\frac{\partial q_x}{\partial x} + \frac{\partial q_y}{\partial y} + \frac{\partial q_z}{\partial z} = Q \tag{10.2}$$

Introducing the gradient operator

$$\mathbf{\nabla} = \left\{ \begin{matrix} \dfrac{\partial}{\partial x} \\[2mm] \dfrac{\partial}{\partial y} \\[2mm] \dfrac{\partial}{\partial z} \end{matrix} \right\} \tag{10.3}$$

we can write the above as

$$\mathbf{\nabla}^T \mathbf{q} - Q = 0 \tag{10.4}$$

Generally the rates of flow will be related to *gradients* of some potential quantity ϕ. This may be temperature in the case of heat flow, etc. A very general relationship will be of a form

$$
\mathbf{q} = \begin{Bmatrix} q_x \\ q_y \\ q_z \end{Bmatrix} = -\mathbf{k} \begin{Bmatrix} \dfrac{\partial \phi}{\partial x} \\ \dfrac{\partial \phi}{\partial y} \\ \dfrac{\partial \phi}{\partial z} \end{Bmatrix} = -\mathbf{k}\, \nabla \phi \tag{10.5}
$$

where $\mathbf{k}$ is a three by three matrix. This is generally of a symmetric form due to energy arguments.

The final governing equation for the 'potential' ϕ is obtained by substitution of Eq. (10.5) into (10.4), leading to

$$
\nabla^T \mathbf{k}\, \nabla \phi + Q = 0 \tag{10.6}
$$

which has to be solved in the domain Ω. On the boundaries of such a domain we shall usually encounter one or other of the following conditions:

1. On Γ_ϕ,

$$
\phi = \bar{\phi} \tag{10.7a}
$$

   i.e., the potential is specified.
2. On Γ_q the normal component of flow, q_n, is given as

$$
q_n = \bar{q} + \alpha \phi \tag{10.7b}
$$

   where α is a transfer or radiation coefficient.
   As

$$
q_n = \mathbf{q}^T \mathbf{n} \qquad \mathbf{n}^T = [n_x, n_y, n_z]
$$

   where $\mathbf{n}$ is a vector of direction cosines of the normal to the surface, this condition can immediately be rewritten as

$$
-(\mathbf{k}\, \nabla \phi)^T \mathbf{n} - \bar{q} - \alpha \phi = 0 \tag{10.7c}
$$

in which $\bar{q}$ and α are given.

10.2.2 *Particular forms.* If we consider the general statement of Eq. (10.5) as being determined for an arbitrary set of coordinate axes x, y, z we shall find that it is always possible to determine locally another set of axes x', y', z' with respect to which the matrix $\mathbf{k}'$ becomes diagonal. With respect to such axes we have

$$
\mathbf{k}' = \begin{bmatrix} k_{x'} & 0 & 0 \\ 0 & k_{y'} & 0 \\ 0 & 0 & k_{z'} \end{bmatrix} \tag{10.8}
$$

and the governing equation [Eq. (10.6)] can be written (now dropping the prime)

$$\frac{\partial}{\partial x}\left(k_x \frac{\partial \phi}{\partial x}\right) + \frac{\partial}{\partial y}\left(k_y \frac{\partial \phi}{\partial y}\right) + \frac{\partial}{\partial z}\left(k_z \frac{\partial \phi}{\partial z}\right) + Q = 0 \qquad (10.9)$$

with a suitable change of boundary conditions.

Lastly, for an isotropic material we can write

$$\mathbf{k} = k\mathbf{I} \qquad (10.10)$$

where $\mathbf{I}$ is an identity matrix. This leads to simple form of Eq. (9.10) which was discussed in detail in Chapter 9.

10.2.3 Weak form of general quasi-harmonic equation [Eq. (10.6)]. Following the principles of Chapter 9, Sec. 9.2, we can obtain the weak form of Eq. (10.6) by writing that

$$\int_\Omega v(\nabla^{\mathrm{T}}\mathbf{k}\,\nabla\phi + Q)\,\mathrm{d}\Omega - \int_{\Gamma_q} v[(\mathbf{k}\,\nabla\phi)^{\mathrm{T}}\mathbf{n} - \bar{q} - \alpha\phi]\,\mathrm{d}\Gamma = 0 \quad (10.11)$$

for all functions v which are zero on Γ_ϕ.

Integration by parts (see Appendix 6) will result in the following weak statement which is equivalent to satisfying the governing equations and the *natural* boundary conditions (10.7b):

$$\int_\Omega \nabla^{\mathrm{T}}v\mathbf{k}\,\nabla\phi\,\mathrm{d}\Omega - \int_\Omega vQ\,\mathrm{d}\Omega + \int_{\Gamma_q} v(\alpha\phi + \bar{q})\,\mathrm{d}\Gamma = 0 \qquad (10.12)$$

The *forced* boundary condition (10.7a) still needs to be imposed.

10.2.4 The variational principle. We shall leave as an exercise to the reader the verification that the functional

$$\Pi = \frac{1}{2}\int_\Omega (\nabla\phi)^{\mathrm{T}}\mathbf{k}\,\nabla\phi\,\mathrm{d}\Omega - \int_\Omega Q\phi\,\mathrm{d}\Omega + \frac{1}{2}\int_{\Gamma_q}\alpha\phi^2\,\mathrm{d}\Gamma + \int_{\Gamma_q}\bar{q}\phi\,\mathrm{d}\Gamma \quad (10.13)$$

gives on minimization [subject to the constraint of Eq. (10.7a)] the satisfaction of the original problem set in Eqs (10.6) and (10.7).

The algebraic manipulations required to verify the above principle follow precisely the lines of Sec. 9.10 of Chapter 9 and can be carried out as an exercise.

10.3 Finite element discretization

This can now proceed on the assumption of a trial function expansion

$$\phi = \sum N_i a_i = \mathbf{N}\mathbf{a} \qquad (10.14)$$

using either the weak formulation of Eq. (10.12) or the variational statement of Eq. (10.13). If, in the first, we take

$$v = N_i \qquad (10.15)$$

according to the Galerkin principle, an identical form will arise with that obtained from the minimization of the variational principle.

Substituting thus Eq. (10.15) into (10.12) we have a typical statement giving

$$\left(\int_{\Omega} \mathbf{\nabla}^T N_i \mathbf{k} \, \mathbf{\nabla N} \, d\Omega + \int_{\Gamma_q} N_i \alpha \mathbf{N} \, d\Gamma \right) \mathbf{a} - \int_{\Omega} N_i Q \, d\Omega + \int_{\Gamma_q} N_i \bar{q} \, d\Gamma = 0$$
$$i = 1, \dots, n \qquad (10.16)$$

or a set of standard discrete equations of the form

$$\mathbf{Ha} + \mathbf{f} = 0 \qquad (10.17)$$

with

$$H_{ij} = \int_{\Omega} \mathbf{\nabla}^T N_i \mathbf{k} \, \mathbf{\nabla} N_j \, d\Omega + \int_{\Gamma_q} N_i \alpha N_j \, d\Gamma$$

$$f_i = - \int_{\Omega} N_i Q \, d\Omega + \int_{\Gamma_q} N_i \bar{q} \, d\Gamma$$

on which prescribed values of $\bar{\phi}$ have to be imposed on boundaries Γ_ϕ.

We note now that an additional 'stiffness' is contributed on boundaries for which a radiation constant α is specified but that otherwise a complete analogy with the elastic-structural problem exists.

Indeed in a computer program the same standard operations will be followed even including an evaluation of quantities analogous to the stresses. These, obviously, are the flow rates

$$q \equiv -\mathbf{k} \, \mathbf{\nabla}\phi = -(\mathbf{k} \, \mathbf{\nabla N})\mathbf{a} \qquad (10.18)$$

and, in accordance with indications of Chapter 12, these should be sampled at the optimal (integration) points according to the expansion order used.

Any of the C_0 expansions, isoparametric transformations, etc., given in Chapters 7 and 8 can again be used.

10.4 Some economic specializations

10.4.1 *Anisotropic and non-homogeneous media.* Clearly material properties defined by the $\mathbf{k}$ matrix can vary from element to element in a discontinuous manner. This is implied in both the weak and variational statements of the problem.

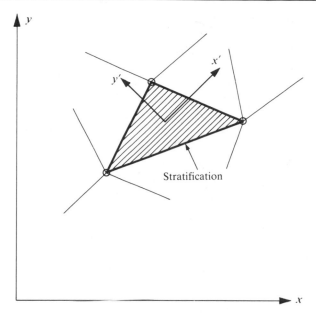

Fig. 10.1 Anisotropic material. Local coordinates coincide with the principal
directions of stratification

The material properties are usually known only with respect to the
principal (or symmetry) axes, and if these directions are constant within
the element it is convenient to use them in the formulation of local axes
specified within each element, as shown in Fig. 10.1.

With respect to such axes only three coefficients k_x, k_y, and k_z need be
specified, and indeed considerable computational economy is achieved as
only a multiplication by a diagonal matrix is needed in formulating the
coefficients of the matrix **H** [Eq. (10.17)].

It is important to note that as the parameters **a** *correspond to scalar
values, no transformation of matrices computed in local coordinates is
necessary before assembly of the global matrices.*

Thus, in most computer programs only a diagonal specification of the
k matrix is used.

10.4.2 *Two-dimensional problem.* The general governing equation (10.9)
specialized to local coordinates becomes, in two dimensions,

$$\frac{\partial}{\partial x}\left(k_x\,\frac{\partial\phi}{\partial x}\right) + \frac{\partial}{\partial y}\left(k_y\,\frac{\partial\phi}{\partial y}\right) + Q = 0 \tag{10.19}$$

On discretization by Eq. (10.16) a slightly simplified form of the
matrices will now be found. Dropping the terms with α and $\bar{q}$ we can

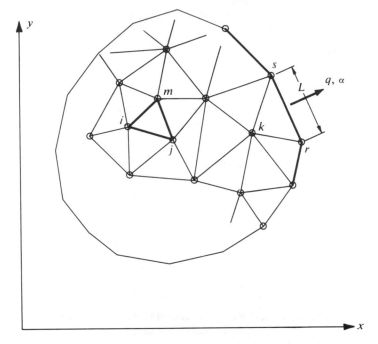

Fig. 10.2   Division of a two-dimensional region into triangular elements

write

$$H_{ij}^e = \int_{V_e} \left(k_x \frac{\partial N_i}{\partial x} \frac{\partial N_j}{\partial x} + k_y \frac{\partial N_i}{\partial y} \frac{\partial N_j}{\partial y} \right) dx \, dy \qquad (10.20)$$

No further discussion at this point appears necessary. However, it may be worth while to particularize here to the most simple yet still useful triangular element (Fig. 10.2).

With

$$N_i = \frac{a_i + b_i x + c_i y}{2\Delta}$$

as in Eq. (3.8) of Chapter 3, we can write down the element 'stiffness' matrix as

$$\mathbf{H}^e = \frac{k_x}{4\Delta} \begin{bmatrix} b_i b_i & b_i b_j & b_i b_m \\ & b_j b_j & b_j b_m \\ \text{symmetric} & & b_m b_m \end{bmatrix} + \frac{k_y}{4\Delta} \begin{bmatrix} c_i c_i & c_i c_j & c_i c_m \\ & c_j c_j & c_j c_m \\ \text{symmetric} & & c_m c_m \end{bmatrix} \qquad (10.21)$$

The load matrices follow a similar simple pattern and thus, for instance, the reader can show that due to Q we have

$$\mathbf{f}^e = -\frac{Q\Delta}{3} \begin{Bmatrix} 1 \\ 1 \\ 1 \end{Bmatrix} \tag{10.22}$$

a very simple (almost 'obvious') result.

Alternatively the equation may be specialized to cylindrical coordinates and used for the solution of axisymmetric situations. Now the differential equation is

$$\frac{\partial}{\partial r}\left(k_r r \frac{\partial \phi}{\partial r}\right) + \frac{\partial}{\partial z}\left(k_z r \frac{\partial \phi}{\partial z}\right) + Q = 0 \tag{10.23}$$

The variational principle could now be again suitably transformed but it is simpler to substitute the values $(k_r r)$ and $(k_z r)$ as modified 'conductivities' and use the previous expressions directly. Integration now will be best carried out numerically as in equivalent problems of Chapter 4.

10.5 Examples—an assessment of accuracy

It is very easy to show that by assembling explicitly worked out 'stiffnesses' of triangular elements for 'regular' meshes shown in Fig. 10.3(a), the discretized equations are *identical* with those that can be derived by well-known finite difference methods.[7]

Obviously the solutions obtained by the two methods will be identical, and so will also be the orders of approximation.†

If an 'irregular' mesh based on a square arrangement of nodes is used a difference between the two approaches will be evident [Fig. 10.3(b)]. This is confined to the 'load' vector $\mathbf{f}^e$. The assembled equations will show 'loads' which differ by small amounts from node to node, but the sum of which is still the same as that due to the finite difference expressions. The solutions therefore differ only locally and will represent the same averages.

In Fig. 10.4 a test comparing the results obtained on an 'irregular' mesh with a relaxation solution of the lowest order finite difference approximation is shown. Both give results of similar accuracy, as indeed would be anticipated. However, it can be shown that in one-dimensional problems the finite element algorithm gives *exact* answers of nodes, while the finite difference method generally does not. In general, therefore, superior accuracy is available with the finite element discretization.

† This is only true in the case where only the boundary values $\bar{\phi}$ are prescribed.

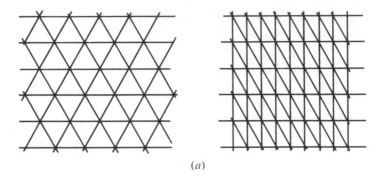

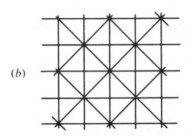

Fig. 10.3   'Regular' and 'irregular' subdivision patterns

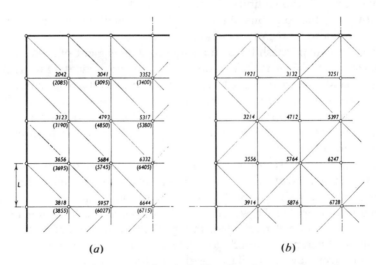

Fig. 10.4   Torsion of a rectangular shaft. Numbers in parentheses show a more accurate solution due to Southwell using a 12 × 16 mesh (values of $\phi/G\theta L^2$)

Further advantages of the finite element process are:

1. It can deal simply with non-homogeneous and anisotropic situations (particularly when the direction of anisotropy is variable).
2. The elements can be graded in shape and size to follow arbitrary boundaries and to allow for regions of rapid variation of the function sought, thus controlling the errors in a most efficient way (viz. Chapter 14).
3. Specified gradient or 'radiation' boundary conditions are introduced naturally and with a better accuracy than in standard finite difference procedures.
4. Higher order elements can be readily used to improve accuracy without complicating boundary conditions—a difficulty always arising with finite difference approximations of a higher order.
5. Finally, but of considerable importance in the computer age, standard (structural) programs may be used for assembly and solution.

Two more sophisticated examples are given at this stage to illustrate the accuracy attainable in practice. The first is the problem of pure torsion of a non-homogeneous shaft illustrated in Fig. 10.5. The basic differential equation here is

$$\frac{\partial}{\partial x}\left(\frac{1}{G}\frac{\partial \phi}{\partial x}\right) + \frac{\partial}{\partial y}\left(\frac{1}{G}\frac{\partial \phi}{\partial y}\right) + 2\theta = 0 \qquad (10.24)$$

in which ϕ is the stress function, G is the shear modulus, and θ the angle of twist per unit length of the shaft.

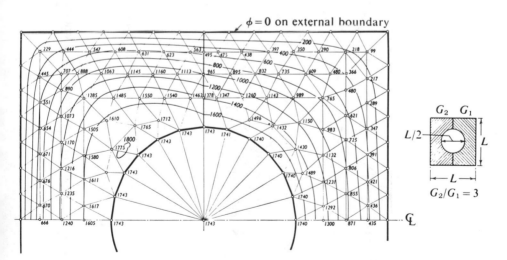

Fig. 10.5 Torsion of a hollow bimetallic shaft. $\phi/G\theta L^2 \times 10^4$

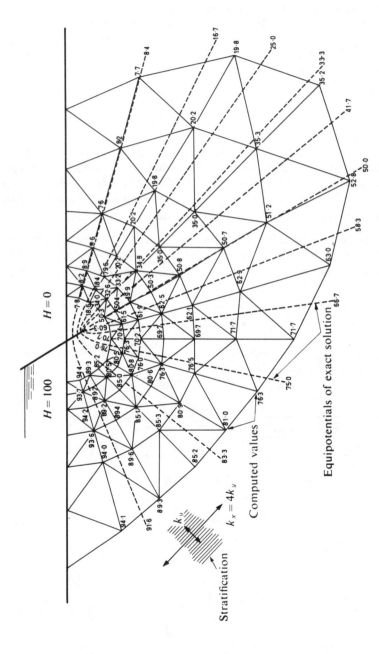

Fig. 10.6 Flow under inclined pile wall in a stratified foundation. A fine mesh near tip of pile is not shown. Comparison with exact solution given by contours

In the finite element solution presented, the hollow section was represented by a material for which G has a value of the order of 10^{-3} compared with the other materials.† The results compare well with the contours derived from an accurate finite difference solution.[8]

An example concerning flow through an anisotropic porous foundation is shown in Fig. 10.6.

Here the governing equation is

$$\frac{\partial}{\partial x}\left(k_x \frac{\partial H}{\partial x}\right) + \frac{\partial}{\partial y}\left(k_y \frac{\partial H}{\partial y}\right) = 0 \qquad (10.25)$$

in which H is the hydraulic head and k_x and k_y represent the permeability coefficients in the direction of the (inclined) principal axes. The answers are here compared against contours derived by an exact solution. The possibilities of the use of a graded size of subdivision are evident in this example.

10.6 Some practical applications

10.6.1 *Anisotropic seepage.* The first of the problems is concerned with the flow through highly non-homogeneous, anisotropic, and contorted strata. The basic governing equation is still Eq. (10.25). However, a special feature has to be incorporated in the computer program to allow for changes of x' and y' principal directions from element to element.

No difficulties are encountered in computation, and the problem together with its solution is given in Fig. 10.7.[3]

17.6.2 *Axisymmetric heat flow.* The axisymmetric heat flow equation can be written in the standard form as

$$\frac{\partial}{\partial r}\left(rk \frac{\partial T}{\partial r}\right) + \frac{\partial}{\partial z}\left(rk \frac{\partial T}{\partial z}\right) = 0 \qquad (10.26)$$

if no heat generation occurs. In the above, T is the temperature and k conductivity. The coordinates x and y are now replaced by r and z, the radial and axial distances.

In Fig. 10.8 the temperature distribution in a nuclear reactor pressure vessel[1] is shown for a steady-state heat conduction when a uniform temperature increase is applied on the inside.

10.6.3 *Hydrodynamic pressures on moving surfaces.* If a submerged surface moves in a fluid with prescribed accelerations and a small amplitude of movement, then it can be shown[9] that if compressibility is

---

† This was done to avoid difficulties due to the 'multiple connection' of the region and to permit the use of a standard program.

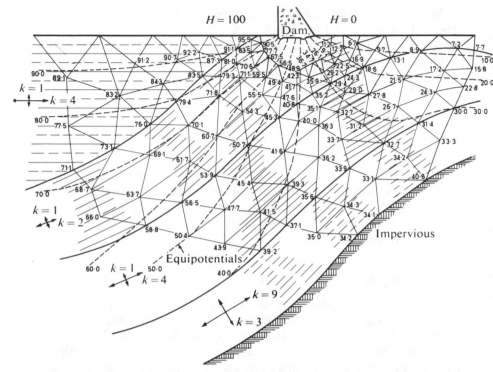

Fig. 10.7   Flow under a dam through a highly non-homogeneous and contorted foundation

ignored the excess pressures that developed obey the Laplace equation

$$\nabla^2 p = 0$$

On moving (or stationary) boundaries the boundary condition is of type 2 [see Eq. (10.7b)] and is given by

$$\frac{\partial p}{\partial n} = -\rho a_n \qquad (10.27)$$

in which ρ is the density of the fluid and a_n is the normal component of acceleration of the boundary.

On free surfaces the boundary condition is (if surface waves are ignored) simply

$$p = 0 \qquad (10.28)$$

The problems clearly therefore come into the category of those discussed in this chapter.

As an example, let us consider the case of a vertical wall in a reservoir, shown in Fig. 10.9, and determine the pressure distribution at points

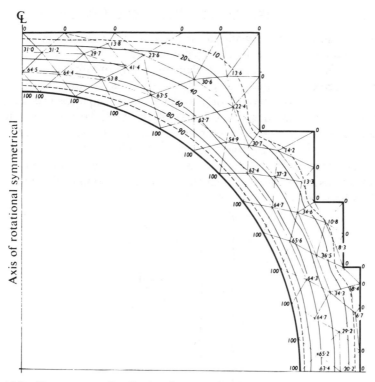

Fig. 10.8 Temperature distribution in a steady-state conduction for an axisymmetrical pressure vessel

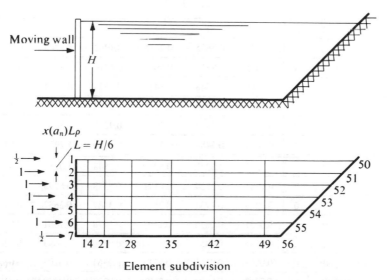

Fig. 10.9 Problem of a wall moving horizontally in a reservoir

273

along the surface of the wall and at the bottom of the reservoir for any prescribed motion of the boundary points 1 to 7.

The division of the region into elements (42 in number) is shown. Here bilinear elements of quadrilateral shape are used. So that results can be made valid for *any* acceleration system, seven separate problems are solved. In each, in turn, the portion of the boundary adjacent to the point in question is given a unit acceleration, resulting in 'loads' $\rho\frac{1}{2}L$, $\rho L, \ldots, \rho L, \rho\frac{1}{2}L$ being applied, in turn, to points 1 to 7. For any arbitrary distribution of acceleration the pressures developed at points 1 to 56 can be listed as a matrix dependent on acceleration of the points 1 to 7. This becomes

$$
\begin{Bmatrix} p_1 \\ \vdots \\ p_7 \\ p_{14} \\ p_{21} \\ p_{28} \\ p_{35} \\ p_{42} \\ p_{49} \\ p_{56} \end{Bmatrix} = \mathbf{M} \begin{Bmatrix} a_1 \\ \vdots \\ a_7 \end{Bmatrix}
\tag{10.29}
$$

in which the matrix $\mathbf{M}$ is given in Table 10.1.

TABLE 10.1

$\mathbf{M} = \rho\,\dfrac{H}{6}$

1	0	0	0	0	0	0	0
2	0	0.7249	0.3685	0.2466	0.1963	0.1743	0.0840
3	0	0.3685	0.9715	0.5648	0.4210	0.3644	0.1744
4	0	0.2466	0.5648	1.1459	0.7329	0.5954	0.2804
5	0	0.1963	0.4210	0.7329	1.3203	0.9292	0.4210
6	0	0.1744	0.3644	0.5954	0.9292	1.5669	0.6489
7	0	0.1680	0.3488	0.5607	0.8420	1.2977	1.1459
14	0	0.1617	0.3332	0.5260	0.7548	1.0285	0.6429
21	0	0.1365	0.2754	0.4171	0.5573	0.6793	0.3710
28	0	0.0879	0.1731	0.2519	0.3187	0.3657	0.1918
35	0	0.0431	0.0838	0.1195	0.1478	0.1661	0.0863
42	0	0.0186	0.0359	0.0150	0.0626	0.0699	0.0362
49	0	0.0078	0.0150	0.0213	0.0261	0.0291	0.0151
56	0	0.0069	0.0134	0.0190	0.0232	0.0259	0.0134

$(L = H/6)$

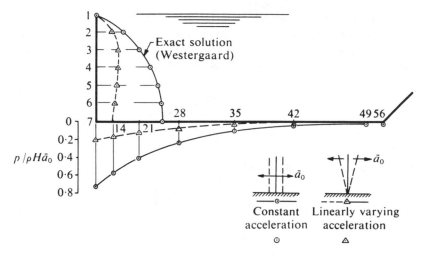

Fig. 10.10   Pressure distribution on moving wall and reservoir bottom

Now for any distribution of accelerations the pressures can be found. For example, if acceleration a is uniform the pressures can be computed taking

$$\begin{Bmatrix} a_1 \\ \vdots \\ a_7 \end{Bmatrix} = \bar{a} \begin{Bmatrix} 1 \\ \vdots \\ 1 \end{Bmatrix} \tag{10.30}$$

The resulting pressure distribution on the wall and the bottom of the reservoir is shown in Fig. 10.10. The results for the pressures on the wall agree to within 1 per cent with the well-known, exact solution derived by Westergaard.[10]

For any other motion the pressures can be similarly derived. If, for instance, the wall is hinged at the base and oscillates around this point with the top (point 1) accelerating by $\bar{a}$, then

$$\begin{Bmatrix} a_1 \\ \vdots \\ a_7 \end{Bmatrix} = \bar{a} \begin{Bmatrix} 1 \\ \frac{5}{6} \\ \frac{4}{6} \\ \vdots \\ 0 \end{Bmatrix} \tag{10.31}$$

Again, the pressure distribution obtained is given by expression, and the results are plotted in Fig. 10.10.

The importance of deriving such an 'influence matrix' is relevant to vibration problems. If the 'wall' oscillates, then in general its acceleration is unknown. From Eq. (10.29) we can write for pressures at points 1 to 7, taking the upper part of matrix $\mathbf{M}$, say $\mathbf{M}_0$,

$$\begin{Bmatrix} p_1 \\ \vdots \\ p_7 \end{Bmatrix} = \mathbf{M}_0 \begin{Bmatrix} a_1 \\ \vdots \\ a_7 \end{Bmatrix} = \mathbf{M}_0 \, \mathbf{a} \qquad (10.32)$$

These pressures result in nodal forces

$$\mathbf{f} = \begin{Bmatrix} \mathbf{f}_1 \\ \vdots \\ \mathbf{f}_7 \end{Bmatrix} = \mathbf{A}\mathbf{M}_0 \begin{Bmatrix} \mathbf{a}_1 \\ \vdots \\ \mathbf{a}_7 \end{Bmatrix} \qquad (10.33)$$

in which $\mathbf{A}$ is a suitable load assignment matrix and $\mathbf{a}$ represents the acceleration of nodal points on the wall. This can be coupled to the dynamic equations of the wall. This and related problems will be discussed in more detail in the second volume of this text.

In Fig. 10.11 the solution of a similar problem in three dimensions is shown.[4] Here simple tetrahedral elements were used and very good accuracy obtained.

In many practical problems the computation of such simplified 'added' masses is sufficient—and the process described here has become widely used in this context.[11-13]

10.6.4 *Electrostatic and magnetostatic problems.* In this area of activity frequent need arises to determine appropriate field strengths and the governing equations are usually of the standard quasi-harmonic type discussed here. Thus the formulations are directly transferable. One of the first applications made as early as 1967[4] was to fully three-dimensional electrostatic field distributions governed by simple Laplace equations (Fig. 10.12).

In Fig. 10.13 a similar use of triangular elements was made in the context of magnetic two-dimensional fields by Winslow[6] in 1966. These early works stimulated a considerable activity in this area and much work has now been published.[14-17]

The magnetic problem is of particular interest as its formulation usually involves the introduction of a *vector potential* with three components which leads to a formulation different from those discussed in this chapter. It is, therefore, worth while to introduce here a variant which allows the standard programs of this section to be utilized for this problem.[18-20]

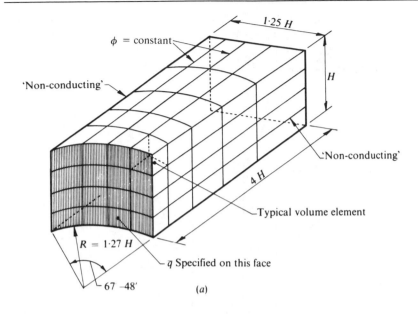

$1.25 H$

$\phi = \text{constant}$

H

'Non-conducting'

'Non-conducting'

$4 H$

Typical volume element

$R = 1.27 H$

q Specified on this face

$67^\circ -48'$

(a)

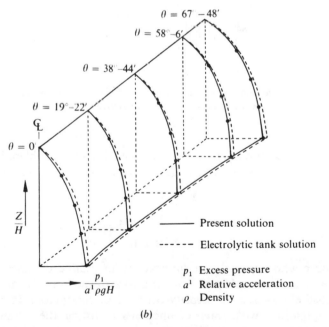

$\theta = 67^\circ -48'$

$\theta = 58^\circ -6'$

$\theta = 38^\circ -44'$

$\theta = 19^\circ -22'$

$\underset{\mathsf{L}}{\mathsf{C}}$

$\theta = 0'$

$\dfrac{Z}{H}$

—— Present solution

------ Electrolytic tank solution

$\dfrac{p_1}{a^1 \rho g H}$

p_1 Excess pressure
a^1 Relative acceleration
ρ Density

(b)

Fig. 10.11 Pressures on an accelerating surface of dam in an incompressible fluid

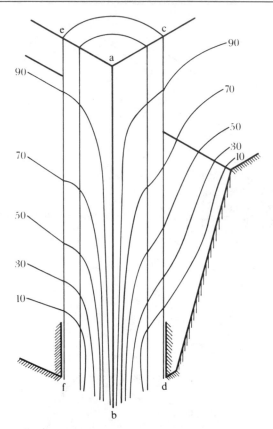

Fig. 10.12 A three-dimensional distribution of electrostatic potential around a porcelain insulator in an earthed trough[9]

In electromagnetic theory for steady-state fields the problem is govern-ed by Maxwell equations which are

$$\mathbf{V}^T \times \mathbf{H} = -\mathbf{J}$$

$$\mathbf{B} = \mu \mathbf{H} \qquad (10.34)$$

$$\mathbf{V}^T \mathbf{B} = 0$$

with the boundary condition specified at an infinite distance from the disturbance, requiring **H** and **B** to tend to zero there. In the above **J** is a prescribed electric current density confined to conductors, **H** and **B** are vector quantities with three components denoting the magnetic field strength and flux density respectively, μ is the magnetic permeability which varies (in an absolute set of units) from unity *in vacuo* to several

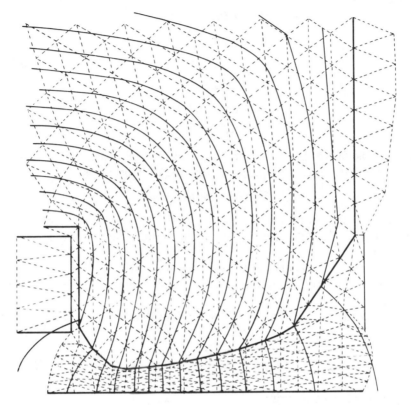

Fig. 10.13   Field near a magnet (after Winslow[6])

thousand in magnetizing materials, and × denotes a vector product, defined in Appendix 5.

The formulation presented here depends on the fact that it is a relatively simple matter to determine a field H_s which exactly solves Eq. (10.34) when $\mu \equiv 1$ everywhere. This is given at any point defined by a vector coordinate r by an integral:

$$H_s = \tfrac{1}{4}\pi \int_\Omega \frac{J \times (r - r')}{(r - r')^2} \, d\Omega \qquad (10.35)$$

In the above, r' refers to the coordinates of $d\Omega$ and obviously the integration domain only involves the electric conductors where $J \neq 0$.

With H_s known we can write

$$H = H_s + H_m$$

and, on substitution into Eq. (17.34), we have a system

$$\mathbf{V}^T \times \mathbf{H}_m = 0$$

$$\mathbf{B} = \mu(\mathbf{H}_s + \mathbf{H}_m) \tag{10.36}$$

$$\mathbf{V}^T\mathbf{B} = 0$$

If we now introduce a *scalar* potential ϕ, defining $\mathbf{H}_m$ as

$$\mathbf{H}_m \equiv \mathbf{V}\phi \tag{10.37}$$

we find the first of Eqs (10.36) to be automatically satisfied and, on eliminating $\mathbf{B}$ in the other two, the governing equation becomes

$$\mathbf{V}^T\mu\mathbf{V}\phi + \mathbf{V}^T\mu\mathbf{H}_s = 0 \tag{10.38}$$

with $\phi \to 0$ at infinity. This is precisely of the standard form discussed in this chapter [Eq. (10.6)] with the second term, which is now specified, replacing Q.

An apparent difficulty exists, however, if μ varies in a discontinuous manner, as indeed we would expect it to do on the interfaces of two materials.

Here the term Q is now undefined and, in the standard discretization of Eq. (10.16) or (10.17), the term

$$\int_\Omega N_i Q \; d\Omega \equiv \int_\Omega N_i \, \mathbf{V}^T\mu\mathbf{H}_s \; d\Omega \tag{10.39}$$

apparently has no meaning.

Integration by parts comes once again to the rescue and we note that

$$\int_\Omega N_i \mathbf{V}^T\mu\mathbf{H}_s \; d\Omega \equiv -\int_\Omega \mathbf{V}^T N_i \, \mu\mathbf{H}_s + \int_\Gamma N_i \mu\mathbf{H}_s \mathbf{n} \; d\Gamma \tag{10.40}$$

As in regions of constant μ, $\mathbf{V}^T\mathbf{H}_s \equiv 0$, the only contribution to the forcing terms comes as a line integral of the second term at discontinuity interfaces.

Introduction of the scalar potential makes both two- and three-dimensional magnetostatic problems solvable by a standard program used for all the problems in this section. Figure 10.14 shows a typical three-dimensional solution for a transformer. Here isoparametric quadratic brick elements were used.[18]

In the typical magnetostatic problems a high non-linearity exists with

$$\mu = \mu(|\mathbf{H}|) \quad \text{where} \quad |\mathbf{H}| = \sqrt{H_x^2 + H_y^2 + H_z^2} \tag{10.41}$$

Treatment of such non-linearities will be discussed in Volume 2.

Considerable economy in this and other problems of infinite extent can be achieved by the use of *infinite* elements discussed in Chapter 8.

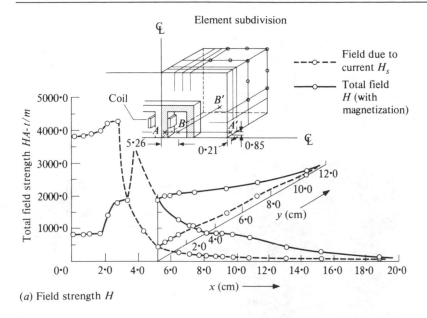

(a) Field strength H

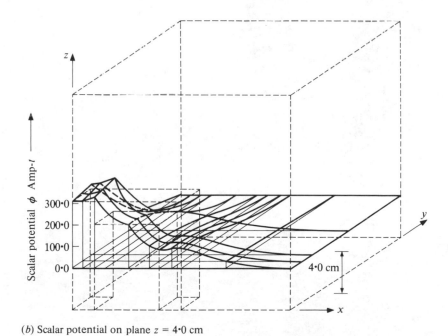

(b) Scalar potential on plane z = 4·0 cm

Fig. 10.14 Three-dimensional transformer

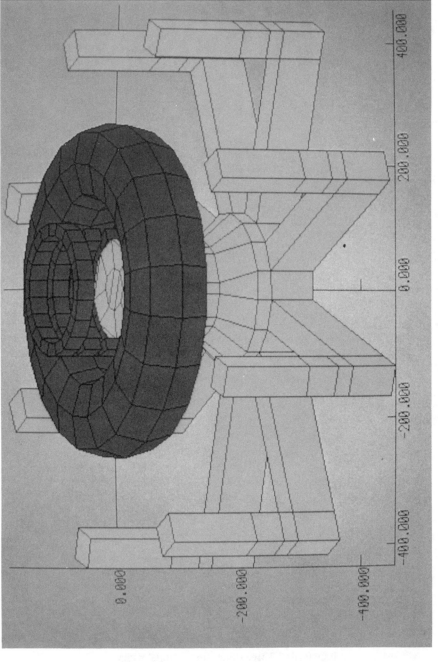

Fig. 10.15 Section of the finite element model of the JET Tokamak (Program TOSCA, courtesy of Vector Fields Ltd, Oxford, UK)

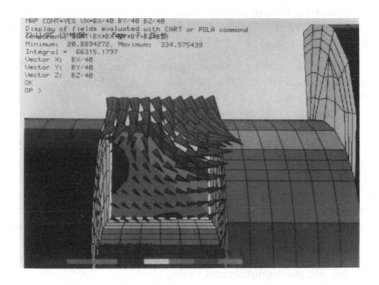

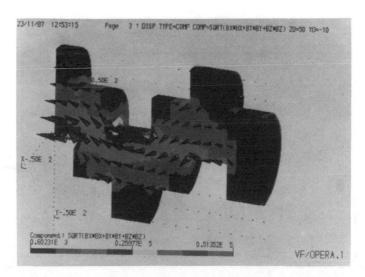

Fig. 10.16 Two views of a camshaft for a magnetic particle inspection application, showing flux density contours and field vectors (Program TOSCA, courtesy of Vector Fields Ltd, Oxford, UK)

Figures 10.15 and 10.16 show more complex applications of procedures outlined above in problems associated with nuclear fusion reactors.[20, 21, 22] Here a two-scalar potential formulation was used. This is computationally somewhat more efficient.[22]

10.6.5 *Lubrication problems.* Once again a standard Poisson type of equation is encountered in the two-dimensional domain of a bearing pad. In the simplest case of constant lubricant density and viscosity the equation to be solved is (Reynolds equation)

$$\frac{\partial}{\partial x}\left(h^3\frac{\partial p}{\partial x}\right) + \frac{\delta}{\partial y}\left(h^3\frac{\partial p}{\partial y}\right) = 6\mu V\frac{\partial h}{\partial x} \qquad (10.42)$$

where h is the film thickness, p the pressure developed, μ the viscosity and V the velocity of the pad in the x direction.

Figure 10.17 shows the pressure distribution in a typical case of a stepped pad.[23] The boundary condition is simply that of zero pressure and it is of interest to note that the step causes an equivalent of a 'line load' on integration of the right-hand side of Eq. (10.42), just as in the case of magnetic discontinuity mentioned above.

More general cases of lubrication problems, including vertical pad movements (squeeze films) and compressibility, can obviously be dealt with, and much work has been done here.[24–31]

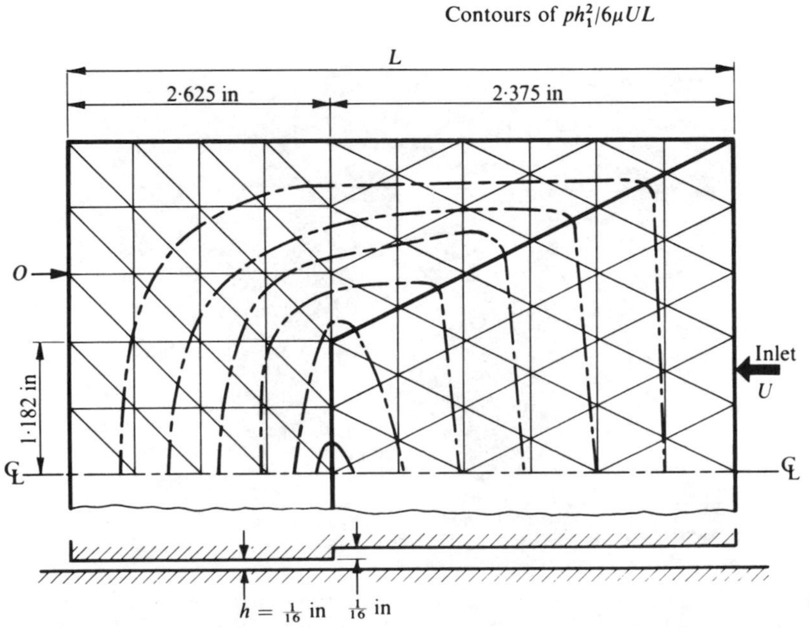

Contours of $ph_1^2/6\mu UL$

Fig. 10.17   A stepped pad bearing. Pressure distribution

Irrotational and free surface flows. The basic Laplace equation which governs the flow of viscous fluid in seepage problems is also applicable in the problem of irrotational fluid flow outside the boundary layer created by viscous effects. The examples already given are adequate to illustrate the general applicability in this context. Further examples are quoted by Martin[32] and others.[33-38]

If no viscous effects exist, then it can be shown that for a fluid starting at rest the motion must be irrotational, i.e.,

$$\omega_z \equiv \frac{\partial u}{\partial y} - \frac{\partial v}{\partial x} = 0, \qquad \text{etc.} \tag{10.43}$$

where u and v are appropriate velocity components.

This implies the existence of a velocity potential, giving

$$u = -\frac{\partial \phi}{\partial x} \qquad v = -\frac{\partial \phi}{\partial y} \tag{10.44}$$

$$(\text{or } \mathbf{u} = -\nabla \phi)$$

If, further, the flow is incompressible the continuity equation [see Eq. (10.2)] has to be satisfied, i.e.,

$$\nabla^T u = 0 \tag{10.45}$$

and therefore

$$\nabla^T \nabla \phi = 0 \tag{10.46}$$

Alternatively, for two-dimensional flow a stream function may be introduced defining the velocities as

$$u = -\frac{\partial \psi}{\partial y} \qquad v = \frac{\partial \psi}{\partial x} \tag{10.47}$$

and this identically satisfies the continuity equation. The irrotationality condition now must ensure that

$$\nabla^T \nabla \psi = 0 \tag{10.48}$$

and thus problems of ideal fluid flow can be posed in one form or the other. As the standard formulation is again applicable, there is little more that needs to be added, and for examples the reader can well consult the literature quoted.

The similarity with problems of seepage flow, which has already been discussed, is obvious.[39,40]

A particular class of fluid flow deserves mention. This is the case when free surface limits the extent of the flow and this surface is not known *a priori*.

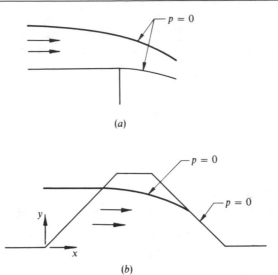

Fig. 10.18   Typical free surface problems with a streamline also satisfying an additional condition of pressure = 0. (*a*) Jet overflow. (*b*) Seepage through an earth dam

The class of problem is typified by two examples—that of a freely overflowing jet [Fig. 10.18(*a*)] and that of a flow through an earth dam [Fig. 10.18(*b*)]. In both, the free surface represents a streamline and in both the position of the free surface is unknown *a priori* but has to be determined so that an *additional condition* on this surface is satisfied. For instance, in the second problem, if formulated in terms of potential H, Eq. (10.25) governs the problem.

The free surface, being a streamline, imposes the condition

$$\frac{\partial H}{\partial n} = 0 \qquad (10.49)$$

to be satisfied there.

In addition, however, the pressure must be zero on the surface as this is exposed to atmosphere. As

$$H = \frac{p}{\gamma} + y \qquad (10.50)$$

where γ is the fluid specific weight, p is the fluid pressure, and y elevation above some (horizontal) datum, we must have on the surface

$$H = y \qquad (10.51)$$

The solution may be approached iteratively. Starting with a prescribed free surface streamline the standard problem is solved. A check is carried out to see if Eq. (10.51) is satisfied and, if not, an adjustment of the surface is carried out to make new y equal to H just found. A few iterations of this kind show that convergence is reasonably rapid. Taylor and Brown[41] show such a process. Alternative methods including special variational principles for dealing with this problem have been devised over the years and interested readers can consult references 42–50.

10.7 Concluding remarks

We have shown how a general formulation for the solution of a steady-state quasi-harmonic problem can be written—and how a single program of such a form can be applied to a wide variety of physical situations. Indeed, the selection of problems dealt with is by no means exhaustive and many other examples of application are of practical interest. Readers will doubtless find appropriate analogies for their own problems.

References

1. O. C. ZIENKIEWICZ and Y. K. CHEUNG, 'Finite elements in the solution of field problems', *The Engineer*, 507–10, Sept. 1965.
2. W. VISSER, 'A finite element method for the determination of non-stationary temperature distribution and thermal deformations', *Proc. Conf. on Matrix Methods in Structural Mechanics*, Air Force Inst. Tech., Wright-Patterson AF Base, Ohio, 1965.
3. O. C. ZIENKIEWICZ, P. MAYER, and Y. K. CHEUNG, 'Solution of anisotropic seepage problems by finite elements', *Proc. Am. Soc. Civ. Eng.*, **92**, EM1, 111–20, 1966.
4. O. C. ZIENKIEWICZ, P. L. ARLETT, and A. K. BAHRANI, 'Solution of three-dimensional field problems by the finite element method', *The Engineer*, 27 October 1967.
5. L. R. HERRMANN, 'Elastic torsion analysis of irregular shapes', *Proc. Am. Soc. Civ. Eng.*, **91**, EM6, 11–19, 1965.
6. A. M. WINSLOW, 'Numerical solution of the quasi-linear Poisson equation in a non-uniform triangle "mesh"', *J. Comp. Phys.*, **1**, 149–72, 1966.
7. D. N. DE G. ALLEN, *Relaxation Methods*, p. 199, McGraw-Hill, 1955.
8. J. F. ELY and O. C. ZIENKIEWICZ, 'Torsion of compound bars—a relaxation solution', *Int. J. Mech. Sci.*, **1**, 356–65, 1960.
9. O. C. ZIENKIEWICZ and B. NATH, 'Earthquake hydrodynamic pressures on arch dams—an electric analogue solution', *Proc. Inst. Civ. Eng.*, **25**, 165–76, 1963.
10. H. M. WESTERGAARD, 'Water pressure on dams during earthquakes', *Trans. Am. Soc. Civ. Eng.*, **98**, 418–33, 1933.

11. O. C. ZIENKIEWICZ and R. E. NEWTON, 'Coupled vibrations of a structure submerged in a compressible fluid', *Proc. Symp. on Finite Element Techniques*, pp. 359–71, Stuttgart, 1969.
12. R. E. NEWTON, 'Finite element analysis of two-dimensional added mass and damping', in *Finite Elements in Fluids* (eds R. H. Gallagher, J. T. Oden, C. Taylor, and O. C. Zienkiewicz), Vol. I, pp. 219–32, Wiley, 1975.
13. P. A. A. BACK, A. C. CASSELL, R. DUNGAR, and R. T. SEVERN, 'The seismic study of a double curvature dam', *Prov. Inst. Civ. Eng.*, **43**, 217–48, 1969.
14. P. SILVESTER and M. V. K. CHARI, 'Non-linear magnetic field analysis of D.C. machines', *Trans. IEEE*, No. 7, 5–89, 1970.
15. P. SILVESTER and M. S. HSIEH, 'Finite element solution of two dimensional exterior field problems', *Proc. IEEE*, **118**, 1971.
16. B. H. MCDONALD and A. WEXLER, 'Finite element solution of unbounded field problems', *Proc. IEEE*, MTT-20, No. 12, 1972.
17. E. MUNRO, 'Computer design of electron lenses by the finite element method', in *Image Processing and Computer Aided Design in Electron Optics*, p. 284, Academic Press, 1973.
18. O. C. ZIENKIEWICZ, J. F. LYNESS, and D. R. J. OWEN, 'Three dimensional magnetic field determination using a scalar potential. A finite element solution, *IEEE, Trans. Magnetics MAG*, **13**, 1649–56, 1977.
19. J. SIMKIN and C. W. TROWBRIDGE, 'On the use of the total scalar potential in the numerical solution of field problems in electromagnets', *Int. J. Num. Meth. Eng.*, **14**, 423–40, 1979.
20. J. SIMKIN and C. W. TROWBRIDGE, 'Three-dimensional non-linear electromagnetic field computations using scalar potentials', *Proc. Inst. Elec. Eng.*, **127**, B(6), 1980.
21. C. W. TROWBRIDGE, 'Low frequency electromagnetic field computations in flow simulation', *Comp. Mech. Appl. Meth. Eng.*, **52**, 653–79, 1985.
22. M. M. SUSSMAN, 'Remarks on computational magnetostatics', *Int. J. Num. Meth. Eng.*, **26**, 987–1000, 1988.
23. D. V. TANESA and I. C. RAO, *Student project report on lubrication*, Royal Naval College, Dartmouth, 1966.
24. M. M. REDDI, 'Finite element solution of the incompressible lubrication problem', *Trans. Am. Soc. Mech. Eng.*, **91** (Ser. F), 524, 1969.
25. M. M. REDDI and T. Y. CHU, 'Finite element solution of the steady state compressible lubrication problem', *Trans. Am. Soc. Mech. Eng.*, **92** (Ser. F), 495, 1970.
26. J. H. ARGYRIS and D. W. SCHARPF, 'The incompressible lubrication problem', *J. Roy. Aero. Soc.*, **73**, 1044–6, 1969.
27. J. F. BOOKER and K. H. HUEBNER, 'Application of finite element methods to lubrication: an engineering approach', *J. Lubr. Techn., Trans. Am. Soc. Mech. Eng.*, **14** (Ser. F), 313, 1972.
28. K. H. HUEBNER, 'Application of finite element methods to thermohydrodynamic lubrication', *Int. J. Num. Meth. Eng.*, **8**, 139–68, 1974.
29. S. M. ROHDE and K. P. OH, 'Higher order finite element methods for the solution of compressible porous bearing problems', *Int. J. Num. Meth. Eng.*, **9**, 903–12, 1975.
30. A. K. TIEU, 'Oil film temperature distributions in an infinitely wide glider bearing: an application of the finite element method', *J. Mech. Eng. Sci.*, **15**, 311, 1973.
31. K. H. HUEBNER, 'Finite element analysis of fluid film lubrication—a survey', in *Finite Elements in Fluids* (eds R. H. Gallagher, J. T. Oden, C. Taylor, and

O. C. Zienkiewicz), Vol. II, pp. 225–54, Wiley, 1975.

32. H. C. Martin, 'Finite element analysis of fluid flows', *Proc. 2nd Conf. on Matrix Methods in Structural Mechanics*, Air Force Inst. Tech., Wright-Patterson AF Base, Ohio, 1968.

33. G. de Vries and D. H. Norrie, *Application of the finite element technique to potential flow problems*, Reports 7 and 8, Dept. Mech. Eng., Univ. of Calgary, Alberta, Canada, 1969.

34. J. H. Argyris, G. Mareczek, and D. W. Scharpf, 'Two and three dimensional flow using finite elements', *J. Roy. Aero. Soc.*, **73**, 961–4, 1969.

35. L. J. Doctors, 'An application of finite element technique to boundary value problems of potential flow', *Int. J. Num. Meth. Eng.*, **2**, 243–52, 1970.

36. G. de Vries and D. H. Norrie, 'The application of the finite element technique to potential flow problems', *J. Appl. Mech., Am. Soc. Mech. Eng.*, **38**, 978–802, 1971.

37. S. T. K. Chan, B. E. Larock, and L. R. Herrmann, 'Free surface ideal fluid flows by finite elements', *Proc. Am. J. Civ. Eng.*, **99**, HY6, 1973.

38. B. E. Larock, 'Jets from two dimensional symmetric nozzles of arbitrary shape', *J. Fluid Mech.*, **37**, 479–83, 1969.

39. C. S. Desai, 'Finite element methods for flow in porous media', in *Finite Elements in Fluids* (ed. R. H. Gallagher), Vol. 1, pp. 157–82, Wiley, 1975.

40. I. Javandel and P. A. Witherspoon, 'Applications of the finite element method to transient flow in porous media', *Trans. Soc. Petrol. Eng.*, **243**, 241–51, 1968.

41. R. L. Taylor and C. B. Brown, 'Darcy flow solutions with a free surface', *Proc. Am. Soc. Civ. Eng.*, **93**, HY2, 25–33, 1967.

42. J. C. Luke, 'A variational principle for a fluid with a free surface', *J. Fluid Mech.*, **27**, 395–7, 1957.

43. K. Washizu, *Variational Methods in Elasticity and Plasticity*, 2nd ed., Pergamon Press, 1975.

44. J. C. Bruch, 'A survey of free-boundary value problems in the theory of fluid flow through porous media', *Advances in Water Resources*, **3**, 65–80, 1980.

45. C. Baiocchi, V. Comincioli, and V. Maione, 'Unconfined flow through porous media', '*Meccanice*', *Ital. Ass. Theor. Appl. Mech.*, **10**, 51–60, 1975.

46. J. M. Sloss and J. C. Bruch, 'Free surface seepage problem', *Proc. ASCE*, **108**, EM5, 1099–1111, 1978.

47. N. Kikuchi, 'Seepage flow problems by variational inequalities', *Int. J. Num. Anal. Meth. geomech.*, **1**, 283–90, 1977.

48. C. S. Desai, 'Finite element residual schemes for unconfined flow', *Int. J. Num. Meth. Eng.*, **10**, 1415–18, 1976.

49. C. S. Desai and G. C. Li, 'A residual flow procedure and application for free surface, and porous media', *Advances in Water Resources*, **6**, 27–40, 1983.

50. K. J. Bathe and M. Koshgoftar, 'Finite elements from surface seepage analysis without mesh iteration', *Int. J. Num. Anal. Meth. geomech.*, **3**, 13–22, 1979.

11

The patch test, reduced integration, and non-conforming elements

11.1 Introduction

We have briefly referred in Chapter 2 to the patch test as a means of assessing convergence of displacement-type elements for elasticity problems in which the shape functions violate continuity requirements. In this chapter we shall deal in more detail with this test which is applicable to all finite element forms and will show that

(a) it is a *necessary* condition for assessing the convergence of any finite element approximation and further that, if properly extended and interpreted, it can provide

(b) a *sufficient* requirement for convergence,

(c) an assessment of the (asymptotic) convergence rate of the element tested,

(d) a check on the robustness of the algorithm, and

(e) a means of developing new and accurate finite element forms which violate compatibility (continuity) requirements.

While for elements which *a priori* satisfy all the continuity requirements, have correct polynomial expansions, and are exactly integrated such a test is superfluous in principle, it is nevertheless useful as its gives

(f) a check that correct programming was achieved.

For all the reasons cited above the patch test has been, since its inception, and continues to be the most important check for practical finite element codes.

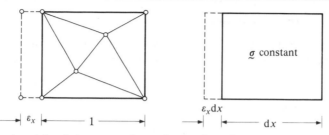

Fig. 11.1 A patch of element and a volume of continuum subject to constant strain ε_x. A physical interpretation of the constant strain or linear displacement field patch test

The original test was introduced by Irons[1-3] in a physical way and could be interpreted as a check which ascertained whether a patch of elements (Fig. 11.1) subject to a constant strain reproduced exactly the constitutive behaviour of the material and resulted in correct stresses when it became infinitesimally small. If it did, it could then be argued that the finite element model represented the real material behaviour and, in the limit, as the size of the elements decreased would therefore reproduce exactly the behaviour of the real structure.

Clearly, although this test would only have to be passed when the size of the element patch became infinitesimal, for most elements in which polynomials are used the patch size did not in fact enter the consideration and the requirement that the patch test be passed for any element size became standard.

Quite obviously a rigid body displacement of the patch would cause no strain, and if the proper constitutive laws were reproduced no stress changes would result. The patch test thus guarantees that no rigid body motion straining will occur.

When curvilinear coordinates are used the patch test still requires to be passed in the limit but generally will not do so for a finite size of the patch. (An exception here is the isoparametric coordinate system in problems discussed in Chapter 8.) Thus for many problems such as shells, where local curvilinear coordinates are used, this test has to be restricted to infinitesimal patch sizes and, on physical grounds alone, appears to be a *necessary and sufficient condition* for convergence.

Numerous publications on the theory and practice of the test have followed the original publications cited[4-6] and mathematical respectability was added to those by Strang.[7,8] Although some authors have cast doubts on its validity[9,10] these have been fully refuted[11,12] and if the test is used as described here it fulfills the requirements (a) to (d) stated above.

11.2 Convergence requirements

We shall consider in the following the patch test as applied to a finite element solution of a set of differential equations

$$A(u) \equiv L(u) + g = 0 \tag{11.1}$$

in the domain Ω together with the conditions

$$B(u) = 0 \tag{11.2}$$

on the boundary of the domain, Γ.

The finite element approximation is given in the form

$$u \approx \hat{u} = \mathbf{N}\mathbf{a} \tag{11.3}$$

where $\mathbf{N}$ are shape functions defined in each element, Ω_e, and $\mathbf{a}$ are unknown parameters.

By applying standard procedures of finite element approximation (viz. Chapters 2 and 9) the problem reduces in a linear case to a set of algebraic equations

$$\mathbf{K}\mathbf{a} = \mathbf{f} \tag{11.4}$$

which when solved give an approximation to the differential equation and its boundary conditions.

What is meant by 'convergence' in the approximation sense is that the approximate solution, $\hat{u}$, should tend to the exact solution u when the size of the elements h approaches zero (with some specified subdivision pattern). Stated mathematically we must find that the error at any point becomes (when h is sufficiently small)

$$|u - \hat{u}| = O(h^q) \leqslant Ch^q \tag{11.5}$$

where $q > 0$ and C is a positive constant, depending on the position.

This must also be true for all the derivatives of u defined in the approximation.

By the order of convergence in the variable u we mean the value of the index q in the above definition.

To ensure convergence it is necessary that the approximation fulfil both consistency and stability conditions.[13]

The *consistency requirement* ensures that as the size of the elements h tends to zero, the approximation equation (11.4) will represent the exact differential equation (11.1) and the boundary conditions (11.2) (at least in the weak sense).

The *stability condition* is simply translated as a requirement that the solution of the discrete equation system (11.4) be unique and avoid spurious mechanisms which may pollute the solution for all sizes of elements. For linear problems in which we solve the system of algebraic equations (11.4) as

$$a = K^{-1}f \tag{11.6}$$

this means simply that the matrix K must be non-singular for all possible element assemblies. However, this requirement may on occasion be too stringent if, for instance, an iterative solution is adopted.

The patch test traditionally has been used as a procedure for verifying the consistency requirement; the stability was checked independently by ensuring non-singularity of matrices.[14] Further, it generally tested only the consistency in satisfaction of the differential equation (11.1) but not of its natural boundary conditions. In what follows we shall show how all the necessary requirements of convergence can be tested by a properly conceived patch test.

A 'weak' singularity of a single element may on occasion be permissible and some elements exhibiting it have been, and still are, successfully used in practice. One such case is given by the eight-node isoparametric element with a 2×2 Gauss quadrature, to which we shall refer later here. This element is on occasion observed to show peculiar behaviour (though its use has many advantages discussed in Chapter 12). An element that occasionally fails is termed *non-robust* and the patch test provides a means of assessing the degree of robustness.

11.3 The simple patch test (forms A and B)—a necessary condition for convergence

We shall first consider the consistency condition which requires that in the limit (as h tends to zero) the finite element approximation of Eq. (11.4) should model exactly the differential equation (11.1) and the boundary conditions (11.2). If we consider a 'small' region of the domain (of size $2h$) we can expand the unknown function u and the essential derivatives entering the weak approximation in a Taylor series. From this we conclude that for convergence of the function and its first derivative in typical problems of a second-order equation and two dimensions, we require that around a point i assumed to be at coordinate origin,

$$u = u_i + \left(\frac{\partial u}{\partial x}\right)_i x + \left(\frac{\partial u}{\partial y}\right)_i y + O(h^p)$$

$$\frac{\partial u}{\partial x} = \left(\frac{\partial u}{\partial x}\right)_i + O(h^{p-1}) \tag{11.7}$$

$$\frac{\partial u}{\partial y} = \left(\frac{\partial u}{\partial y}\right)_i + O(h^{p-1})$$

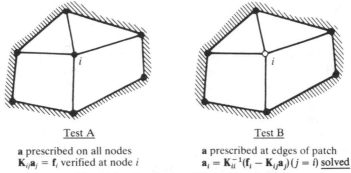

Test A | Test B

a prescribed on all nodes
$K_{ij}a_j = f_i$ verified at node i

a prescribed at edges of patch
$a_i = K_{ii}^{-1}(f_i - K_{ij}a_j)(j = i)$ solved

Fig. 11.2   Patch test of forms A and B

with $p \geqslant 2$. The finite element approximation should therefore reproduce exactly the problem posed for *any linear forms* of u as h tends to zero. Similar conditions can obviously be written for higher order problems. This requirement is tested by the current interpretation of the patch test illustrated in Fig. 11.2.

In this we compute first an arbitrary linear solution of the differential equation and the corresponding set of parameters **a** [viz. Eq. (11.3)] at all 'nodes' of a *patch* which assembles completely the nodal variable a_i (i.e., provides all the equation terms corresponding to it).

In *test A* we simply insert the exact value of the parameters **a** into the ith equation and verify that

$$K_{ij}a_j - f_i \equiv 0 \tag{11.8}$$

In *test B* only the values of **a** corresponding to the boundaries of the 'patch' are inserted and a_i is found as

$$a_i = K_{ii}^{-1}(f_i - K_{ij}a_j) \qquad j \neq i \tag{11.9}$$

and compared against the exact value.

Both patch tests verify only the satisfaction of the basic differential equation and not of the boundary approximations, as these have been explicitly excluded here.

We mentioned earlier that the test is, in principle, required only for an infinitesimally small patch of elements; however, for differential equations with constant coefficients and with a mapping involving constant jacobian the size of the patch is immaterial and the test can be carried out on a patch of arbitrary dimensions.

Indeed, if the coefficients are not constant the same size independence exists providing that a constant set of such coefficients is used in the formulation of the test. (This applies, for instance, in axisymmetric problems where coefficients of the type 1/radius enter the equations and when

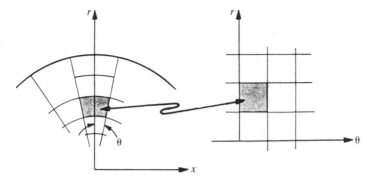

Fig. 11.3 Polar coordinate mapping

the patch test is here applied, it is simply necessary to enter the computation with such quantities assumed constant.)

If mapped, curvilinear, elements are used it is not obvious that the patch test posed in global coordinates needs to be satisfied. Here, in general, convergence in the mapping coordinates may exist but a finite patch test may not be satisfied. However, once again if we specify the nature of the subdivision without changing the mapping function, in the limit the jacobian becomes locally constant and the previous remarks apply. To illustrate this point consider, for instance, a set of elements in which local coordinates are simply the polar coordinates as shown in Fig. 11.3. With shape functions using polynomial expansions in the $r - \theta$ terms the patch test of the kind we have described above will not be satisfied with elements of finite size—nevertheless in the limit as the element size tends to zero it will become true. Thus it is evident that the patch test satisfaction is a *necessary condition* which has always to be achieved *providing the size of the patch is infinitesimal.*

This proviso which we shall call *weak patch test satisfaction* is not always simple to verify, particularly if the element coding does not easily permit the insertion of constant coefficients or jacobian. In Sec. 11.9 we shall discuss in some detail its implementation, which, however, is only necessary in very special element forms. It is indeed fortunate that the standard isoparametric element form reproduces exactly the linear polynomial in global coordinates (viz. Chapter 8) and for this reason does not require the special treatment unless some other *crime* (such as selective or reduced integration) is introduced.

11.4 Generalized patch test (test C) and the single-element test

The patch test described in the preceding section was shown to be a necessary condition for convergence of the formulation but did not

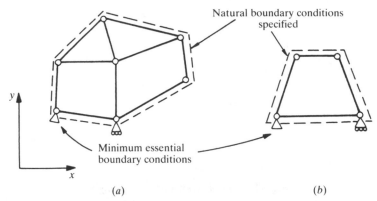

Fig. 11.4   (a) Patch test of form C. (b) The single-element test

establish sufficient conditions for it. In particular, it omitted the testing of the boundary 'load' approximation for the case when the 'natural' (e.g. 'traction of elasticity') conditions are specified. Further it did not verify the stability of the approximation. A test including a check on the above conditions is easily constructed. We show this in Fig. 11.4 as the *test C*. In this the patch of elements is assembled as before but subject to prescribed natural boundary conditions (or tractions around its perimeter). The assembled matrix of the whole patch is written as

$$\mathbf{Ka} = \mathbf{f}$$

Fixing only the minimum number of parameters **a** necessary to obtain a physically valid solution (e.g., eliminating the rigid body motion in an elasticity example or a single value of temperature in a heat conduction problem) a solution is sought for remaining **a** values and compared with the exact basic solution assumed.

Now any singularity of the **K** matrix will be immediately observed and, as the vector **f** includes all necessary source and boundary traction terms, the formulation will be completely tested (providing of course a sufficient number of test states are used). The test described is now not only *necessary* but *sufficient* for convergence.

With boundary traction included it is of course possible to reduce the size of the patch to a single element and an alternative form of test C is illustrated in Fig. 11.4(b), which is termed the single-element test.[11] This test is indeed one requirement of good finite element formulation as, on occasion, a larger patch may not reveal the inherent instabilities of a single element. This happens in the well-documented case of the plane strain–stress eight-noded isoparametric element with (reduced) four-point Gauss integration (i.e., where the singular deformation mode of a single element (viz. Fig. 11.5) disappears when several elements are

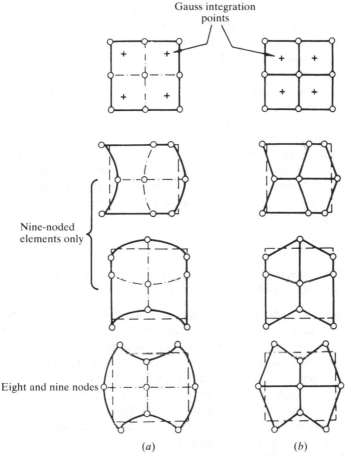

Fig. 11.5 (a) Zero energy (singular) modes for eight- and nine-noded quadratic elements and (b) for a patch of bilinear elements with single integration points

assembled.† *It should be noted, however, that satisfaction of a single element test is not a sufficient condition for convergence. For sufficiency we require at least one internal element boundary to test that consistency of a patch solution is maintained between elements.*

11.5 Higher order patch tests[6,8]

While the patch tests discussed in the last two sections ensure (when satisfied) that convergence will occur, they did not test the order of this

---

† This figure shows also similar singularity for a patch of four bilinear elements with single-point quadrature, and we note the similar shape of zero energy modes (viz. Chapter 8, Sec. 8.11.3).

convergence, beyond assuring us that in the case of Eq. (11.7) the errors were, at least, of order $O(h^2)$ in u. It is an easy matter to determine the actual highest asymptotic rate of convergence of a given element by simply imposing instead of a linear solution exact higher order polynomial solutions. The highest value of such polynomials for which complete satisfaction of the patch test is achieved automatically evaluates the corresponding convergence rate. It goes without saying that for such exact solutions generally non-zero source terms in the original equation (11.1) will need to be involved.

In addition, test C in conjunction with a higher order patch test may be used to illustrate any tendency for 'locking' to occur (see Chapter 12). Accordingly, element robustness with regard to various parameters (e.g., Poisson's ratios near one-half for elasticity problems in plane strain) may be established.

11.6 Application of the patch test to plane elasticity elements with 'standard' and 'reduced' quadrature

In the next few sections we consider several applications of the patch test in the evaluation of finite element models. In each case we consider only one of the necessary tests which need to be implemented. For a complete evaluation of a formulation it is necessary to consider all possible independent essential polynomial solutions as well as a variety of patch configurations which test the effects of element distortion or alternative meshing interconnections which will be commonly used in analysis. As we shall emphasize, it is important that both consistency and stability be evaluated in a properly conducted test.

In Chapter 8 (Sec. 8.11) we have discussed the minimum required order of numerical integration for various finite element problems which results in no loss of convergence rate. However, it was also shown that for some problems such a minimum integration order must result in singular matrices. If we define the *standard* integration as one which evaluates the stiffness of an element exactly (at least in the undistorted form) then any lower order of integration is generally called *reduced*.

Such *reduced* integration has some merits in certain problems for reasons which we shall discuss in the next chapter, but it can cause singularities which should be discovered by a patch test (which supplements and verifies the arguments of Sec. 8.11.3). Application of the patch test to some typical problems will now be shown.

We consider first a plane stress problem on the patch shown in Fig. 11.6(a). The material is linear, isotropic elastic with properties $E = 1000$ and $v = 0.3$. The finite element procedure used is based on the

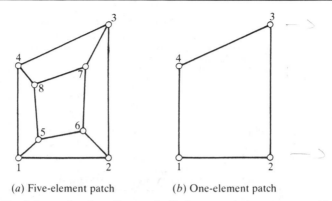

(a) Five-element patch   (b) One-element patch

Fig. 11.6   Patch for evaluation of numerically integrated plane stress problems

displacement form using four-noded isoparametric shape functions, and numerical integration. Analyses are conducted using the plane element and program described in Chapter 15. Since the stiffness computation includes only first derivatives of displacements, the formulation converges provided that the patch test is satisfied for all linear polynomial solutions of displacements. Here we consider only one of the six independent linear polynomial solutions necessary to verify satisfaction of the patch test. The solution considered is

$$u = 0.002x$$
$$v = -0.0006y \qquad (11.10)$$

which produces zero body forces and zero stresses except for

$$\sigma_x = 2 \qquad (11.11)$$

The solution given in Table 11.1 is obtained for the nodal displacements and satisfies Eq. (11.10) exactly.

The patch test is performed first using 2×2 gaussian, 'standard', quadrature to compute each element stiffness and resulting reaction forces at nodes. For patch test A all nodes are restrained and nodal displacement values are specified according to Table 11.1. Stresses are computed at specified Gauss points (1×1, 2×2, and 3×3 Gauss points were sampled) and all are exact to within round-off error (64 bit precision was used on a VAX 11/750 which produced round-off errors less than 10^{-15} in the quantities computed). Reactions were also computed at all nodes and again produced the force values shown in Table 11.1 to within round-off limits. This above approximation satisfies all conditions required for a finite element procedure (i.e., conforming shape functions and normal-order quadrature). Accordingly, the patch

TABLE 11.1
PATCH SOLUTION FOR FIG. 11.6

Node i	Coordinates		Computed displacements		Forces	
	x_i	y_i	u_i	v_i	F_{x_i}	F_{y_i}
1	0.0	0.0	0.0	0.0	−2	0
2	2.0	0.0	0.0040	0.0	3	0
3	2.0	3.0	0.0040	−0.00186	2	0
4	0.0	2.0	0.0	−0.00120	−3	0
5	0.4	0.4	0.0008	−0.00024	0	0
6	1.4	0.6	0.0028	−0.00036	0	0
7	1.5	2.0	0.0030	−0.00120	0	0
8	0.3	1.6	0.0006	−0.00096	0	0

test merely verifies that the programming steps used contain no errors. Patch test A does not require explicit use of the stiffness matrix to compute results; consequently the above patch test was repeated using patch test B where only nodes 1 to 4 are restrained with their displacements specified according to Table 11.1. This tests accuracy of the stiffness matrix and, as expected, exact results are once again recovered to within round-off errors (i.e., errors of order 10^{-15}). Finally, patch test C was performed with node 1 fully restrained and node 4 restrained only in the x direction. Nodal forces were applied to nodes 2 and 3 in accordance with the values generated through the boundary tractions by σ_x (i.e., nodal forces shown in Table 11.1). This test also produced exact solutions for all other nodal quantities in Table 11.1 and recovered σ_x of 2 at all Gauss points in each element.

The above test was repeated for patch tests A, B, and C but using a 1×1 'reduced' Gauss quadrature to compute the element stiffness and nodal force quantities. Patch test C indicated that the global stiffness matrix contained two global 'zero energy modes' (i.e., the global stiffness matrix was rank deficient by 2), thus producing incorrect nodal displacements whose results depend solely on the round-off errors in the calculations. These in turn produced incorrect stresses except at the 1×1 Gauss point used in each element to compute the stiffness and forces. Thus, based upon stability considerations, the use of 1×1 quadrature on four-noded elements produces a failure in the patch test. The element does satisfy consistency requirements, however, and provided a proper stabilization scheme is employed (e.g., stiffness or viscous methods are used in practice) this element may be used for practical calculations.[15,16]

It should be noted that a one-element patch test may be performed using the mesh shown in Fig. 11.6(b). The results are given by nodes 1 to 4 in Table 11.1. For the one-element patch, patch tests A and B coincide and neither evaluates the accuracy or stability of the stiffness matrix. On

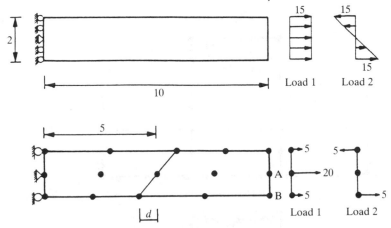

Fig. 11.7   Patch test for eight- and nine-noded isoparametric quadrilaterals

the other hand, patch test C leads to the conclusions reached using the five-element patch: namely, 2×2 gaussian quadrature passes a patch test whereas 1×1 quadrature fails the stability part of the test (as indeed we would expect by arguments of Chapter 8, Sec. 8.11).

A simple test on cancellation of a diagonal during the triangular decomposition step is sufficient to warn of rank deficiencies in the stiffness matrix. In a profile method, described in Chapter 15, this is easily monitored as compact elimination converts the initial value of a diagonal element to the final value in one step. Thus only one extra scalar variable is needed to test the initial and final values.

In Fig. 11.7 we consider a two-element patch of quadratic, isoparametric, quadrilaterals. Both eight-noded serendipity and nine-noded lagrangian types are considered and a basic patch test type C is performed for load case 1. For the eight-noded element both 2×2 ('reduced') and 3×3 ('standard') gaussian quadrature satisfy the patch test, whereas for the nine-noded element only 3×3 quadrature is satisfactory, with 2×2 reduced quadrature leading to failure in rank of the stiffness matrix. However, if we perform a one-element test for the eight-noded and 2×2 quadrature element, we discover the spurious zero-energy mode shown in Fig. 11.5 and thus the one-element test is failed. We consider such elements suspect and to be used only with the greatest of care. To illustrate what can happen in practice we consider the simple problem shown in Fig. 11.8(a). In this example the 'structure' modelled by a single element is considered rigid and interest is centred on the 'foundation' response. Accordingly only one element is used to model the structure. Use of 2×2 quadrature throughout leads to answers

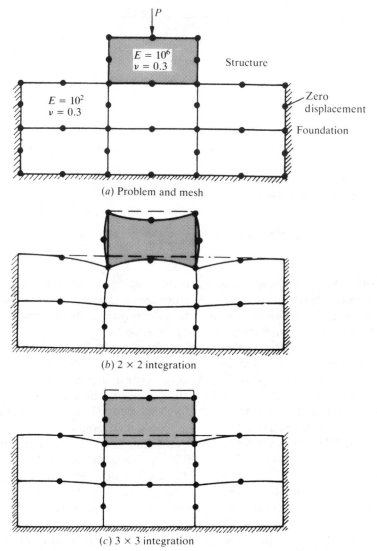

(a) Problem and mesh

(b) 2 × 2 integration

(c) 3 × 3 integration

Fig. 11.8   A propagating spurious mode from a single unsatisfactory element

shown in Fig. 11.8(b) while results for 3 × 3 quadrature are shown in Fig. 11.8(c). It should be noted that no zero-energy mode exists since more than one element is used. There is, however, here a spurious response due to the large modulus variation between structure and foundation. This suggests that problems in which non-linear response may lead to a large variation in material parameters could also induce such

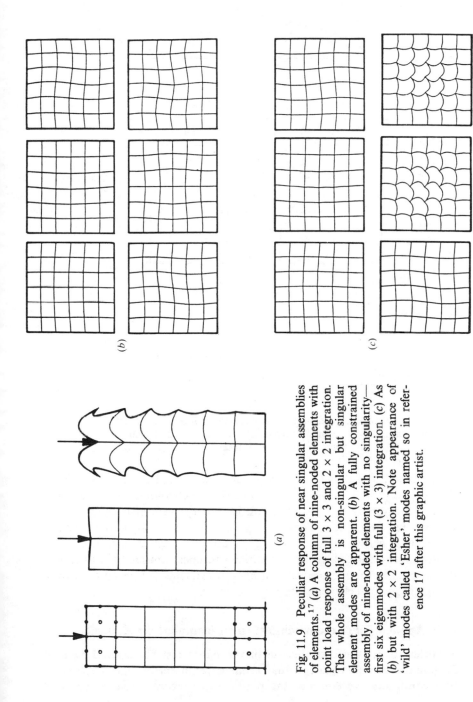

Fig. 11.9 Peculiar response of near singular assemblies of elements.[17] (a) A column of nine-noded elements with point load response of full 3 × 3 and 2 × 2 integration. The whole assembly is non-singular but singular element modes are apparent. (b) A fully constrained assembly of nine-noded elements with no singularity—first six eigenmodes with full (3 × 3) integration. (c) As (b) but with 2 × 2 integration. Note appearance of 'wild' modes called 'Esher' modes named so in reference 17 after this graphic artist.

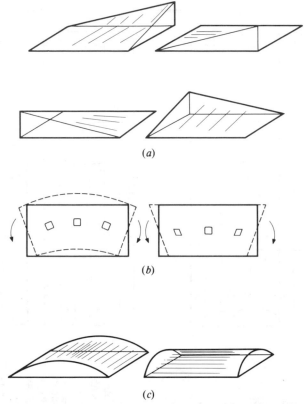

Fig. 11.10 (a) Linear quadrilateral with auxiliary incompatible shape functions; (b) pure bending and linear displacements causing shear; (c) auxiliary 'bending' shape functions with internal variables

performance, and thus use of the eight-noded 2×2 integrated element should always be closely monitored to detect such anomalous behaviour.

Indeed, support or loading conditions may themselves induce very suspect responses for elements in which near singularity occurs. Figure 11.9 shows some amusing peculiarities which can occur for reduced integration elements and which disappear entirely if full integration is used.[17] In all cases the *assembly* of elements is non-singular even though individual elements are rank deficient.

11.7 Application of the patch test to an incompatible element

In order to demonstrate the use of the patch test for a finite element formulation which violates usually stated requirements for shape function continuity, we consider the plane strain incompatible modes first

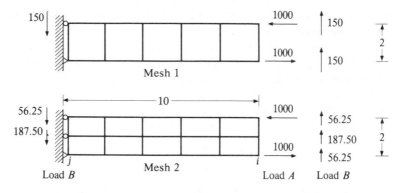

Fig. 11.11   Performance of the non-conforming quadrilateral in beam bending treated as plane stress: (a) conforming linear quadrilateral, (b) non-conforming quadrilateral

		Displacement at i		Bending stress at j	
		Load A	Load B	Load A	Load B
	Beam theory	10.00	103.0	300.0	4050
(a)	Mesh 1	6.81	70.1	218.2	2945
	Mesh 2	7.06	72.3	218.8	2954
(b)	Mesh 1	10.00	101.5	300.0	4050
	Mesh 2	10.00	101.3	300.0	4050

introduced by Wilson et al.[18] and discussed further by Taylor et al.[19] The specific incompatible formulation considered uses the element displacement approximations:

$$\hat{\mathbf{u}} = N_I(\xi, \eta)\mathbf{a}_I + (1 - \xi^2)\boldsymbol{\alpha}_1 + (1 - \eta^2)\boldsymbol{\alpha}_2 \qquad (11.12)$$

where $N_I(I = 1 - 4)$ are the usual conforming bilinear shape functions and the last two terms are incompatible modes of deformation defined as hierarchical functions and independently for each element.

The shape functions used are illustrated in Fig. 11.10. The first, a set of standard Lagrange type, gives a displacement pattern which, as shown in Fig. 11.10(b), introduces shear strains even in pure bending. The second, in which the parameters α_1 and α_2 are strictly associated with a specific element, therefore introduces incompatibility but assures correct bending behaviour. The excellent performance of this element in the bending situation is illustrated in Fig. 11.11.

In reference 19 the finite element approximation is computed by summing the potential energies of each element and computing the nodal loads due to boundary tractions from the conforming part of the displacement field only. Thus for the purposes of conducting patch tests we compute the strains using all parts of the displacement field leading to a generalization of (11.4) which may be written as

$$\begin{bmatrix} \mathbf{K}_{11} & \mathbf{K}_{12} \\ \mathbf{K}_{21} & \mathbf{K}_{22} \end{bmatrix} \begin{Bmatrix} \mathbf{a} \\ \alpha \end{Bmatrix} = \begin{Bmatrix} \mathbf{f}_1 \\ \mathbf{f}_2 \end{Bmatrix} \qquad (11.13)$$

Here $\mathbf{K}_{11}$ and $\mathbf{f}_1$ are the stiffness and loads of the four-noded (conforming) bilinear element, $\mathbf{K}_{12}$ and $\mathbf{K}_{21}$ ($=\mathbf{K}_{12}^{\mathrm{T}}$) are coupling stiffnesses between the conforming and non-conforming displacements, and $\mathbf{K}_{22}$ and $\mathbf{f}_2$ are the stiffness and loads of the non-conforming displacements. We note that, according to the algorithm of reference 18, $\mathbf{f}_2$ must vanish from the patch test solutions.

For a patch test in plane strain or plane stress, only linear polynomials need be considered for which all non-conforming displacements must vanish. Thus for a successful patch test we must have

$$\mathbf{K}_{11}\mathbf{a} = \mathbf{f}_1 \qquad (11.14a)$$

and

$$\mathbf{K}_{21}\mathbf{a} - \mathbf{f}_2 = 0 \qquad (11.14b)$$

If we carry out a patch test for the mesh shown in Fig. 11.12(a) we find that all three forms (i.e., patch tests A, B, and C) satisfy these conditions and thus pass the patch test. If we consider the patch shown in Fig. 11.12(b), however, the patch test is not satisfied. The lack of satisfaction shows up in different ways for each form of the patch test. Patch test A produces non-zero $\mathbf{f}_2$ values when α is set to zero and $\mathbf{a}$ according to the displacements considered. In form B the value of the nodal displacements $\mathbf{a}_5$ are in error and α are non-zero, leading also to erroneous stresses in each element. In form C all unspecified displacements are in error as well as stresses.

It is interesting to note that when a patch is constructed according to Fig. 11.12(c) in which all elements are parallelograms all three forms of the patch test are once again satisfied. Accordingly we can note that if any mesh is systematically refined by subdivision of each element into four elements whose sides are all along ξ, η lines in the original element with values of -1, 0, or 1 (i.e., by bisections) the mesh converges to constant jacobian approximations of the type shown in Fig. 11.12(e). Thus, in this special case the incompatible mode element satisfies a weak

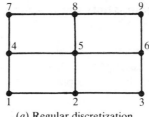

(a) Regular discretization

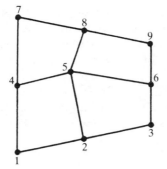

(b) Irregular discretization about node 5

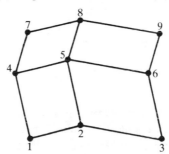

(c) Constant jacobian discretization
about node 5

Fig. 11.12   Patch test for an incompatible element form

patch test and thus will converge. In general, however, it may be necessary to use a very fine discretization to achieve sufficient accuracy, and thus the element probably has no practical (nor efficient) engineering use.

The simple artifice of ensuring that the jacobian is taken as constant in each element is to evaluate it only once at $\eta = \xi = 0$ (the element centre). This ensures convergence for all element shapes—and with this alteration of the algorithm the incompatible element proves convergent though not extremely accurate.[19]

11.8 Generation of incompatible shape functions which satisfy the patch test

In the previous section we have shown how an incompatible element can, on occasion, produce superior results despite its violation of the rules generally postulated. In solving plates and shells we deal with problems requiring C_1 continuity; the use of such incompatible functions is widespread not only because these produce superior results but also due to the difficulty of developing functions which satisfy not only the continuity of the functions but also their slope. In this section we address the problem of how to generate incompatible shape functions in a manner that will automatically ensure the satisfaction of the patch test and hence convergence. The rules for doing this have been recently developed[20,21] and applied to derivation of new plate bending elements. We derive these rules here in a simple example of a second-order partial differential equation problem but the results are easy to generalize to other situations.

Consider the finite element solution to the following equation:

$$A(u) \equiv -T\nabla^2 u + ku - q = 0 \qquad \text{in the domain } \Omega \qquad (11.15)$$

with boundary conditions

$$u = \bar{u} \qquad \text{on } \Gamma_u$$

and $\qquad\qquad\qquad\qquad\qquad\qquad\qquad\qquad\qquad\qquad\qquad$ (11.16)

$$T\frac{\partial u}{\partial n} = \bar{t} \qquad \text{on } \Gamma_t$$

This may represent the displacement u of elastic membrane on an elastic foundation with an initial tension T and spring constant k. Let the unknown u be approximated by two sets of (hierarchic) expansions

$$u = u^c + u^n \qquad (11.17a)$$

$$u^c = \mathbf{N}^c \mathbf{a}^c \qquad \text{and} \qquad u^n = \mathbf{N}^n \mathbf{a}^n \qquad (11.17b)$$

in which $\mathbf{N}^c$ and $\mathbf{N}^n$ are respectively compatible and incompatible shape functions. It must be stressed that these are linearly independent as otherwise stability conditions (i.e., the non-singularity of matrices) would be violated as was the case in the counterexample of Stummel.[9]

When a patch of elements is subject to a linear variation of u such that Eq. (11.15) is satisfied, the approximation u^c is capable of yielding this solution and satisfying all the patch test requirements. (Now, of course, $q = -ku$ has to be assumed.)

It follows therefore that in the patch test u^n will be zero. However, it is important to consider here a single element test in which the constant traction $\bar{t}$ (corresponding to $u^c = u$) is applied. The Galerkin equation corresponding to the incompatible mode yields now

$$\int_{\Gamma_e} N_i^n k \frac{\partial u}{\partial n} \, d\Gamma \equiv \int_{\Gamma_e} N_i^n k \left(n_x \frac{\partial u}{\partial x} + n_y \frac{\partial u}{\partial y} \right) = \int_{\Gamma_e} N_i^n \bar{t} \, d\Gamma \qquad (11.18)$$

and this equation has to be satisfied identically with $\bar{t}$, k, and $\partial u / \partial n$ being constants. In the above Γ_e represents the total boundary of the element and n_x; n_y and $\mathbf{n}$ are the boundary normal vectors (viz. Appendix 6).

The above condition can be easily achieved by ensuring that

$$\int_{\Gamma_e} N_i^n \mathbf{n} \, d\Gamma = 0 \qquad (11.19)$$

A similar condition has been called an 'interpolation test' by Specht[21] and also introduced earlier by Samuelsson.[20] Using this a non-conforming triangular plate bending element has been developed which satisfies the patch test and also demonstrates excellent results in sample applications.

In the present example we can use alternatively a weaker patch requirement stating that

$$\int_{\Gamma_e} N_i^n n_x \, d\Gamma = 0 \qquad (11.20a)$$

and

$$\int_{\Gamma_e} N_i^n n_y \, d\Gamma = 0 \qquad (11.20b)$$

for each element, thus imposing the constraint

$$\int_{\Gamma_e} N_i^n \bar{t} \, d\Gamma = 0 \qquad (11.21)$$

which implies (as originally suggested by Wilson) that boundary loads from the incompatible displacements must vanish or be ignored. On the other hand, nodal contribution from body forces (i.e., q) should be computed using both the conforming and non-conforming parts of the displacements.

In order to illustrate the use of the above procedure in developing incompatible mode shape functions, we consider the case of a non-conforming four-noded quadrilateral element which in the special case of a rectangle reproduces the non-conforming element of reference 18. The convergence of this non-conforming element for the rectangular or constant jacobian case has been illustrated in the previous section.

We take the conforming part of the shape function for each displacement component as the four-noded isoparametric functions

$$u^c = N_I u_I \qquad (11.22)$$

where

$$N_I = \tfrac{1}{4}(1 + \xi_I \xi)(1 + \eta_I \eta) \qquad (11.23)$$

and ξ, η are natural coordinates on the interval $(-1, 1)$ with values at each corner node I given by ξ_I, η_I. The non-conforming functions will be constructed from the remaining four shape functions for the eight-noded isoparametric serendipity element (Chapter 8). Accordingly we take for the non-conforming field

$$u^n = \tfrac{1}{2}(1 - \xi^2)(1 - \eta)\alpha_1 + \tfrac{1}{2}(1 + \xi)(1 - \eta^2)\alpha_2$$
$$+ \tfrac{1}{2}(1 - \xi^2)(1 + \eta)\alpha_3 + \tfrac{1}{2}(1 - \xi)(1 + \eta^2)\alpha_4 \qquad (11.24)$$

Substitution into the constraint conditions (11.20) yields the two scalar conditions

$$\sum_{i=1}^{4} a_i \alpha_i = 0 \qquad (11.25)$$

and

$$\sum_{i=1}^{4} b_i \alpha_i = 0 \qquad (11.26)$$

where a_i, b_i depend on the geometry of the element through

$$a_i = x_i - x_j$$

and

$$b_i = y_j - y_i \qquad (11.27)$$

with

$$j = \mathrm{mod}\,(i, 4) + 1$$

The two constraint conditions may be used to express two of the α_i in terms of the other two. The result gives two incompatible displacement modes which may be added to the conforming field with the satisfaction of a strong patch test still ensured. For elements which are rectangular the two resulting modes are identical to those proposed and used in Eq. (11.12).

Other possibilities exist for constructing non-conforming or incompatible functions.[5]

11.9 The weak patch test—example

The problems described above yield exact solutions for the patch tests performed and accordingly satisfy strong conditions. In order to illustrate the performance of an element which only satisfies a weak patch test we consider an axisymmetric linear elastic problem modelled by four-noded isoparametric elements. The material is assumed isotropic and the finite element stiffness and reaction force matrices are computed using a selective integration method where terms associated with the bulk modulus are evaluated by a single-point Gauss quadrature, whereas all other terms are computed using a 2×2 (normal) gaussian quadrature. It may be readily verified that the stiffness matrix is of proper rank and thus stability of solutions is not an issue. On the other hand, consistency must still be evaluated.

In order to assess the performance of a selective reduced quadrature formulation we consider the patch of elements shown in Fig. 11.13. The patch is not as generally shaped as desirable and is only used to illustrate performance of an element that satisfies a weak patch test. The polynomial solution considered is

$$u = 2r$$
$$w = 0$$
(11.28)

and material constants $E = 1$ and $v = 0$ are used in the analysis. The resulting stress field is given by

$$\sigma_r = \sigma_\theta = 2$$
(11.29)

with other components identically zero. The exact solution for the nodal quantities of the mesh shown in Fig. 11.13 are summarized in Table 11.2. Patch tests have been performed for this problem using the selective reduced integration scheme described above and values of h of 0.8, 0.4, 0.2, 0.1, and 0.05. The result for the radial displacement at nodes 2 and 5

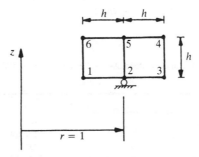

Fig. 11.13   Patch for selective, reduced quadrature on axisymmetric four-noded elements

TABLE 11.2
EXACT SOLUTION FOR PATCH

Node	Radius	Displacement		Force	
I	r_I	U_I	W_I	F_{rI}	F_{zI}
1, 4	$1 - h$	$2(1 - h)$	0	$-(1 - h)h$	0
2, 5	1	2	0	0	0
3, 6	$1 + h$	$2(1 + h)$	0	$(1 + h)h$	0

(reported to six digits) is given in Table 11.3. All other quantities (displacements, strains, and stresses) have a similar performance with convergence rates of at least $O(h)$ or more.

11.10 Higher order patch test example—robustness

In order to demonstrate a higher order patch test we consider the two-element plane stress problem shown in Fig. 11.7 and subjected to bending loading as shown. As above, two different types of elements are considered: (*a*) an eight-noded serendipity quadrilateral element and (*b*) a nine-noded lagrangian quadrilateral element. In our test we wish to demonstrate a feature for nine-noded element mapping discussed in Chapter 8 (viz. Sec. 8.7) and first shown by Wachspress.[22] In particular we restrict the mapping into the xy plane to be that produced by the four-noded isoparametric bilinear element, but permit the dependent variable to assume the full range of variations permitted by the eight- or nine-noded shape functions. In Chapter 8 we showed that the nine-noded element can approximate a complete quadratic displacement function in x, y whereas the eight-noded element cannot. Thus we expect that the nine-noded element when restricted to the isoparametric mappings of the four-noded element will pass a higher order patch test for all arbitrary quadratic displacement fields. The pure bending solution in elasticity is composed of polynomial terms up to quadratic order.

TABLE 11.3
RADIAL DISPLACEMENT AT NODES 2 AND 5

h	u
0.8	2.01114
0.4	2.00049
0.2	2.00003
0.1	2.00000
0.05	2.00000

TABLE 11.4
BENDING LOAD CASE ($E = 100$, $v = 0.3$)

Element	Quadrature	d	v_A	u_B	v_B
Eight-node	3×3		0.750	0.150	0.75225
Eight-node	2×2	0	0.750	0.150	0.75225
Nine-node	3×3		0.750	0.150	0.75225
Eight-node	3×3		0.7448	0.1490	0.74572
Eight-node	2×2	1	0.750	0.150	0.75100
Nine-node	3×3		0.750	0.150	0.75225
Eight-node	3×3		0.6684	0.1333	0.66364
Eight-node	2×2	2	0.750	0.150	0.75225
Nine-node	3×3		0.750	0.150	0.75225
Exact	—	—	0.750	0.150	0.75225

Furthermore, no body force loadings are necessary to satisfy the equilibrium equations. For the mesh considered the nodal loadings are equal and opposite on the top and bottom nodes as shown. The results for the two elements are shown in Table 11.4 for the indicated quadratures with $E = 100$ and $v = 0.3$.

From this test we observe that the nine-noded element does pass the higher order test performed. Indeed, provided the mapping is restricted to the four-noded shape it will always pass a patch test for displacements with terms no higher than quadratic. On the other hand, the eight-noded element passes the higher order patch test performed only for rectangular element (or constant jacobian) mappings. Moreover, the accuracy of the eight-noded element deteriorates very rapidly with increased distortions defined by the parameter d.

The use of 2×2 reduced quadrature improves results for the higher order patch test performed. Indeed, two of the points sampled give exact results and the third is only slightly in error. As noted previously, however, a single element test for the 2×2 integrated eight-noded element will fail the stability part of the patch test and it should thus be used with great care.

The use of a higher order patch test may also be used to assess element 'robustness'. An element is termed robust if its performance is not sensitive to physical parameters of the differential equation. For example, the performance of many elements for solution of plane strain linear elasticity problems is sensitive to Poisson's ratio values near 0.5 (called 'near incompressibility'). Indeed, for Poisson's ratios near 0.5 the energy stored by a unit volumetric strain is many orders larger than the energy stored by a unit deviatoric strain. Accordingly finite elements which exhibit a strong coupling between volumetric and deviatoric strains often produce poor results in the nearly incompressible range, a problem discussed further in Chapter 12.

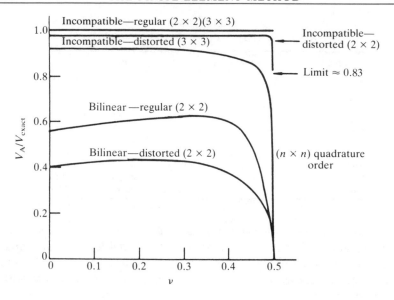

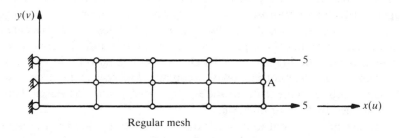

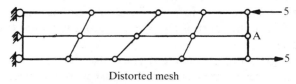

Fig. 11.14   Plane strain four-noded quadrilaterals with and without incompatible modes (higher order patch test for performance evaluation)

This may be observed using a four-noded element to solve a problem with a quadratic displacement field (i.e., a higher order patch test). If we again consider the pure bending example and an eight-element mesh shown in Fig. 11.14 we can clearly observe the deterioration of results as Poisson's ratio approaches a value of one-half. Also shown in Fig. 11.14 are results for the incompatible modes derived in Sec. 11.8. It is evident

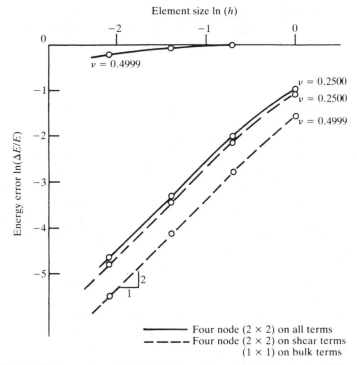

Fig. 11.15   Higher order patch test on element robustness (see Fig. 11.14) (convergence test under subdivision of elements)

that the response is considerably improved by adding these modes, especially if 2×2 quadrature is used.

If we consider the regular mesh and four-noded elements and further keep the domain constant and successively refine the problem using meshes of 8, 32, 128, and 512 elements, we observe that the answers do converge as guaranteed by the patch test. However, as shown in Fig. 11.15, the rate of convergence in energy for Poisson's ratio values of 0.25 and 0.4999 is quite different. For 0.25 the rate of convergence is nearly a straight line for all meshes, whereas for 0.4999 the rate starts out quite low and approaches an asymptotic value of 2 as h tends towards zero. For v near 0.25 the element is called robust, whereas for v near 0.5 it is not. If we use selective reduced integration (which for plane strain passes strong patch tests) and repeat the experiment, both values of v produce a similar response and thus the element becomes robust for all values of Poisson's ratio less than 0.5.

The use of higher order patch tests can thus be very important to separate robust elements from non-robust elements. For methods which seek to automatically refine a mesh adaptively in regions with high errors, as discussed in Chapter 14, it is extremely important to use robust elements.

11.11 Closure

In the preceding sections we have described the patch test and its use in practice by considering several example problems. The patch test described has two essential parts: (a) a consistency evaluation and (b) a stability check. In the consistency test a set of linearly independent essential polynomials (i.e., all independent terms up to the order needed to describe the finite element model) is used as a solution to the differential equations and boundary conditions, and in the limit as the size of a patch tends to zero the finite element model must exactly satisfy each solution. We presented three forms to perform this portion of the test which we call forms A, B, and C.

The use of form C, where all boundary conditions are the natural ones (e.g., tractions for elasticity) except for the minimum number of essential conditions needed to ensure a unique solution to the problem (e.g., rigid body modes for elasticity), is recommended to test consistency and stability simultaneously. Both one-element and more-than-one-element tests are necessary to ensure that the patch test is satisfied. With these conditions and assuming that the solution procedure used can detect any possible rank deficiencies the stability of solution is also tested. If no such condition is included in the program a stability test must be conducted independently. This can be performed by computing the number of zero eigenvalues in the coefficient matrix for methods that use a solution of linear equations to compute the finite element parameters, $\mathbf{a}$. Alternatively, the loading used for the patch solution may be perturbed at one point by a small value (say square root of the round-off limit—e.g., by 10^{-6} for round-offs of order 10^{-12}) and the solution tested to ensure that it does not change by a large amount.

Once an element has been shown to pass all of the essential patch tests for both consistency and stability, convergence is assured as the size of elements tends to zero. However, in some situations (e.g., the nearly incompressible elastic problem) convergence may be very slow until a very large number of elements is used. Accordingly, we recommend that higher order patch tests be used to establish element robustness. Higher order patch tests involve the use of polynomial solutions of the differential equation and boundary conditions with the order of terms larger than the basic polynomials used in a patch test. Indeed, the order of polynomials used should be increased until the patch test is satisfied only in a weak sense (i.e., as h trends to zero). The advantage of using a higher order patch test, as opposed to other boundary value problems, is that the exact solution may be easily computed everywhere in the model.

We have tested in some of the examples the use of incompatible function and inexact numerical integration procedures (reduced and selective integration). Some of these violations of the rules previously stipulated have proved justified not only by yielding improved performance but by providing methods for which convergence is guaranteed. We shall discuss in the next chapter (Chapter 12) some of the reasons for such improved performance.

References

1. B. M. IRONS, 'Numerical integration applied to finite element methods', *Conf. on Use of Digital Computers in Structural Engineering*, Univ. of Newcastle, 1966.
2. G. P. BAZELEY, Y. K. CHEUNG, B. M. IRONS, and O. C. ZIENKIEWICZ, 'Triangular elements in plate bending. Conforming and nonconforming solutions', *Proc. 1st Conf. on Matrix Methods in Structural Mechanics*, pp. 547–76, AFFDLTR-CC-80, Wright-Patterson AF Base, Ohio, 1966.
3. B. M. IRONS and A. RAZZAQUE, 'Experience with the patch test for convergence of finite element method', in *Mathematical Foundations of the Finite Element Method* (ed. A. K. Aziz), pp. 557–87, Academic Press, 1972.
4. B. FRAEIJS DE VEUBEKE, 'Variational principles and the patch test', *Int. J. Num. Meth. Eng.*, **8**, 783–801, 1974.
5. G. SANDER and P. BECKERS, 'The influence of the choice of connectors in the finite element method', *Int. J. Num. Meth. Eng.*, **11**, 1491–505, 1977.
6. E. R. DE ARANTES OLIVEIRA, 'The patch test and the general convergence criteria of the finite element method', *Int. J. Solids Struct.*, **13**, 159–78, 1977.
7. G. STRANG, 'Variational crimes and the finite element method', in *Proc. Foundations of the Finite Element Method* (ed. A. K. Aziz), pp. 689–710, Academic Press, 1972.
8. G. STRANG and G. J. FIX, *An Analysis of the Finite Element Method*, Prentice-Hall, 1973.
9. F. STUMMEL, 'The limitations of the patch test', *Int. J. Num. Meth. Eng.*, **15**, 177–88, 1980.
10. J. ROBINSON *et al.*, 'Correspondence on patch test', *Finite Element News*, **1**, 30–4, 1982.
11. R. L. TAYLOR, O. C. ZIENKIEWICZ, J. C. SIMO, and A. H. C. CHAN, 'The patch test—a condition for assessing f.e.m. convergence', *Int. J. Num. Meth. Eng.*, **22**, 39–62, 1986.
12. R. E. GRIFFITHS and A. R. MITCHELL, 'Non-conforming elements', in *Mathematical Basis of Finite Element Methods*, Inst. Math. and Appl. Conference series, pp. 41–69, Clarendon Press, Oxford, 1984.
13. A. RALSTON, *A First Course in Numerical Analysis*, McGraw-Hill, New York, 1965.
14. B. M. IRONS and S. AHMAD, *Techniques of Finite Elements*, Horwood, Chichester, 1980.
15. D. KOSLOFF and G. A. FRASIER, 'Treatment of hour glass patterns in low order finite element codes', *Int. J. Num. Anal. Meth. Geomechanics*, **2**, 57–72, 1978.

16. T. BELYTCHKO and W. E. BACHRACH, 'The efficient implimentation of quadrilaterals with high coarse mesh accuracy', *Comp. Meth. Appl. Mech. Eng.*, **54**, 276–301, 1986.
17. N. BIĈANIĈ and E. HINTON, 'Spurious modes in two dimensional isoparametric elements', *Int. J. Num. Meth. Eng.*, **14**, 1545–57, 1979.
18. E. L. WILSON, R. L. TAYLOR, W. P. DOHERTY, and J. GHABOUSSI, 'Incompatible displacement models', in *Num. and Comp. Meth. in Struct. Mech.* (eds S. T. Fenves *et al.*), pp. 43–57, Academic Press, 1973.
19. R. L. TAYLOR, P. J. BERESFORD, and E. L. WILSON, 'A non-conforming element for stress analysis', *Int. J. Num. Meth. Eng.*, **10**, 1211–20, 1976.
20. A. SAMUELSSON, 'The global constant strain condition and the patch test', Chapter 3 of *Energy Methods in Finite Element Methods* (eds R. Glowinski, E. Y. Rodin, and O. C. Zienkiewicz), pp. 47–58, Wiley, 1979.
21. B. SPECHT, 'Modified shape functions for the three-node plate bending element passing the patch test', *Int. J. Num. Mech. Eng.*, **26**, 705–15, 1988.
22. E. L. WACHSPRESS, 'High-order curved finite elements', *Int. J. Num. Meth. Eng.*, **17**, 735–45, 1981.

Mixed formulation and constraints—complete field methods

12.1 Introduction

The set of differential equations from which we start the discretization process will determine whether we shall refer to the formulation as *mixed* or *irreducible*. Thus if we consider an equation system with several dependent variables **u** written as [viz. Eqs (9.1) and (9.2)]

$$\mathbf{A}(\bar{\mathbf{u}}) = 0 \quad \text{in domain } \Omega$$

and

$$\mathbf{B}(\mathbf{u}) = 0 \quad \text{on boundary } \Gamma$$

$$(12.1)$$

in which none of the components of **u** can be eliminated still leaving a well-defined problem, then the formulation will be termed *irreducible*. If this is not the case the formulation shall be called *mixed*.

This definition is not the only one possible[1] but appears to the authors to be the most widely applicable[2,3] if in the elimination process referred to we are allowed to introduce penalty functions. Further, for any given physical situation we shall find that more than one irreducible forms are usually possible.

As an example we shall consider the simple problem of heat conduction (or quasi-harmonic equations) to which we have referred in Chapter 9 and 10. In this we start by a physical constitutive relation defining the fluxes [viz. Eq. (10.5)] in terms of the potential (temperature) gradients, i.e.,

$$\mathbf{q} = -\mathbf{k} \, \nabla \phi \qquad \mathbf{q} = \begin{Bmatrix} q_x \\ q_y \end{Bmatrix} \qquad (12.2)$$

The conservation equation can be written as [viz. Eq. (10.7)]

$$\mathbf{V}^T \mathbf{q} \equiv \frac{\partial q_x}{\partial x} + \frac{\partial q_y}{\partial y} = Q \tag{12.3}$$

If the above equations are satisfied in Ω and the boundary conditions

$$\phi = \tilde{\phi} \text{ on } \Gamma_\phi \quad \text{or} \quad q_n = \tilde{q}_n \text{ on } \Gamma_q \tag{12.4}$$

are obeyed then the problem is solved.

Clearly elimination of the vector $\mathbf{q}$ is possible and simple substitution of Eq. (12.2) into Eq. (12.3) leads to

$$\mathbf{V}^T(\mathbf{k} \, \mathbf{V}\phi) + Q = 0 \quad \text{in } \Omega \tag{12.5}$$

with appropriate boundary conditions expressed in terms of ϕ or its gradient.

In Chapter 10 we have shown discretized solutions starting from this point and clearly, as no further elimination of variables is possible, the formulation was *irreducible*.

On the other hand, if we were to start the discretization from Eqs. (12.2) to (12.4) the formulation would be *mixed*.

An alternative irreducible form is also possible in terms of the variables $\mathbf{q}$. Here we have to introduce a penalty form and write in place of Eq. (12.3)

$$\mathbf{V}^T \mathbf{q} - Q = \frac{\phi}{\alpha} \quad \alpha \to \infty \tag{12.6}$$

where α is a penalty number which tends to infinity. Clearly in the limit both equations are the same and in general if α is very large but finite the solutions should be approximately the same.

Now substitution into Eq. (12.2) gives the single governing equation

$$\mathbf{k}\mathbf{V} \, \mathbf{V}^T \mathbf{q} + \frac{\mathbf{q}}{\alpha} = \mathbf{k} \, \mathbf{V}Q \tag{12.7}$$

which again could be used for the start of a discretization process as a possible irreducible form.[4]

The reader should observe that, by the definition given, the formulations so far used in this book were *irreducible*. In subsequent sections we will show how elasticity problems can be dealt with in *mixed* form and indeed will show how such formulations are essential in certain problems typified by the incompressible elasticity example to which we have referred in Chapter 3. In Chapter 9 (Sec. 9.10.2) we have shown how discretization of a mixed problem can be accomplished.

Before proceeding to discussion of such discretization (which will reveal advantages and disadvantages of mixed methods) it is important to observe that if the operator specifying the mixed form is *symmetric* or *self-adjoint* (viz. Sec. 9.11.1) the formulation can proceed from the basis of a *variational principle* which can be directly obtained for linear problems. We invite the reader to prove by using the methods of Chapter 9 that stationarity of the *variational principle* given below is equivalent to the differential equations (12.2) and (12.3) together with the boundary conditions (12.4):

$$\Pi = \tfrac{1}{2} \int_\Omega \mathbf{q}^{\mathrm{T}} \mathbf{k}^{-1} \mathbf{q} \ d\Omega - \int_\Omega \phi (\nabla^{\mathrm{T}} \mathbf{q} - Q) \ d\Omega + \int_{\Gamma_q} \phi (q_{\mathrm{n}} - \tilde{q}_{\mathrm{n}}) \ d\Gamma$$

for

$$\phi = \bar{\phi} \text{ on } \Gamma_\phi \tag{12.8}$$

The establishment of such variational principles is a worthy academic pursuit and has led to many famous forms given in the classical work of Washizu.[5] However, we also know (viz. Sec. 9.9) that if in a linear problem symmetry of weighted residual matrices is obtained then a variational principle exists and can be determined. As such symmetry can be established by inspection we shall, in what follows, proceed with such weighting directly and thus avoid some unwarranted complexity.

12.2 Discretization of mixed forms—some general remarks

We shall demonstrate the discretization process on the basis of the mixed form of the heat conduction equations (12.2) and (12.3). Here we start by assuming that each of the unknowns is approximated in the usual manner by appropriate shape functions and corresponding unknown parameters. Thus

$$\mathbf{q} \cong \hat{\mathbf{q}} = \mathbf{N}_q \bar{\mathbf{q}} \qquad \text{and} \qquad \phi \cong \hat{\phi} = \mathbf{N}_\phi \bar{\phi} \tag{12.9}$$

where $\bar{\mathbf{q}}$ and $\bar{\phi}$ stand for nodal or more generally element parameters that have to be determined.

Assuming that the boundary conditions for $\phi = \tilde{\phi}$ are satisfied by the choice of the expansion, the weighted statement of the problem is, for Eq. (12.2),

$$\int_\Omega \mathbf{W}_q^{\mathrm{T}} (\mathbf{k}^{-1} \hat{\mathbf{q}} + \nabla \hat{\phi}) \ d\Omega = 0 \tag{12.10}$$

and, for Eq. (12.3) and the 'natural' boundary conditions,

$$\int_\Omega \mathbf{W}_\phi^{\mathrm{T}} (\nabla^{\mathrm{T}} \hat{\mathbf{q}} - Q) \ d\Omega - \int_{\Gamma_q} \mathbf{W}_\phi^{\mathrm{T}} (\hat{q}_{\mathrm{n}} - \tilde{q}_{\mathrm{n}}) \ d\Gamma = 0 \tag{12.11}$$

The reason we have premultiplied Eq. (12.2) by $\mathbf{k}^{-1}$ is now evident as the choice

$$\mathbf{W}_q = \mathbf{N}_q \qquad \mathbf{W}_\phi = \mathbf{N}_\phi \qquad\qquad (12.12)$$

will yield symmetric equations [using Green's theorem to perform integration by parts on the gradient term in Eq. (12.11)] of the form

$$\begin{bmatrix} \mathbf{A} & \mathbf{C} \\ \mathbf{C}^\mathrm{T} & \mathbf{O} \end{bmatrix} \begin{Bmatrix} \bar{\mathbf{q}} \\ \bar{\boldsymbol{\phi}} \end{Bmatrix} = \begin{Bmatrix} \mathbf{f}_1 \\ \mathbf{f}_2 \end{Bmatrix} \qquad\qquad (12.13)$$

with

$$\mathbf{A} = \int_\Omega \mathbf{N}_q^\mathrm{T} \mathbf{k}^{-1} \mathbf{N}_q \, d\Omega$$

$$\mathbf{C} = \int_\Omega \mathbf{N}_q^\mathrm{T} \nabla \mathbf{N}_\phi \, d\Omega$$

$$\mathbf{f}_1 = \mathbf{0}$$

$$(12.14)$$

$$\mathbf{f}_2 = - \int_\Omega \mathbf{N}_\phi^\mathrm{T} Q \, d\Omega + \int_{\Gamma_q} \mathbf{N}_\phi^\mathrm{T} \bar{q} \, d\Gamma$$

This problem, which we shall consider as typifying a large number of mixed approximations, illustrates the main features of the mixed formulation, including its advantages and disadvantages. We note that

1. The continuity requirements on the shape functions chosen are reduced. It is easily seen that while the irreducible form [viz. Eq. (12.5)] requires C_0 continuity of the shape functions, the integrals of Eq. (12.14) allow $\mathbf{N}_q$ to be discontinuous, as no derivatives of this are present. Alternatively, this discontinuity can be transferred to $\mathbf{N}_\phi$ (using Green's theorem on the integral in $\mathbf{C}$) while maintaining C_0 continuity for $\mathbf{N}_q$.

   This relaxation of continuity is of particular importance in plate and shell bending problems (viz. Vol. 2) and indeed the most important and earliest uses of mixed forms have been made in that context.[6-9]

2. If interest is focused on the variable $\mathbf{q}$ rather than ϕ, use of improved approximation for this may result in higher accuracy than possible with the irreducible form previously discussed. Indeed, in Sec. 12.8.3 we shall show how a C_0-continuous stress approximation can improve dramatically results of standard displacement analysis. *However, we must note that if the approximation function for $\mathbf{q}$ is capable of reproducing precisely the same type of variation as that determinable from the irreducible form then no additional accuracy will result and, indeed, the two approximations will yield identical answers.*

   Thus, for instance, if we consider the mixed approximation to the field problems discussed using a linear triangle to determine $\mathbf{N}_\phi$ and

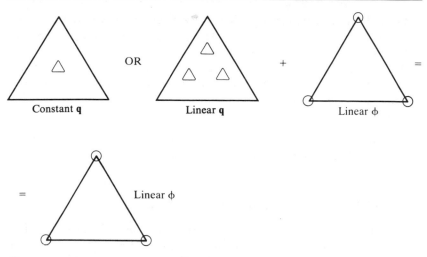

Fig. 12.1 A mixed approximation to the heat conduction problem yielding identical results as corresponding irreducible form (constant **k** assumed in each element)

piecewise constant $\mathbf{N}_q$, as shown in Fig. 12.1, we will obtain precisely the same results as those obtained by the irreducible formulation with the same $\mathbf{N}_\phi$ applied directly to Eq. (12.5), *providing* **k** *is constant within each element*. This is evident as the second of equations (12.13) is precisely the weighted continuity statement used in deriving the irreducible formulation in which the first of the equations is identically satisfied.

Indeed, should we choose to use a linear but discontinuous approximation form of $\mathbf{N}_q$ in the interior of such a triangle, we would still obtain precisely the same answers—with the additional coefficients becoming zero. This discovery was made by Fraeijs de Veubeke[10] and is called the *principle of limitation*, showing that under some circumstances no additional accuracy is to be expected from a mixed formulation. In a more general case where **k** is, for instance, discontinuous and variable within an element, the results of the mixed approximation will be different and on occasion superior.[2] Note that a C_0-continuous approximation for **q** could improve results as it is not capable of reproducing the discontinuous ones.

3. The equations resulting from mixed formulation frequently have zero-diagonal terms as indeed in the case of Eq. (12.13).

We noted in Chapter 9 that this is a characteristic of problems constrained by a lagrangian variable. Indeed, this is the origin of the problem which adds some difficulty to the standard gaussian elimination processes used in equation solving (viz. Chapter 15). As the form of Eq. (12.13) is typical of many two-field problems we shall refer to the first variable (here $\bar{q}$) as the *primary variable* and the second (here $\bar{\phi}$) as the *constraint variable*.

4. The added number of variables means that generally larger size algebraic problems have to be dealt with. However, in Sec. 12.8 we shall show how such difficulties can often be avoided by a suitable iterative solution.

The characteristics so far discussed did not mention one vital point which we elaborate in the next section.

12.3 Stability of mixed approximation. The patch test

12.3.1 *General requirements.* Despite the relaxation of shape function continuity requirements in mixed approximation for certain choices of the individual shape functions the mixed approximation will not yield meaningful results. This limitation is indeed much more severe than in *irreducible* formulation where a very simple, 'constant gradient' (or constant strain), condition sufficed to ensure a convergent form once continuity requirements were satisfied.

The mathematical reasons for this difficulty are discussed by Babuska[11] and Brezzi,[12] who formulated a rather complex criterion associated with their names. However, some sources of the difficulties (and hence ways of avoiding them) follow from quite simple reasoning.

If we consider the equation system (12.13) to be typical of many mixed systems in which $\bar{q}$ is the *primary variable* and $\bar{\phi}$ is the *constraint variable* (equivalent to a lagrangian multiplier), we note that the solution can proceed by eliminating $\bar{q}$ from the first equation and by substituting into the second to obtain

$$(\mathbf{C}^T \mathbf{A}^{-1} \mathbf{C})\bar{\phi} = -\mathbf{f}_2 + \mathbf{C}^T \mathbf{A}^{-1} \mathbf{f}_1 \qquad (12.15)$$

providing the matrix $\mathbf{A}$ is non-singular (or $\mathbf{A}\bar{q} \neq 0$ for all $\bar{q} \neq 0$). To calculate $\bar{\phi}$ it is necessary to ensure that the bracketed matrix, i.e.

$$\mathbf{H} = \mathbf{C}^T \mathbf{A}^{-1} \mathbf{C} \qquad (12.16)$$

is non-singular.

Singularity of the $\mathbf{H}$ matrix will always occur if the number of unknowns in the vector $\bar{q}$, which we call n_q, is less than the number of unknowns n_ϕ in the vector $\bar{\phi}$. Thus for avoidance of singularity

$$n_q \geqslant n_\phi \qquad (12.17)$$

is needed.

The reason for this is evident as the rank of the matrix (12.16), which needs to be n_ϕ, cannot be greater than n_q, i.e., the rank of A^{-1}.†

The same condition (12.17) ensures that non-zero answers for the variables $\bar{q}$ are possible. If it is violated *locking* or non-convergent results will occur in the formulation, giving near-zero answers for $\bar{q}$ in the case of 'locking' [viz. Chapter 9, Eq. (9.165) on].

To show this we shall replace Eq. (12.13) by its penalized form:

$$\begin{bmatrix} A & C \\ C^T & -I/\alpha \end{bmatrix} \begin{Bmatrix} \bar{q} \\ \bar{\phi} \end{Bmatrix} = \begin{Bmatrix} f_1 \\ f_2 \end{Bmatrix} \quad \begin{array}{l} \text{with } \alpha \to \infty \\ \text{and } I = \text{identity matrix} \end{array} \tag{12.18}$$

Elimination of $\bar{\phi}$ leads to

$$(A + \alpha CC^T)\bar{q} = f_1 + \alpha Cf_2 \tag{12.19}$$

As $\alpha \to \infty$ the above becomes simply

$$(CC^T)\bar{q} = Cf_2 \tag{12.20}$$

Non-zero answers for $\bar{q}$ should exist even when $f_2 = 0$ and hence the matrix CC^T *must be singular*. This singularity will always exist if $n_q > n_\phi$.

However, it is possible for this singularity to exist even if $n_\phi > n_q$, providing that the relationships $C^T\bar{q}$ are linearly dependent. Now the 'locking' will not occur and solution for $\bar{q}$ is still available, although H in Eq. (12.16) is still singular and thus no unique solution for $\bar{\phi}$ exists.

The stability conditions derived on the particular example of Eq. (12.13) are generally valid for any problem exhibiting the standard lagrangian constrained form. In particular the necessary condition of Eq. (12.17) will in most cases suffice to determine element acceptability.

---

† In some problems the matrix A may well be singular. It can normally be made non-singular by addition of a multiple of the second equation, thus changing it to $\bar{A} = A + \gamma CC^T$ and $\bar{f}_1 = f_1 + \gamma Cf_2$, where γ is an arbitrary number.

Although both the matrices A and CC^T are singular their combination $\bar{A}$ is not, providing we ensure that for all vectors $q \neq 0$ *either*

$$A\bar{q} \neq 0 \quad or \quad C^T\bar{q} \neq 0$$

(In mathematics terminology above this means that A is non-singular in the null space of C^T.)

The requirements of Eq. (12.17) is a necessary but not sufficient condition for non-singularity of the matrix H. An additional requirement evident from Eq. (12.15) is

$$C\bar{\phi} \neq 0 \quad \text{for all } \bar{\phi} \neq 0$$

If this is not the case the solution would not be unique.

The above requirements are inherent in the Babuska–Brezzi condition previously mentioned, but can always be verified algebraically.

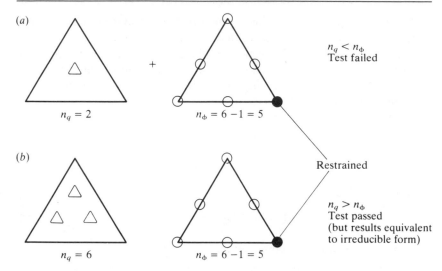

Fig. 12.2   Single element patch test for mixed approximations to the heat conduction problem with discontinuous flux **q** assumed: (*a*) quadratic C_0, ϕ; constant **q**; (*b*) quadratic C_0, ϕ; linear **q**

12.3.2 *The patch test.* The patch test for mixed elements can be carried out in exactly the way we have described in the previous chapter for irreducible elements. As *consistency* is easily assured by taking a polynomial approximation space; only *stability* needs generally to be investigated and *most answers* to this can be obtained by ensuring that condition (12.17) is satisfied for an isolated patch on the boundaries of which we constrain the maximum number of **q** and minimum number of ϕ variables.[13]

In Fig. 12.2 we illustrate a single element test for two possible formulations with C_0 continuous N_ϕ (quadratic) and discontinuous N_q, assumed to be either constant or linear within an element of triangular form. As no values of $\bar{q}$ can here be specified on the boundaries,† we shall fix a single value of $\bar{\phi}$ only, as is necessary to ensure uniqueness, on the patch boundary—which here is simply that of a single element. A count shows that only one of the formulations, i.e., that with linear flux variation, satisfies condition (12.17) and is therefore acceptable.

In Fig. 12.3 we illustrate a similar patch test on the same element but with identical C_0 continuous shape functions specified for both $\bar{q}$ and $\bar{\phi}$ variables. This example shows satisfaction of the basic condition of Eq. (12.17) and therefore is apparently a permissible formulation.

---

† In the formulation given, the specification of boundary fluxes is not imposed as a constraint on $\bar{q}$ variables.

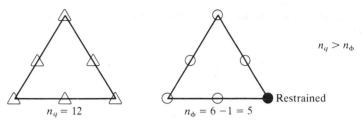

$$n_q > n_\phi$$

Restrained

$$n_q = 12$$ $$n_\phi = 6 - 1 = 5$$

Fig. 12.3   As Fig. 12.2 but with C_0 continuous **q**

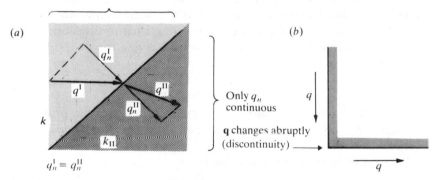

(a)

(b)

$$\left.\begin{array}{l}\text{Only } q_n \\ \text{continuous}\end{array}\right.$$

q changes abruptly
(discontinuity)

$$q_n^{\rm I} = q_n^{\rm II}$$

Fig. 12.4   Some situations for which C_0 continuity of flux **q** is inappropriate: (a)
discontinuous change of material properties; (b) singularity

Even if the patch test is satisfied occasional difficulties can arise, and
these are indicated mathematically by the Babuska–Brezzi condition
already referred to[14] (see footnote page 325). These difficulties can be due
to *excessive continuity* imposed on the problem by requiring, for
instance, the flux condition to be of C_0 continuity class. In Fig. 12.4 we
illustrate some cases in which the imposition of such continuity is *physically incorrect* and therefore can be expected to produce erroneous (and
usually highly oscillating) results. In all such problems we recommend
that the continuity be relaxed at least locally.

In Sec. 12.4.4 we shall discuss this problem further.

12.4 Mixed formulation in elasticity

12.4.1 *General.* In all the previous formulation of elasticity problems in
this book we have used an irreducible formulation, using the displacement **u** as the primary variable. The virtual work principle was used to

establish the equilibrium conditions which were written as (viz. Chapter 2)

$$\int_\Omega \delta\varepsilon^T\sigma \; d\Omega - \int_\Omega \delta u^T b \; d\Omega - \int_{\Gamma_t} \delta u^T \bar{t} \; d\Gamma = 0 \qquad (12.21)$$

where $\bar{t}$ are the tractions prescribed on Γ_t and with

$$\sigma = D\varepsilon \qquad (12.22)$$

as the constitutive relation (omitting here initial strains and stresses for clarity).

We recall that such statements as Eq. (12.21) are equivalent to weighted residual forms (viz. Chapter 9) and in what follows we shall use these frequently. In the above the strains are related to displacement by the matrix operator S introduced in Chapter 2, giving

$$\varepsilon = Su \qquad (12.23)$$

$$\delta\varepsilon = S \; \delta u \qquad (12.24)$$

with the displacement expansions constrained to satisfy the prescribed displacements on Γ_u. This is, of course, equivalent to Galerkin-type weighting.

With the displacement u approximated as

$$u \cong \hat{u} = N_u \bar{u} \qquad (12.25)$$

the required stiffness equations were obtained in terms of the unknown displacement vector $\bar{u}$ and solution obtained.

It is possible to use mixed forms in which either σ or ε, or, indeed, both these variables are approximated independently. We shall discuss such formulations below.

12.4.2 *The u–σ mixed form.* In this we shall assume that Eq. (12.21) is valid but that we approximate σ independently as

$$\hat{\sigma} = N_\sigma \bar{\sigma} \qquad (12.26)$$

and satisfy approximately the constitutive relation

$$\sigma = DSu \qquad (12.27)$$

which replaces (12.22) and (12.23). The approximate integral form is written as

$$\int_\Omega \delta\sigma^T(Su - D^{-1}\sigma) \; d\Omega = 0 \qquad (12.28)$$

where the expression in the brackets is simply Eq. (12.27) premultiplied by D^{-1} to establish symmetry and $\delta\sigma$ is introduced as a weighting function.

Indeed, Eqs (12.21) and (12.28) which now define the problem are equivalent to the stationarity of the functional

$$\Pi_{HR} = \tfrac{1}{2} \int_{\Omega} \sigma^T D^{-1} \sigma \; d\Omega + \int_{\Omega} u^T (S^T \sigma + b) - \int_{\Gamma_t} u^T (G\sigma - \tilde{t}) \; d\Gamma \quad (12.29)$$

where the boundary tractions are

$$t \equiv G\sigma$$

and

$$u = \tilde{u}$$

is enforced on Γ_u, as the reader can readily verify. This is the well-known Hellinger–Reissner[15,16] variational principle, but, as we have remarked earlier, it is unnecessary in deriving approximate equations. Using

$$N_u \text{ in place of } \delta u$$

$$B \equiv S N_u \text{ in place of } \delta\varepsilon$$

$$N_\sigma \text{ in place of } \delta\sigma$$

we write the approximate equations (12.28) and (12.21) in the standard form [viz. Eq. (12.13)]

$$\begin{bmatrix} A & C \\ C^T & 0 \end{bmatrix} \begin{Bmatrix} \bar{\sigma} \\ \bar{u} \end{Bmatrix} - \begin{Bmatrix} f_1 \\ f_2 \end{Bmatrix} \qquad (12.30)$$

with

$$A = - \int_{\Omega} N_\sigma^T D^{-1} N_\sigma \; d\Omega$$

$$C = + \int_{\Omega} N_\sigma^T B \; d\Omega$$

$$f_1 = 0$$

$$f_2 = + \int_{\Omega} N_u^T b \; d\Omega + \int_{\Gamma_t} N_u^T \tilde{t} \; d\Gamma$$

(12.31)

In the form given above the N_u shape functions have still to be of C_0 continuity, though N_σ can be discontinuous. However, integration by parts of the expression for C allows a reduction of such continuity and indeed this form has been used by Herrmann[6,17,18] for the problems of plates and shells.

A u–ε mixed form can be obtained in an exactly analogous manner and we shall leave the derivation to the reader as a simple exercise.

12.4.3 *The* **u**–**σ**–ε *mixed form. A three-field problem.* It is, of course, possible to use independent approximation to all the essential variables entering the elasticity problem. We can then write the three equations (12.23), (12.22), and (12.21) in their weak form as

$$\int_\Omega \delta \varepsilon^{\mathrm{T}} (\mathbf{D}\varepsilon - \boldsymbol{\sigma}) \, d\Omega = 0$$

$$\int_\Omega \delta \boldsymbol{\sigma}^{\mathrm{T}} (\mathbf{S}\mathbf{u} - \varepsilon) \, d\Omega = 0 \qquad (12.32)$$

$$\int_\Omega \delta (\mathbf{S}\mathbf{u})^{\mathrm{T}} \boldsymbol{\sigma} \, d\Omega - \int_\Omega \delta \mathbf{u}^{\mathrm{T}} \mathbf{b} \, d\Omega - \int_{\Gamma_t} \delta \mathbf{u}^{\mathrm{T}} \bar{\mathbf{t}} \, d\Gamma = 0$$

with a corresponding variational principle requiring the stationarity of

$$\Pi_{\mathrm{HW}} = \int_\Omega \tfrac{1}{2} \varepsilon^{\mathrm{T}} \mathbf{D}\varepsilon \, d\Omega - \int_\Omega \mathbf{u}^{\mathrm{T}} \mathbf{b} \, d\Omega - \int_\Omega \boldsymbol{\sigma}^{\mathrm{T}} (\varepsilon - \mathbf{S}\mathbf{u}) \, d\Omega - \int_{\Gamma_t} \mathbf{u}^{\mathrm{T}} \bar{\mathbf{t}} \, d\Gamma$$

$$(12.33)$$

where $\mathbf{u} \equiv \tilde{\mathbf{u}}$ on Γ_u is enforced.† This principle is known by the name of Hu–Washizu.[5] However, again we can proceed directly, using Eq. (12.32), taking the following approximations

$$\mathbf{u} \cong \hat{\mathbf{u}} = \mathbf{N}_u \bar{\mathbf{u}} \qquad \boldsymbol{\sigma} \cong \hat{\boldsymbol{\sigma}} = \mathbf{N}_\sigma \bar{\boldsymbol{\sigma}} \qquad \text{and} \qquad \varepsilon \cong \hat{\varepsilon} = \mathbf{N}_\varepsilon \bar{\varepsilon}$$

with corresponding 'variations' and writing the approximating equations in a similar fashion as we have in the previous section. This yields an equation system of the following form:

$$\begin{bmatrix} \mathbf{A} & \mathbf{C} & \mathbf{0} \\ \mathbf{C}^{\mathrm{T}} & \mathbf{0} & \mathbf{E} \\ \mathbf{0} & \mathbf{E}^{\mathrm{T}} & \mathbf{0} \end{bmatrix} \begin{Bmatrix} \bar{\varepsilon} \\ \bar{\boldsymbol{\sigma}} \\ \bar{\mathbf{u}} \end{Bmatrix} = \begin{Bmatrix} \mathbf{f}_1 \\ \mathbf{f}_2 \\ \mathbf{f}_3 \end{Bmatrix} \qquad (12.34)$$

where

$$\mathbf{A} = \int_\Omega \mathbf{N}_\varepsilon^{\mathrm{T}} \mathbf{D} \mathbf{N}_\varepsilon \, d\Omega$$

$$\mathbf{E} = \int_\Omega \mathbf{N}_\sigma^{\mathrm{T}} \mathbf{B} \, d\Omega$$

$$\mathbf{C} = - \int_\Omega \mathbf{N}_\varepsilon^{\mathrm{T}} \mathbf{N}_\sigma \, d\Omega \qquad (12.35)$$

$$\mathbf{f}_1 = \mathbf{f}_2 = 0$$

$$\mathbf{f}_3 = \int_\Omega \mathbf{N}_u^{\mathrm{T}} \mathbf{b} \, d\Omega + \int_{\Gamma_t} \mathbf{N}_u^{\mathrm{T}} \bar{\mathbf{t}} \, d\Gamma$$

---

† It is possible to include the displacement boundary conditions in Eq. (12.33) as a natural rather than imposed constraint; however, most finite element applications of the principle are in the form shown.

The reader will have observed again that in this section we have quoted the variational principles purely as a matter of interest and that all the approximations have been made directly.

12.4.4 *Stability of two-field approximation in elasticity* ($\mathbf{u}$–$\boldsymbol{\sigma}$; $\mathbf{u}$–$\boldsymbol{\varepsilon}$). Before attempting to formulate practical mixed approach approximations in detail, identical stability problems to those discussed in Sec. 12.3 have to be considered.

For the $\mathbf{u}$–$\boldsymbol{\sigma}$ forms it is clear that $\boldsymbol{\sigma}$ is the *primary variable* and $\mathbf{u}$ the *constraint variable* (viz. Sec. 12.2), and for the total problem as well as for element patches we must have as a necessary, but not sufficient condition

$$n_\sigma \geqslant n_u \tag{12.36}$$

where n_σ and n_u stand for numbers of degrees of freedom in appropriate variables.

In Fig. 12.5 we consider a two-dimensional plane problem and show a

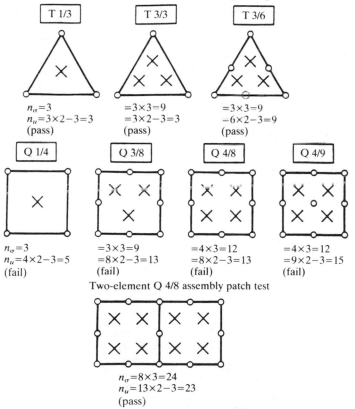

T 1/3	T 3/3	T 3/6
$n_\sigma = 3$ $n_u = 3 \times 2 - 3 = 3$ (pass)	$= 3 \times 3 = 9$ $= 3 \times 2 - 3 = 3$ (pass)	$= 3 \times 3 = 9$ $- 6 \times 2 - 3 = 9$ (pass)

Q 1/4	Q 3/8	Q 4/8	Q 4/9
$n_\sigma = 3$ $n_u = 4 \times 2 - 3 = 5$ (fail)	$= 3 \times 3 = 9$ $= 8 \times 2 - 3 = 13$ (fail)	$= 4 \times 3 = 12$ $= 8 \times 2 - 3 = 13$ (fail)	$= 4 \times 3 = 12$ $= 9 \times 2 - 3 = 15$ (fail)

Two-element Q 4/8 assembly patch test

$n_\sigma = 8 \times 3 = 24$
$n_u = 13 \times 2 - 3 = 23$
(pass)

Fig. 12.5   Elasticity by the mixed $\boldsymbol{\sigma}$–$\mathbf{u}$ formulation. Discontinuous stress approximation. Single element patch test. No restraint on $\bar{\boldsymbol{\sigma}}$ variables but three $\bar{\mathbf{u}}$ degrees of freedom restrained on patch. Test condition $n_\sigma \geqslant n_u$ (X denotes $\bar{\boldsymbol{\sigma}}$ (3 DOF) and o the $\bar{\mathbf{u}}$ (2 DOF) variables)

series of elements in which $\mathbf{N}_\sigma$ is discontinuous while $\mathbf{N}_u$ has C_0 continuity. We note again, by invoking the 'principle of limitation', that all the elements that pass the single-element test here will in fact yield identical results to those obtained by using the equivalent irreducible form, providing the $\mathbf{D}$ matrix is constant within each element. They are therefore of little interest. However, we note in passing that the Q 4/8, which fails in a single-element test, passes the patch test for assemblies of two or more elements, and is therefore usable, performing well in many circumstances. We shall see later that this is equivalent to using four-point Gauss, *reduced* integration (viz. Sec. 12.7), and as we have mentioned in Chapter 11 such elements will not always be robust.

It is of interest to note that if a higher order of interpolation is used for σ than for $\mathbf{u}$ the patch test is still satisfied, but in general the results will not be improved.

We do not show the similar patch test for the C_0 continuous $\mathbf{N}_\sigma$ assumption but state simply that, similarly to the example of Fig. 12.3, identical interpolation of $\mathbf{N}_\sigma$ and $\mathbf{N}_u$ is acceptable from the point of view of stability. However, as in Fig. 12.4, restriction of *excessive continuity* for stresses has to be avoided at singularities and at abrupt material property change interfaces, where only the normal and tangential tractions are continuous. We shall show in Sec. 12.8.3 that such continuous stress interpolations often lead to much improved accuracy.

The disconnection of stress variables at corner nodes can only be accomplished for all the variables. For this reason an alternative set of elements with continuous stress nodes at element interfaces can be introduced (viz. Fig. 12.6).[19]

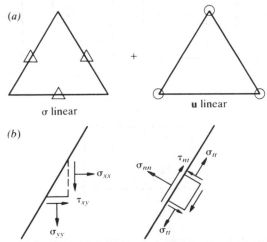

Fig. 12.6   Elasticity by the mixed σ–$\mathbf{u}$ formulation. Partially continuous σ (continuity at nodes only). (*a*) σ linear, $\mathbf{u}$ linear; (*b*) possible transformation of interface stresses with σ_{tt} disconnected

In such elements excessive continuity can easily be avoided by disconnecting only the direct stress components parallel to an interface at which material changes occur. It should be noted that even in the case when all stress components are connected at a mid-side node such elements do not ensure stress continuity along the whole interface. Indeed, the amount of such discontinuity can be useful as an error measure. However, we observe that for the linear element [Fig. 12.6(a)] the inter-element stresses are continuous *in the mean*.

It is, of course, possible to derive elements that exhibit complete continuity of the appropriate components along interfaces and indeed this was achieved by Raviart and Thomas[20] in the case of the heat conduction problem discussed previously. Extension to the full stress problem is difficult[21] and as yet such elements have not been successfully used.

12.4.5 *Stability condition of three-field approximation* (ε–σ–u). The stability condition derived in Sec. 12.3 [Eq. (12.17)] for two-field problems, which we later used in Eq. (12.36) for the simple mixed elasticity form, needs to be modified when three-field approximations of the form given in Eq. (12.34) are considered.

Many other problems fall into a similar category (for instance, plate bending) and hence the conditions of stability are generally useful. The requirement now is that

$$n_\varepsilon + n_u \geqslant n_\sigma$$

$$n_\sigma \geqslant n_u \tag{12.37}$$

This was first stated in reference 22 and follows directly from the two-field criterion as shown below.

The system of Eq. (12.34) can be 'regularized' by adding γE times the third equation to the second, with γ being an arbitrary constant. We now have

$$\begin{bmatrix} A & C & 0 \\ C^T & \gamma EE^T & E \\ 0 & E^T & 0 \end{bmatrix} \begin{Bmatrix} \bar{\varepsilon} \\ \bar{\sigma} \\ \bar{u} \end{Bmatrix} = \begin{Bmatrix} f_1 \\ f_2 + \gamma Ef_3 \\ f_3 \end{Bmatrix}$$

On elimination of ε using the first of above we have

$$\begin{bmatrix} \gamma EE^T - C^T A^{-1}C, & E \\ E^T, & 0 \end{bmatrix} \begin{Bmatrix} \bar{\sigma} \\ \bar{u} \end{Bmatrix} = \begin{Bmatrix} f_2 + \gamma Ef_3 - C^T A^{-1}f_1 \\ f_3 \end{Bmatrix}$$

From the two-field requirement [Eq. (12.17)] it follows that we require for no singularity

$$n_\sigma \geqslant n_u$$

Rearranging Eq. (12.34) we can write

$$\begin{bmatrix} \mathbf{A} & \mathbf{0} & \mathbf{C} \\ \mathbf{0} & \mathbf{0} & \mathbf{E}^T \\ \mathbf{C}^T & \mathbf{E} & \mathbf{0} \end{bmatrix} \begin{Bmatrix} \bar{\boldsymbol{\varepsilon}} \\ \bar{\mathbf{u}} \\ \bar{\boldsymbol{\sigma}} \end{Bmatrix} = \begin{Bmatrix} \mathbf{f}_1 \\ \mathbf{f}_3 \\ \mathbf{f}_2 \end{Bmatrix}$$

This again can be regularized by adding multiples $\gamma \mathbf{C}$ and $\gamma \mathbf{E}^T$ of the third of the above equations to the first and second respectively obtaining

$$\begin{bmatrix} \mathbf{A} + \gamma \mathbf{CC}^T, & \gamma \mathbf{CE} & \vdots & \mathbf{C} \\ \gamma \mathbf{E}^T \mathbf{C}^T, & \gamma \mathbf{E}^T \mathbf{E} & \vdots & \mathbf{E}^T \\ \hline \mathbf{C}^T, & \mathbf{E} & \vdots & \mathbf{0} \end{bmatrix} \begin{Bmatrix} \bar{\boldsymbol{\varepsilon}} \\ \bar{\mathbf{u}} \\ \bar{\boldsymbol{\sigma}} \end{Bmatrix} = \begin{Bmatrix} \mathbf{f}_1 + \gamma \mathbf{Cf}_2 \\ \mathbf{f}_3 + \gamma \mathbf{E}^T \mathbf{f}_2 \\ \mathbf{f}_2 \end{Bmatrix}$$

By partitioning as above it is evident that we require

$$n_\varepsilon + n_u \geqslant n_\sigma$$

We shall not discuss in detail any of the possible approximations to the $\boldsymbol{\varepsilon}$–$\boldsymbol{\sigma}$–$\mathbf{u}$ formulation or their corresponding patch tests as the arguments are similar to those of the two-field problems.

In some practical applications of the three-field form the approximation of the second and third equations in (12.32) is used directly to eliminate all but the displacement terms. This leads to a special form of the displacement method which has been called a $\bar{\mathbf{B}}$ (**B**-bar) form.[23,24] In the $\bar{\mathbf{B}}$ form the shape function derivatives are replaced by approximations resulting from the mixed form. We shall illustrate this concept with an example of a nearly incompressible material in Sec. 12.5.2.

12.5 Incompressible (or nearly incompressible) elasticity

12.5.1 *Two-field approximation* ($\mathbf{u}$–p). We have noted earlier that the standard displacement formulation of elastic problems fails when Poisson's ratio v approaches 0.5 or when the material becomes incompressible. Indeed, problems arise even when the material becomes nearly incompressible with $v > 0.4$ and the simple linear approximation with triangular elements gives highly oscillatory results in such cases.

The application of a mixed formulation for such problems avoids the difficulties and is of great practical interest as *nearly* incompressible elastic behaviour is encountered in a variety of real engineering problems ranging from soil mechanics to aerospace engineering. Identical problems arise when the flow of incompressible fluids is considered.

The main problem in application of 'standard' displacement formulation to incompressible or nearly incompressible problems lies in the determination of the mean stress or pressure which is related to the volumetric part of the strain. For this reason it is convenient to separate this from the total stress field and treat it as an independent variable. Using the 'vector' notation of stresses the mean stress or pressure is given by

$$p = \frac{\sigma_x + \sigma_y + \sigma_z}{3} = \frac{\sigma^T m}{3} \tag{12.38}$$

where m for the general three-dimensional state of stress is given by

$$m^T = [1,\ 1,\ 1,\ 0,\ 0,\ 0]$$

The 'pressure' is related to the volumetric strain, ε_v, by the bulk modulus of the material K, for isotropic behaviour. Thus

$$\varepsilon_v = \varepsilon_x + \varepsilon_y + \varepsilon_z = m^T \varepsilon \tag{12.39}$$

$$\varepsilon_v = \frac{p}{K} \tag{12.40}$$

For an incompressible material $K = \infty$ and the volumetric strain is simply zero.

The deviatoric strain ε_d defined by

$$\varepsilon_d = \varepsilon - \frac{m\varepsilon_v}{3} \equiv (I - \tfrac{1}{3}mm^T)\varepsilon \tag{12.41}$$

is related in isotropic elasticity to the deviatoric stress σ_d by the stress modulus G as

$$\sigma_d = \sigma - mp = G D_0 \varepsilon_d = G(D_0 - \tfrac{2}{3}mm^T)\varepsilon \tag{12.42}$$

where

$$D_0 = \begin{bmatrix} 2 & & & & & 0 \\ & 2 & & & & \\ & & 2 & & & \\ & & & 1 & & \\ & & & & 1 & \\ 0 & & & & & 1 \end{bmatrix}$$

is introduced because of the vector notation.

The above relationships are but an alternate way of determining the stress–strain relations shown in Chapters 2 and 3, and the reader can verify that

$$G = \frac{E}{2(1 + v)} \qquad K = \frac{E}{3(1 - 2v)} \tag{12.43}$$

and indeed Eqs (12.42) and (12.40) can be used to define the standard D matrix in an alternate manner.

In the mixed form considered next we shall use as the variables the displacement $\mathbf{u}$ and the pressure p.

Now the equilibrium equation (12.21) is rewritten using (12.42), treating p as an independent variable as

$$\int_\Omega \delta\varepsilon^T[G(\mathbf{D}_0 - \tfrac{2}{3}\mathbf{mm}^T)\varepsilon + \mathbf{m}p]\ d\Omega$$

$$-\int_\Omega \delta\mathbf{u}^T\mathbf{b}\ d\Omega - \int_{\Gamma_t} \delta\mathbf{u}^T\tilde{\mathbf{t}}\ d\Gamma = 0 \quad (12.44)$$

and in addition we shall impose a weak form of Eq. (12.40), i.e.,

$$\int_\Omega \delta p^T\left[\mathbf{m}^T\varepsilon - \frac{p}{K}\right]d\Omega = 0 \qquad \text{with } \varepsilon = \mathbf{Su} \qquad (12.45)$$

Independent approximation of $\mathbf{u}$ and p as

$$\mathbf{u} \cong \hat{\mathbf{u}} = \mathbf{N}_u\bar{\mathbf{u}} \qquad \text{and} \qquad p \cong \hat{p} = \mathbf{N}_p\bar{\mathbf{p}} \qquad (12.46)$$

immediately gives the mixed approximation in the form

$$\begin{bmatrix} \mathbf{A} & \mathbf{C} \\ \mathbf{C}^T & -\mathbf{V} \end{bmatrix}\begin{Bmatrix} \bar{\mathbf{u}} \\ \bar{\mathbf{p}} \end{Bmatrix} = \begin{Bmatrix} \mathbf{f}_1 \\ \mathbf{f}_2 \end{Bmatrix} \qquad (12.47)$$

where

$$\mathbf{A} = \int_\Omega \mathbf{B}^T G(\mathbf{D}_0 - \tfrac{2}{3}\mathbf{mm}^T)\mathbf{B}\ d\Omega$$

$$\mathbf{C} = \int_\Omega \mathbf{B}^T\mathbf{m}\mathbf{N}_p\ d\Omega$$

$$\mathbf{V} = \int_\Omega \left(\mathbf{N}_p^T\mathbf{N}_p\frac{d\Omega}{K}\right) \qquad (12.48)$$

$$\mathbf{f}_1 = \int_\Omega \mathbf{N}_u^T\mathbf{b}\ d\Omega + \int_{\Gamma_t} \mathbf{N}_u^T\tilde{\mathbf{t}}\ d\Gamma$$

$$\mathbf{f}_2 = 0$$

We note that for incompressible situations the equations are of the 'standard' form [viz. Eq. (12.13)] with $\mathbf{V} = 0$ (as $K \to \infty$), but the formulation is practically useful when K has a high value (or $v \to 0.5$).

A formulation similar to that above and using the corresponding variational theorem has been first proposed by Herrmann[25] and later generalized by Key[26] for anisotropic elasticity.

The arguments concerning stability (or singularity) of the matrices which we outlined in Sec. 12.3 are again of great importance in this problem.

Clearly the condition about the number of degrees of freedom now yields [viz. Eq. (12.17)]

$$n_u \geqslant n_p \tag{12.49}$$

and has to be observed for locking (or instability) prevention with the pressure acting now as the constraint variable or lagrangian multiplier enforcing incompressibility.

In the form of a *patch test* this condition is most critical and we show in Figs 12.7 and 12.8 a series of such patch tests on elements with C_0 continuous interpolation of **u** and either discontinuous or continuous interpolation of p. For each we have included all combinations of constant, linear, and quadratic functions.

In the test we prescribe *all* the displacements on the boundaries of the patch and one pressure variable (as it is well known that in fully incompressible situations pressure will be indeterminate by a constant).

The single-element test is very stringent and eliminates most continuous pressure approximations whose performance is known to be acceptable in many situations. For this reason we attach more importance to the assembly test and it would appear that the following elements are permissible according to the criteria of Eq. (12.49) (indeed all pass the B–B condition fully):

Triangles: T 6/1　T 10/3　T 6/C3
Quadrilaterals: Q 9/3　Q 8/C4　Q 9/C4

We note, however, that in practical applications quite adequate answers have been reported with Q 4/1, Q 8/3, and Q 9/4 quadrilaterals, although severe oscillations of p may occur. If robustness is sought the choice of the elements is limited.[2]

It is unfortunate that in the present 'acceptable' list the linear triangle and quadrilateral are missing. This appreciably restricts the use of these simplest elements. A possible and indeed effective procedure here is to apply the pressure constraint not at the level of a single element but on an assembly. This was done by Herrmann in his original presentation[25] where four elements were chosen for such a constraint as shown in Fig. 12.9(a). This 'element' passes the single-element (and multiple-element) patch tests but so do apparently several others fitting into this category. In Fig. 12.9(b) we show how a single triangle can be internally subdivided into three parts by introduction of a central node. This coupled with constant pressure on the assembly allows the necessary condition to be satisfied and a standard element procedure applies to the original triangle treating the central node as an internal variable. Indeed, the same effect could be achieved by introduction of any other internal element function which gives zero value on the main triangle perimeter.

Such a bubble function can simply be written in terms of the area coordination (viz. Chapter 7) as

$$L_1 L_2 L_3$$

However, as we have stated before, the degree of freedom count is a necessary but not sufficient condition for stability and other tests are necessary. In particular it can be verified by algebra that the conditions stated in the footnote to page 325 are not fulfilled for this triple subdivision of a linear triangle (or the case with the bubble function) and thus

$$\mathbf{Cp} = 0 \text{ for some non-zero values of } \mathbf{p}$$

thus retaining instability.

In Fig. 12.9(c) we show, however, that the same concept can be used with good effect for C_0 continuous p. Similar internal subdivision into quadrilaterals or introduction of bubble functions can be used, as shown in Fig. 12.9(d), with success.

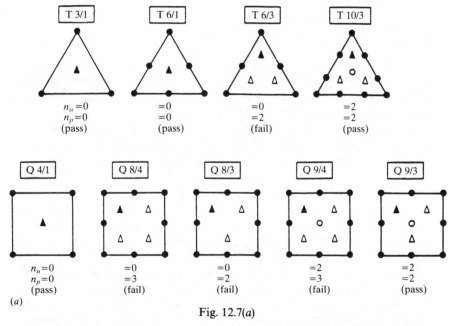

Fig. 12.7(a)

Fig. 12.7 Incompressible elasticity $\mathbf{u}$–p formulation. Discontinuous pressure approximation. (a) Single-element patch tests. (b) Multiple-element patch tests

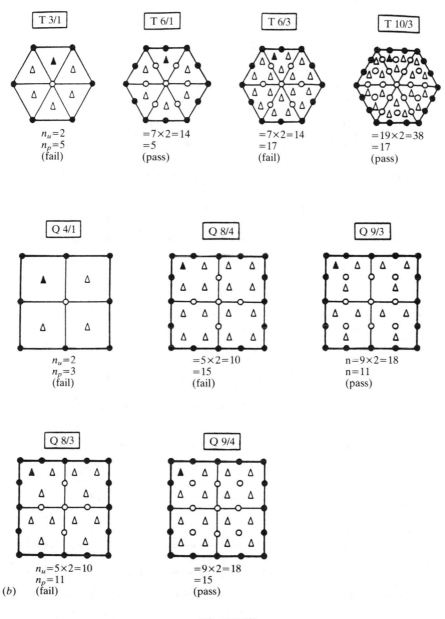

Fig. 12.7(b)

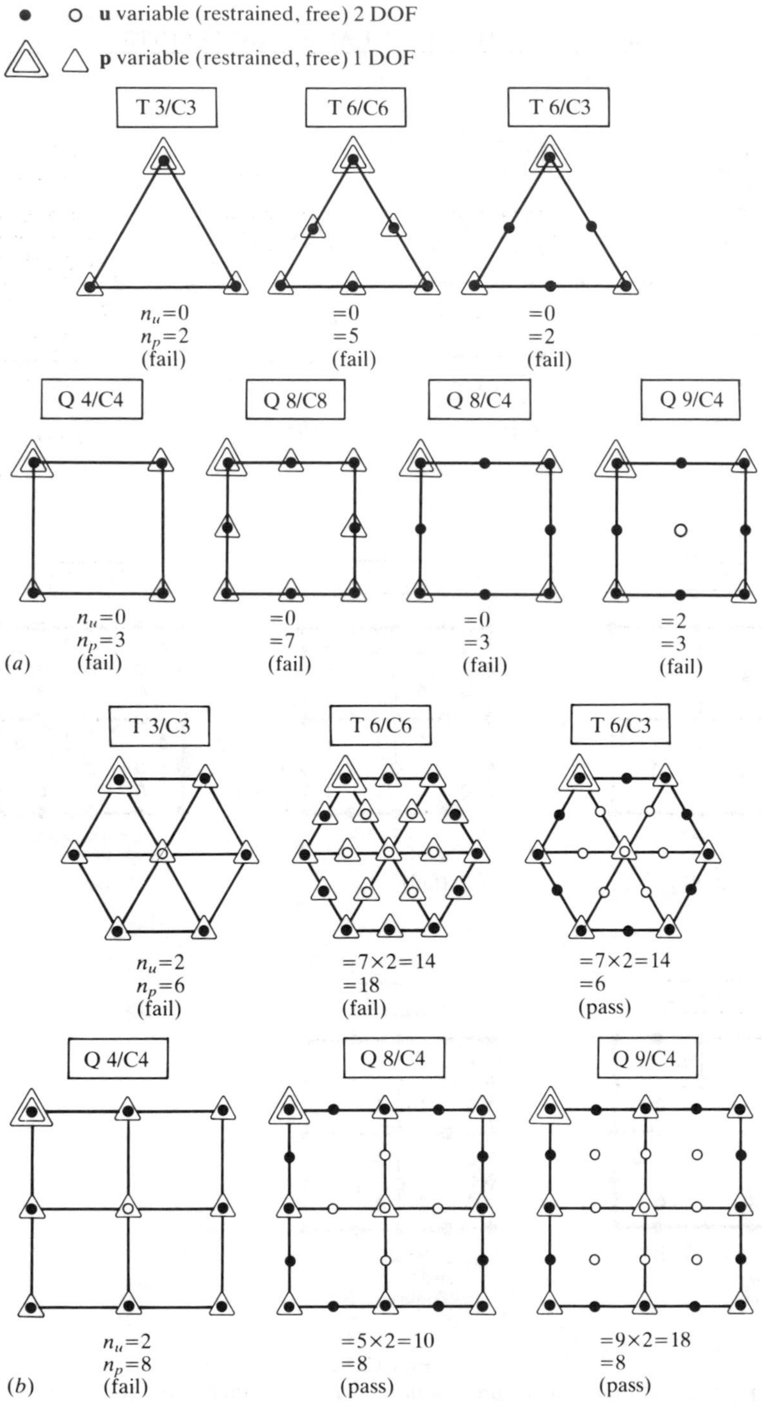

Fig. 12.8 Incompressible elasticity **u**–p formulation. Continuous (C_0) pressure approximation. (a) Single-element patch tests. (b) Multiple-element patch tests

340

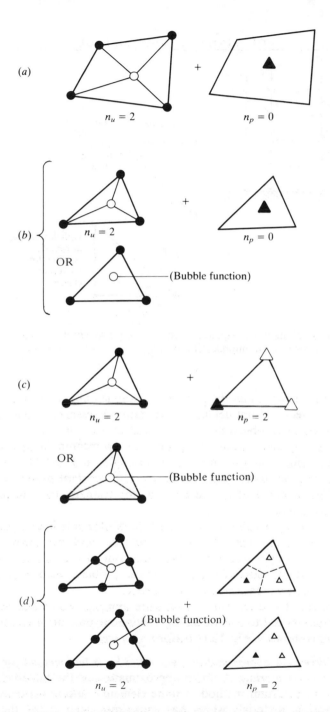

Fig. 12.9 Some simple combinations of linear triangles and quadrilaterals that pass the necesary patch test counts. Combinations (a), (c), and (d) are successful but (b) is still singular and not usable

341

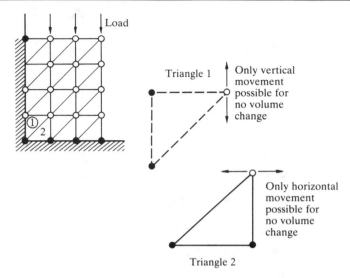

Fig. 12.10   Locking (zero displacements) of a simple assembly of linear triangles for which incompressibility is fully required ($n_p = n_u = 24$)

The performance of all the elements mentioned above has been extensively discussed[27-32] but detailed comparative assessment of merit is difficult. As we have observed, it is essential to have $n_u \geqslant n_p$ but if near equality is only obtained in a large problem no meaningful answers will be obtained for **u** as we observe, for example, in Fig. 12.10 in which linear triangles for **u** are used with the element constant p. Here the only permissible answer is of course $\mathbf{u} = 0$ as the triangles have to preserve constant volumes.

The ratio n_u/n_p which occurs as the field of elements is enlarged gives some indication of the relative performance, and we show this in Fig. 12.11. This approximates to the behaviour of a very large element assembly, but of course for any practical problem such a ratio will depend on the boundary conditions imposed.

We see that for discontinuous pressure approximation this ratio for 'good' elements is 2 to 3 while for C_0 continuous pressure it is 6 to 8. All the elements shown in Fig. 12.11 perform very well.

12.5.2 *Three-field approximation* ($\mathbf{u}$–p–ε_v). *The* **B**-*bar method for nearly incompressible materials*. A direct approximation of the three-field form leads to an important method in finite element solution procedures for incompressible materials which has sometimes been called the **B**-bar method. The methodology can be illustrated for the (nearly) incompressible problem. The usual irreducible form (displacement method) has been

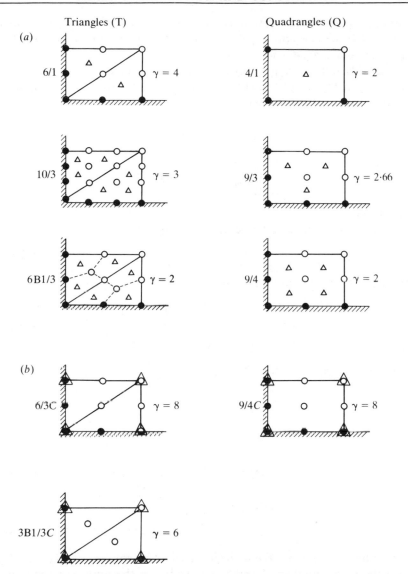

Fig. 12.11 The *freedom index* or *infinite patch ratio* for various **u**–*p* elements for incompressible elasticity ($\gamma = n_u/n_p$). (*a*) Discontinuous pressure. (*b*) Continuous pressure

shown to 'lock' for the nearly incompressible problem. As shown in Sec. 12.5.1, the use of a mixed method can avoid this locking phenomenon when properly implemented (e.g., using the Q 9/3 two-field form). Below we present an alternative which leads to an efficient and accurate implementation in many situations. For the development shown we shall assume that the material is isotropic linear elastic but it may be extended easily to include anisotropic materials.

Assuming independent approximation to ε_v and p we can formulate the problem by use of Eq. (12.44) and the weak statement of relations (12.39) and (12.40) written as

$$\int_\Omega \delta p[\varepsilon_v - \mathbf{m}^T \mathbf{Su}] \, d\Omega = 0 \tag{12.50}$$

$$\int_\Omega \delta\varepsilon_v[K\varepsilon_v - p] \, d\Omega = 0 \tag{12.51}$$

and (12.44).

If we approximate $\mathbf{u}$ and p fields by (12.46) and

$$\varepsilon_v \approx \hat{\varepsilon}_v = \mathbf{N}_v \bar{\varepsilon}_v \tag{12.52}$$

we obtain a mixed approximation of the form

$$\begin{bmatrix} \mathbf{A} & \mathbf{C} & \mathbf{0} \\ \mathbf{C}^T & \mathbf{0} & -\mathbf{E} \\ \mathbf{0} & -\mathbf{E}^T & \mathbf{H} \end{bmatrix} \begin{Bmatrix} \bar{\mathbf{u}} \\ \bar{\mathbf{p}} \\ \bar{\varepsilon}_v \end{Bmatrix} = \begin{Bmatrix} \mathbf{f}_1 \\ \mathbf{f}_2 \\ \mathbf{f}_3 \end{Bmatrix} \tag{12.53}$$

where $\mathbf{A}$, $\mathbf{C}$, $\mathbf{f}_i$ are given by (12.48) and

$$\mathbf{E} = \int_\Omega \mathbf{N}_p^T \mathbf{N}_v \, d\Omega \qquad \mathbf{f}_3 = 0 \tag{12.54}$$

with

$$\mathbf{H} = \int_\Omega \mathbf{N}_v^T K \mathbf{N}_v \, d\Omega \tag{12.55}$$

The second of (12.53) has the solution

$$\bar{\varepsilon}_v = \mathbf{E}^{-1} \mathbf{C}^T \bar{\mathbf{u}} = \mathbf{W}\bar{\mathbf{u}} \tag{12.56}$$

In the above we assume $\mathbf{E}$ may be inverted which implies that $\mathbf{N}_v$ and $\mathbf{N}_p$ have the same number of terms. Furthermore, the approximations for the volumetric strain and pressure are constructed for each element individually and are not continuous across element boundaries. Thus, the solution of Eq. (12.56) may be performed for each individual element. In practice $\mathbf{N}_v$ normally is assumed identical to $\mathbf{N}_p$ so that $\mathbf{E}$ is symmetric positive definite. The solution of the third of (12.53) yields the pressure parameters in terms of the volumetric strain parameters and is given by

$$\bar{\mathbf{p}} = (\mathbf{E}^T)^{-1} \mathbf{H}\bar{\varepsilon}_v \tag{12.57}$$

Substitution of (12.56) and (12.57) into the first of (12.53) gives a solution that is in terms of displacements only. Accordingly,

$$\bar{\mathbf{A}}\bar{\mathbf{u}} = \mathbf{f}_1 \tag{12.58}$$

where for isotropy

$$\bar{\mathbf{A}} = \int_\Omega \mathbf{B}^T G(\mathbf{D}_0 - \tfrac{2}{3}\mathbf{mm}^T)\mathbf{B} \ d\Omega + \mathbf{W}^T\mathbf{H}\mathbf{W} \tag{12.59}$$

The solution of (12.58) yields the nodal parameters for the displacements. Use of (12.56) and (12.57) then gives the approximations for the volumetric strain and pressure.

The result given by (12.59) may be further modified to obtain a form that is similar to the standard displacement method. Accordingly we write

$$\bar{\mathbf{A}} = \int_\Omega \bar{\mathbf{B}}^T\mathbf{D}\bar{\mathbf{B}} \ d\Omega \tag{12.60}$$

where the strain displacement matrix is now

$$\bar{\mathbf{B}} = (I - \tfrac{1}{3}\mathbf{mm}^T)\mathbf{B} + \tfrac{1}{3}\mathbf{m}\mathbf{N}_v\mathbf{W} \tag{12.61}$$

For isotropy the modulus matrix is

$$\mathbf{D} = G(\mathbf{D}_0 - \tfrac{2}{3}\mathbf{mm}^T) + K\mathbf{mm}^T \tag{12.62}$$

We note that the above form is identical to a standard displacement model except that $\mathbf{B}$ is replaced by $\bar{\mathbf{B}}$. The method has been discussed more extensively in references 23 and 24.

The equivalence of (12.59) and (12.60) can be verified by simple matrix multiplication.

The formulation shown above has been implemented into an element included as part of the program given in Chapter 15. The elegance of the method is more fully utilized when considering non-linear problems, such as plasticity and finite deformation elasticity. For a more complete discussion the reader is referred to reference 24.

We note that of course elimination of p could be accomplished also in the $\mathbf{u}$–p two-field form using K as a penalty number, but if K is a real variable with spatial variation the new form given here is much more convenient as the $\mathbf{E}$ matrix is independent of it.

Of course, precisely the same stability criteria operate here as in the two-field approximation discussed earlier.

12.6 Stress smoothing/optimal sampling

12.6.1 *Stress projection.* We have observed in the earlier chapters that the displacement formulation has frequently resulted in an unrealistic stress prediction, giving interelement stress 'jumps' even if the true stresses were continuous. Resort to nodal averaging of element stresses is frequently made in practice to make the results more palatable to the user. However, it is possible to obtain a better stress picture by a *projection* or *variational recovery* process which in itself is another possible mixed formulation.[33-36]

In this we obtain displacements $\mathbf{u}$ (or in an equivalent field problem such as discussed in Sec. 12.3, the temperature ϕ) by an irreducible formulation. Instead of reporting, however, the stresses $\hat{\boldsymbol{\sigma}}$ so approximated, i.e.,

$$\hat{\boldsymbol{\sigma}} = \mathbf{DB\bar{u}} \tag{12.63}$$

we compute a set that is interpolated by

$$\boldsymbol{\sigma}^* = \mathbf{N}_\sigma \bar{\boldsymbol{\sigma}} \tag{12.64}$$

and in the weak sense approximates to $\hat{\boldsymbol{\sigma}}$. We write this approximation as

$$\int_\Omega \mathbf{N}_\sigma^T (\boldsymbol{\sigma}^* - \hat{\boldsymbol{\sigma}}) \, d\Omega = 0 \tag{12.65}$$

or

$$\left(\int_\Omega \mathbf{N}_\sigma^T \mathbf{N}_\sigma \, d\Omega \right) \bar{\boldsymbol{\sigma}} = \left(\int_\Omega \mathbf{N}_\sigma^T \mathbf{DB\bar{u}} \right) d\Omega \tag{12.66}$$

This is *almost* identical to the first of (12.30) used in the $\boldsymbol{\sigma}$–$\mathbf{u}$ mixed approximation and would be identical if the integrand had been premultiplied by $\mathbf{D}^{-1}$. We have not done so in the present case as the structure of the matrix on the left-hand side of Eq. (12.66) is of particular interest. Indeed, we shall encounter its form in problems of dynamics as the 'mass matrix'. If we call this matrix

$$\mathbf{M} = \int_\Omega \mathbf{N}_\sigma^T \mathbf{N}_\sigma \, d\Omega \tag{12.67}$$

we find that it is an easy matter to approximately 'diagonalize' it as $\mathbf{M}_L$, a lumped matrix. Several diagonalization procedures are described in Appendix 8 and we shall find that one of the most useful ways is to apply an approximate integration rule with integrating points being confined to nodes i of $\mathbf{N}_\sigma$. If $\mathbf{N}_\sigma$ functions are written in a 'standard' rather than hierarchic form a purely diagonal matrix will result.

If we write Eq. (12.66) as

$$\mathbf{M}\bar{\boldsymbol{\sigma}} = \mathbf{P} \tag{12.68}$$

we can approximate to its solution as

$$\bar{\sigma} = M_L^{-1} P \tag{12.69}$$

where the inversion is trivial. Successively better solutions can be obtained by an iteration

$$\bar{\sigma}^n = \bar{\sigma}^{n-1} + M_L^{-1}(M\bar{\sigma}^{n-1} - P) \tag{12.70}$$

The most obvious projection is to use continuous N_σ such that

$$N_\sigma = N_u \tag{12.71}$$

obtaining thus an approximation of higher accuracy.

It is interesting to note that the 'projection' of Eq. (12.66) is equivalent to the 'least square fit' or minimization of

$$\Pi = \int_\Omega (\sigma^* - \hat{\sigma})^2 \, d\Omega \tag{12.72}$$

with respect to $\bar{\sigma}$ of Eq. (12.64).

12.6.2 *Optimal sampling*. The fact that the smoothed and thus more accurate stresses are a least square fit of computed stresses provides a clue as to the location of points at which the sampling or evaluation of stress is optimal.

Consider Fig. 12.12 in which we show piecewise discontinuous linear element stresses (gradients) approximating the exact distribution. If the latter is a least square fit of the former it is evident that at some points the two will be equal. Indeed, this equality will generally occur within each element. If we knew in advance where such points were located we would always find there the exact solution; clearly this is a dream hardly to be realized.

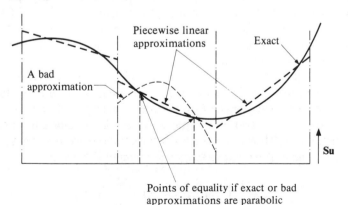

Points of equality if exact or bad
approximations are parabolic

Fig. 12.12   Piecewise linear least square fit to curve **Su** of stress or strain

However, a useful property of numerical integration (Gauss–Legendre) points can help us here. *This property can be stated as follows: if we devise a numerical integration formula with a minimum number of sampling points which just integrates precisely a polynomial of degree $2M + 1$, then generally at such points a polynomial of order $M + 1$ is equal to its least square approximations by a polynomial of order M.*

This proposition is exactly true in the case of one-dimensional Gauss point integration and approximately satisfied for other two- and three-dimensional integration expressions.[37]

It is immediately obvious in the example shown that if the exact curve were a parabola, then two Gauss points would define uniquely a straight line which is the least square approximation to it. Conversely, if we sampled the approximation **Su** at these points we would thus obtain an accuracy of one order greater than that available elsewhere by the approximation. Clearly such points are optimal for sampling the quantity **Su** (or the strains and stresses in an elasticity problem).

For smooth solutions we can state quite generally that the approximation to **Su** is always of the order $O(h^{p-m+1})$, where p is the complete polynomial in the approximating shape function and m the order of the operator **S** (Chapter 9, Sec. 9.8). Therefore, *at numerical integration points which just integrate exactly a polynomial of order $2(p - m) + 1$* [i.e., with an error of order $O(h^{2(p-m)+2})$], *the approximation to* **Su** *will be nearly one order better, i.e.,* $O(h^{p-m+2})$.

Obviously in any finite element computation it therefore pays to sample the strains of such integration points, as has been realized by many investigators.[34,37–39]

Figure 12.13 shows some such optimal points for sampling for various C_0-type elements ($m = 1$).

The results for linear triangles and quadrilaterals are physically obvious (and we have already remarked in Chapter 3, Sec. 3.2.9, that 'obviously' the stresses would be best represented at centroids). For higher order C_0 elements the results are by no means self-evident—they turn out, however, to be true.

The concept of least square fit has additional justification in self-adjoint problems in which an energy functional is minimized. In such cases, typical of displacement formulation of elasticity, it can be readily shown that the minimization is equivalent to a least square fit of stresses to the exact ones. Thus quite generally we can start from a theory which states that *minimization of an energy functional* Π *defined as*

$$\Pi = \frac{1}{2} \int_\Omega (\mathbf{Su})^{\mathrm{T}} \mathbf{A}(\mathbf{Su}) \, d\Omega + \int_\Omega \mathbf{u}^{\mathrm{T}} \mathbf{p} \, d\Omega \tag{12.73}$$

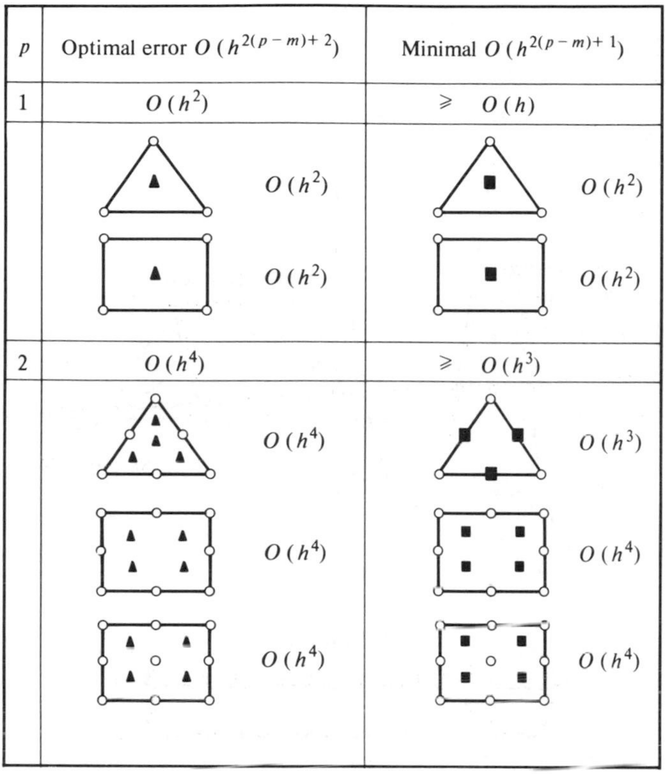

p	Optimal error $O(h^{2(p-m)+2})$	Minimal $O(h^{2(p-m)+1})$
1	$O(h^2)$	$\geqslant \quad O(h)$

	$O(h^2)$
	   $O(h^2)$
		$O(h^2)$
		   $O(h^2)$

2	$O(h^4)$	$\geqslant \quad O(h^3)$

	$O(h^4)$
	   $O(h^4)$
	   $O(h^4)$
		$O(h^3)$
		   $O(h^4)$
		   $O(h^4)$

Fig. 12.13 Optimal sampling and minimum integration points for some C_0 elements

which gives the exact solution $\mathbf{u} = \bar{\mathbf{u}}$ is equivalent to minimization of another functional Π^* defined as

$$\Pi^* = \frac{1}{2} \int_\Omega [\mathbf{S}(\mathbf{u} - \bar{\mathbf{u}})]^T \mathbf{A}\mathbf{S}(\mathbf{u} - \bar{\mathbf{u}}) \, d\Omega \qquad (12.74)$$

In the above $\mathbf{S}$ is a self-adjoint linear operator and $\mathbf{A}$ and $\mathbf{p}$ are pre-scribed matrices of position. The above quadratic form [Eq. (12.73)] is such as arises in the majority of linear, self-adjoint, problems.

This theorem is given in different forms by Herrmann,[40] Moan,[37] and Brauchli and Oden,[38] and shows that the approximate solution for $\mathbf{Su}$ approaches the exact one $\mathbf{S\bar{u}}$ as a weighted least square approximation.

In the context of elastic analysis, for instance, we can state that the minimization of total potential energy is equivalent to finding a weighted least square fit of the exact strains by those assumed approximately.

The proof of the above theorem is given at the end of this section.

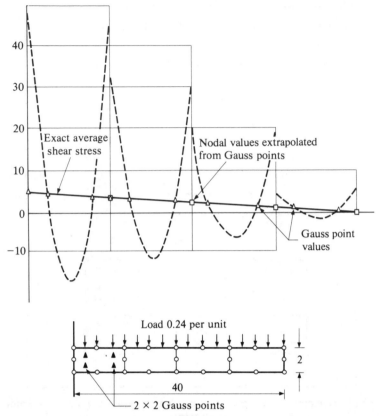

Fig. 12.14   A cantilever beam with four quadratic (Q8) elements. Stress sampling at cubic order (2 × 2) Gauss points with extrapolation to nodes

We shall find that on some occasions individual components of the strains or stresses exhibit a very bad approximation locally due to higher order spurious terms present in quadrilateral elements. Here optimal sampling points again come to the rescue.

In Fig. 12.14 we show, for instance, an analysis of a cantilever beam using four quadratic, 'serendipity'-type, elements. While the results for deflections and axial stresses are excellent, the shear stresses show a parabolic 'variation' in each element which provides an extremely poor representation of the actual stresses. However, the values sampled at the Gauss points are an excellent representation of the correct mean shear stresses.

Similar improvements can be shown in the context of other elements and problems, although (fortunately) the discrepancies are not always so large.

The example just quoted suggests that in quadratic C_0 elements, whether two- or three-dimensional, the stresses (or similar quantities) *should never be calculated at nodes.* If nodal values are desired, then a projection or a simple bilinear extrapolation from Gauss points should be made. Such values are again shown to be excellent in Fig. 12.14. Further examples of such extrapolations are given by Hinton and Campbell.[34] Hinton *et al.*[41] and Barlow[39] give a very simple extrapolation algorithm for use in such problems.

Proof of the theorem

Variation of Π defined in Eq. (12.73) gives, at $\mathbf{u} = \bar{\mathbf{u}}$ (the exact solution),

$$\delta\Pi = \frac{1}{2}\int_\Omega (\mathbf{S}\,\delta\mathbf{u})^\mathrm{T}\mathbf{AS}\bar{\mathbf{u}}\,d\Omega + \frac{1}{2}\int_\Omega (\mathbf{S}\bar{\mathbf{u}})^\mathrm{T}\mathbf{AS}\,\delta\mathbf{u}\,d\Omega + \int_\Omega \delta\mathbf{u}^\mathrm{T}\mathbf{p}\,d\Omega = 0$$

or, if $\mathbf{A}$ is symmetric,

$$\delta\Pi = \int_\Omega (\mathbf{S}\,\delta\mathbf{u})^\mathrm{T}\mathbf{AS}\bar{\mathbf{u}}\,d\Omega + \int_\Omega \delta\mathbf{u}^\mathrm{T}\mathbf{p}\,d\Omega = 0$$

in which $\delta\mathbf{u}$ is any arbitrary variation. Thus we can write

$$\delta\mathbf{u} = \mathbf{u}$$

and

$$\int_\Omega (\mathbf{S}\mathbf{u})^\mathrm{T}\mathbf{AS}\bar{\mathbf{u}}\,d\Omega + \int_\Omega \mathbf{u}^\mathrm{T}\mathbf{p}\,d\Omega = 0$$

Subtracting the above from Eq. (12.73) and noting the symmetry of the $\mathbf{A}$ matrix, we can write

$$\Pi = \frac{1}{2}\int_\Omega [\mathbf{S}(\mathbf{u} - \bar{\mathbf{u}})]^\mathrm{T}\mathbf{AS}(\mathbf{u} - \bar{\mathbf{u}})\,d\Omega - \frac{1}{2}\int_\Omega (\mathbf{S}\bar{\mathbf{u}})^\mathrm{T}\mathbf{AS}\bar{\mathbf{u}}\,d\Omega$$

where the last term is not subject to variation. Thus

$$\Pi^* = \Pi + \text{constant}$$

and its stationarity is equivalent to the stationarity of Π.

12.7 Reduced and selective integration and its equivalence to penalized mixed problems

In Chapter 8 we have mentioned the lowest order numerical integration rules that still preserve the required convergence order for various elements, but at the same time pointed out the possibility of singularity in the resulting element matrices. In Chapter 11 we again referred to such low order integration rules, introducing the name of 'reduced integration' for those that did not evaluate the stiffness exactly for simple elements and pointed out some dangers of its indiscriminate use due to resulting instability. Nevertheless, such reduced integration and selective integration (where low order approximation is only applied to certain parts of the matrix) has proved its worth in practice yielding often much more accurate results than the use of more precise integration rules. This was particularly noticeable in nearly incompressible elasticity (or Stokes flow which is similar)[42-44] and in problems of plate and shell flexure dealt with as a case of degenerate solid.[45,46]

The success of these procedures derived initially by heuristic arguments proved quite spectacular—though some considered it somewhat verging on immorality to obtain improved results while doing less work! Obviously fuller justification of such processes was sought[47] and one of the obvious reasons was that frequently the reduced integration formulae sample at precisely the same points that we have identified in the previous section as optimal. However, the main reason for success does not lie here but is associated with the fact that it provides the necessary singularity of the constraint part of the matrix [viz. Eqs (12.18) to (12.20)] which avoids locking. Such singularity can be deduced from a count of integration points,[47,48] but it is simpler to show that there is a complete equivalence between reduced (or selective) integration procedures and the mixed formulation already discussed. This equivalence was first shown by Malkus and Hughes[49] and later in a general context by Zienkiewicz and Nakazawa.[50]

We shall demonstrate this equivalence on the basis of the nearly incompressible elasticity problem for which the mixed weak integral statement is given by Eqs (12.44) and (12.45).

The corresponding irreducible form can be written by satisfying the second of these equations exactly by implying

$$p = K\mathbf{m}^T\boldsymbol{\varepsilon} \tag{12.75}$$

and substituting above into (12.44) as

$$\int_\Omega \delta\boldsymbol{\varepsilon}^T G(\mathbf{D}_0 - \tfrac{2}{3}\mathbf{mm}^T)\boldsymbol{\varepsilon}\ d\Omega + \int_\Omega \delta\boldsymbol{\varepsilon}^T \mathbf{m}K\mathbf{m}^T\boldsymbol{\varepsilon}\ d\Omega$$
$$- \int_\Omega \delta\mathbf{u}^T\mathbf{b}\ d\Omega - \int_{\Gamma_t} \delta\mathbf{u}^T\bar{\mathbf{t}}\ d\Gamma = 0 \tag{12.76}$$

On substituting

$$\mathbf{u} \cong \hat{\mathbf{u}} = \mathbf{N}_u\bar{\mathbf{u}} \quad \text{and} \quad \boldsymbol{\varepsilon} \approx \hat{\boldsymbol{\varepsilon}} = \mathbf{SN}\bar{\mathbf{u}} = \mathbf{B}\bar{\mathbf{u}} \tag{12.77}$$

we have

$$(\mathbf{A} + \bar{\mathbf{A}})\bar{\mathbf{u}} = \mathbf{f}_1 \tag{12.78}$$

where $\mathbf{A}$ and $\mathbf{f}_1$ are exactly as given in Eq. (12.48) and

$$\bar{\mathbf{A}} = \int_\Omega \mathbf{B}^T \mathbf{m} K \mathbf{m}^T \mathbf{B} \, d\Omega \qquad (12.79)$$

The solution of Eq. (12.78) for $\bar{\mathbf{u}}$ allows the pressures to be determined at all points by Eq. (12.75). In particular, if we have used an integration scheme for evaluating (12.79) which sampled at points (ξ) we can write

$$p(\xi) = K\mathbf{m}^T \boldsymbol{\varepsilon}(\xi) = K\mathbf{m}^T \mathbf{B}(\xi) \bar{\mathbf{u}} \qquad (12.80)$$

Now if we turn our attention to the penalized mixed form of Eqs (12.44) to (12.48) we note that the second of equations (12.47) is explicitly

$$\int_\Omega \mathbf{N}_p^T \left(\mathbf{m}^T \mathbf{B} \bar{\mathbf{u}} - \frac{\mathbf{N}_p \bar{\mathbf{p}}}{K} \right) d\Omega = 0 \qquad (12.81)$$

If a numerical integration is applied to above sampling at the pressure nodes located at coordinate (ξ), previously defined in Eq. (12.80), we can write for each scalar component of $\mathbf{N}_p$

$$\sum \mathbf{N}_{p,j}(\xi) \left(\mathbf{m}^T \mathbf{B}(\xi) \bar{\mathbf{u}} - \frac{\mathbf{N}_p(\xi) \bar{\mathbf{p}}}{K} \right) W_\xi = 0 \qquad (12.82)$$

in which the summation is over all integration points (ξ) and W_ξ are the appropriate weighting function and jacobian determinants.

Now as

$$N_{p,j}(\xi) = 1$$

if ξ is at the node j and zero at other nodes Eq. (12.82) reduces simply to the requirement that at all pressure nodes

$$\mathbf{m}^T \mathbf{B}(\xi) \bar{\mathbf{u}} = \frac{\mathbf{N}_p(\xi) \bar{\mathbf{p}}}{K} \qquad (12.83)$$

This is precisely the same condition as that given by Eq. (12.80) and the equivalence of the procedures is proved, *providing the integrating scheme used for evaluating $\bar{\mathbf{A}}$ gives an exact integral of the mixed form of Eq. (12.81)*.

This is true in most cases and for those the reduced integration–mixed equivalence is exact. In all other cases this equivalence exists for a mixed problem in which an inexact rule of integration has been used in evaluating such equations as (12.81).

For curved isoparametric elements the equivalence is in fact inexact, and slightly different results can be obtained using reduced integration and mixed forms. This is illustrated in examples given in reference 51.

We can conclude without detailed proof that this type of equivalence is quite general and that with any problem of a similar type the application of numerical quadrature at n_p points in evaluating the matrix $\bar{\mathbf{A}}$ within each element is equivalent to a mixed problem in which the variable p is interpolated element by element using as nodal values the same integrating points.

The equivalence is only complete for the selective integration process, i.e., application of numerical quadrature only to the matrix $\bar{\mathbf{A}}$, and ensures that this matrix is singular, i.e., no locking occurs if we have satisfied the previously stated conditions ($n_u > n_p$).

The full use of reduced integration on the remainder of the matrix determining $\mathbf{u}$, i.e., $\mathbf{A}$, is only permissible if that remains non-singular— the case which we have discussed previously for the Q 8/4 element.

It can therefore be concluded that all the elements with discontinuous interpolation of p which we have verified as applicable to the mixed problem (viz. Fig. 12.7, for instance) can be implemented for nearly incompressible situations by a penalized irreducible form using corresponding selective integration.†

In Fig. 12.15 we show an example which clearly indicates the improvement of displacements achieved by such reduced integration as the compressibility modulus K increases (or the Poisson's ratio tends to 0.5). We note also in this example the dramatically improved performance of such points for stress sampling.

For problems in which the p (constraint) variable is continuously interpolated (C_0) the arguments given above fail as such quantities as $\mathbf{m}^T\varepsilon$ are not interelement continuous in the irreducible form.

A very interesting corollary of the equivalence just proved for (nearly) incompressible behaviour is observed if we note the rapid increase of order of integrating formulae with the number of quadrature points (viz. Chapter 8). *For high order elements the number of quadrature points equivalent to the p constraint permissible for stability rapidly reaches that required for exact integration and hence their performance in nearly incompressible situations is excellent, even if exact integration is used.* This was observed on many occasions[52-54] and Sloan and Randolph[55] have shown good performance with the quintic triangle. Unfortunately such high order elements pose other difficulties and are seldom used in practice.

A final remark concerns the use of 'reduced' integration in particular and of penalized, mixed, methods in general. As we have pointed out in Sec. 12.3.1 it is possible in such forms to obtain sensible results for the

† The Q 9/3 element would involve a three-point quadrature which is somewhat unnatural for quadrilaterals. It is therefore better to simply use the mixed form here.

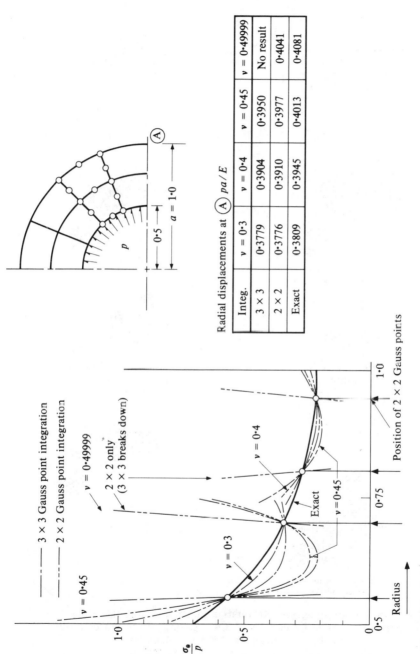

Radial displacements at Ⓐ pa/E

Integ.	$v = 0.3$	$v = 0.4$	$v = 0.45$	$v = 0.49999$
3 × 3	0·3779	0·3904	0·3950	No result
2 × 2	0·3776	0·3910	0·3977	0·4041
Exact	0·3809	0·3945	0·4013	0·4081

Fig. 12.15 Sphere under internal pressure. Effect of numerical integration rules on results with different Poisson's ratios

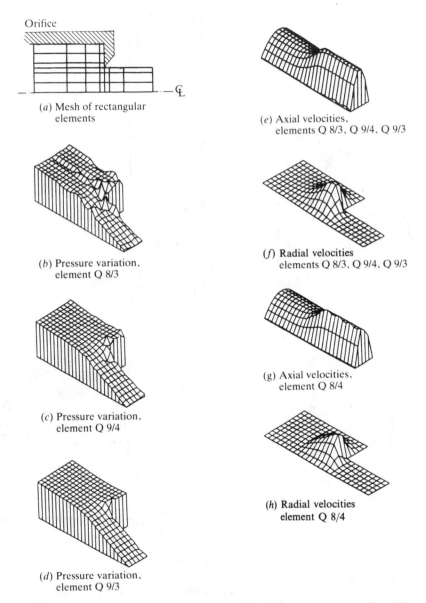

(a) Mesh of rectangular
    elements

(b) Pressure variation,
    element Q 8/3

(c) Pressure variation,
    element Q 9/4

(d) Pressure variation,
    element Q 9/3

(e) Axial velocities,
    elements Q 8/3, Q 9/4, Q 9/3

(f) Radial velocities
    elements Q 8/3, Q 9/4, Q 9/3

(g) Axial velocities,
    element Q 8/4

(h) Radial velocities
    element Q 8/4

Fig. 12.16 Steady-state, low Reynolds number flow through an orifice. Note that pressure variation for element Q 8/4 is so large it can not be plotted. Solution with **u**/p elements Q 8/3, Q 8/4, Q 9/3, Q 9/4

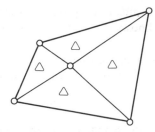

Fig. 12.17 A quadrilateral with intersecting diagonals forming an assembly of four T 3/1 elements. This allows displacements to be determined for nearly incompressible behaviour but does not yield pressure results

primary variable (**u** in the present example) even though the general stability conditions are violated, providing some of the *constraint equations* are linearly dependent. Now of course the *constraint variable* (*p* in the present example) is not determinate in the limit.

This situation occurs with some elements that are occasionally used for solution of incompressible problems but which do not pass our patch test, such as Q 8/4 and Q 9/4 of Fig. 12.7. If we take the latter number to correspond to the integrating points these will yield acceptable **u** fields, though not *p*.

Figure 12.16 illustrates the point on an application involving slow viscous flow through an orifice—a problem that obeys identical equations to those of incompressible elasticity. Here elements Q 8/4, Q 8/3, Q 9/4, and Q 9/3 are compared although only the last satisfies the stability requirements completely. All elements are found to give a reasonable velocity (**u**) field but pressures are acceptable only for the last one, with element Q 8/4 failing to give plottable results.[2]

It is of passing interest to note that a similar situation develops if four triangles of the T 3/1 type are assembled to form a quadrilateral in the manner of Fig. 12.17. Although the original element locks, as we have previously demonstrated, a linear dependence of the constraint equation allows the assembly to be used quite effectively in many incompressible situations, as shown in reference 53.

12.8 A simple iterative solution process for mixed problems

12.8.1 *General.* In the general remarks on the algebraic solution of mixed problems characterized by equations of the type [viz. Eq. (12.13)]

$$\begin{bmatrix} \mathbf{A} & \mathbf{C} \\ \mathbf{C}^{\mathrm{T}} & \mathbf{O} \end{bmatrix} \begin{Bmatrix} \mathbf{x} \\ \mathbf{y} \end{Bmatrix} = \begin{Bmatrix} \mathbf{f}_1 \\ \mathbf{f}_2 \end{Bmatrix} \tag{12.84}$$

we have remarked on the difficulties posed by the zero diagonal and the increased number of unknowns $(n_x + n_y)$ as compared with the irreducible form (n_x).

A general iterative form of solution is, however, possible which substantially reduces the cost.[56] In this we solve successively

$$\mathbf{y}^{n+1} = \mathbf{y}^n + \mathbf{\rho}\mathbf{r}^n \tag{12.85}$$

where $\mathbf{r}^n$ is the residual of the second equation computed as

$$\mathbf{r}^n = \mathbf{C}^T\mathbf{x}^n - \mathbf{f}_2 \tag{12.86}$$

and follow with solution of the first equation, i.e.,

$$\mathbf{x}^{n+1} = \mathbf{A}^{-1}(\mathbf{f}_1 - \mathbf{C}\mathbf{y}^{n+1}) \tag{12.87}$$

In the above $\mathbf{\rho}_n$ is a 'convergence accelerator matrix' and is chosen to be efficient and simple to use.

The algorithm is similar to that described initially by Uzawa[57] and has been widely applied in an optimization context.[58-62]

Its relative simplicity can best be grasped when particular examples are considered.

12.8.2 *Iterative solution for incompressible elasticity.* In this case we start from Eq. (12.47) now written with $V = 0$, i.e., complete incompressibility assumed. The various matrices are defined in (12.48), resulting in the form

$$\begin{bmatrix} \mathbf{A} & \mathbf{C} \\ \mathbf{C}^T & \mathbf{O} \end{bmatrix} \begin{Bmatrix} \bar{\mathbf{u}} \\ \bar{\mathbf{p}} \end{Bmatrix} = \begin{Bmatrix} \mathbf{f}_1 \\ \mathbf{0} \end{Bmatrix} \tag{12.88}$$

Now, however, the matrix $\mathbf{A}$ is singular (as volumetric changes are not restrained) and it is necessary to *augment* it to make it non-singular. We can do this in the manner described in the footnote to Sec. 12.3, or more conveniently by addition of a fictitious compressibility matrix, thus replacing $\mathbf{A}$ by

$$\bar{\mathbf{A}} = \mathbf{A} + \int_\Omega \mathbf{B}^T(\lambda G\mathbf{m}\mathbf{m}^T)\mathbf{B}\, d\Omega \tag{12.89}$$

If the second matrix uses an integration consistent with the number of discontinuous pressure parameters assumed, then this is precisely equivalent to writing

$$\bar{\mathbf{A}} = \mathbf{A} + \lambda G\mathbf{C}\mathbf{C}^T \tag{12.90}$$

and is simpler to evaluate. Clearly this addition does not change the equation system.

The iteration of the algorithm (12.85) to (12.87) is now conveniently taken with the 'convergence accelerator', being simply defined as

$$\rho = \lambda G \mathbf{I} \qquad (12.91)$$

We now have the iterative system given as

$$\bar{\mathbf{p}}^{n+1} = \bar{\mathbf{p}}^n + \rho \mathbf{r}^n \qquad (12.92a)$$

$$\mathbf{r}^n = \mathbf{C}^T \bar{\mathbf{u}}^n \qquad \text{(the residual of} \atop \text{incompressible constraint)} \qquad (12.92b)$$

and

$$\bar{\mathbf{u}}^{n+1} = \bar{\mathbf{A}}^{-1}(\mathbf{f}_1 - \mathbf{C}\bar{\mathbf{p}}^{n+1}) \qquad (12.92c)$$

In this $\bar{\mathbf{A}}$ can be interpreted as the stiffness matrix of a compressible material with a bulk modulus

$$K = \lambda G$$

and the process of iteration as that of successive addition of volumetric 'initial' strains designed to reduce the volumetric strain to zero. Indeed, this simple approach first led to the first realization of this algorithm.[63-65] Alternatively the process can be visualized as an amendment of the original equation (12.88) by subtracting the term $\mathbf{p}/\rho$ from each side of the second to give (this is often called an augmented lagrangian form)[56, 62]

$$\begin{bmatrix} \mathbf{A} & \mathbf{C} \\ \mathbf{C}^T & -\mathbf{I}/\rho \end{bmatrix} \begin{Bmatrix} \bar{\mathbf{u}} \\ \bar{\mathbf{p}} \end{Bmatrix} = \begin{Bmatrix} \mathbf{f}_1 \\ -\bar{\mathbf{p}}/\rho \end{Bmatrix} \qquad (12.93)$$

and adopting an iteration

$$\begin{bmatrix} \mathbf{A} & \mathbf{C} \\ \mathbf{C}^T & -\mathbf{I}/\rho \end{bmatrix} \begin{Bmatrix} \bar{\mathbf{u}} \\ \bar{\mathbf{p}} \end{Bmatrix}^{n+1} = \begin{Bmatrix} \mathbf{f}_1 \\ -\bar{\mathbf{p}}^n/\rho \end{Bmatrix} \qquad (12.94)$$

With this, on elimination a sequence similar to Eq. (12.92) will be obtained provided that $\bar{\mathbf{A}}$ is defined by Eq. (12.90).

Starting the iteration from

$$\mathbf{u}^0 = 0 \qquad \text{and} \qquad \bar{\mathbf{p}}^0 = 0$$

in Fig. 12.18 we show the convergence of the maximum divergence of $\mathbf{u}$ computed at any of the integrating points used. We note that this convergence becomes quite rapid for large values of $\lambda = (10^3$ to $10^4)$.

For smaller λ values the process can be accelerated by using different ρ values[56] but for practical purposes the simple algorithm suffices. Clearly much better satisfaction of the incompressibility constraint can now be obtained than by the simple use of a 'large enough' bulk

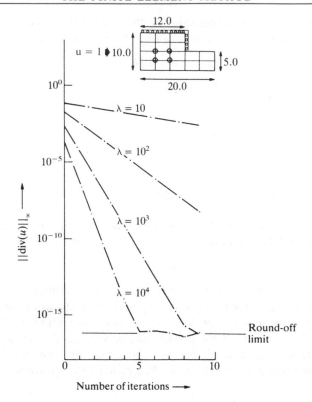

Fig. 12.18   Convergence of iterations in an extrusion problem for different
values of the penalty ratio $\lambda = \gamma/\mu$

modulus (or penalty) parameter. With $\lambda = 10^4$, for instance, in five iterations the initial divergence of **u** is reduced from the value $\sim 10^{-4}$ to 10^{-16}, which is at the limit of precision of the particular computer here used.

The reader should note that the iteration often allows a single precision operation to be used in a problem which otherwise would require a double precision operation.

12.8.3 *Iterative solution for* **u**, **σ**, **ε** *mixed formulation.* In this example we shall show how the mixed solution procedure can be used to improve the results of an irreducible, displacement formulation.

The starting point is the formulation of Eq. (12.34) in which we assume that identical, C_0 continuous, shape functions approximate all variables.

We thus have, assuming 'standard' shape functions,

$$\mathbf{N}_u = \mathbf{N}_\sigma = \mathbf{N}_\varepsilon = \mathbf{N} \tag{12.95}$$

$$\begin{bmatrix} \mathbf{0} & \mathbf{C} & \mathbf{E} \\ \mathbf{C}^T & \mathbf{A} & \mathbf{0} \\ \mathbf{E}^T & \mathbf{0} & \mathbf{0} \end{bmatrix} \begin{Bmatrix} \bar{\boldsymbol{\sigma}} \\ \bar{\boldsymbol{\varepsilon}} \\ \bar{\mathbf{u}} \end{Bmatrix} = \begin{Bmatrix} 0 \\ 0 \\ \mathbf{f}_3 \end{Bmatrix} \tag{12.96}$$

with

$$\mathbf{A} = \int_\Omega \mathbf{N}^T \mathbf{D} \mathbf{N} \, d\Omega$$

$$\mathbf{C} = -\int_\Omega \mathbf{N}^T \mathbf{N} \, d\Omega = \mathbf{C}^T \tag{12.97}$$

$$\mathbf{E} = \int_\Omega \mathbf{N}^T \mathbf{B} \, d\Omega$$

where $\bar{\boldsymbol{\sigma}}$, $\bar{\boldsymbol{\varepsilon}}$, and $\bar{\mathbf{u}}$ define nodal parameters.

Further we shall assume that the second equation giving the constitutive law is evaluated using *nodal quadrature* and thus simply leads to the specification of the constitutive law at such nodes, giving there

$$(\bar{\boldsymbol{\sigma}})_i = (\mathbf{D}\bar{\boldsymbol{\varepsilon}})_i \tag{12.98}$$

The iterative process starts, as in the general case of Eq. 12.84, from the constraint variable and as we know that good solutions can be obtained with the standard irreducible form we specify

$$\boldsymbol{\rho} = \mathbf{K}^{-1} \equiv \left(\int_\Omega \mathbf{B}^T \mathbf{D} \mathbf{B} \, d\Omega \right)^{-1} \tag{12.99}$$

The iteration becomes

$$\bar{\mathbf{u}}^{n+1} = \bar{\mathbf{u}}^n - \mathbf{K}^{-1} \mathbf{r}^n \tag{12.100a}$$

with

$$\mathbf{r}^n = \mathbf{E}^T \bar{\boldsymbol{\sigma}}^n - \mathbf{f}_3 \tag{12.100b}$$

$$\bar{\boldsymbol{\varepsilon}}^{n+1} = -\mathbf{C}^{-1} \mathbf{E} \bar{\mathbf{u}}^{n+1} \tag{12.100c}$$

$$(\bar{\boldsymbol{\sigma}}^{n+1})_i = (\mathbf{D}\bar{\boldsymbol{\varepsilon}}^{n+1})_i \tag{12.100d}$$

Starting from $\bar{\boldsymbol{\sigma}}^0 = \bar{\boldsymbol{\varepsilon}}^0 = \mathbf{u}^0 = 0$ we note immediately that the first computation of $\mathbf{u}^1$ represents simply the standard displacement solution.

The second step of the computation is precisely the smoothing of discontinuous strains $\mathbf{B}\bar{\mathbf{u}}^1$ to obtain nodally continuous values. The oper-

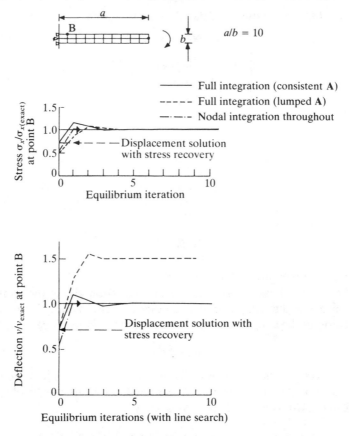

Fig. 12.19   Iterative solution of the mixed σ/u formulation for a beam. Bilinear **u**
and σ

ations are identical to those used in Sec. 12.6.1 and again a saving of computation can be made by lumping (diagonalizing) the matrix **C** and using the iterations of Eq. (12.70).

Alternatively, excellent results can often be obtained by evaluating both **C** and $E\bar{u}^{n+1}$ by a nodal quadrature and dispensing with the iterative process.[56,66-68]

In Figs 12.19 and 12.20 we show some examples which illustrate the very considerable improvement of performance obtainable by this mixed procedure. The cost of solution is not much larger than that of irreducible form and convergence is rapid.

The process is similar to the one suggested by Cantin *et al.*[69,70] as an *iterative improvement* of standard irreducible, displacement-type, solution.

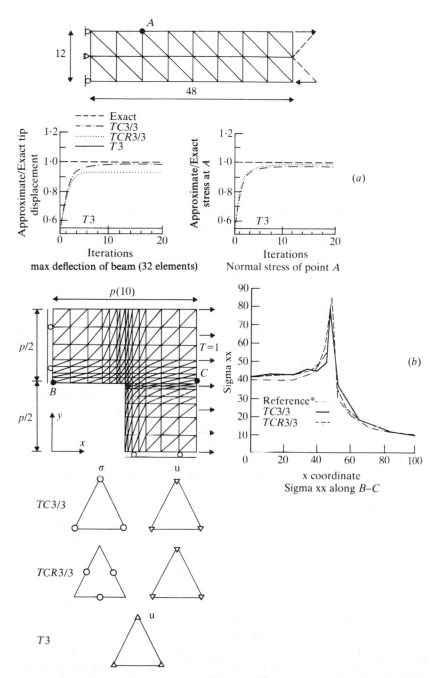

Fig. 12.20 Iterative solution of the mixed **σ/u** formulation using two triangular element forms TC 3/3 and TCR 3/3. (a) A beam showing convergence with iterations. (b) An L-shaped domain showing the improved results of stress distribution when no continuity of stress is imposed at singularity (element TCR 3/3)

The sequence can be recognized simply as that of

- (a) smoothing of stresses and strains to continuous form (by the averaging procedure) and
- (b) adjusting the displacement solution to account for out-of-balance forces created by the smoothing.

If interpreted in this way we note that in fact it is not necessary to proceed to full convergence of the mixed method. *Each solution with a fixed number of iterations offers an acceptable approximation, usually of better quality than the pure displacement solution.*

It is important once again to stress that continuity of strains/stresses should not be imposed on singularities and abrupt material changes. If this is done the mixed solution can deteriorate the result of the simple displacement approach.

The simple iteration is a worthwhile addition to 'standard' displacement programs to improve accuracy.

12.9 Complementary forms with direct constraint

12.9.1 *General forms.* In the introduction to this chapter we defined the irreducible and mixed forms and indicated that on occasion it is possible to obtain more than one 'irreducible' form. To illustrate this in the problem of heat transfer given by Eqs (12.2) and (12.3) we introduced a penalty function α in Eq. (12.6) and derived a corresponding single governing equation (12.7) given in terms of q. This penalty function here has no obvious physical meaning and served simply as a device to obtain a *close enough* approximation to the satisfaction of the continuity of flow equations. Its role is precisely the same as that of the physically meaningful penalty we introduced in the incompressible problem and *selective integration* procedures could be justified. In the context of elasticity a similar penalty function was used successfully to solve the σ–u problem of Sec. 12.4.2, eliminating u and dealing entirely with stresses as the problem variable.[4] Selective integration was here found essential in the manner indicated in Sec. 12.7.

On occasion it is possible to solve the problem as an irreducible one assuming *a priori* that the choice of the variable satisfies one of the equations. We call such forms *directly constrained* and obviously the choice of the shape function becomes difficult.

We shall consider two examples.

The complementary heat transfer problem. In this we assume *a priori* that the choice of q is such that it satisfies Eq. (12.3) and the natural boundary conditions

$$\mathbf{V}^T\mathbf{q} = Q \text{ in } \Omega \quad \text{and} \quad q_n = \bar{q} \text{ on } \Gamma_q \qquad (12.101)$$

Thus we only have to satisfy the constitutive relation (12.2), i.e.,

$$\mathbf{k}^{-1}\mathbf{q} + \mathbf{V}\phi = \mathbf{0} \text{ in } \Omega \quad \text{with } \phi = \bar{\phi} \text{ on } \Gamma_\phi \qquad (12.102)$$

A weak statement of the above is

$$\int_\Omega \delta\mathbf{q}^T(\mathbf{k}^{-1}\mathbf{q} + \nabla\phi)\, d\Omega - \int_{\Gamma_\phi} \delta q_n(\phi - \tilde{\phi})\, d\Gamma = 0 \qquad (12.103)$$

in which δq_n represents the variation of normal flux on the boundary.

Use of Green's theorem transforms the above into

$$\int_\Omega \delta\mathbf{q}^T\mathbf{k}^{-1}\mathbf{q}\, d\Omega - \int_\Omega \nabla^T\delta\mathbf{q}\phi\, d\Omega + \int_\Gamma \delta q_n \tilde{\phi}\, d\Gamma + \int_{\Gamma_q} \delta q_n \phi\, d\Gamma = 0$$
$$(12.104)$$

If we further assume that $\nabla^T\delta\mathbf{q} \equiv 0$ in Ω and $\delta q_n = 0$ on Γ_q, i.e., that the weighting functions are simply the variations of $\mathbf{q}$ and satisfy the continuity conditions of Eq. (12.101), the equation reduces to

$$\int_\Omega \delta\mathbf{q}^T\mathbf{k}^{-1}\mathbf{q}\, d\Omega + \int_{\Gamma_\phi} \delta q_n \tilde{\phi}\, d\Gamma = 0 \qquad (12.105)$$

This is in fact the variation of a complementary flux principle

$$\Pi = \int_\Omega \tfrac{1}{2}\mathbf{q}^T\mathbf{k}^{-1}\mathbf{q}\, d\Omega + \int_{\Gamma_\phi} q_n \tilde{\phi}\, d\Gamma \qquad (12.106)$$

Numerical solutions can obviously be started from either of the above equations but the difficulty is the choice of the trial function satisfying the constraints. We shall return to this problem later.

The complementary elastic energy principle. In the elasticity problem specified in Sec. 12.4 we can proceed similarily, assuming stress fields which satisfy the equilibrium conditions both on the boundary Γ_t and in the domain Ω.

Thus in an analogous manner to that of the previous example we impose on the permissible stress field the constraints which we assume to be satisfied by the approximation identically, i.e.,

$$\mathbf{S}^T\boldsymbol{\sigma} + \mathbf{b} = 0 \text{ in } \Omega \quad \text{and} \quad \mathbf{t} = \tilde{\mathbf{t}} \text{ on } \Gamma_t, \qquad (12.107)$$

Thus only the constitutive relations and displacement boundary conditions remain to be satisfied, i.e.,

$$\mathbf{D}^{-1}\boldsymbol{\sigma} - \mathbf{Su} = 0 \text{ in } \Omega \quad \text{and} \quad \mathbf{u} = \tilde{\mathbf{u}} \text{ on } \Gamma_u \qquad (12.108)$$

The weak statement of the above can be written as

$$\int_\Omega \delta\boldsymbol{\sigma}^T(\mathbf{D}^{-1}\boldsymbol{\sigma} - \mathbf{Su})\, d\Omega + \int_{\Gamma_u} \delta\mathbf{t}^T(\mathbf{u} - \tilde{\mathbf{u}})\, d\Gamma = 0 \qquad (12.109)$$

which on integration by Green's theorem gives

$$\int_\Omega \delta\boldsymbol{\sigma}^T\mathbf{D}^{-1}\boldsymbol{\sigma}\, d\Omega + \int_\Omega (\mathbf{S}^T\delta\boldsymbol{\sigma})^T\mathbf{u}\, d\Omega - \int_{\Gamma_u} \delta\mathbf{t}^T\tilde{\mathbf{u}}\, d\Gamma - \int_{\Gamma_t} \delta\mathbf{t}^T\mathbf{u}\, d\Gamma = 0$$
$$(12.110)$$

Again assuming that the test functions are complete variations satisfying the homogeneous equilibrium equation, i.e.,

$$\mathbf{S}^T \delta\boldsymbol{\sigma} = 0 \text{ in } \Omega \qquad \text{and} \qquad \delta\mathbf{t} = 0 \text{ on } \Gamma_t \qquad (12.111)$$

we have as the weak statement

$$\int_\Omega \delta\boldsymbol{\sigma}^T \mathbf{D}^{-1}\boldsymbol{\sigma} \, d\Omega - \int_{\Gamma_u} \delta\mathbf{t}\tilde{\mathbf{u}} \, d\Gamma = 0 \qquad (12.112)$$

The corresponding complementary energy variational principle is

$$\Pi = \frac{1}{2} \int_\Omega \boldsymbol{\sigma}^T \mathbf{D}^{-1}\boldsymbol{\sigma} \, d\Omega - \int_{\Gamma_u} \mathbf{t}^T\tilde{\mathbf{u}} \, d\Gamma \qquad (12.113)$$

Once again in practical use the difficulties connected with the choice of the approximating function arise but on occasion a direct choice is possible.[70]

12.9.2 *Solution using auxiliary function.* Both the complementary forms can be solved using auxiliary functions to ensure the satisfaction of the constraints.

In the *heat transfer problem* it is easy to verify that the homogeneous equation

$$\mathbf{V}^T\mathbf{q} \equiv \frac{\partial}{\partial x}\, q_x + \frac{\partial}{\partial y}\, q_y = 0 \qquad (12.114)$$

is automatically satisfied by defining a function ψ such that

$$q_x = \frac{\partial\psi}{\partial y} \qquad q_y = -\frac{\partial\psi}{\partial x} \qquad (12.115)$$

Thus if we define

$$\mathbf{q} = \mathbf{L}\psi + \mathbf{q}_0 \qquad \text{and} \qquad \delta\mathbf{q} = \mathbf{L}\,\delta\psi \qquad (12.116)$$

where $\mathbf{q}_0$ is any flux chosen so that

$$\mathbf{V}^T\mathbf{q}_0 = Q \qquad (12.117)$$

and

$$\mathbf{L}^T = \left[\frac{\partial}{\partial y},\ -\frac{\partial}{\partial x}\right] \qquad (12.118)$$

the formulation of Eqs (12.105) and (12.106) can be used without any constraints and, for instance, the stationarity

$$\Pi = \int_\Omega \tfrac{1}{2}(\mathbf{L}\psi + \mathbf{q}_0)^T\mathbf{k}^{-1}(\mathbf{L}\psi + \mathbf{q}_0) \, d\Omega - \int_{\Gamma_\phi} \left(\frac{\partial\psi}{\partial t}\right)^T \tilde{\phi} \, d\Gamma \quad (12.119)$$

will suffice to so formulate the problem (here t is the tangented direction to boundary).

The above form will require shape functions for ψ satisfying C_0 continuity.

In the corresponding elasticity problem a similar two-dimensional form can be obtained by the use of the so-called Airy stress function ψ.[71]

Now the equilibrium equations

$$\mathbf{S}^T\boldsymbol{\sigma} + \mathbf{b} \equiv \begin{bmatrix} \dfrac{\partial}{\partial x}\, \sigma_x + \dfrac{\partial}{\partial y}\, \tau_{xy} \\[2mm] \dfrac{\partial}{\partial y}\, \sigma_y + \dfrac{\partial}{\partial x}\, \tau_{xy} \end{bmatrix} + \begin{Bmatrix} b_x \\ b_y \end{Bmatrix} = 0 \tag{12.120}$$

are identically solved by choosing

$$\boldsymbol{\sigma} = \mathbf{L}\psi + \boldsymbol{\sigma}_0 \tag{12.121}$$

where

$$L^T = \begin{bmatrix} \dfrac{\partial^2}{\partial y^2}, & \dfrac{\partial^2}{\partial x^2}, & -\dfrac{\partial^2}{\partial x\,\partial y} \end{bmatrix} \tag{12.122}$$

and $\boldsymbol{\sigma}_0$ is an arbitrary stress so chosen that

$$\mathbf{S}^T\boldsymbol{\sigma}_0 + \mathbf{b} = 0 \tag{12.123}$$

Again the substitution of (12.121) into the weak statement (12.112) or the complementary variational principle (12.113) will yield a direct formulation to which no additional constraints need be applied.

The use of this two-dimensional stress function formulation in two-dimensional context was first made by De Veubeke and Zienkiewicz[72] and Elias,[73] but the reader should note that now with second-order operators present, C_1 continuity of shape functions is needed in a similar manner to the problems which we have to consider in plate bending (viz. Vol. 2).

Incidentally, analogies with plate bending go further here and indeed it can be shown that some of these can be usefully employed for other problems.[74]

12.10 Concluding remarks—mixed formulation or a test of element 'robustness'

The mixed form of finite element formulation outlined in this chapter opens a new range of possibilities, some with potentially higher accuracy than those offered by irreducible forms. Further, it allows direct treatment of constrained problems without the introduction of penalty forms. Typical in the last context is the development of direct approximations for incompressible elasticity.

However, an additional advantage arises even in situations where, by the *principle of limitation*, the irreducible and mixed forms yield identical results. Here the study of the behaviour of the mixed form can frequently reveal weaknesses or lack of 'robustness' in the irreducible form which otherwise would be difficult to determine. To illustrate this we can consider the case of compressible elasticity which can always be cast as a mixed form (viz. Sec. 12.5.1) or an equivalent irreducible form in which the variation of pressure is identical.

This equivalence in the case of a linear triangle (which we have used in the early chapters of the book) or the linear tetrahedron immediately reveals why such elements are 'not robust' when compressibility is small (or in isotropic elasticity Poisson's ratio tends to 0.5). The reason is clearly the inadmissible form of such elements when compressibility is at the zero limit.

By the same argument we can justify the use of 'reduced' (selective) integration in quadratic elements.

Such study of 'limiting' cases can often throw light on an otherwise mysterious element behaviour. The mixed approximation of properly understood expands the potential of the finite element method and presents almost limitless possibilities of detailed improvement. Some of these will be discussed in the next chapter—others in Vol. 2.

An interesting survey is given by Crisfield.[75]

References

1. S. N. ATLURI, R. H. GALLAGHER, and O. C. ZIENKIEWICZ (eds), *Hybrid and Mixed Finite Element Methods*, Wiley, 1983.
2. O. C. ZIENKIEWICZ, R. L. TAYLOR, and J. A. W. BAYNHAM, 'Mixed and irreducible formulations in finite element analysis', Chapter 21 of *Hybrid and Mixed Finite Element Methods* (eds S. N. Atluri, R. H. Gallagher, and O. C. Zienkiewicz), pp. 405–31, Wiley, 1983.
3. I. BABUSKA and J. E. OSBORN, 'Generalized finite element methods and their relations to mixed problems', *SIAM J. Num. Anal.*, **20**, 510–36, 1983.
4. R. L. TAYLOR and O. C. ZIENKIEWICZ, 'Complementary energy with penalty function in finite element analysis', Chapter 8 of *Energy Methods in Finite Element Analysis* (eds R. Glowinski, E. Y. Rodin, and O. C. Zienkiewicz), Wiley, 1979.
5. K. WASHIZU, *Variational Methods in Elasticity and Plasticity*, 2nd ed., Pergamon Press, 1975.
6. L. R. HERRMANN, 'Finite element bending analysis of plates', *Proc. 1st Conf. Matrix Methods in Structural Mechanics*, AFFDL-TR-80, Wright-Patterson AF Base, Ohio, 1965.
7. K. HELLAN, 'Analysis of plates in flexure by a simplified finite element method', *Acta Polytechnica Scandinavia*, Civ. Eng. Series 46, Trondheim, 1967.
8. R. S. DUNHAM and K. S. PISTER, 'A finite element application of the Hellinger Reissner variational theorem, *Proc. 1st Conf. Matrix Methods in Structural Mechanics*, Wright-Patterson AF Base, Ohio, 1965.

9. R. L. TAYLOR and O. C. ZIENKIEWICZ, 'Mixed finite element solution of fluid flow problems', Chapter 1 of *Finite Elements in Fluid*, Vol. 4 (eds R. H. Gallagher, D. N. Norrie, J. T. Oden, and O. C. Zienkiewicz), pp. 1–20, Wiley, 1982.

10. B. FRAEIJS DE VEUBEKE, 'Displacement and equilibrium models in finite element method', Chapter 9 of *Stress Analysis* (eds O. C. Zienkiewicz and C. S. Holister), pp. 145–97, Wiley, 1965.

11. I. BABUŠKA, 'The finite element method with Lagrange multipliers', *Num. Math.*, **20**, 179–92, 1973; also 'Error bounds for finite element methods', *Num. Math.*, **16**, 322–33, 1971.

12. F. BREZZI, 'On the existence, uniqueness and approximation of saddle point problems arising from lagrangian multipliers', *RAIRO*, 8-R2, 129–151, 1974.

13. O. C. ZIENKIEWICZ, S. QU, R. L. TAYLOR, and S. NAKAZAWA, 'The patch test for mixed formulation', *Int. J. Num. Meth. Eng.*, **23**, 1873–83, 1986.

14. J. T. ODEN and N. KIKUCHI, 'Finite element methods for constrained problems of elasticity', *Int. J. Num. Mech. Eng.*, **18**, 701–25, 1982.

15. E. HELLINGER, 'Die allgemeine Aussetze der Mechanik der Kontinua', in *Encyclopedia der Matematischen Wissenschaften*, Vol. 4 (eds F. Klein and C. Muller), Tebner, Leipzig, 1914.

16. E. REISSNER, 'On a variational theorem in elasticity', *J. Math. Phys.*, **29**, 90–5, 1950.

17. L. R. HERRMANN, 'Finite element bending analysis of plates', *Proc. Am. Soc. Civ. Eng.*, **94**, EM5, 13–25, 1968.

18. L. R. HERRMANN and D. M. CAMPBELL, 'A finite element analysis for thin shells, *JAIAA*, **6**, 1842–7, 1968.

19. O. C. ZIENKIEWICZ and D. LEFEBVRE, 'Mixed methods for FEM and the patch test. Some recent developments', in *Analyse Mathematique of Application* (eds F. Murat and O. Pirenneau), Gauthier Villars, Paris, 1988.

20. P. A. RAVIART and J. M. THOMAS, 'A mixed finite element method for second order elliptic problems', *Lect. Notes in Math.*, no. 606, pp. 292–315, Springer Verlag, 1977.

21. D. ARNOLD, F. BREZZI, and J. DOUGLAS, 'PEERS, a new mixed finite element for plane elasticity', *Japan J. Appl. Math.*, **1**, 347–67, 1984.

22. O. C. ZIENKIEWICZ and D. LEFEBVRE, 'Three field mixed approximation and the plate bending problem', *Comm. Appl. Num. Math.*, **3**, 301–9, 1987.

23. T. J. R. HUGHES, 'Generalization of selective integration procedures to anisotropic and non-linear media', *Int. J. Num. Meth. Eng.*, **15**, 1413–1418, 1980.

24. J. C. SIMO, R. L. TAYLOR and K. S. PISTER, 'Variational and projection methods for the volume constraint in finite deformation plasticity', *Comp. Meth. App. Mech. Eng.*, **51**, 177–208, 1985.

25. L. R. HERRMANN, 'Elasticity equations for incompressible and nearly incompressible materials by a variational theorem', *JAIAA*, **3**, 1896–900, 1965.

26. S. W. KEY, 'A variational principle for incompressible and nearly incompressible anisotropic elasticity', *Int. J. Solids Struct.*, 1970.

27. M. FORTIN and N. FORTIN, 'Newer and newer elements for incompressible flow', Chapter 7 of *Finite Elements in Fluids*, Vol. 6 (eds R. H. Gallagher, G. F. Carey, J. T. Oden, and O. C. Zienkiewicz), pp. 171–88, Wiley, 1985.

28. J. T. ODEN, 'R.I.P. methods for Stokesian flow', Chapter 15 of *Finite Elements in Fluids*, Vol. 4 (eds R. H. Gallagher, D. N. Norrie, J. T. Oden, and O. C. Zienkiewicz), pp. 305–18, Wiley, 1982.

29. M. Crouzcix and P. A. Raviart, 'Conforming and non-conforming finite element methods for solving stationary Stokes equations', *RAIRO*, 7-R3, 33–76, 1973.
30. D. S. Malkus, 'Eigenproblems associated with the discrete LBB condition for incompressible finite elements', *Int. J. Eng. Sci.*, **19**, 1299–370, 1981.
31. M. Fortin, 'Old and new finite elements for incompressible flow', *Int. J. Num. Meth. Fluids*, **1**, 347–64, 1981.
32. C. Taylor and P. Hood, 'A numerical solution of the Navier–Stokes equations using the finite element technique', *Computers in Fluids*, **1**, 73–100, 1973.
33. J. T. Oden and J. N. Reddy, 'Note on approximation method for computing consistent conjugate stresses in finite elements, *Int. J. Num. Meth. Eng.*, **6**, 55–61, 1973.
34. E. Hinton and J. Campbell, 'Local and global smoothing of discontinuous finite element function using a least squares method', *Int. J. Num. Meth. Eng.*, **8**, 461–80, 1974.
35. O. C. Zienkiewicz and S. Nakazawa, 'The penalty function method and its application to numerical solution of boundary value problems', *Am. Soc. Mech. Eng. AMD*, **51**, 157–79, 1982.
36. R. L. Sani, P. M. Gresho, R. L. Lee, and D. F. Griffiths, 'The cause and cure of the spurious pressures generated by certain FEM solutions of the Navier–Stokes equations, Part 1', *Int. J. Num. Meth. Fluids*, **1**, 17, 1981.
37. T. Moan, 'Orthogonal polynomials and "best" numerical integration formulas on a triangle', *ZAMM*, **54**, 501–8, 1974.
38. H. J. Brauchli and J. T. Oden, 'On the calculation of consistent stress distribution in finite element applications', *Int. J. Num. Meth. Eng.*, **3**, 317–25, 1971.
39. J. Barlow, 'Optimal stress locations in finite element models', *Int. J. Num. Meth. Eng.*, **10**, 243–51, 1976.
40. L. R. Herrmann, 'Interpretation of finite element procedure in stress error minimisation', *Proc. Am. Soc. Civ. Eng.*, **98**, EM5, 1331–6, 1972.
41. E. Hinton, F. C. Scott, and R. E. Ricketts, 'Local least squares stress smoothing for parabolic isoparametric elements', *Int. J. Num. Meth. Eng.*, **9**, 235–56, 1975.
42. D. J. Naylor, 'Stresses in nearly incompressible materials for finite elements with application to the calculation of excess pore pressures', *Int. J. Num. Meth. Eng.*, **8**, 443–60, 1974.
43. O. C. Zienkiewicz and P. N. Godbote, 'Viscous incompressible flow with special reference to non-Newtonian (plastic) flows', Chapter 2 of *Finite Elements in Fluids*, Vol. 1 (eds R. H. Gallagher *et al.*), pp. 25–55, Wiley, 1975.
44. T. J. R. Hughes, R. L. Taylor, and J. F. Levy, 'High Reynolds number, steady, incompressible flows by a finite element method', in *Finite Elements in Fluids*, Vol. 3 (eds R. H. Gallagher *et al.*), Wiley, Chichester, 1978.
45. O. C. Zienkiewicz, R. L. Taylor, and J. M. Too, 'Reduced integration techniques in general analysis of plates and shells', *Int. J. Num. Meth. Eng.*, **3**, 275–90, 1971.
46. S. F. Pawsey and R. W. Clough, 'Improved numerical integration of thick shell finite elements', *Int. J. Num. Meth. Eng.*, **3**, 545–86, 1971.
47. O. C. Zienkiewicz and E. Hinton, 'Reduced integration, function smoothing and nonconformity in finite element analysis', *J. Franklin Inst.*, **302**, 443–61, 1976.

48. O. C. ZIENKIEWICZ, *The Finite Element Method*, 3rd ed., McGraw-Hill, 1977.
49. D. S. MALKUS and T. J. R. HUGHES, 'Mixed finite element methods in reduced and selective integration techniques: a unification of concepts', *Comp. Meth. App. Mech. Eng.*, **15**, 1978.
50. O. C. ZIENKIEWICZ and S. NAKAZAWA, 'On variational formulation and its modification for numerical solution', *Comp. Struct.*, **19**, 303–13, 1984.
51. M. S. ENGLEMAN, R. L. SANI, P. M. GRESHO, and H. BERCOVIER, 'Consistent v. reduced integration penalty methods for incompressible media using several old and new elements', *Int. J. Num. Meth. Fluids*, **2**, 25–42, 1982.
52. D. N. ARNOLD, 'Discretization by finite elements of a model parameter dependent problem', *Num. Meth.*, **37**, 405–21, 1981.
53. J. C. NAGTEGAAL, D. M. PARKS, and J. R. RICE, 'On numerically accurate finite element solutions in the fully plastic range', *Comp. Meth. Appl. Mech. Eng.*, **4**, 153–78, 1974.
54. M. VOGELIUS, 'An analysis of the *p*-version of the finite element method for nearly incompressible materials; uniformly optimal error estimates', *Num. Math.*, **41**, 39–53, 1983.
55. S. W. SLOAN and M. F. RANDOLPH, 'Numerical prediction of collapse loads using finite element methods', *Int. J. Num. Anal. Meth. Geomechanics*, **6**, 47–76, 1982.
56. O. C. ZIENKIEWICZ, J. P. VILOTTE, S. TOYOSHIMA, and S. NAKAZAWA, 'Iterative method for constrained and mixed approximation. An inexpensive improvement of FEM performance', *Comp. Meth. Appl. Mech. Eng.*, **51**, 3–29, 1985.
57. K. J. ARROW, L. HURWICZ, and H. UZAWA, *Studies in Non-Linear Programming*, Stanford University Press, 1958.
58. M. R. HESTENS, 'Multiplier and gradient methods', *J. Optim. Theory Appl.*, **4**, 303–20, 1969.
59. M. J. D. POWELL, 'A method for non-linear constraints in optimization problems', in *Optimization* (ed. R. Fletcher), pp. 283–98, Academic Press, London, 1969.
60. C. A. FELLIPPA, 'Iterative procedure for improving penalty function solutions of algebraic systems', *Int. J. Num. Meth. Eng.*, **12**, 165–85, 1978.
61. M. FORTIN and F. THOMASSET, 'Mixed finite element methods for incompressible flow problems', *J. Comp. Physics*, **31**, 113–45, 1973.
62. M. FORTIN and R. GLOWINSKI, *Methodes de Lagrangien Augmente*, Dunod, Paris, 1982.
63. J. ARGYRIS, 'Three-dimensional anisotropic and inhomogeneous media. Matrix analysis of small and large displacements', *Ingr. Arch.*, **34**, 33–5, 1965.
64. O. C. ZIENKIEWICZ and S. VALLIAPPAN, 'Analysis of real structures for creep, plasticity and other complex constitutive laws, *Proc. Conf. Civil Engineering Materials*, Southampton, 1969, in *Structure of Solid Mechanics and Engineering Design*, Part 1 (ed. M. Te'eni), pp. 27–48, Wiley, London, 1971.
65. O. C. ZIENKIEWICZ, *The Finite Element Methods in Engineering Science*, pp. 404–5, McGraw-Hill, 1971.
66. O. C. ZIENKIEWICZ, XI KUI LI and S. NAKAZAWA, 'Iterative solution of mixed problems and stress recovery procedures', *Comm. Appl. Num. Meth.*, **1**, 3–9, 1985.
67. S. NAKAZAWA, 'Mixed finite elements and iterative solution procedures', in *Innovative Methods in Non-linear Problems* (eds W. K. Liu *et al.*), Pineridge Press, 1984.

68. O. C. ZIENKIEWICZ, XI KUI LI and S. NAKAZAWA, 'Dynamic transient analysis by a mixed iterative method', *Int. J. Num. Meth. Eng.*, **23**, 1343–53, 1986.
69. C. CANTIN, C. LOUBIGNAC, and C. TOUZOT, 'An iterative scheme to build continuous stress and displacement solutions', *Int. J. Num. Meth. Eng.*, **12**, 1493–506, 1978.
70. C. LOUBIGNAC, C. CANTIN, and C. TOUZOT, 'Continuous stress fields in finite element analysis', *AIAA J.*, **15**, 1645–7, 1978.
71. S. TIMOSHENKO and J. N. GOODIER, *Theory of Elasticity*, 2nd ed., McGraw-Hill, 1951.
72. B. FRAEIJS DE VEUBEKE and O. C. ZIENKIEWICZ, 'Strain energy bounds in finite element analysis', *J. Strain Analysis*, **2**, 265–71, 1967.
73. Z. M. ELIAS, 'Duality in finite element methods', *Proc. Am. Soc. Civ. Eng.*, **94**, EM4, 931–46, 1968.
74. R. V. SOUTHWELL, 'On the analogues relating flexure and displacement of flat plates', *Quart. J. Mech. Appl. Math.*, **3**, 257–70, 1950.
75. M. A. CRISFIELD, *Finite Elements and Solution Procedures for Structural Analysis*, Vol. 1, Linear Analysis, Pineridge Press, Swansea, UK, 1986.

13

Mixed formulation and constraints—incomplete (hybrid) field methods

13.1 General

In the previous chapter we have assumed in the mixed approximation that all the variables were defined and approximated in the same manner throughout the domain of the analysis. This process can, however, be conveniently abandoned on occasion with different formulations adopted in different subdomains and with some variables being only approximated on surfaces joining such subdomains. In this part we shall discuss such *incomplete* or *partial field* approximations which include various so-called *hybrid* formulations.

In all the examples given here we shall consider elastic solid body approximations only, but extension to the heat transfer or other field problems, etc., can be readily made as a simple exercise following the procedures outlined.

13.2 Interface traction link of two (or more) irreducible form subdomains

One of the most obvious and frequently used examples of an 'incomplete field' approximation is the subdivision of a problem into two (or more) subdomains in each of which an irreducible (displacement) formulation is used and the use of independently defined Lagrange multipliers (tractions) on the interface to join the subdomains, as in Fig. 13.1(a).

In this problem we formulate the approximation in domain Ω^1 in terms of displacements u^1 and the interface tractions $t^1 = \lambda$. With the weak form using the standard virtual work expression [viz. Eqs (12.21)

373

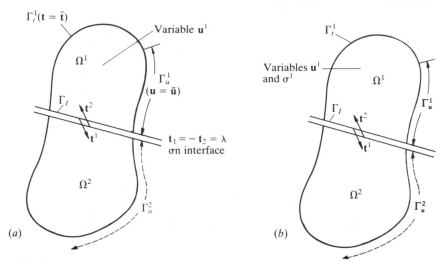

Fig. 13.1 Linking of two (or more) domains by traction variables defined only on the interfaces. (a) Variables in each domain are displacements **u** (internal irreducible form). (b) Variables in each domain are displacements and stresses **σ–u** (mixed form)

to (12.24)] we have

$$\int_{\Omega^1} \delta(\mathbf{Su}^1)^{\mathrm{T}} \mathbf{D}^1 \mathbf{Su}^1 \, d\Omega - \int_{\Gamma_I} \delta\mathbf{u}^{1\mathrm{T}}\boldsymbol{\lambda} \, d\Gamma$$

$$- \int_{\Omega^1} \delta\mathbf{u}^{1\mathrm{T}}\mathbf{b} \, d\Omega - \int_{\Gamma_t^1} \delta\mathbf{u}^{1\mathrm{T}}\bar{\mathbf{t}} \, d\Gamma = 0 \qquad (13.1)$$

in which as usual we assume that the satisfaction of prescribed displacement on Γ_{u^1} is implied by the approximation for $\mathbf{u}^1$. Similarly in domain Ω^2 we can write, now putting the interface traction as $\mathbf{t}^2 = -\boldsymbol{\lambda}$ to ensure equilibrium between the two domains,

$$\int_{\Omega^2} \delta(\mathbf{Su}^2)^{\mathrm{T}} \mathbf{D}^2 \mathbf{Su}^2 \, d\Omega + \int_{\Gamma_I} \delta\mathbf{u}^{2\mathrm{T}}\boldsymbol{\lambda} \, d\Gamma$$

$$- \int_{\Gamma^2} \delta\mathbf{u}^{2\mathrm{T}}\mathbf{b} \, d\Omega - \int_{\Gamma_t^2} \delta\mathbf{u}^{2\mathrm{T}}\bar{\mathbf{t}} \, d\Gamma = 0 \qquad (13.2)$$

The two subdomain equations are completed by a weak statement of displacement continuity on the interface between the two domains, i.e.,

$$\int_{\Gamma_I} \delta\boldsymbol{\lambda}^{\mathrm{T}}(\mathbf{u}^2 - \mathbf{u}^1) \, d\Gamma = 0 \qquad (13.3)$$

Discretization of displacements in each domain and of the tractions on the interface yields the final system of equations. Thus putting independent approximations as

$$\mathbf{u}^1 = \mathbf{N}_{u^1}\,\bar{u}^1 \tag{13.4}$$

$$\mathbf{u}^2 = \mathbf{N}_{u^2}\bar{u}^2 \tag{13.5}$$

$$\lambda = \mathbf{N}_\lambda\,\bar{\lambda} \tag{13.6}$$

we have

$$\begin{bmatrix} \mathbf{K}^1 & \mathbf{Q}^1 & \mathbf{O} \\ \mathbf{Q}^{1T} & \mathbf{O} & \mathbf{Q}^{2T} \\ \mathbf{O} & \mathbf{Q}^2 & \mathbf{K}^2 \end{bmatrix} \begin{Bmatrix} \bar{u}^1 \\ \bar{\lambda} \\ \bar{u}^2 \end{Bmatrix} = \begin{Bmatrix} \mathbf{f}^1 \\ 0 \\ \mathbf{f}^2 \end{Bmatrix} \tag{13.7a}$$

where

$$\mathbf{K}^1 = \int_{\Omega^1} \mathbf{B}^{1T}\mathbf{D}^1\mathbf{B}^1 \, d\Omega$$

$$\mathbf{Q}^1 = -\int_{\Gamma_I} \mathbf{N}_{u^1}^T\mathbf{N}_\lambda \, d\Gamma$$

$$\mathbf{K}^? = \int_{\Omega^2} \mathbf{B}^{2T}\mathbf{D}^2\mathbf{B}^2 \, d\Omega$$

$$\mathbf{Q}^2 = \int_{\Gamma_I} \mathbf{N}_{u^2}^T\mathbf{N}_\lambda \, d\Gamma \tag{13.7b}$$

$$\mathbf{f}_1 = \int_{\Omega^1} \mathbf{N}_{u^1}^T\mathbf{b} \, d\Omega + \int_{\Gamma_t^1} \mathbf{N}_{u^1}^T\tilde{\mathbf{t}} \, d\Gamma$$

$$\mathbf{f}_2 = \int_{\Omega^2} \mathbf{N}_{u^2}^T\mathbf{b} \, d\Omega + \int_{\Gamma_t^2} \mathbf{N}_{u^2}^T\tilde{\mathbf{t}} \, d\Gamma$$

We note that in the derivation of the above matrices the shape function $\mathbf{N}_\lambda$ and hence λ itself are only specified along the interface line—hence complying with our definition of partial field approximation.

The formulation just outlined can obviously be extended to many subdomains and in many cases of practical analysis is useful in ensuring a better matrix conditioning and allowing solution to be obtained with reduced computational effort.[1]

The variables $\mathbf{u}^1$ and $\mathbf{u}^2$, etc., appear as internal variables within each subdomain (or superelement) and can be eliminated locally providing matrices $\mathbf{K}^1$, $\mathbf{K}^2$, etc., are non-singular. Such non-singularity presupposes, however, that each of the subdomains has enough prescribed displacements to eliminate rigid body modes. If this is not the case partial elimination is always possible, retaining the rigid body modes till the complete solution is achieved.

The formulation just used can, of course, be applied to a single field displacement formulation in which we are required to specify displacement on the boundaries in a weak sense (rather than imposing these directly on the displacement shape functions).

This problem can be approached directly or can be derived simply via the first equation of (13.7a) in which we put $u^2 = \tilde{u}$, the specified displacement on Γ_I.

Now the equation system is simply

$$\begin{bmatrix} \mathbf{K}^1 & \mathbf{Q}^1 \\ \mathbf{Q}^{1\mathrm{T}} & \mathbf{0} \end{bmatrix} \begin{Bmatrix} \bar{\mathbf{u}}^1 \\ \bar{\lambda} \end{Bmatrix} = \begin{Bmatrix} \mathbf{f}_1 \\ \mathbf{f}_\lambda \end{Bmatrix} \tag{13.8}$$

where

$$\mathbf{f}_\lambda = - \int_{\Gamma_I} \mathbf{N}_\lambda^\mathrm{T} \tilde{\mathbf{u}} \, \mathrm{d}\Gamma \tag{13.9}$$

This formulation is often convenient for imposing a prescribed displacement on a displacement element field when the boundary values cannot fit the shape function field.

We have approached in the above the formulation directly via weak forms or weighted residuals. Of course, a variational principle could be given here simply as the minimization of total potential energy (viz. Chapter 2) subject to a Lagrange multiplier λ imposing subdomain continuity. The stationarity of

$$\Pi = \tfrac{1}{2} \int_\Omega (\mathbf{S}\mathbf{u})^\mathrm{T} \mathbf{D}(\mathbf{S}\mathbf{u}) \, \mathrm{d}\Omega - \int_\Omega \mathbf{u}^\mathrm{T} \mathbf{b} \, \mathrm{d}\Omega - \int_{\Gamma_t} \mathbf{u}^\mathrm{T} \bar{\mathbf{t}} \, \mathrm{d}\Gamma$$
$$+ \int_{\Gamma_I} \lambda^\mathrm{T} (\mathbf{u}^1 - \mathbf{u}^2) \, \mathrm{d}\Gamma \tag{13.10}$$

would result in the equation set (13.1) to (13.3).

13.3 Interface traction link of two or more mixed form subdomains

The problem discussed in the previous section could of course be tackled by assuming a mixed type of two-field approximation ($\sigma/\mathbf{u}$) in each subdomain, as illustrated in Fig. 13.1(b).

Now in each subdomain variables $\mathbf{u}$ and σ will appear, but the linking will be carried out again with the interface traction λ.

We now have, using the formulation of Sec. 12.4.2 for domain Ω^1 [viz. Eqs (12.28) and (12.21)],

$$\int_{\Omega^1} \delta\boldsymbol{\sigma}^{1\mathrm{T}}[(\mathbf{D}^1)^{-1}\boldsymbol{\sigma}^1 - \mathbf{S}\mathbf{u}^1] \, d\Omega = 0 \qquad (13.11\text{a})$$

$$\int_{\Omega^1} \delta(\mathbf{S}\mathbf{u}^1)^{\mathrm{T}}\boldsymbol{\sigma}^1 \, d\Omega - \int_{\Gamma_I} \delta\mathbf{u}^{1\mathrm{T}}\boldsymbol{\lambda} \, d\Gamma$$

$$- \int_{\Omega^1} \delta\mathbf{u}^{1\mathrm{T}}\mathbf{b} \, d\Omega - \int_{\Gamma_t^1} \delta\mathbf{u}^{1\mathrm{T}}\bar{\mathbf{t}} \, d\Gamma = 0 \qquad (13.11\text{b})$$

and for domain Ω^2 similarly

$$\int_{\Omega^2} \delta\boldsymbol{\sigma}^{2\mathrm{T}}[(\mathbf{D}^2)^{-1}\boldsymbol{\sigma}^2 - \mathbf{S}\mathbf{u}^2] \, d\Omega = 0 \qquad (13.12\text{a})$$

$$\int_{\Omega^2} \delta(\mathbf{S}\mathbf{u}^2)^{\mathrm{T}}\boldsymbol{\sigma}^2 \, d\Omega + \int_{\Gamma_I} \delta\mathbf{u}^{2\mathrm{T}}\boldsymbol{\lambda} \, d\Gamma$$

$$- \int_{\Omega^2} \delta\mathbf{u}^{2\mathrm{T}}\mathbf{b} \, d\Omega - \int_{\Gamma_t^2} \delta\mathbf{u}^{2\mathrm{T}}\bar{\mathbf{t}} \, d\Gamma = 0 \qquad (13.12\text{b})$$

With interface tractions in equilibrium the restoration of continuity demands that

$$\int_{\Gamma_I} \delta\boldsymbol{\lambda}^{\mathrm{T}}(\mathbf{u}^2 - \mathbf{u}^1) \, d\Gamma = 0 \qquad (13.13)$$

On discretization we now have

$$\mathbf{u}^1 = \mathbf{N}_{u^1}\bar{\mathbf{u}}^1 \qquad \mathbf{u}^2 = \mathbf{N}_{u^2}\bar{\mathbf{u}}^2$$

$$\boldsymbol{\sigma}^1 = \mathbf{N}_{\sigma^1}\bar{\boldsymbol{\sigma}}^1 \qquad \boldsymbol{\sigma}^2 = \mathbf{N}_{\sigma^2}\bar{\boldsymbol{\sigma}}^2$$

$$\boldsymbol{\lambda} = \mathbf{N}_\lambda\bar{\boldsymbol{\lambda}}$$

and

$$\begin{bmatrix} \mathbf{A}^1 & \mathbf{C}^1 & 0 & 0 & 0 \\ \mathbf{C}^{1\mathrm{T}} & 0 & \mathbf{Q}^1 & 0 & 0 \\ 0 & \mathbf{Q}^{1\mathrm{T}} & 0 & 0 & \mathbf{Q}^{2\mathrm{T}} \\ 0 & 0 & 0 & \mathbf{A}^2 & \mathbf{C}^2 \\ 0 & 0 & \mathbf{Q}^2 & \mathbf{C}^{2\mathrm{T}} & 0 \end{bmatrix} \begin{Bmatrix} \bar{\boldsymbol{\sigma}}^1 \\ \bar{\mathbf{u}}^1 \\ \bar{\boldsymbol{\lambda}} \\ \bar{\boldsymbol{\sigma}}^2 \\ \bar{\mathbf{u}}^2 \end{Bmatrix} = \begin{Bmatrix} \mathbf{f}_1^1 \\ \mathbf{f}_2^1 \\ 0 \\ \mathbf{f}_1^2 \\ \mathbf{f}_2^2 \end{Bmatrix} \qquad (13.14)$$

with $\mathbf{A}$, $\mathbf{C}$, $\mathbf{f}_1$, and $\mathbf{f}_2$ defined similarly to Eq. (12.31) with appropriate subdomain subscripts and $\mathbf{Q}^1$ and $\mathbf{Q}^2$ given as in (13.7b).

All the remarks made in the previous section apply here once again—though use of the above form does not appear frequent.

13.4 Interface displacement 'frame'

13.4.1 *General.* In the preceding examples we have used traction as the interface variable linking two or more subdomains. Due to lack of rigid body constraints the elimination of local subdomain displacements has generally been impossible. For this and other reasons it is convenient to accomplish the linking of subdomains via a displacement field *defined only on the interface* [Fig. 13.2(a)] and to eliminate all the interior variables so that this linking can be accomplished via a standard stiffness matrix procedure using only the interface variables.

The *displacement frame* can be made to surround the subdomain completely and if all internal variables are eliminated will yield a stiffness matrix of a new 'element' which can be used directly in coupling with any other element with similar displacement assumptions on the interface, irrespective of the procedure used for deriving such an element [Fig. 13.2(b)].

In all the examples of this section we shall approximate the frame displacements as

$$\mathbf{v} = \mathbf{N}_v \bar{\mathbf{v}} \qquad \text{or} \qquad \Gamma_I \qquad (13.15)$$

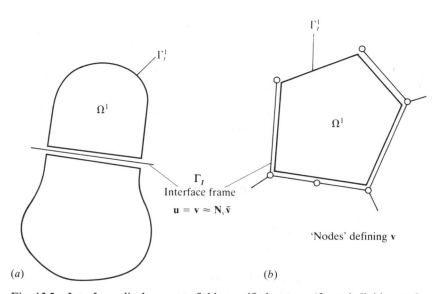

(a)   (b)

Fig. 13.2 Interface displacement field specified on a 'frame' linking subdomains: (a) two-domain link; (b) a 'superelement' (hybrid) which can be linked to many other similar elements

and consider the 'nodal forces' contributed by a single subdomain Ω^1 to the 'nodes' on this frame. Using virtual work (or weak) statements we have with discretization

$$\int_{\Gamma_I^1} \mathbf{N}_v^T \mathbf{t} = \mathbf{q}^1 \tag{13.16}$$

where $\mathbf{t}$ are the tractions the interior exerts on the imaginary frame. The balance of the nodal forces contributed by each subdomain provides now the weak condition for traction continuity.

As finally the tractions $\mathbf{t}$ can be expressed in terms of the frame parameters $\bar{\mathbf{v}}$ only, we shall arrive at

$$\mathbf{q}^1 = \mathbf{K}^1 \bar{\mathbf{v}} + \mathbf{f}_0^1 \tag{13.17}$$

where $\mathbf{K}^1$ is the stiffness matrix of the subdomain Ω^1 and $\mathbf{f}_0^1$ its internally contributed 'forces'.

From this point onwards the standard assembly procedures are valid and the subdomain can be treated as a standard element which can be assembled with others by ensuring that

$$\sum \mathbf{q} = 0 \tag{13.18}$$

where the sum includes all subdomains (elements!). We thus have only to consider a single subdomain in what follows.

13.4.2 *Linking two or more mixed form subdomains.* We shall assume as in Sec. 13.3 that in each subdomain, now labelled e for generality, the stresses $\boldsymbol{\sigma}^e$ and displacements $\mathbf{u}^e$ are independently approximated. The equations (13.11) are rewritten adding to the first weak statement of displacement continuity.

We now have in place of (13.11a) and (13.13) (dropping superscripts)

$$\int_{\Omega^e} \delta\boldsymbol{\sigma}^T (\mathbf{D}^{-1}\boldsymbol{\sigma} - \mathbf{S}\mathbf{u}) \, d\Omega - \int_{\Gamma_I^e} \delta\mathbf{t}^T(\mathbf{u} - \mathbf{v}) \, d\Gamma = 0 \tag{13.19}$$

Equation (13.11b) will be rewritten as the weighted statement of the equilibrium relation, i.e.,

$$-\int_{\Omega^e} \delta\mathbf{u}^T(\mathbf{S}^T\boldsymbol{\sigma} + \mathbf{b}) \, d\Omega + \int_{\Gamma_t^e} \delta\mathbf{u}^T(\mathbf{t} - \tilde{\mathbf{t}}) \, d\Gamma = 0$$

or, after integration by parts

$$\int_{\Omega^e} \delta(\mathbf{S}\mathbf{u})^T\boldsymbol{\sigma} \, d\Omega - \int_{\Omega^e} \delta\mathbf{u}^T\mathbf{b} \, d\Omega - \int_{\Gamma_I^e} \delta\mathbf{u}^T\mathbf{t} \, d\Gamma - \int_{\Gamma_t^e} \delta\mathbf{u}^T\tilde{\mathbf{t}} \, d\Gamma = 0 \tag{13.20}$$

In the above, t are the tractions corresponding to the stress field σ [viz. Eq. (12.29)]:

$$t = G\sigma \qquad (13.21)$$

In what follows Γ_t^e, i.e., the boundary with prescribed tractions, will generally be taken as zero.

On approximating Eqs (13.19) and (13.20) and (13.16), with

$$\mathbf{u} = \mathbf{N}_u \bar{\mathbf{u}} \qquad \sigma = \mathbf{N}_\sigma \bar{\sigma} \qquad \text{and} \qquad \mathbf{v} = \mathbf{N}_v \bar{\mathbf{v}}$$

we can write using galerkin weighting and limiting the variables to the 'element' e

$$\begin{bmatrix} \mathbf{A}^e & \mathbf{C}^e & \mathbf{Q}^e \\ \mathbf{C}^{eT} & 0 & 0 \\ \mathbf{Q}^{eT} & 0 & 0 \end{bmatrix} \begin{Bmatrix} \bar{\sigma}^e \\ \bar{\mathbf{u}}^e \\ \bar{\mathbf{v}} \end{Bmatrix} = \begin{Bmatrix} 0 \\ \mathbf{f}^e \\ \mathbf{q}^e \end{Bmatrix} \qquad (13.22a)$$

where

$$\mathbf{A}^e = \int_{\Omega^e} \mathbf{N}_\sigma^T \mathbf{D}^{-1} \mathbf{N}_\sigma \; d\Omega$$

$$\mathbf{C}^e = \int_{\Omega^e} \mathbf{N}_\sigma^T \mathbf{B} \; d\Omega - \int_{\Gamma_{t^e}} (\mathbf{G}\mathbf{N}_\sigma)^T \mathbf{N}_u \; d\Gamma \qquad (13.22b)$$

$$\mathbf{Q}^e = \int_{\Gamma_{t^e}} (\mathbf{G}\mathbf{N}_\sigma)^T \mathbf{N}_v \; d\Gamma$$

$$\mathbf{f}^e = \int_{\Omega^e} \mathbf{N}_u^T \mathbf{b} \; d\Omega$$

Elimination of σ^e and $\mathbf{u}^e$ from above yields the stiffness matrix of the element and the internally contributed force [viz. Eq. (13.17)].

Once again we can note that the simple stability criteria discussed in Chapter 12 will help in choosing the number of σ, $\mathbf{u}$, and $\mathbf{v}$ parameters. As the final stiffness matrix of an element should be singular for three rigid body displacements we must have [by Eq. (12.17)]

$$n_\sigma \geqslant n_u + n_v - 3 \qquad (13.23)$$

in two-dimensional applications.

Various alternative variational forms of the above formulation exist. A particularly useful one is developed by Pian et al.[2–4] In this the full mixed representation can be written completely in terms of a single variational principle (for zero body forces) and no boundary of type Γ_t

present:

$$\Pi_\Omega = -\int_\Omega \tfrac{1}{2}\boldsymbol{\sigma}\mathbf{D}^{-1}\boldsymbol{\sigma}\ d\Omega - \int_\Omega (\mathbf{S}^T\boldsymbol{\sigma})^T\mathbf{u}_I\ d\Omega + \int_\Omega \boldsymbol{\sigma}^T\mathbf{S}\mathbf{v}\ d\Omega \quad (13.24)$$

In the above it is assumed that the compatible field of $\mathbf{v}$ is *specified throughout the element* domain and not only on its interfaces and $\mathbf{u}_I$ stands for the incompatible field defined only inside the element domain.†

We note that in the present definition

$$\mathbf{u} = \mathbf{u}_I + \mathbf{v} \quad (13.25)$$

To show the validity of this variational principle, which is convenient as no interface integrals need to be evaluated, we shall derive the weak statement corresponding to Eqs (13.19) and (13.20) using the condition (13.25).

We can now write in place of (13.19) (noting that for interelement compatibility we have to ensure that $\mathbf{u}_I = 0$ on the interfaces)

$$\int_{\Omega^e} \delta\boldsymbol{\sigma}^T(\mathbf{D}^{-1}\boldsymbol{\sigma} - \mathbf{S}\mathbf{v})\ d\Omega - \int_{\Omega^e} \delta\boldsymbol{\sigma}^T\mathbf{S}\mathbf{u}_I\ d\Omega + \int_{\Gamma_I^e} \delta\mathbf{t}^T\mathbf{u}_I\ d\Gamma = 0 \,(13.26)$$

After use of Green's theorem the above becomes simply

$$\int_{\Omega^e} \delta\boldsymbol{\sigma}^T(\mathbf{D}^{-1}\boldsymbol{\sigma} - \mathbf{S}\mathbf{v})\ d\Omega + \int_{\Omega^e} (\mathbf{S}^T\delta\boldsymbol{\sigma})^T\mathbf{u}_I\ d\Gamma = 0 \quad (13.27)$$

In place of (13.20) we write (in the absence of body forces $\mathbf{b}$ and Γ_t boundary)

$$\int_{\Omega^e} \delta\mathbf{u}_I^T(\mathbf{S}^T\boldsymbol{\sigma})\ d\Omega + \int_{\Omega^e} \delta\mathbf{v}^T(\mathbf{S}^T\boldsymbol{\sigma})\ d\Omega = 0 \quad (13.28)$$

and again after use of Green's theorem

$$\int_{\Omega^e} \delta\mathbf{u}_I^T\mathbf{S}^T\boldsymbol{\sigma}\ d\Omega - \int_{\Omega^e} \delta(\mathbf{S}\mathbf{v})^T\boldsymbol{\sigma}\ d\Omega = 0 \qquad (\text{if } \delta\mathbf{v} = 0 \quad \text{on } \Gamma_I) \quad (13.29)$$

These equations are precisely the variations of the functional (13.24).

The variational principle developed in Eq. (13.24) has been applied effectively to derive several simple, new, elements by Pian and Sumihara.[2] We shall derive one such quadrilateral element in detail below.

Of course, the procedure developed in this section can be applied to other mixed or irreducible representations with 'frame' links. Tong and

---

† In this form, of course, the element could well fit into Chapter 12 and the subdivision of hybrid and mixed forms is not unique here.

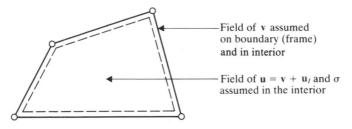

Field of **v** assumed
on boundary (frame)
and in interior

Field of **u** = **v** + **u**$_I$ and σ
assumed in the interior

Fig. 13.3   The Pian–Sumihara quadrilateral

Pian[5,6] developed several alternative element forms by using this procedure.

The Pian–Sumihara hybrid quadrilateral (Fig. 13.3). The derivation of this four-noded element is separated into two parts. In the first the approximation for stress is established in terms of linear polynomials in the *natural* (ξ, ζ) coordinates. The first term in Eq. (13.28) is used to reduce the stress parameters to a minimum number. Subsequent computations then are in terms of the reduced stress interpolation and the compatible displacement field only. In the second part of the development the stress interpolations are modified to reduce computational effort needed to establish the element 'stiffness' matrix.

The development given below is for plane stress and plane strain applications. The approximation for the stress in each element is taken as

$$\boldsymbol{\sigma} = \boldsymbol{\sigma}_0 + \boldsymbol{\sigma}^1(\xi, \eta)$$

where $\boldsymbol{\sigma}_0$ is a constant set of three parameters and

$$\boldsymbol{\sigma}^1 = \begin{Bmatrix} \sigma_x^1 \\ \sigma_y^1 \\ \sigma_{xy}^1 \end{Bmatrix} = \begin{Bmatrix} \alpha_1 \xi + \alpha_2 \eta \\ \beta_1 \xi + \beta_2 \eta \\ \gamma_1 \xi + \gamma_2 \eta \end{Bmatrix} = \mathbf{N}_\sigma^1 \bar{\boldsymbol{\sigma}}^1$$

Thus nine parameters define the variation of stress over the element.

The incompatible displacement field is defined by four parameters of $\bar{\mathbf{u}}_\lambda$ as

$$\mathbf{u}_I = \mathbf{N}_\lambda(\xi, \eta)\bar{\mathbf{u}}_\lambda$$

with

$$\mathbf{N}_\lambda = \begin{bmatrix} (1 - \xi^2) & 0 & (1 - \eta^2) & 0 \\ 0 & (1 - \xi^2) & 0 & (1 - \eta^2) \end{bmatrix}$$

and the compatible frame displacements in the element are assumed as

$$\mathbf{v} = N_i(\xi, \eta)\bar{\mathbf{v}}_i$$

where $N_i(\xi, \eta)$ are the usual, bilinear, four-noded isoparametric shape functions (viz. Chapter 8). Equation (13.28) leads to the requirement that in each element

$$\int_{\Omega_e} \begin{Bmatrix} (1 - \xi^2) \\ (1 - \eta^2) \end{Bmatrix} (\sigma^1_{x,x} + \sigma^1_{xy,y}) \, d\Omega = 0$$

and

$$\int_{\Omega_e} \begin{Bmatrix} (1 - \xi^2) \\ (1 - \eta^2) \end{Bmatrix} (\sigma^1_{xy,x} + \sigma^1_{y,y}) \, d\Omega = 0$$

It may easily be verified that substitution of the stress approximation into the above relations will yield only two independent equations:[2]

$$b_3 \alpha_1 - b_1 \alpha_2 - a_3 \gamma_1 + a_1 \gamma_2 = 0$$

and

$$-a_3 \beta_1 + a_1 \beta_2 + b_3 \gamma_1 - b_1 \gamma_2 = 0$$

where

$$a_1 = \sum_1^4 x_i \xi_i \qquad b_1 = \sum_1^4 y_i \xi_i$$

$$a_2 = \sum_1^4 x_i \xi_i \eta_i \qquad b_2 = \sum_1^4 y_i \xi_i \eta_i$$

$$a_3 = \sum_1^4 x_i \eta_i \qquad b_3 = \sum_1^4 y_i \eta_i$$

Thus, from Eq. (13.28) the stress approximation may be expressed in terms of seven independent parameters. Use of this seven-parameter stress field in the functional of Eq. (13.24) will eliminate the term involving the incompatible displacements from further considerations.

Proper rank of a four-noded plane stress–strain element is achieved with a five-term stress approximation. In reference 2 the seven-term approximation is reduced to five independent terms by perturbing the geometric shape of the element. Alternatively, the same result may be achieved through integration by parts of the incompatible term in Eq. (13.28). The result is a volume integral and a boundary integral. If we require the volume and boundary terms to be satisfied independently, we obtain four constraint equations to reduce the original nine-term approximation to the desired five terms. After integration by parts the volume term is

$$\int_{\Omega_e} (\mathbf{S}\delta\mathbf{u}_I)^T \boldsymbol{\sigma}^1 \, d\Omega$$

Setting this to zero yields upon substitution of the approximating fields

$$\int_{\Omega_e} \mathbf{B}_\lambda^T \boldsymbol{\sigma}^1 \, d\Omega = 0$$

where $\mathbf{B}_\lambda$ is the strain-displacement matrix deduced from the incompatible displacements. This relationship when combined with the original two independent equations yields the four equations

$$b_3\alpha_1 = a_3\gamma_1$$

$$b_1\alpha_2 = a_1\gamma_2$$

$$a_3\beta_1 = a_1\gamma_1$$

$$a_1\beta_2 = b_1\gamma_2$$

which may be used to construct the five-term stress approximation. The boundary term will yield constraint equations which necessarily are linear combinations of the above four equations.

The four constraint equations may be used to write a five-term stress approximation. We note that any of the a_i, b_i may be zero; hence the stress parameters are redefined to avoid division by these geometric parameters. The result is

$$\boldsymbol{\sigma}^1 = \begin{bmatrix} a_3^2\xi & a_1^2\eta \\ b_3^2\xi & b_1^2\eta \\ a_3b_3\xi & a_1b_1\eta \end{bmatrix} \begin{Bmatrix} \sigma_1^1 \\ \sigma_2^1 \end{Bmatrix}$$

The above stress approximation may now be used in (13.24) to deduce an element 'stiffness' matrix. As stated previously, the construction of the stiffness matrix requires a solution of the matrix resulting from the first term in (13.24), i.e., inversion of a 5×5 matrix. In the present case this would lead to a significant number of numerical operations which may be avoided in situations where the material property matrix is constant over the element. This step is achieved by slightly modifying the stress interpolation to

$$\boldsymbol{\sigma} = \bar{\boldsymbol{\sigma}}_0 + \bar{\mathbf{N}}_\sigma \bar{\boldsymbol{\sigma}}^1$$

where

$$\bar{\mathbf{N}}_\sigma = \begin{bmatrix} a_3^2(\xi - \xi_0) & a_1^2(\eta - \eta_0) \\ b_3^2(\xi - \xi_0) & b_1^2(\eta - \eta_0) \\ a_3b_3(\xi - \xi_0) & a_1b_1(\eta - \eta_0) \end{bmatrix}$$

This modification merely involves a rescaling of the parameters defining the constant part of the stress approximation. We may now deduce values for the parameters ξ_0, η_0 such that

$$\int_{\Omega^e} \boldsymbol{\sigma}^{\mathrm{T}} \mathbf{D}^{-1} \boldsymbol{\sigma} \, d\Omega = \langle \bar{\boldsymbol{\sigma}}_0^{\mathrm{T}}, \bar{\boldsymbol{\sigma}}^{1\mathrm{T}} \rangle \begin{bmatrix} \mathbf{D}^{-1}\Omega^e & 0 \\ 0 & \bar{\mathbf{A}} \end{bmatrix} \begin{Bmatrix} \bar{\boldsymbol{\sigma}}_0 \\ \bar{\boldsymbol{\sigma}}^1 \end{Bmatrix}$$

Accordingly, the inverse is

$$\begin{bmatrix} \Omega^e\mathbf{D}^{-1} & 0 \\ 0 & \bar{\mathbf{A}} \end{bmatrix}^{-1} = \begin{bmatrix} \dfrac{1}{\Omega^e}\mathbf{D} & 0 \\ 0 & \bar{\mathbf{A}}^{-1} \end{bmatrix}$$

where $\bar{\mathbf{A}}$ is a 2×2 matrix defined as

$$\bar{\mathbf{A}} = \int_{\Omega^e} \bar{\mathbf{N}}_\sigma^{\mathrm{T}} \mathbf{D}^{-1} \bar{\mathbf{N}}_\sigma \, d\Omega$$

The appropriate values of ξ_0, η_0 are

$$\xi_0 = \frac{J_1}{3J_0} \qquad \text{and} \qquad \eta_0 = \frac{J_2}{3J_0}$$

where J_i are the parameters in the jacobian determinant for the four-noded element with

$$J = J_0 + J_1\xi + J_2\eta$$

The above steps have been implemented in an element routine included in Chapter 15. The resulting element is very efficient numerically, compared to a four-noded isoparametric displacement element. The efficiency in computation results from the fact that

$$\int_{\Omega^e} \sigma^{\mathrm{T}}(\mathbf{S}\mathbf{v}) \, d\Omega = \langle \bar{\sigma}_0^{\mathrm{T}}, \bar{\sigma}^{1\mathrm{T}} \rangle \begin{bmatrix} \mathbf{B}_0 \, \Omega^e \\ \mathbf{C}^1 \end{bmatrix} \bar{\mathbf{v}}$$

where $\mathbf{B}_0$ is the four-noded isoparametric element strain-displacement matrix evaluated at the origin of the natural coordinates and thus is identical to a one-point Gauss quadrature value. Also

$$\mathbf{C}^1 = \int_{\Omega^e} \bar{\mathbf{N}}_\sigma \mathbf{B} \, d\Omega$$

and hence the element stiffness matrix is given by

$$\mathbf{k} = \mathbf{B}_0^{\mathrm{T}} \mathbf{D} \mathbf{B}_0 \, \Omega^e + \mathbf{C}^{1\mathrm{T}} \bar{\mathbf{A}}^{-1} \mathbf{C}^1$$

The first term is identical to a one-point quadrature evaluation of the four-noded displacement element. The second term is a rank 2 stabilization matrix whose terms may be easily computed analytically. Accordingly the element may be implemented with considerably fewer numerical operations than a 2×2 Gauss quadrature derivation of the displacement element.

After solution of the global problem for the displacement parameters $\bar{\mathbf{v}}$, the stress parameters $\bar{\sigma}_0$ may be deduced from

$$\bar{\sigma}_0 = \mathbf{D}\mathbf{B}_0 \bar{\mathbf{v}}$$

Comparing this with the stress interpolation, it may be observed that the $\bar{\sigma}_0$ are the mean of the stresses over each element. These are accurate values which are easy to compute and report in an element. It should be noted that the mean stresses are associated with the point ξ_0, η_0 and not the origin, where $\mathbf{B}_0$ is computed. This result from the mixed formulation is in contrast with the usual statement of the displacement solution. Also note that it is not necessary to compute the $\bar{\sigma}^1$ to determine the mean values in the element.

The element described above is not only very efficient to implement but is probably the most accurate four-noded element tested to date on a wide range of plane stress and plane strain problems. The element performs well in problems involving bending (e.g., involving the $\bar{\mathbf{N}}_\sigma^1$ terms) and in nearly incompressible applications. Moreover, the sensitivity to distortion is less

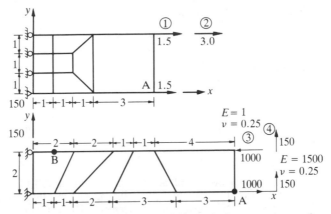

Fig. 13.4 The Pian–Sumihara quadrilateral. Comparative performance on examples shown

Results

	Case 1	Case 2	Case 3		Case 4	
Elements	u_A	$-v_A$	v_A	σ_{x_B}	v_A	σ_{x_B}
Bilinear iso-parametric Q 4	6.00	17.00	45.7	−1761	50.7	−2448
Incompatible (Taylor, Wilson) Q 6 (viz. Chapter 11)	6.00	17.61	—	—	—	—
Pian–Sumihara	6.00	17.64	96.18	−3014	98.19	−4137
Exact	6.00	18.00	100	−3000	102.6	−4050

than other four-noded isoparametric elements. Typical results are obtained in reference 2 and are shown in Figs 13.4 and 13.5

13.4.3 *Linking of equilibrating form subdomains.* In this form we shall assume *a priori* that the stress field expansion is such that

$$\boldsymbol{\sigma}_T = \boldsymbol{\sigma} + \boldsymbol{\sigma}_0 \tag{13.30}$$

and that the equilibrium equations are identically satisfied. Thus

$$\mathbf{S}^T\boldsymbol{\sigma} \equiv 0; \ \mathbf{S}^T\boldsymbol{\sigma}_0 \equiv b \ \text{in} \ \Omega \quad \text{and} \quad \mathbf{G}\boldsymbol{\sigma} = 0; \ \mathbf{G}\boldsymbol{\sigma}_0 = \bar{\mathbf{t}} \ \text{on} \ \Gamma_t^e$$

In the absence of Γ_t^e, Eq. (13.20) is identically satisfied and we write (13.19) as (viz. Chapter 12, Sec. 12.9)

$$\int_{\Omega^e} \delta\boldsymbol{\sigma}^T(\mathbf{D}^{-1}\boldsymbol{\sigma}_T - \mathbf{S}\mathbf{u}) \, d\Omega + \int_{\Gamma_I^e} \delta\mathbf{t}^T(\mathbf{u} - \mathbf{v}) \, d\Gamma \equiv$$

$$\equiv \int_{\Omega^e} \delta\boldsymbol{\sigma}^T\mathbf{D}^{-1}(\boldsymbol{\sigma} + \boldsymbol{\sigma}_0) \, d\Omega - \int_{\Gamma_I^e} (\mathbf{G}\delta\boldsymbol{\sigma})^T\mathbf{v} \, d\Gamma = 0 \tag{13.31}$$

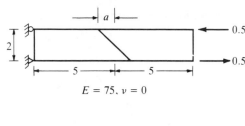

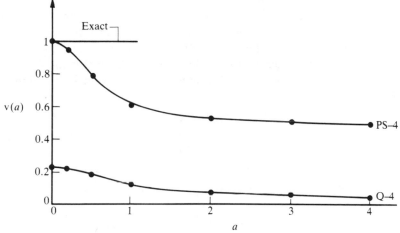

Fig. 13.5   The Pian–Sumihara quadrilateral (P–S 4) compared with displacement quadrilateral. Effect of element distortion (Q 4)

On discretization, noting that the field **u** does not enter the problem

$$\boldsymbol{\sigma} = \mathbf{N}_\sigma \bar{\boldsymbol{\sigma}} \qquad \mathbf{v} = \mathbf{N}_v \bar{\mathbf{v}}$$

we have, on including Eq. (13.16)

$$\begin{bmatrix} \mathbf{A}^e & \mathbf{Q}^e \\ \mathbf{Q}^{eT} & 0 \end{bmatrix} \begin{Bmatrix} \bar{\boldsymbol{\sigma}} \\ \bar{\mathbf{v}} \end{Bmatrix} = \begin{Bmatrix} \mathbf{f}_1^e \\ \mathbf{q}^e - \mathbf{f}_2^e \end{Bmatrix} \tag{13.32}$$

when

$$\mathbf{A}^e = \int_{\Omega^e} \mathbf{N}_\sigma \mathbf{D}^{-1} \mathbf{N}_\sigma \, d\Omega$$

$$\mathbf{f}_1 = \int_{\Omega^e} \mathbf{N}_\sigma \mathbf{D}^{-1} \boldsymbol{\sigma}_0 \, d\Omega$$

$$\mathbf{Q}^e = \int_{\Gamma_1^e} (\mathbf{G}\mathbf{N}_\sigma)^T \mathbf{N}_v \, d\Gamma$$

and

$$\mathbf{f}_2^e = \int_{\Gamma_1^e} \mathbf{N}_v \mathbf{G} \boldsymbol{\sigma}_0 \, d\Gamma$$

Here elimination of $\bar{\sigma}$ is simple and we can write directly

$$\mathbf{K}^e \bar{\mathbf{v}} = \mathbf{q}^e - \mathbf{f}_2^e - \mathbf{Q}^{eT} \mathbf{A}^{e-1} \mathbf{f}_1^e$$

and   (13.33)

$$\mathbf{K}^e = \mathbf{Q}^{eT} \mathbf{A}^{e-1} \mathbf{Q}^e$$

In Sec. 12.9 we have discussed the possible equilibration fields and have indicated the difficulties in choosing such fields for a finite element, subdivided, field. In the present case, on the other hand, the situation is quite simple as the parameters describing the equilibrating stresses inside the element can be chosen arbitrarily in a polynomial expression.

For instance, if we use a simple polynomial expression in two dimensions:

$$\sigma_x = \alpha_0 + \alpha_1 x + \alpha_2 y$$
$$\sigma_y = \beta_0 + \beta_1 x + \beta_2 y \qquad (13.34)$$
$$\sigma_{xy} = \gamma_0 + \gamma_1 x + \gamma_2 y$$

we note that to satisfy the equilibrium we require

$$\mathbf{S}^T \boldsymbol{\sigma} = \begin{bmatrix} \dfrac{\partial}{\partial x} & 0 & \dfrac{\partial}{\partial y} \\ 0 & \dfrac{\partial}{\partial y} & \dfrac{\partial}{\partial x} \end{bmatrix} \boldsymbol{\sigma} = \begin{bmatrix} \alpha_1 + \gamma_2 \\ \beta_2 + \gamma_1 \end{bmatrix} = 0 \qquad (13.35)$$

and this simply means

$$\gamma_2 = -\alpha_1$$
$$\gamma_1 = \beta_2$$

Thus a linear expansion in terms of $6 - 2 = 4$ independent parameters is easily achieved. Similar expansions can of course be used with higher order terms.

It is interesting to observe that:

1. $n_\sigma \geqslant n_v - 3$ is needed to preserve stability.
2. By the principle of limitation, accuracy of such approximation cannot be better than that achieved by a simple displacement formulation with compatible expansion of $\mathbf{v}$ throughout the element, providing similar polynomial expressions arise in stress component variations.

However, in practice two advantages of such elements, known as hybrid-stress elements, are obtained. In the first place it is not necessary to construct compatible displacement fields throughout the element (a point useful in their application to, say, a plate bending problem). In the second for distorted (isoparametric) elements it is easy to use stress fields varying with the global coordinates and thus achieve higher order accuracy.

The first use of such elements was made by Pian[7] and many successful variants are in use today.[8-22]

13.5 Linking of boundary (or Trefftz)-type solution by 'frame' of specified displacements

We have already referred to boundary (Trefftz)-type solutions[23] earlier (Chapter 9). Here the chosen displacement/stress fields are such that *a priori* the homogeneous equations of equilibrium and constitutive relation are satisfied in the domain under consideration (and indeed on occasion some prescribed boundary traction or displacement conditions).

Thus in Eqs (13.19) and (13.20) the subdomain (element *e*) integral terms disappear and, as the internal δt and δu variations are linked, we combine all into a single statement (in the absence of body force terms) as

$$-\int_{\Gamma_I e} \delta t(u - v)\, d\Gamma + \int_{\Gamma_t e} \delta u^T(t - \tilde{t})\, d\Gamma + d\Omega = 0 \qquad (13.36)$$

This coupled with the boundary statement (13.16) provides the means of devising stiffness matrix statements of such subdomains.

For instance, if we express the approximate fields as

$$u = N\bar{a} \qquad (13.37)$$

implying

$$\sigma = D(SN)\bar{a} \qquad \text{and} \qquad t = G\sigma = GD(SN)\bar{a}$$

we can write in place of (13.22)

$$\begin{bmatrix} H^e & Q^e \\ Q^{eT} & 0 \end{bmatrix} \begin{Bmatrix} \bar{a} \\ \bar{v} \end{Bmatrix} = \begin{Bmatrix} f_1^e \\ q \end{Bmatrix} \qquad (13.38)$$

where

$$H^e = \int_{\Gamma_I e} [GD(SN)]^T N\, d\Gamma + \int_{\Gamma_t e} N^T GD(SN)\, d\Gamma$$

$$Q^e = \int_{\Gamma_I e} [GD(SN)]^T N_v\, d\Gamma \qquad (13.39)$$

$$f_1^e = \int_{\Gamma_t e} N^T \tilde{t}\, d\Gamma$$

In Eqs (13.38) and (13.39) we have omitted the domain integral of the particular solution σ_0 corresponding to the body forces $\mathbf{b}$ but have allowed a portion of the boundary Γ_t^e to be subject to prescribed tractions. Full expressions including the particular solution can easily be derived.

Equation (13.38) is immediately available for solution of a single boundary solution problem in which $\mathbf{v}$ and $\tilde{\mathbf{t}}$ are described on portions of the boundary. More importantly, however, it results in a very simple stiffness matrix for a full element enclosed by the frame. We now have

$$\mathbf{K}^e\bar{\mathbf{v}} = \mathbf{q} - \mathbf{f}^e \qquad (13.40)$$

in which

$$\mathbf{K}^e = \mathbf{Q}^{e\mathrm{T}}\mathbf{H}^{e-1}\mathbf{Q}^e$$
$$\mathbf{f}^e = \mathbf{Q}^{e\mathrm{T}}\mathbf{H}^{e-1}\mathbf{f}_1^e \qquad (13.41)$$

This form is very similar to that of Eq. (13.33) except that now only integrals on the boundaries of the subdomain element need to be evaluated.

It is not immediately apparent that the matrix $\mathbf{H}^e$ of Eq. (13.39) must be symmetric. This symmetry follows from the uniqueness of strain energy and thus

$$\int_{\Gamma^e} \mathbf{t}^\mathrm{T}\mathbf{u} \, d\Gamma \equiv \int_{\Gamma^e} \mathbf{u}^\mathrm{T}\mathbf{t} \, d\Gamma$$

and hence

$$\int_{\Gamma^e} \mathbf{N}^\mathrm{T}\mathbf{GD}(\mathbf{SN}) \, d\Gamma \equiv \int_{\Gamma^e} [\mathbf{GD}(\mathbf{SN})]^\mathrm{T}\mathbf{N} \, d\Gamma \qquad (13.42)$$

It is thus convenient to write $\mathbf{H}^e$ in an obviously symmetric form as

$$\mathbf{H}^e = \frac{1}{2}\int_{\Gamma_f e} \{[\mathbf{GD}(\mathbf{SN})]^\mathrm{T}\mathbf{N} + \mathbf{N}^\mathrm{T}\mathbf{GD}(\mathbf{SN})\} \, d\Gamma$$

$$- \frac{1}{2}\int_{\Gamma_f e} \{[\mathbf{GD}(\mathbf{SN})]^\mathrm{T}\mathbf{N} + \mathbf{N}^\mathrm{T}[\mathbf{GD}(\mathbf{SN})]\} \, d\Gamma \qquad (13.43)$$

Much has been written about so-called 'boundary elements' and their merits and disadvantages.[24-26] Very frequently singular solutions (Green's functions) are used to accomplish the purpose of satisfying the governing field equations and these involve complex integration procedures. However, it is possible to derive complete sets of functions satisfying the governing equations without introducing singularities[27-31] and simple integration then suffices.

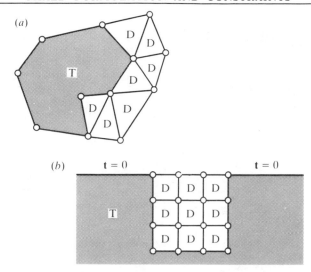

Fig. 13.6 Boundary–Trefftz-type elements (T) with complex-shaped 'frames' allowing combination with standard, displacement elements (D): (a) an *interior* element; (b) an *exterior* element

While boundary solutions are confined to linear homogeneous domains these give very accurate solutions for a limited range of parameters, and their combination with 'standard' finite elements has been occasionally described. Several coupling procedures have been developed in the past,[32–35] but the form given here coincides with more recent work of Zielinski and Zienkiewicz,[36] Jirousek[37–40] and Piltner.[41] The latter has developed very general two-dimensional elasticity and plate bending elements which can be enclosed by a many-sided polygonal domain (element) that can be directly coupled to standard elements providing that same-displacement interpolation along the edges is involved, as shown in Fig. 13.6. Here both *interior* elements with a frame enclosing an element volume and *exterior* elements satisfying tractions at free surface and infinity are illustrated.

Such elements can include in the arsenal of the 'shape functions' N^e [viz. Eq. (13.37)] solutions, which are exact solutions to singularities or which satisfy automatically traction boundary conditions on internal boundaries, e.g., circles or ellipses inscribed within large elements as shown in Fig. 13.7.

Clearly such elements can perform very well when compared with standard ones, as the nature of the analytical solution has been essentially included. Figure 13.8 shows excellent results which can be obtained using such complex elements. The number of degrees of freedom is here much smaller than with a standard displacement solution but, of course, the bandwidth is much larger.[39]

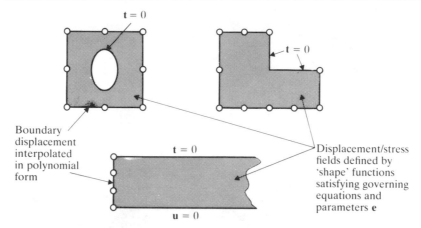

Fig. 13.7   Boundary–Trefftz-type elements. Some useful general forms[39]

Two points come out clearly in the general formulation of Eqs (13.36) to (13.39).

First, the displacement field, **a**, can only be determined excluding any rigid body modes. These can only give strains **SN** identically equal to zero and hence make no contribution to the **H** matrix.

Second, stability conditions require that (in two dimensions)

$$n_a \geqslant n_u - 3$$

and thus the minimum n_a can be readily found. Once again there is little point in increasing the number of internal parameters substantially above the minimum number as additional accuracy may not be gained.

We have said earlier that the 'translation' of the formulation discussed to problems governed by the quasi-harmonic equations is almost evident. Now identical relations will hold if we replace

$$\mathbf{u} \rightarrow \phi$$

$$\boldsymbol{\sigma} \rightarrow \mathbf{q} \tag{13.44}$$

$$\mathbf{t} \rightarrow q_n$$

$$\mathbf{S} \rightarrow \nabla$$

For the Poisson equation

$$\nabla^2 \phi = Q \tag{13.45}$$

a complete series of analytical solutions in two dimensions can be written as

$$\mathrm{Re}\,(z^n) = 1,\, x,\, x^2 - y^2,\, x^3 - 3xy^3,\, \dots \qquad \text{for } z = x + iy \tag{13.46}$$

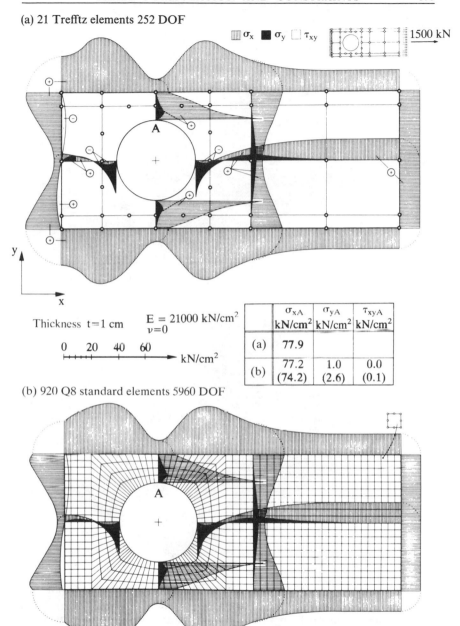

(a) 21 Trefftz elements 252 DOF

	σ_{xA} kN/cm^2	σ_{yA} kN/cm^2	τ_{xyA} kN/cm^2
(a)	77.9		
(b)	77.2 (74.2)	1.0 (2.6)	0.0 (0.1)

Thickness $t=1$ cm

$E = 21000$ kN/cm^2
$\nu=0$

0   20   40   60   kN/cm^2

(b) 920 Q8 standard elements 5960 DOF

Fig. 13.8   Application of Trefftz-type elements to a problem of a plane-stress tension bar with a circular hole. (*a*) Trefftz element solution. (*b*) Standard displacement element solution. (Numbers in parenthesis indicate standard solution with 230 elements, 1600 DOF)

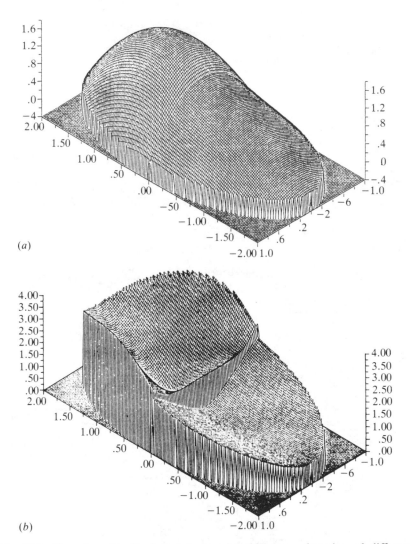

Fig. 13.9 Boundary–Trefftz-type 'elements' linking two domains of different materials in an elliptic bar subject to torsion (Poisson equations).[36] (a) Stress function given by internal variables showing almost complete continuity. (b) x component of shear stress (gradient of stress function showing abrupt discontinuity of material junction)

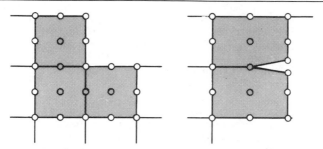

Fig. 13.10 'Superelements' built from assembly of standard displacement elements with global functions eliminating singularities confined to the assembly

A simple solution involving two subdomains with constant but different values of Q and a linking on the boundary is shown in Fig. 13.9, indicating the accuracy of the linking procedures.

13.6 Subdomains with 'standard' elements and global functions

The procedure just described can be conveniently used with approximations made internally with standard (displacement) elements and global functions helping to deal with singularities or other internal problems. Now simply an additional term will arise inside nodes placed internally in the subdomain but effect of global functions can be contained inside the subdomain. The formulation is somewhat simpler as complicated Trefftz-type functions need not be used.

We leave details to the reader and in Fig. 13.10 show some possible, useful subdomain assemblies.

13.7 Concluding remarks

The possibilities of elements or 'superelements' constructed by the mixed–incomplete field methods of this chapter are very large. Many have found practical use in existing computer codes as 'hybrid elements'; others are only now being made widely available. Much further research will elucidate the advantages of some of the forms discovered and we expect the use of such developments to increase in the future.

References

1. N. E. WIBERG, 'Matrix structural analysis with mixed variables', *Int. J. Num. Meth. Eng.*, **8**, 167–94, 1974.
2. T. H. H. PIAN and K. SUMIHARA, 'Rational approach for assumed stress finite elements', *Int. J. Num. Meth. Eng.*, **20**, 1685–95, 1984.

3. T. H. H. PIAN and D. P. CHEN, 'Alternative ways for formulation of hybrid elements', *Int. J. Num. Meth. Eng.*, **18**, 1679–84, 1982.
4. T. H. H. PIAN, D. P. CHEN, and D. KONG, 'A new formulation of hybrid/mixed finite elements', *Comp. Struct.*, **16**, 81–7, 1983.
5. P. TONG, 'A family of hybrid elements', *Int. J. Num. Meth. Eng.*, **18**, 1455–68, 1982.
6. T. H. H. PIAN and P. TONG, 'Relations between incompatible displacement model and hybrid strain model', *Int. J. Num. Meth. Eng.*, **22**, 173–181, 1986.
7. T. H. H. PIAN, 'Derivation of element stiffness matrices by assumed stress distributions', *JAIAA*, **2**, 1333–5, 1964.
8. S. N. ATLURI, R. H. GALLAGHER, and O. C. ZIENKIEWICZ (Eds), *Hybrid and Mixed Finite Element Methods*, Wiley, 1983.
9. T. H. H. PIAN, 'Element stiffness matrices for boundary compatibility and for prescribed boundary stresses', *Proc. Conf. Matrix Methods in Structural Mechanics*, AFFDL-TR-66-80, pp. 457–78, 1966.
10. R. D. COOK and J. AT-ABDULLA, 'Some plane quadrilateral "hybrid" finite elements', *JAIAA*, **7**, 1969.
11. T. H. H. PIAN and P. TONG, 'Basis of finite element methods for solid continua', *Int. J. Num. Meth. Eng.*, **1**, 3–28, 1969.
12. S. N. ATLURI, 'A new assumed stress hybrid finite element model for solid continua', *JAIAA*, **9**, 1647–9, 1971.
13. R. D. HENSHELL, 'On hybrid finite elements', in *The Mathematics of Finite Elements and Applications* (ed. J. R. Whiteman), pp. 299–312, Academic Press, 1973.
14. R. DUNGAR and R. T. SEVERN, 'Triangular finite elements of variable thickness', *J. Strain Analysis*, **4**, 10–21, 1969.
15. R. J. ALLWOOD and G. M. M. CORNES, 'A polygonal finite element for plate bending problems using the assumed stress approach', *Int. J. Num. Meth. Eng.*, **1**, 135–49, 1969.
16. T. H. H. PIAN, 'Hybrid models', in *Numerical and Computer Methods in Applied Mechanics* (eds S. J. Fenves *et al.*), Academic Press, 1971.
17. R. ALI, S. GOPALACHARYULU, and P. W. SHARMAN, 'The development of a series of hybrid-stress finite elements', *Proc. World Congress Finite Element Methods in Structural Mechanics*, **2**, 13.1–13.27, 1978.
18. Y. YOSHIDA, 'A hybrid stress element for thin shell analysis', in *Finite Element Methods in Engineering*, (eds V. Pulmano and A. Kabaila), pp. 271–86, University of New South Wales, Australia, 1974.
19. R. D. COOK and S. G. LADKANY, 'Observations regarding assumed-stress hybrid plate elements', *Int. J. Num. Meth. Eng.*, **8** (3), 513–20, 1974.
20. J. P. WOLF, 'Generalized hybrid stress finite element models', *JAIAA*, **11**, 1973.
21. P. L. GOULD and S. K. SEN, 'Refined mixed method finite elements for shells of revolution', *Proc. 3rd Air Force Conf. Matrix Methods in Structural Mechanics*, Wright-Patterson AF Base, Ohio, 1971.
22. P. TONG, 'New displacement hybrid finite element models for solid continua', *Int. J. Num. Meth. Eng.*, **2**, 73–83, 1970.
23. E. TREFFTZ, 'Ein Gegenstruck zum Ritz'schem Verfohren', *Proc. 2nd Int. Cong. Appl. Mech.*, Zurich, 1926.
24. P. K. BANERJEE and R. BUTTERFIELD, *Boundary Element Methods in Engineering Science*, McGraw-Hill, London and New York, 1981.
25. J. A. LIGGET and P. L-F. LIU, *The Boundary Integral Equation Method for Porous Media Flow*, Allen and Unwin, London, 1983.

26. C. A. BREBBIA and S. WALKER, *Boundary Element Technique in Engineering*, Newnes-Butterworth, London, 1980.
27. I. HERRERA, 'Boundary methods: a criteria for completeness', *Proc. Nat. Acad. Sci. USA*, **77**(8), 4395–8, August 1980.
28. I. HERRERA, 'Boundary methods for fluids', Chapter 19 of *Finite Elements in Fluids*, Vol. 4 (eds R. H. Gallagher, H. D. Norrie, J. T. Oden, and O. C. Zienkiewicz), Wiley, New York, 1982.
29. I. HERRERA, 'Trefftz method', in *Progress in Boundary Element Methods*, Vol. 3 (ed. C. A. Brebbia), Wiley, New York, 1983.
30. I. HERRERA and H. GOURGEON, 'Boundary methods, C-complete system for Stokes problems', *Comp. Meth. Appl. Mech. Eng.*, **30**, 225–44, 1982.
31. I. HERRERA and F. J. SABINA, 'Connectivity as an alternative to boundary integral equations: construction of bases', *Proc. Nat. Acad. Sci. USA*, **75**(5), 2059–63, May 1978.
32. O. C. ZIENKIEWICZ, D. W. KELLY, and P. BETTESS, 'The coupling of the finite element method and boundary solution procedures', *Int. J. Num. Meth. Eng.*, **11**, 355–75, 1977.
33. O. C. ZIENKIEWICZ, D. W. KELLY, and P. BETTESS, 'Marriage a la mode—the best of both worlds (finite elements and boundary integrals)', Chapter 5 of *Energy Methods in Finite Element Analysis* (eds R. Glowinski, E. Y. Rodin, and O. C. Zienkiewicz), pp. 81–107, Wiley, London and New York, 1979.
34. O. C. ZIENKIEWICZ and K. MORGAN, *Finite Elements and Approximation*, Wiley, London and New York, 1983.
35. O. C. ZIENKIEWICZ, 'The generalized finite element method—state of the art and future directions', *J. Appl. Mech.*, 50th anniversary issue, 1983.
36. A. P. ZIELINSKI and O. C. ZIENKIEWICZ, 'Generalized finite element analysis with T complete boundary solution functions', *Int. J. Num. Mech. Eng.*, **21**, 509–28, 1985.
37. J. JIROUSEK, 'A powerful finite element for plate bending', *Comp. Meth. Appl. Mech. Eng.*, **12**, 77–96, 1977.
38. J. JIROUSEK, 'Basis for development of large finite elements locally satisfying all field equations', *Comp. Meth. Appl. Mech. Eng.*, **14**, 65–92, 1978.
39. J. JIROUSEK and P. TEODORESCU, 'Large finite elements for the solution of problems in the theory of elasticity', *Comp. Struct.*, **15**, 575–87, 1982.
40. J. JIROUSEK and LAN GUEX, 'The hybrid Trefftz finite element model and its application to plate bending', *Int. J. Num. Mech. Eng.*, **23**, 651–93, 1986.
41. R. PILTNER, 'Special elements with holes and internal cracks', **21**, 1471–85, 1985.

Error estimates and adaptive finite element refinement

14.1 Introduction

Throughout this book we have stressed the fact that the finite element method offers only *an approximation* to the exact solution of a mathematically posed problem. We have ascertained that the differences between the *exact* and *approximate* solutions, e.g., the errors in displacements

$$\mathbf{e}_u = \mathbf{u} - \hat{\mathbf{u}} \tag{14.1a}$$

or the errors in stresses

$$\mathbf{e}_\sigma = \boldsymbol{\sigma} - \hat{\boldsymbol{\sigma}} \tag{14.1b}$$

decrease as the size of the subdivision 'h' gets smaller or as 'p', the order of the polynomial in the trial function used, increases. This established convergence and the acceptability or otherwise of various finite element forms. However, the central question of determining the magnitude of the error at a given, finite, stage of subdivision was so far not addressed (other than comparing in some examples the 'finite element' with 'exact' solutions).

In this chapter we shall be concerned with determining approximately:

(*a*) the error that has occurred in a particular finite element analysis carried out (*a posteriori* error estimate);

(*b*) how best to refine the approximation to achieve results of a given desired accuracy economically.

In general the interaction between (*a*) and (*b*) will be *adaptive* and many steps may be required to achieve results that are optimal.

,We shall discuss the principles of such fully adaptive analysis but also will indicate more direct approaches that can be used in practice.

In regenerating the mesh the user has often the choice of saving some data preparation efforts by retaining the original mesh and refining locally, either by

(a) introducing new elements of the type used originally but of smaller size (h) or

(b) using same-element definitions but increasing the order of polynomials used (p) involving new 'nodes' placed in such elements or

(c) using a combination of (a) and (b).

In Fig. 14.1 we illustrate the first two possibilities, often called 'h' or 'p' refinements respectively.

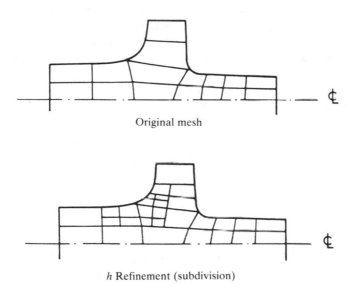

Original mesh

h Refinement (subdivision)

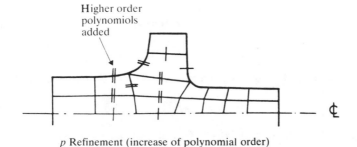

Higher order polynomiols added

p Refinement (increase of polynomial order)

Fig. 14.1   Possible refinements of an inaccurate mesh

In Chapter 8 dealing with hierarchic shape functions we have indicated how easily higher polynomial terms may be added to existing elements. Here we shall again note the merits of such forms in adaptive analysis.

14.2 Error norms and convergence rates

The specification of local error in the manner given in Eqs (14.1) is generally not convenient and occasionally misleading. For instance, under a point load both errors in displacements and stresses will be locally infinite but the overall solution may well be acceptable. Similar situations will exist near reentrant corners where, as is well known, stress singularities exist in elastic analysis and gradient singularities develop in field problems. For this reason various 'norms' representing some integral scalar quantity are often introduced to measure the error or indeed the function itself.

If, for instance, we are concerned with a general linear equation of the form of Eq. (9.6) (cf. Chapter 9), i.e.,

$$\mathbf{Lu} + \mathbf{p} = 0 \quad \text{in } \Omega \tag{14.2}$$

we can define an 'energy norm' written for the error as

$$\|e\| = \left(\int_\Omega \mathbf{e}^\mathrm{T} \mathbf{Le} \; \mathrm{d}\Omega \right)^{1/2} \equiv \left[\int_\Omega (\mathbf{u} - \hat{\mathbf{u}})^\mathrm{T} \mathbf{L}(\mathbf{u} - \hat{\mathbf{u}}) \; \mathrm{d}\Omega \right]^{1/2} \tag{14.3}$$

This scalar measure corresponds in fact to the square root of the quadratic functional such as we have discussed in Sec. 9.11 of Chapter 9 and sought its minimum in the case of a self-adjoint operator $\mathbf{L}$.

For elasticity problems the energy norm is identically defined and yields, on integration by parts,

$$\|e\|^2 = \int_\Omega (\mathbf{Se})^\mathrm{T} \mathbf{D}(\mathbf{Se}) \; \mathrm{d}\Omega \tag{14.4}$$

(with symbols as used in Chapter 2).

Here $\mathbf{e}$ is given by Eq. (14.1a) and the operator $\mathbf{S}$ defines the strains as

$$\boldsymbol{\varepsilon} = \mathbf{Su} \quad \text{and} \quad \hat{\boldsymbol{\varepsilon}} = \mathbf{S\hat{u}} \tag{14.5}$$

and $\mathbf{D}$ is the elasticity matrix (viz. Chapter 2), giving stresses as

$$\boldsymbol{\sigma} = \mathbf{D}\boldsymbol{\varepsilon} \quad \text{and} \quad \hat{\boldsymbol{\sigma}} = \mathbf{D}\hat{\boldsymbol{\varepsilon}} \tag{14.6}$$

The energy norm of Eq. (14.4) can thus be written alternatively as

$$\|e\| = \left[\int_\Omega (\boldsymbol{\varepsilon} - \hat{\boldsymbol{\varepsilon}})^\mathrm{T} \mathbf{D}(\boldsymbol{\varepsilon} - \hat{\boldsymbol{\varepsilon}}) \, d\Omega \right]^{1/2}$$

$$= \left[\int_\Omega (\boldsymbol{\varepsilon} - \hat{\boldsymbol{\varepsilon}})(\boldsymbol{\sigma} - \hat{\boldsymbol{\sigma}}) \, d\Omega \right]^{1/2}$$

$$= \left[\int_\Omega (\boldsymbol{\sigma} - \hat{\boldsymbol{\sigma}}) \mathbf{D}^{-1}(\boldsymbol{\sigma} - \hat{\boldsymbol{\sigma}}) \, d\Omega \right]^{1/2} \tag{14.7}$$

and its relation to strain energy is evident.

Other scalar norms can easily be devised. For instance, the so-called L_2 norm of displacement or stress error can be written as

$$\|e_u\|_{L_2} = \left[\int_\Omega (\mathbf{u} - \hat{\mathbf{u}})^\mathrm{T}(\mathbf{u} - \hat{\mathbf{u}}) \, d\Omega \right]^{1/2} \tag{14.8}$$

$$\|e_\sigma\|_{L_2} = \left[\int_\Omega (\boldsymbol{\sigma} - \hat{\boldsymbol{\sigma}})^\mathrm{T}(\boldsymbol{\sigma} - \hat{\boldsymbol{\sigma}}) \, d\Omega \right]^{1/2} \tag{14.9}$$

Such norms allow us to focus on the particular quantity of interest and indeed it is possible to evaluate 'root mean square' values of its error. For instance, the RMS error in stress, $\Delta\sigma$, becomes for the domain Ω

$$|\Delta\sigma| = \left(\frac{\|e_\sigma\|_{L_2}^2}{\Omega} \right)^{1/2} \tag{14.10}$$

Any of the above norms can be evaluated over the whole domain or over subdomains or even individual elements.

We note that

$$\|e\|^2 = \sum_{i=1}^m \|e\|_i^2 \tag{14.11}$$

where i refers to individual elements Ω_i such that their union is Ω.

We note further that the energy norm given in terms of stresses, the L_2 stress norm and the RMS stress error have a very similar structure and that these are similarly approximated.

At this stage it is of interest to invoke the discussion of Chapter 2 (Sec. 2.6) concerning the rates of convergence. We noted from it that with trial functions in the displacement formulation of degree p, the errors in stresses were of the order $O(h)^p$. This order of error should therefore apply to the energy norm, $\|e\|$, error. While the arguments are correct for well-behaved problems with no singularity, it is of interest to see how the above rule is violated when singularities exist.

In Figs 14.2 and 14.3 we show two similar stress analysis problems, in the first of which a strong singularity is, however, present. In both figures

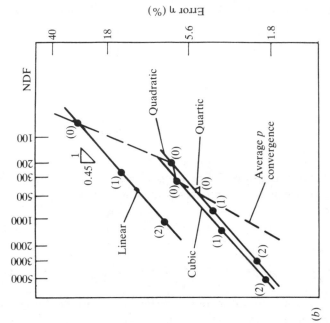

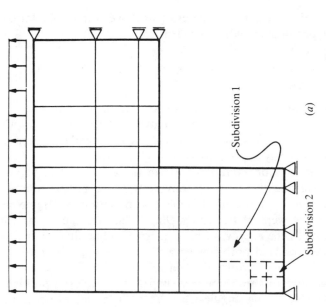

Fig. 14.2   Analysis of an L-shaped domain with singularity. h convergence for various p values. Plane stress $E = 1$, $v = 0.3$. (a) Initial mesh (0) and uniform subdivisions (i). (b) % error (η) versus NDF (number of degrees of freedom)

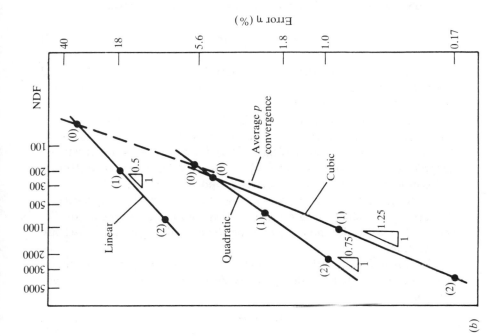

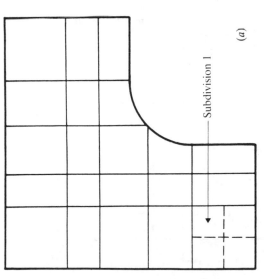

Fig. 14.3 Analysis of an L-shaped domain without singularity. h convergence for various p values and uniform subdivision. Plane stress $E = 1$, $v = 0.3$. (a) Initial mesh (0). (b) % error (η) versus NDF (number of degrees of freedom)

we show the variation of the *relative energy norm error* (percentage):

$$\eta = \frac{\|e\|}{\|u\|} \times 100\% \tag{14.12}$$

for an h refinement proceeding by a uniform subdivision of the initial mesh and for a p refinement in which polynomial order is increased throughout the original mesh.

We note two interesting facts. Firstly, the h convergence rates for various polynomial orders of the shape functions are nearly the same in the example with singularity (Fig. 14.2) and are well below the predicted theoretical order $O(h)^p$, [or $O(\text{NDF})^{-p/2}$ as the NDF (number of degrees of freedom) is approximately inversely proportional to h^2].

Secondly, in the case shown in Fig. 14.3, where the singularity is avoided by rounding the corner, the convergence rates improve for elements of higher order though again the theoretical (asymptotic) rates are not achieved.

The reason for this behaviour is clearly the singularity, and in general it can be shown that the rate of convergence for problems with singularity is

$$O(\text{NDF})^{-[\min(\lambda,\,p)]/2} \tag{14.13}$$

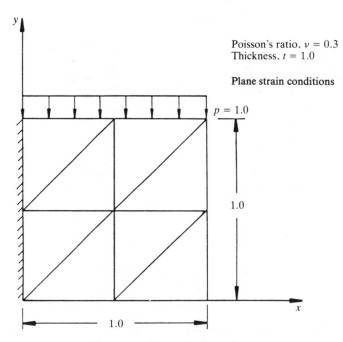

Poisson's ratio, $v = 0.3$
Thickness, $t = 1.0$

Plane strain conditions

$p = 1.0$

Fig. 14.4　Short cantilever beam

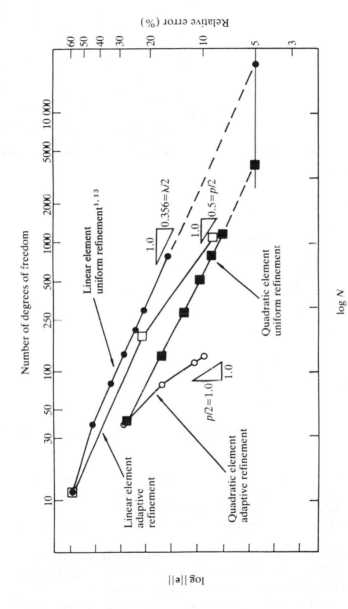

Fig. 14.5 Experimental rates of convergence for short cantilever beam.

$\lambda/2 = 0.356$, theoretical rate of convergence for uniform refinement

$p/2$, maximum rate of convergence

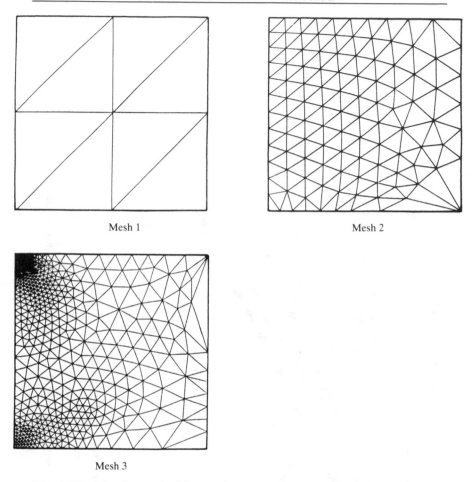

Mesh 1

Mesh 2

Mesh 3

Fig. 14.6   Adaptive mesh of linear triangular elements for short cantilever beam

where λ is a number associated with the intensity of the singularity. For elasticity problems λ ranges from 0.5 for a nearly closed crack to 0.71 for a 90° corner.[1] The rate of convergence illustrated in Fig. 14.2 approaches the value controlled by the singularity for all values of p used in the elements.

In Fig. 14.4 we show another problem with a strong singularity influence taken from the same reference.[1] Here for uniform subdivision of linear and quadratic triangles the convergence shown in Fig. 14.5 is almost the same at the singularity dominated limit unless the sequence of meshes shown in Fig. 14.6 is used. For this sequence we attempt to achieve an *optimal mesh* in which the error in each element is (in the limit) constant. For such meshes the theoretical convergence rate dependent on p is achievable as shown in Fig. 14.5.

How such mesh subdivision can be achieved will be discussed later.

A final remark evident from examining the results of Fig. 14.2 or Fig. 14.3 shows that p convergence has in general a higher rate. These results are simply shown by examining the answers given for the same mesh using different polynomials.

Once again it is possible to show that the p convergence rate reaches very high values if an optimally subdivided mesh is used.[2]

The fact that it is impractical to determine the convergence rate *a priori* means that, with the exception of the simplest problems, the Richardson type of extrapolation to the exact answer (described in Chapter 2) is generally of little use in estimating errors. For this reason we shall use more sophisticated procedures in what follows.

14.3 Error estimates—a simple and effective procedure for h refinement

In irreducible (displacement) approximations to linear elasticity and field problems discussed earlier we have generally assumed a C_0 continuous approximation for $\hat{\mathbf{u}}$ (the displacement) and this resulted in discontinuous stresses $\hat{\boldsymbol{\sigma}}$. In Chapter 12 we have shown how, by averaging or 'projection', a continuous set of stresses $\boldsymbol{\sigma}^*$ can be obtained. Thus if $\boldsymbol{\sigma}^*$ is such a C_0 continuous field interpolated by the same shape functions as are used for representing the $\hat{\mathbf{u}}$ field, i.e.,

$$\hat{\mathbf{u}} = \mathbf{N}\bar{\mathbf{u}} \tag{14.14}$$

and

$$\boldsymbol{\sigma}^* = \mathbf{N}\bar{\boldsymbol{\sigma}}^* \tag{14.15}$$

the approximating equation is achieved by a weighted residual requirement for equality between $\boldsymbol{\sigma}^*$ and $\hat{\boldsymbol{\sigma}}$, i.e.,

$$\int_{\Omega} \mathbf{N}^{\mathrm{T}}(\boldsymbol{\sigma}^* - \hat{\boldsymbol{\sigma}}) \, d\Omega = 0 \tag{14.16}$$

giving

$$\bar{\boldsymbol{\sigma}}^* = \mathbf{A}^{-1}\left(\int_{\Omega} \mathbf{N}^{\mathrm{T}}\hat{\boldsymbol{\sigma}} \, d\Omega \right) \tag{14.17a}$$

with

$$\hat{\boldsymbol{\sigma}} = \mathbf{D}\mathbf{B}\bar{\mathbf{u}} \tag{14.17b}$$

and

$$A = \int_{\Omega} N^T N \, d\Omega \tag{14.17c}$$

The above computation is made particularly simple if lumped or diagonal approximation is made to **A** and an iterative process followed (viz. Chapter 12 Sec. 12.6.1).

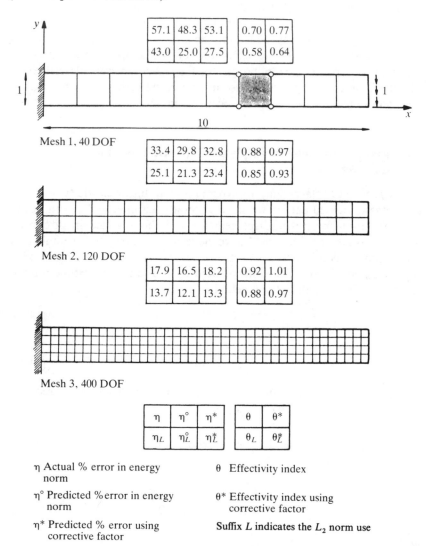

η	η°	η^*	θ	θ^*
η_L	η_L°	η_L^*	θ_L	θ_L^*

η Actual % error in energy norm

η° Predicted % error in energy norm

η^* Predicted % error using corrective factor

θ Effectivity index

θ^* Effectivity index using corrective factor

Suffix L indicates the L_2 norm use

Fig. 14.7 Bilinear elements. Cantilever beam, plane stress, $E = 10^5$, $v = 0.3$. Analysis and error estimates for uniform subdivision

With both $\hat{\boldsymbol{\sigma}}$ and $\boldsymbol{\sigma}^*$ fields available we find that the error in stresses can be *estimated with good accuracy* as

$$\boxed{\mathbf{e}_\sigma = \boldsymbol{\sigma}^* - \hat{\boldsymbol{\sigma}}} \tag{14.18}$$

Insertion of this quantity into Eqs (14.7), (14.9), or (14.10) *after the solution of the problem* (i.e., *a posteriori*) allows the error to be computed in any of the norms discussed so far.

The estimates so obtained are excellent, as comparisons with the exact values of error indicate. Figures 14.7 to 14.15[3] show a series of problems in which, for various meshes and element types, the effectivity index of the error estimates is presented. This index is defined as

$$\theta = \frac{\|e\|_{\text{estimated}}}{\|e\|_{\text{exact}}} \tag{14.19}$$

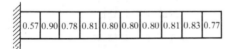

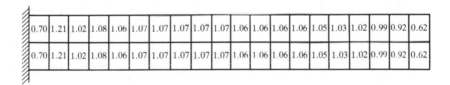

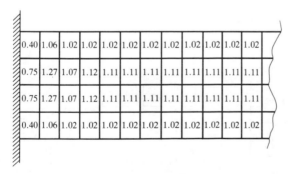

Fig. 14.8 Local effectivity indices for the problem of Fig. 14.7 (mesh 1 to 3) (θ^* energy norm); in L_2 norm results are very similar)

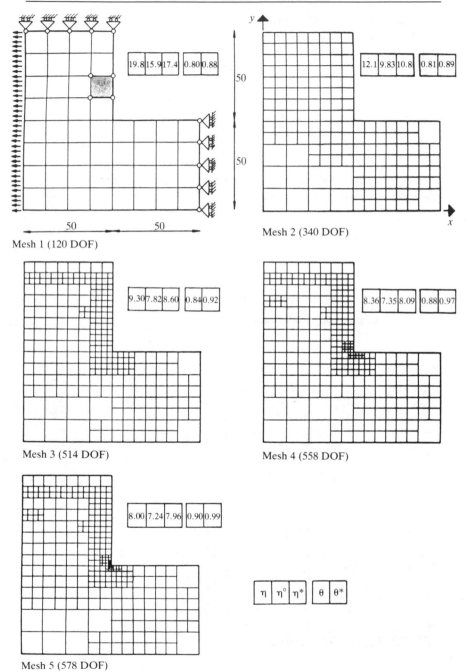

Mesh 1 (120 DOF)

Mesh 2 (340 DOF)

Mesh 3 (514 DOF)

Mesh 4 (558 DOF)

Mesh 5 (578 DOF)

Fig. 14.9 Bilinear elements. An L-shaped region in plane stress. Sequences of mesh refinement

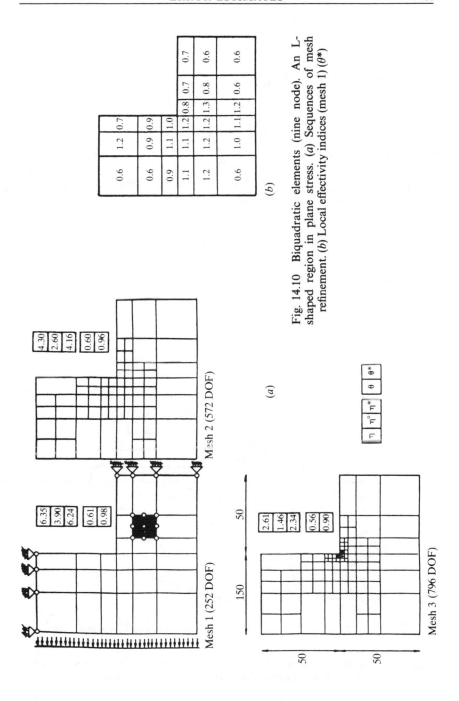

Fig. 14.10 Biquadratic elements (nine node). An L-shaped region in plane stress. (a) Sequences of mesh refinement. (b) Local effectivity indices (mesh 1) (θ^*)

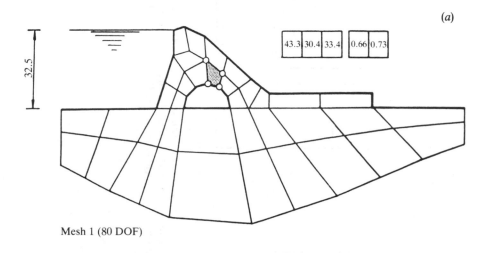

Mesh 1 (80 DOF)

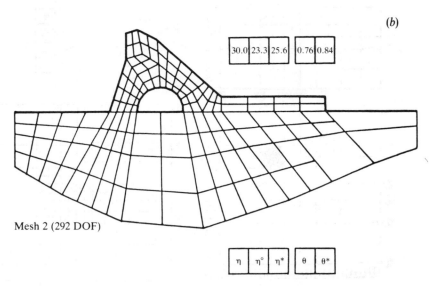

Mesh 2 (292 DOF)

η	η°	η^{*}	θ	θ^{*}

Fig. 14.11 Bilinear elements. Plane strain analysis of a dam with perforation. Water loading only. Various stages of refinement (a)–(d)

(c)

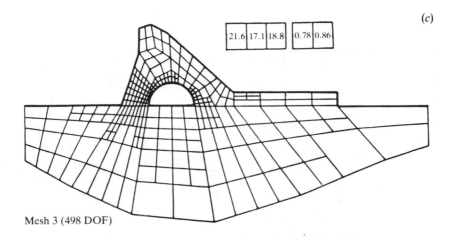

Mesh 3 (498 DOF)

(d)

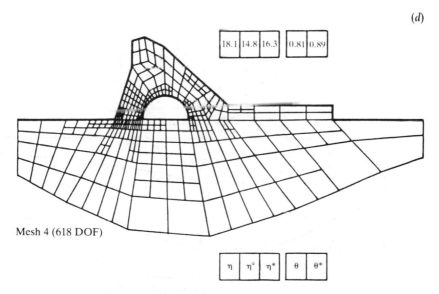

Mesh 4 (618 DOF)

η	η°	η^*	θ	θ^*

Fig. 14.11 (continued)

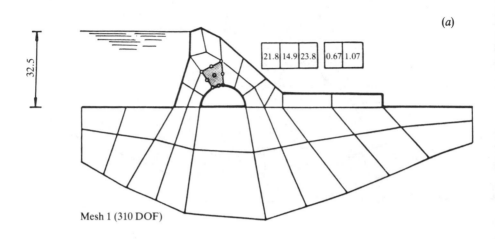

(a)

Mesh 1 (310 DOF)

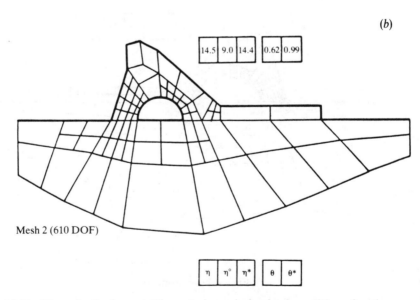

(b)

Mesh 2 (610 DOF)

η	η°	η^*	θ	θ^*

Fig. 14.12   Biquadratic element. Plane strain analysis of a dam with perforation. Water loading only. Various stages of refinement (a)–(d)

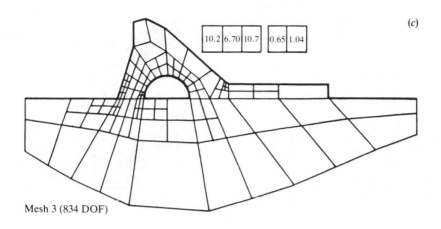

Mesh 3 (834 DOF)

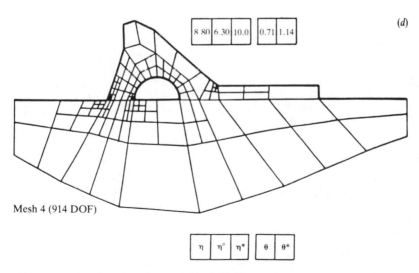

Mesh 4 (914 DOF)

Fig. 14.12 (continued)

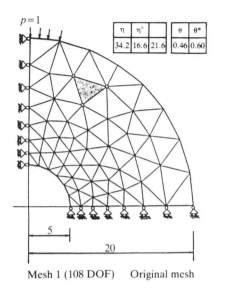

$p=1$

η	η°		θ	θ^*
34.2	16.6	21.6	0.46	0.60

5

20

Mesh 1 (108 DOF)   Original mesh   (*a*)

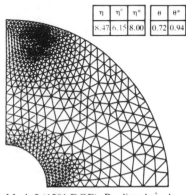

η	η°	η^*	θ	θ^*
8.47	6.15	8.00	0.72	0.94

Mesh 2 (1201 DOF)   Predicted mesh

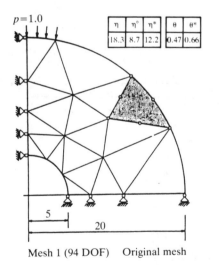

$p=1.0$

η	η°	η^*	θ	θ^*
18.3	8.7	12.2	0.47	0.66

5

20

Mesh 1 (94 DOF)   Original mesh   (*b*)

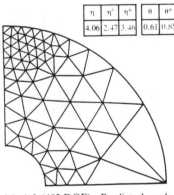

η	η°	η^*	θ	θ^*
4.06	2.47	3.46	0.61	0.85

Mesh 2 (482 DOF)   Predicted mesh

Fig. 14.13 Automatic mesh generation to achieve 5 per cent accuracy: thick cylinder under diametral load. (*a*) Linear elements. (*b*) Quadratic elements

Empirically it is found that this index (and hence the error estimate) can be improved by correcting the direct estimate by a constant factor. This factor turns out to be for elasticity problems:

1.1 for bilinear quadrilateral elements
1.3 for linear triangular elements
1.6 for biquadratic, nine-noded elements
1.4 for quadratic triangular elements

The values θ^* and η^* shown in the figures are so obtained. However, even without this empirical device the effectivity indices are good.

While it is intuitively 'obvious' that the 'smoothed' stresses σ^* are more accurate than the discontinuous ones and hence that the estimates based on Eq. (14.18) should be good ones, a more mathematical proof of the correctness of the estimate is needed. A simple and valid explanation is available in Fig. 14.16 in which we show a solution for a one-dimensional elastic problem with a constant modulus using linear and quadratic elements. In this case σ values are proportional to du/dx and it is evident that the 'smoothing' or projection process is simply a *higher order difference approximation to the nodal derivative* than that used to calculate $\hat{\sigma}$. In two and three dimensions the arguments are similar.†

The error estimator introduced above is one of the simplest to evaluate and hence to use in practice. Its accuracy compares well with others in which the computation involves the evaluation of 'residuals' obtained by insertion of the approximate solution $\hat{u}$ into the equation governing the problem [viz. Eq. (14.2)], i.e.,

$$\mathbf{r} = \mathbf{L}\hat{\mathbf{u}} + \mathbf{p} \tag{14.20}$$

This residual is easily enough evaluated within each element but on the element interfaces becomes infinite if, say, we have a second-order differential equation and a C_0 continuous approximation. The integrated effect of the residual near an interface I as $d\Omega \to 0$ becomes equivalent to a line integral of the discontinuity in gradients of $\mathbf{u}$ (or tractions in an elastic problem) which we shall call $\mathbf{J}$ (jump). Thus

$$\int_\Omega \mathbf{r}\, d\Omega \equiv \int_I \mathbf{J}\, dI \tag{14.21}$$

The error estimators using energy norm derived by various authors[4-10] have a general form (with $r^2 \equiv \mathbf{r}^T\mathbf{r}$, etc.)

$$\|e\|^2 = C_1 \int_\Omega r^2\, d\Omega + C_2 \int_I J^2\, dI \tag{14.22}$$

where Ω is the total domain and I the total interface between elements.

† A more formal proof of validity of the error estimation presented here has been given very recently.[17]

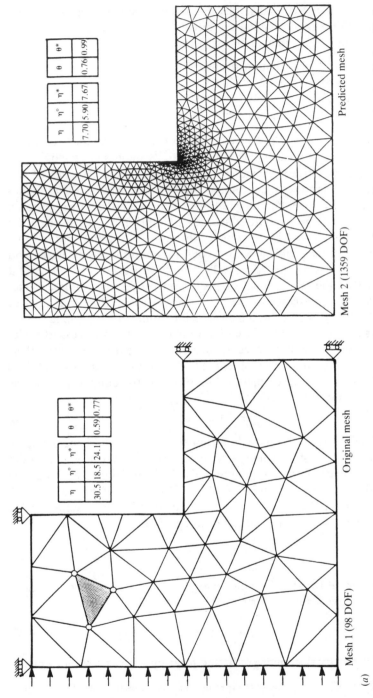

Fig. 14.14 Automatic mesh generation to achieve 5 per cent. L-shaped region in plane stress. (a) Linear triangle. (b) Quadratic triangle. With refinement strategy using Eq. (14.32), accuracy of 4.61 per cent is reached in the operation with 358 DOF[14]

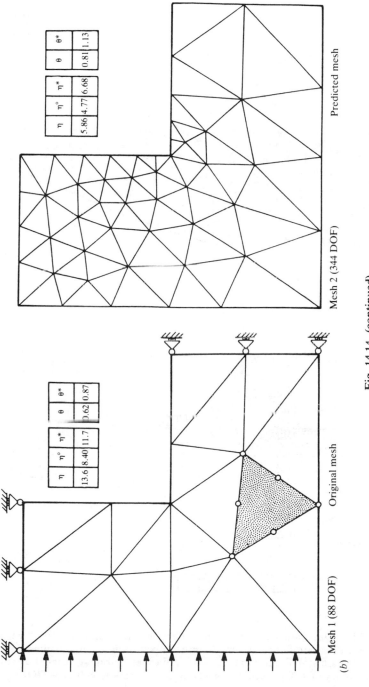

Fig. 14.14 (continued)

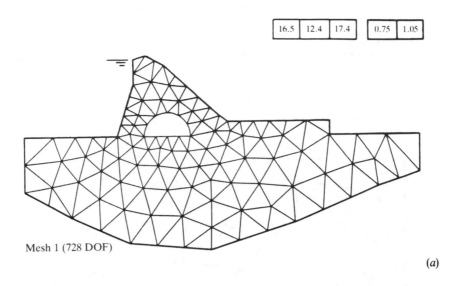

Mesh 1 (728 DOF)

(a)

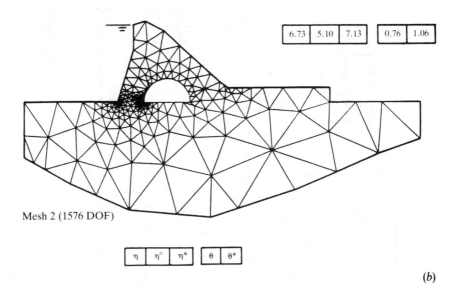

Mesh 2 (1576 DOF)

(b)

Fig. 14.15 Quadratic triangle. Automatic mesh generation to achieve 5 per cent accuracy. Plane strain analysis of a dam with perforation water loading only. (a) Original mesh. (b) Refined mesh. With refinement strategy using Eq. (14.32) accuracy of 4.88 per cent is reached in one operation with 1764 DOF[14]

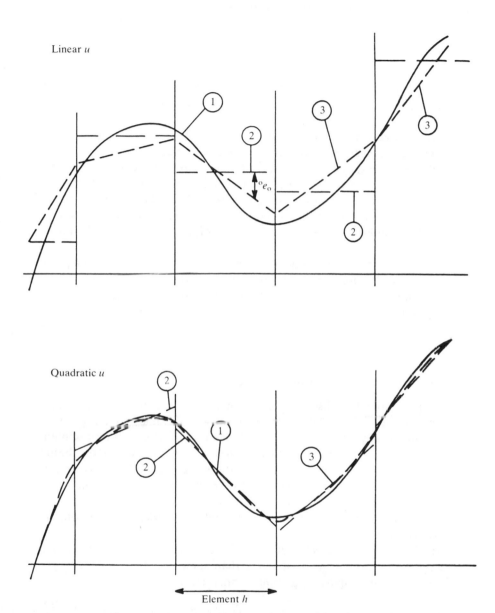

Fig. 14.16   Why projection gives higher order approximation (1) σ Exact solution. (2) $\hat{\sigma}$ FEM solution = least square approximation to σ (discontinuous). (3) σ^* projected solution = least square approximation to $\hat{\sigma}$ (continuous high order)

For a single element a particular expression derived in references 5 and 6 for two-dimensional problems gives an element contribution as

$$\|e\|_i^2 = \frac{h^2}{24kp^2} \int_{\Omega^e} r^2 \, d\Omega + \frac{h}{24kp} \int_{I_e} J^2 \, dI \qquad (14.23)$$

where k is dependent on the problem (being $E/(1 - v)$ for plane stress elasticity), p is the polynomial order of the approximations used, and h the element size.

We shall return to an estimator of this type later in the chapter when discussing p convergence problems, but we note that

(a) the estimator is more complex to use than that previously derived as it involves evaluation of line integrals of the discontinuities;

(b) the simple estimator using stress smoothing can be shown to be equivalent to that derived and proved earlier by Babuska and Rheinboldt[5], at least in the case of bilinear (four-noded) elements;[5,11]

(c) that for low order elements the major contribution to the integrals of (14.23) is from the term involving traction (stress) discontinuities.

These have frequently been taken by practitioners of finite element analysis as a direct measure of error and indeed there is a theoretical justification for this as shown above.

14.4 The h refinement process—adaptivity

The error estimators discussed in the preceding section allow the global energy (or similar) norm of the error to be determined and the errors occurring locally (at the element level) are usually also well represented, as shown in Figs 14.8 and 14.10. If these errors are within the limits prescribed by the analyst then clearly the work is completed. More frequently these limits are exceeded and refinement is necessary. The question which this section addresses is how best to effect this refinement. Here obviously many strategies are possible and much depends on the *objectives* to be achieved.

In the simplest case we shall seek, for instance, to make the relative energy norm percentage error η less than some specified value $\bar{\eta}$ (say 5 per cent in many engineering applications). Thus

$$\eta \leq \bar{\eta} \qquad (14.24)$$

is to be achieved.

In an 'optimal mesh' it is desirable that the distribution of energy norm error (i.e., $\|e\|_i$) should be equal between all elements. Thus if the total permissible error is determined (assuming that it is given by the result of the approximate analysis) as

$$\bar{\eta}(\|\hat{u}\|^2 + \|e\|^2)^{1/2} \tag{14.25}$$

we could pose a requirement that error in any element i should be

$$\|e\|_i < \bar{\eta}\left(\frac{\|\hat{u}\|^2 + \|e\|^2}{m}\right)^{1/2} \equiv \bar{e}_m \tag{14.26}$$

where m is the number of elements involved.

Elements in which the above is not satisfied are obvious candidates for refinement. Thus if we define the ratio

$$\frac{\|e\|_i}{\bar{e}_m} = \xi_i \tag{14.27}$$

we shall refine wherever†

$$\xi_i > 1 \tag{14.28}$$

The refinement could be carried out progressively by refining only a certain number of elements in which ξ is higher than a certain limit and at each time halve the size of such elements. This type of process known as *enrichment of mesh* is illustrated in Figs 14.9 to 14.12. The process of refinement shown above, though ultimately leading to a satisfactory solution being obtained with a relatively small number of total degrees of freedom, is in general not economical as the total number of trial solutions may be excessive (as seen from the examples).

A more efficient procedure is to try to design a completely new mesh which satisfies the requirement that

$$\xi_i \leqslant 1 \tag{14.29}$$

One possibility here is to invoke the asymptotic convergence rate criteria at the element level (although we have seen that these are not realistic in the presence of singularities) and to predict the element size distribution. For instance, if we assume

$$\|e\|_i \propto h_i^p \tag{14.30}$$

where h_i is the current element size and p the polynomial order of approximation, then to satisfy the requirement of Eq. (14.25) the element size should be no larger than

$$h = \xi_i^{-1/p} h_i \tag{14.31}$$

Mesh generation programs in which the local element size can be specified are available today and these can be used to predict a new mesh for which the reanalysis is carried out.[12] In Fig. 14.6 and Figs 14.13 to 14.15 we show how starting from a relatively coarse solution a

† We can indeed 'derefine' or use a larger element spacing where $\xi_i < 1$ if computational economy is desired.

single mesh prediction allows a solution (almost) satisfying the specified accuracy requirement to be achieved.

The reason of success of the mesh regeneration based on the simple assumption of asymptotic convergence rate implied in Eq. (14.30) is the fact that with refinement the mesh tends to be 'optimal' and the localized singularity influence no longer affects the overall convergence. We have shown this effect already in Fig. 14.6.

Of course the effects of singularity will still remain present in the elements adjacent to it—and improved mesh subdivision can be obtained if in such elements we use the appropriate convergence and write, in place of Eq. (14.13),[13]

$$h = \xi_i^{-1/\lambda} h_i \tag{14.32}$$

in which λ is the singularity strength. A convenient number to use here is $\lambda = 0.5$ as most singularity parameters lie in the range 0.5–1.0. With this procedure added to the refinement strategy we achieve accuracies better than 5 per cent in one remeshing for problems of Fig. 14.14b and 14.15.

In the examples illustrated so far we have shown in general a process of refinement with the total number of degrees of freedom increasing with each stage, even though the mesh is redesigned. This need not necessarily be the case as a fine but badly structured mesh can show much greater error than in a near-optimal one. To illustrate this point we show in Fig. 14.17 the one-stage refinement designed to reach a 5 per cent accuracy in one step starting from uniform mesh subdivisions. The problem here is the same as that illustrated in Figs 14.4 to 14.6 and in the refinement process we use both the mesh criteria of Eqs (14.31) and (14.32).[14]

We note that now, in at least one refinement, a decrease of total error occurs with a reduction of total degrees of freedom (starting for uniform 8×8 subdivision with NDF = 544 and $\eta = 9.8$ per cent to NDF = 460 and $\eta = 3.1$ per cent).

Neither of the strategies suggested is 'optimal' in the sense that a uniform mesh refinement is by no means necessary if the problem shows 'directional features' (i.e., the function u may be varying rapidly along some local coordinate x' and at a slower rate along y'). For such situations 'correction indicators' which tell us in which direction the refinement is more effective could be introduced. This procedure we shall discuss in a section dealing with p convergence. Alternatively, different 'directional' error norms could be introduced. We shall allude to such refinements later.

As we mentioned earlier, the energy norm is not necessarily the best criterion for practical refinement. Local stress error [viz. Eq. (14.10)] can be used effectively and, though such local estimates by the procedures

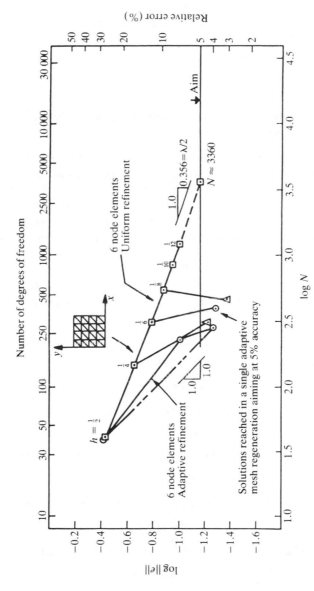

Fig. 14.17 The influence of initial mesh to convergence rates in *h* version. Adaptive refinement using quadratic triangular elements

indicated are not necessarily precise, these have been used with good effect.[3]

Very precise local information can be obtained with regard to stresses, displacements, stress intensity factors, etc., by suitable recasting of the problem and necessitates generally an auxiliary solution.[15] Discussion of such processes is referred to the published literature as it is somewhat too complex to include here.

14.5 Error estimates for hierarchic formulations. A basis for p-adaptive refinement

When deriving an approximation to the total error [such as we have introduced in Eq. (14.7)] various alternative possibilities exist. One of these is, of course, to obtain a complete solution by, say, halving the existing mesh. Another is to introduce an additional, complete, polynomial to the approximation. The difference between two such consecutive solutions gives an estimate of the local error albeit at a very considerable cost.

If a hierarchic form of approximation is used this difference can be obtained at more reasonable cost, utilizing the original approximation. In such a hierarchic form the discretized equations are (viz. Chapter 7), for $\hat{\mathbf{u}} = \mathbf{N}\bar{\mathbf{u}}$,

$$\mathbf{K}\bar{\mathbf{u}} = \mathbf{f} \tag{14.33a}$$

and, with a hierarchically added variable $\bar{\mathbf{u}}^h$,

$$\hat{\mathbf{u}} = \mathbf{N}\bar{\mathbf{u}}^n + \mathbf{N}^h\bar{\mathbf{u}}^h$$

$$\begin{bmatrix} \mathbf{K} & \mathbf{K}^{nh} \\ \mathbf{K}^{hn} & \mathbf{K}^{hh} \end{bmatrix} \begin{Bmatrix} \bar{\mathbf{u}}^n \\ \bar{\mathbf{u}}^h \end{Bmatrix} = \begin{Bmatrix} \mathbf{f}_1 \\ \mathbf{f}_2 \end{Bmatrix} \tag{14.33b}$$

Solution of Eqs (14.33) will, of course, yield the error approximation as

$$\mathbf{e} = \mathbf{N}(\bar{\mathbf{u}}^n - \bar{\mathbf{u}}) + \mathbf{N}^h\bar{\mathbf{u}}^h \tag{14.34}$$

A reasonable approximation, however, can be made by putting

$$\bar{\mathbf{u}}^n \approx \bar{\mathbf{u}} \tag{14.35}$$

into the second equation, i.e., using the original approximation. Now

$$\bar{\mathbf{u}}^h = (\mathbf{K}^{hh-1})(\mathbf{f}_2 - \mathbf{K}^{hn}\bar{\mathbf{u}}) \tag{14.36}$$

and to get this only the inversion of the matrix $\mathbf{K}^{hh}$ is needed.

Even a cruder approximation will often suffice by introducing hierarchical refinements one by one. Now for each degree of freedom introduced at a 'node' i, $\bar{u}_i^h$ is a scalar and

$$(K_{ii}^{hh})^{-1} \equiv \frac{1}{K_{ii}^{hh}} \tag{14.37}$$

To translate the local error estimate $e_i^h = N_i^h \bar{u}_i^h$ into an energy norm we can proceed again in various ways.

One possibility is to evaluate the corresponding stress and strain changes, i.e., we obtain [viz. Eq. (14.7)]

$$\|e\|_i^2 = \bar{u}_i^{hT} \int (SN_i^h)^T D(SN_i^h)\, d\Omega \bar{u}_i^h = \bar{u}_i^{hT} K_{ii}^{hh} \bar{u}_i^h \tag{14.38}$$

with the total estimate being

$$\|e\| = \left(\sum_{i=1}^{s} \|e\|_i^2 \right)^{1/2} \tag{14.39}$$

for a large number of degrees of freedom introduced.

Another procedure looks at the residual and notes that

$$\|e\|_i = \left(\int_\Omega e^T L e\, d\Omega \right)^{1/2} = \left(-\int_\Omega e^T r\, d\Omega \right)^{1/2} \tag{14.40}$$

for any differential equation, as

$$L\hat{u} + p = r$$
$$Lu + p = 0 \tag{14.41}$$

and

$$Le = L(u - \hat{u}) = -r$$

Now r can be evaluated for a given $\hat{u}^h$ both inside the element and on the interfaces [viz. Eqs (14.20) and (14.21), and if we note that in fact the second of equations (14.33b) is a weighted residual form

$$K^{hh}\bar{u}^h = -\int_\Omega N^h r\, d\Omega \tag{14.42}$$

we can write for a single variable introduced using Eqs (14.33) and (14.40) to (14.42),

$$\|e\|_i^2 = \frac{(\int_\Omega N_i^h r\, d\Omega)^2}{K_{ii}^{hh}} = C_i^2 \tag{14.43}$$

The above quantity is useful as a *correction indicator* giving quite correctly the amount of error that can be corrected by introducing a particular degree of freedom and is, of course, identical to Eq. (14.38).

However, it is not a good error estimator as on occasion the new shape function N_i^h introduced can be orthogonal to the residual **r**.

An improved error estimator is derived using the Schwartz inequality to replace the integral in Eq. (14.43) as[8]

$$\left(\int_\Omega N_i^h \mathbf{r} \, d\Omega \right)^2 \leqslant \left(\int_\Omega N_i^{h2} \, d\Omega \right) \left(\int_\Omega r^2 \, d\Omega \right) \qquad (14.44)$$

It is found that this form gives an estimator that is efficient, particularly if a correction factor of $\sqrt{2}$ is introduced. We now evaluate

$$\|e\|_i = \frac{1}{\sqrt{2}} \left[\frac{\sum \left(\int_\Omega N_i^{h2} \, d\Omega \int_\Omega r^2 \, d\Omega + \int_I N_i^{h2} \, dI \int_I J^2 \, dI \right)}{K_{ii}^{hh}} \right]^{1/2} \qquad (14.45)$$

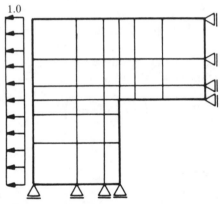

Step 1. Initial mesh geometry, boundary and
load conditions. $\eta = 25.8\%$. 72 DOF

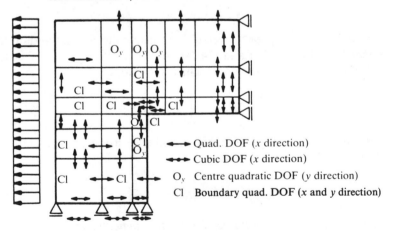

Step 5. Adaptive p version. $\eta = 7.2\%$. 144 DOF

Fig. 14.18   A p refinement solution of the L-shaped domain problem

recognizing as before the need to deal with interelement jumps. The summation is conducted through all new degrees of freedom which can be introduced at the next hierarchical stage.

Obviously the procedure is advantageous for p refinement when we consider the next higher polynomial order in an element in assessing the correction indicators and the error estimate.

During the refining process the polynomial order can be variable from element to element.

With the availability of error indicators for all candidate degrees of freedom that can be introduced the adaptive strategy can directly introduce those that provide the greatest error correction [viz. Eq. (14.43)].

A technique sometimes adopted[9] is to include in the next analysis all those degrees of freedom which give

$$C_i \geqslant \gamma C_{max} \qquad (14.46)$$

where C_{max} is the maximum value of any correction indicator and γ a number in the range 0.1–0.5.

Obviously $\gamma = 0$ simply corresponds to complete refinement of all elements but any reasonable value greater than zero will tend to give an optimal refinement, in which the error in each element tends to constant after many adaptive steps.

Figure 14.18 shows again the problem previously discussed but now solved using hierarchic p-type refinements. The figure shows the final distribution of the degrees of freedom.

The adaptive process indicated here has to be continued until required accuracy is reached—and though leading to an economy in degrees of freedom used finally is less effective than that described previously in which we attempt to reach directly the solution of desired accuracy.

14.6 Concluding remarks

The methods of estimating errors and adaptive refinement which are described in this chapter constitute a very important tool for practical application of finite element methods. The range of applications is large and we have only touched here upon the relatively simple range of linear elasticity and similar self-adjoint problems. A recent survey shows many more areas of application[4] and the reader is referred to this publication for interesting details. At this stage we would like to reiterate that many different norms or measures of error can be used and that for some problems the energy norm is not in fact 'natural'. A good example here is presented by problems of high-speed gas flow, where very steep gradients (shocks) can develop. The formulation of such problems is complex, but this is not necessary for the present argument. In this

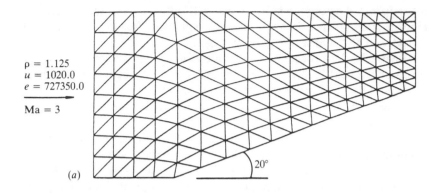

ρ = 1.125
u = 1020.0
e = 727350.0

———

Ma = 3

20°

(*a*)

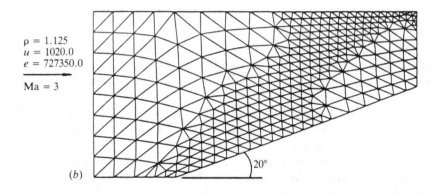

ρ = 1.125
u = 1020.0
e = 727350.0

———

Ma = 3

20°

(*b*)

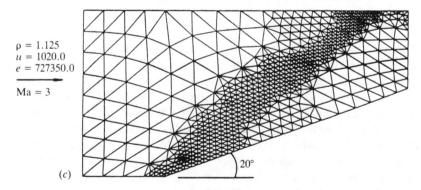

ρ = 1.125
u = 1020.0
e = 727350.0

———

Ma = 3

20°

(*c*)

Fig. 14.19 Mesh enrichment. Supersonic flow past wedge (Mach number 3). (*a*)
Initial configuration of mesh. (*b*) After 101 steps. (*c*) After 201 steps (ρ = density;
u = velocity; *e* = specific energy)

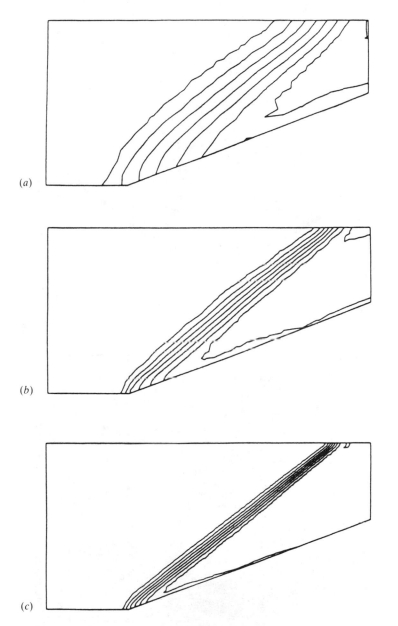

Fig. 14.20   Problem of Fig. 14.14. Density after: (a) 100 steps; (b) 200 steps; (c) 250 steps each using mesh refinements shown in Fig. 14.19

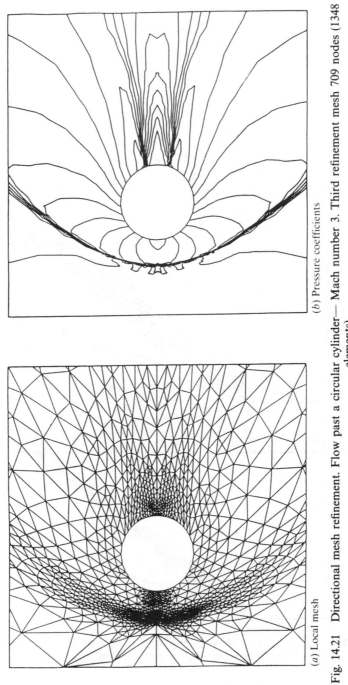

(b) Pressure coefficients

(a) Local mesh

Fig. 14.21   Directional mesh refinement. Flow past a circular cylinder— Mach number 3. Third refinement mesh 709 nodes (1348 elements)

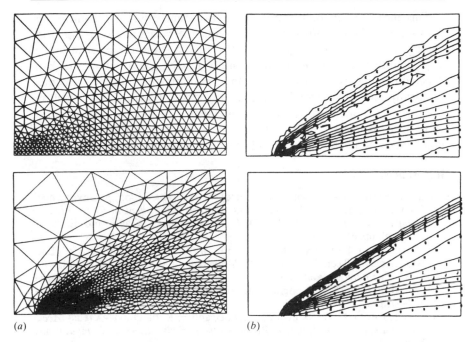

Fig. 14.22   Viscous flow past a flat plate Mach number 3 (a) Initial and regener-
ated meshes. (b) Computed density contours

analysis attention focuses on the function **u** itself rather than on its derivatives (stresses). In such cases we can proceed to use the L_2 norm of the
function **u** itself as the error measure. In a one-dimensional case such as
that illustrated in Fig. 14.16 the nodal derivatives are found to be
approximated well by projection and hence the second derivative (or
curvature) $\mathrm{d}^2\hat{u}/\mathrm{d}x^2$ is approximated within each element closely.

This provides an estimate of the error e as

$$e = u - \bar{u} = \tfrac{1}{2}x_1(h - x_1)\left|\frac{\mathrm{d}^2\hat{u}}{\mathrm{d}x^2}\right| \tag{14.47}$$

if x_1 is the coordinate measured from the side of the element and if we
assume $\hat{u}$ to be nodally exact. Immediately, the L_2 norm of error is given
for element i as

$$\|e\|_i \approx \tfrac{1}{8}h_i^2\left|\frac{\mathrm{d}^2\hat{u}}{\mathrm{d}x^2}\right| \tag{14.48}$$

which can be used in the control of a mesh refinement if we aim to limit
this error to a fixed value in each element.

This measure can be extended to a two-dimensional problem either providing uniform refinement *or* applying the measure along the principal axes of curvature. Figures 14.19 to 14.22 show the application of such adaptive refinements in successfully tracing the areas of shock in supersonic flow problems. In the first example the mesh is reformed uniformly—in the second directionally,[12,16] showing the resulting optimal mesh. Such non-uniform refinement is of interest also in problems such as plastic deformation where directional slip lines are of importance, and here lies yet another field of its application.

References

1. B. A. Szabo, 'Estimation and control of error based on *p*-convergence', Chapter 3 in *Accuracy Estimates and Adaptive Refinement in Finite Element Computations* (eds I. Babuska, O. C. Zienkiewicz, J. Gago, and E. R. de Oliveira), pp. 61–73, Wiley, 1986.
2. B. Guo and I. Babuska, 'The *h-p* version of the finite element method', Lab. Num. Analysis Tech. Note BN 1043, University of Maryland, USA, 1985.
3. O. C. Zienkiewicz and J. Z. Zhu, 'A simple error estimator and adaptive procedure for practical engineering analysis', *Int. J. Num. Meth. Eng.*, **24**, 337–57, 1987.
4. I. Babuska, O. C. Zienkiewicz, J. Gago, and E. R. de Oliveira (eds), *Accuracy Estimates and Adaptive Refinements in Finite Element Computations*, Wiley, 1986.
5. I. Babuska and W. C. Rheinboldt, '*A-posteriori* error estimates for the finite element method', *Int. J. Num. Meth. Eng.*, **11**, 1597–615, 1978.
6. I. Babuska and W. C. Rheinboldt, 'Adaptive approaches and reliability estimates in finite element analysis', *Comp. Meth. Appl. Mech. Eng.*, **17/18**, 519–40, 1979.
7. D. W. Kelly, J. P. de S. R. Gago, O. C. Zienkiewicz, and I. Babuska, '*A-posteriori* error analysis and adaptive processes in the finite element method. Part I—Error analysis, Part II—Adaptive mesh refinement', *Int. J. Num. Meth. Eng.*, **19**, 1593–619, 1621–56, 1983.
8. O. C. Zienkiewicz, J. P. de S. R. Gago, and D. W. Kelly, 'The hierarchical concept in finite element analysis', *Comp. Struct.*, **16**, 53–65, 1981.
9. O. C. Zienkiewicz and A. Craig, 'Adaptive refinement, error estimates, multigrid solution and hierarchic finite element concepts', Chapter 2 of *Accuracy Estimates and Adaptive Refinements in Finite Element Computations* (eds. I. Babuska, O. C. Zienkiewicz, J. Gago, and E. R. de Oliveira), pp. 25–55, Wiley, 1986.
10. B. A. Szabo, 'Some recent developments in finite element analysis', *Comp. Math. Appl.*, **5**, 99–115, 1979.
11. E. Rank and O. C. Zienkiewicz, 'A simple error estimator for the finite element method', *Comm. Appl. Num. Meth.*, **3**, 243–50, 1987.
12. J. Peraire, M. Vahdati, K. Morgan, and O. C. Zienkiewicz, 'Adaptive remeshing for compressible flow computations', *J. Comp. Phys.*, **72**, 449–66, 1987.
13. J. Z. Zhu, O. C. Zienkiewicz, and A. W. Craig, 'Adaptive techniques in finite element analysis', *Proc. NUMETA*, Conference S3/1 to S3/10, Martinus Nijhoff Publ., 1987.

14. J. Z. ZHU and O. C. ZIENKIEWICZ, 'Adaptive techniques in the finite element method', *Comm. Appl. Num. Math*, **4**, 197–204, 1988.
15. I. BABUSKA and A. MILLER, 'The post processing approach in the finite element method: Part 1—Calculation of displacements, stresses and other higher derivatives of displacements. Part 2—The calculation of stress intensity factors. Part 3—*A posteriori* error estimates and adaptive mesh selection', *Int. J. Num. Meth. Eng.*, **20**, 1085–109, 1111–29, 2311–24, 1984.
16. R. LOHNER, K. MORGAN, and O. C. ZIENKIEWICZ, 'An adaptive finite element procedure for compressible high speed flows', *Comp. Meth. Appl. Mech. Eng.*, **51**, 441–65, 1985.
17. M. AINSWORTH, J. Z. ZHU, A. W. CRAIG, and O. C. ZIENKIEWICZ, 'Analysis of a simple a-posteriori error estimator in the finite element method', *Int. J. Num. Meth. Eng.* To Appear.

Computer procedures for finite element analysis

15.1 Introduction

In this chapter we shall consider some of the steps that are involved in the development of a finite element computer program to carry out analyses for the theory presented in previous chapters. The computer program that is discussed here may be used to solve any of the one-, two-, or three-dimensional problems previously discussed, provided the global coefficient matrix of the resulting algebraic equations is symmetric. Modifications to the program may easily be made to permit extension to the unsymmetric case. Although the theory discussed in this volume is restricted to linear applications, the program may also be used to solve non-linear and/or transient problems, which will be discussed in detail in the next volume.

The computer program presented is an extension of the work originally contained in the 3rd edition.[1] Several extensions are provided to permit a wider range of computer implementations and problem classes to be considered (some limitations have also been necessary due to space requirements). The version included in this chapter is called PCFEAP and is specifically intended for use on a personal computer. The program name is an acronym for a Personal Computer Finite Element Analysis Program. In the personal computer version a primary limitation is the available memory, and it is not possible to retain all of the necessary arrays in core at one time. Accordingly, an out-of-core frontal system is adopted to solve linear algebraic equations.[2,3] A simple memory management system utilizing the main memory and disk files is used to store large arrays resulting from the global coefficient matrix and any element history terms.

For implementation on larger computers with virtual memory management, it is more efficient to avoid using disk files as much as possible. Accordingly, for these systems an in-core variable band solver is included as an equation-solving option.[4] It is necessary to select one of these systems when installing the program. The equation-handling systems included are both compatible with the personal computer version, PCFEAP. Some increase in efficiency may be obtained by slight reprogramming to avoid all of the disk input/output operations and by retaining the arrays in the main memory.

The current version of PCFEAP permits both 'batch' and 'interactive' problem solution. The finite element model of the problem is given as an input file and may be prepared using instructions provided later in this chapter. Included also is a simple graphics option which permits display of one- and two-dimensional finite element models in either their undeformed or deformed configurations. Again, limitations in space have precluded the inclusion of other options; however, experienced programmers may easily add other features. In developing the program, library plot subprograms from the Graphics Development Toolkit[5] are used. Options for other plotting systems may be substituted as needed.

Finite element programs can be separated into two basic parts:
(a) data input module and preprocessor, and
(b) solution and output modules to carry out the actual analysis (see Fig. 15.1 for a simplified schematic).
Each of the modules can in practice be very complex. In the subsequent sections we shall discuss in some detail the programming aspects for each of the modules. It will be assumed that the reader is familiar with the finite element principles presented in this book, linear algebra, prog-

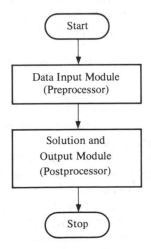

Fig. 15.1 Simplified schematic of finite element program

ramming, and in particular Fortran. Readers who merely intend to use
the program listed at the end of the chapter may wish to skip the following sections and go to the user instructions included in Secs 15.3, 15.4,
and 15.8.

The chapter is divided into eight sections. Sections 15.2 and 15.3
discuss the procedure adopted for data input, describing a finite element
problem and the instructions for data file preparation respectively. Basi-
cally the data consists of nodal quantities (e.g., coordinates, boundary
condition data, loading, etc.) and element quantities (e.g., connection
data, material properties, etc.).

Section 15.4 discusses solution algorithms for various classes of finite
element analyses. In order to have a computer program that can solve
many types of finite element problems a *macro* programming language is
introduced. The macro programming language is associated with a set of
compact subprograms each designed to compute one or at most a few
basic steps in a finite element solution process. Examples in the macro
language are commands to form the global stiffness matrix, as well as
commands to solve equations, print results, etc. The macro programming
concept permits inclusion of a wide class of solution algorithms in the
computer program presented here.

In Sec. 15.5 we discuss a methodology commonly used to develop
element matrices. In particular, numerical integration is used to derive
the element 'stiffness', 'mass', and 'load' matrices for problems in linear
heat transfer and elasticity. The concept of using basic shape function
routines is exploited in these developments (see Chapters 7 and 8).

In Sec. 15.6 methods for solving the large set of algebraic equations
resulting from finite element formulations is presented. The methods
adopted for the computer program are a frontal method which is based
upon Gauss elimination[2,3] and a variable-band profile method which is
based upon the Crout method.[4] The basic method consists of a factor-
ization of the coefficient matrix into a product of a lower triangular
matrix and an upper triangular matrix. The use of this scheme with
either the variable-band or frontal algorithms leads to a very compact
program and allows for inclusion of a resolve capability (i.e., new load
cases) without any significant additional programming effort. Use of the
resolve capability can substantially reduce costs for analysing subsequent
load cases. Included are implementations of the subprograms for solution of symmetric coefficient matrices only. For finite element formulations which do not lead to a symmetric matrix it is possible to extend the
frontal solution system as discussed by Hood.[6] The variable-band system
is capable of solving some unsymmetric problems; however, it is necessary to modify the storage allocation and assembly process (e.g., see references 1 or 4).

In present-day practice there are many complex and efficient system programs for finite element analysis capable of dealing with very large numbers of variables and formulations. The very complexity of such systems means that it is difficult to update them in order to introduce new developments of technology. The program presented here is written specifically as a research and educational tool in which the various 'modules' can be changed or added to as desired. Indeed, quite different combinations of the subroutines for purposes that may even today not be obviously needed are possible.

To provide a program for small computers all equations are not retained in core; thus, the capacity of the program is limited by the memory and the disk size of the computer used. With the core size available in some current personal computers, the program can only handle realistic engineering problems with several hundred unknowns (maximum front width is currently set to 120). At the expense of some complexity and efficiency, the program can be extended to larger problems, and we shall deal with this matter in Sec. 15.7.

Finally, Sec. 15.8 contains a complete listing of the program discussed in this chapter. Included also are element routines to carry out analyses for two-dimensional problems of linear elasticity. Using the format of these routines the reader should, after mastering this chapter, be able to program additional routines for other problems and greatly extend the capabilities of the program.

15.2 Data input module

The data input module shown in Fig. 15.1 must transmit sufficient information to the other modules so that each problem can be solved. In the program discussed here the data input module is used to read from an input file the necessary geometric, material, and loading data so that all subsequent finite element arrays can be established. In the program a set of dimensioned arrays are established which store nodal coordinates, element connections, material properties, boundary restraint codes, prescribed nodal forces and displacements, nodal temperatures, etc. Table 15.1 lists the array names (and their dimensions) which are used to store these quantities.

The notation used for the arrays is at variance with that used in the text. For example, in the text it was found convenient to refer to nodal coordinates as x_i, y_i, z_i, whereas in the program these are called $X(1, i)$, $X(2, i)$, $X(3, i)$, respectively. This change was made so that all arrays used in the program can be dynamically allocated; thus, if a two-dimensional problem is analysed, space will not be reserved for the $X(3, i)$ coordinates, and likewise for $X(2, i)$ in one-dimensional problems.

TABLE 15.1
FORTRAN VARIABLE NAMES USED FOR DATA STORAGE

Variable name (dimension)	Description
D(18,NUMMAT)	Material property data sets, limited to 18 words per set
F(NDF,NUMNP)	Nodal forces and displacements
ID(NDF,NUMNP)	Boundary restraint conditions after input of data changed to equation numbers in global arrays
IE(8,NUMMAT)	Element type for each material set
IX(NEN1,NUMEL)	Element nodal connections and material set numbers
T(NUMNP)	Nodal temperatures
X(NDM,NUMNP)	Nodal coordinates
NDF	Maximum number of degrees of freedom at any node (maximum of 6)
NDM	Spatial dimension of problem (maximum is 3)
NEN	Maximum number of nodes connected to any element
NEN1	NEN + 3
NUMEL	Number of elements
NUMMAT	Number of material sets
NUMNP	Number of nodes

In addition, the nodal displacements in the text were called a_i; in the program these are called either U(i) or U(1, i), U(2, i), etc., where the first subscript refers to the degrees of freedom at a node (from 1 to NDF).

15.2.1 *Storage allocation.* A single array is partitioned to store all the data arrays, as well as some global arrays, e.g., residuals, displacements, loads, etc. Each array indicated in Table 15.1 is dynamically dimensioned to the size and precision required for each problem by using a set of pointers established in the control program (see Fig. 15.2). In this way no space is wasted in data storage and a maximum amount of space is reserved to store the global arrays. Since this automatic dimensioning method is used it is not possible to establish absolute values for maximum numbers of material sets, nodes, or elements.

The program uses different precision (integer, real, double precision) to maintain maximum capacity. Byte alignment is also maintained for maximum efficiency and utility on various computer systems. The program will check that sufficient space exists to solve each problem and, if not, an error message will be printed. The total capacity of the program is controlled by the dimension of the array in the blank common of the main program and the corresponding value of MAXM.

A second limit in the problem size is the resulting size of the global coefficient matrix. The terms are stored in the array contained in labelled common ADATA. In the version given here the frontal solver is limited to a maximum front width of 120 and the profile solver to an array with a non-zero profile of less than 8000 terms. If the capacity is exceeded an error is printed and the program stops.

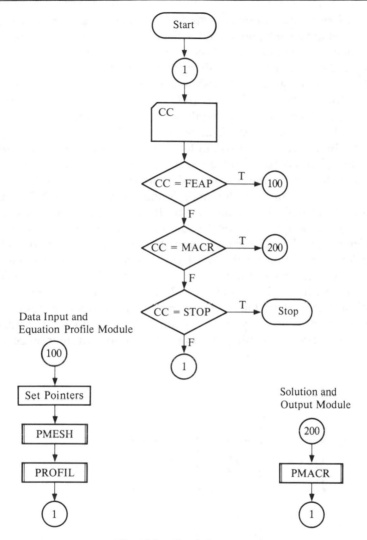

Fig. 15.2   Control program

The front width is dependent on the order of introducing the elements. Renumbering elements may reduce the maximum front to an acceptable size. In the profile solution the number of terms needed depends on the node numbering adopted; a renumbering of the nodes may reduce the size. This issue is discussed more in Sec. 15.6.

In general for the current dimensioning limits, the mesh limit will occur first with the frontal solution system, whereas the equation size limit will occur first for the variable-band solution system.

15.2.2 *Element and coordinate data.* Once a mesh for a problem has been established data can be prepared for the computer program (see Sec. 15.3 for formats). As an example, consider the specification of the nodal coordinate and element connection data for the sample two-dimensional (i.e., NDM = 2) rectangular region shown in Fig. 15.3, where a mesh of 9 four-noded rectangular elements (NUMEL = 9 and NEN = 4) and 16 nodes (NUMNP) has been established. To describe the nodal and element data, values must be assigned to each $X(i, j)$ for $i = 1, 2$ and $j = 1$ to 16 and for each $IX(k, n)$ for $k = 1$ to 4 and $n = 1$ to 9. In the definition of the coordinate array X, the 'i' subscript indicates coordinate direction and the 'j' subscript defines the node number. Thus the value of $X(1, 3)$ is the x coordinate for node 3 and the value of $X(2, 3)$ is the y coordinate for node 3. In a similar way for the element connection array IX the 'k' subscript is the local node number of the element and 'n' is the element number. The value of any $IX(k, n)$ is the number of a global node. The convention for the first local node number is somewhat arbitrary. The local node number 1 for element 3 in Fig. 15.3 could be associated with the global node 3, 4, 7, or 8. Once the first local node is established the others follow according to the convention adopted for each particular element type. For example, the four-noded quadrilateral can be numbered according to Fig. 15.4. If we consider once again

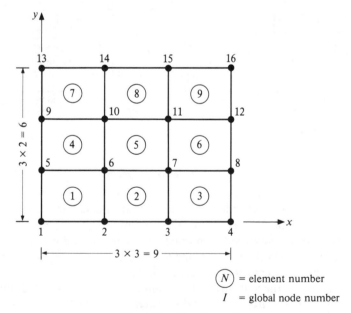

Fig. 15.3   Simple mesh

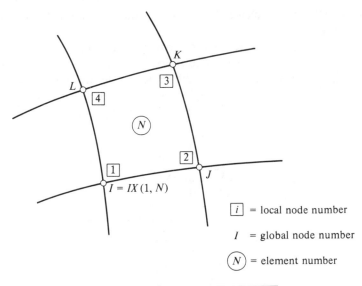

$\boxed{i}$ = local node number

I = global node number

$\left(\,N\,\right)$ = element number

Option number	Local node number			
	1	2	3	4
a	3	4	8	7
b	4	8	7	3
c	8	7	3	4
d	7	3	4	8

Fig. 15.4   Typical four-noded element

element 3 we have four possibilities for specifying the IX(k, 3) array. These are shown in Fig. 15.4. The computation of the element matrices from any of the above descriptions must produce the same coefficients for the global arrays.

For a very large mesh the preparation of each piece of mesh data would be very tedious; consequently, a program should provide capabilities to generate much of the data. A simple scheme for nodal generation is to input end points of any line and generate by some scheme the interior points, e.g., linear interpolation is used in the program given here. Thus for the mesh in Fig. 15.3 one could input coordinates of nodes 1 and 4 and generate coordinates for nodes 2 and 3. Even for this simple problem nodal coordinate data preparation is reduced by half.

For elements a pattern usually exists from which elements can be generated. Again consider the mesh in Fig. 15.3. The nodal values for element 2 are those of element 1 incremented by 1; and nodes of element 3 are those of element 2 incremented by 1. Thus, one could input the nodal connections for elements 1, 4, and 7 and generate the rest using a specified increment.

More sophisticated generation schemes can be developed. For example, for regions that are regular blocks of elements an alternative is to describe a superelement that is subsequently subdivided into elements and nodes.[7] The procedure adopted to input the mesh data must be consistent with the particular class of problems the user wishes to

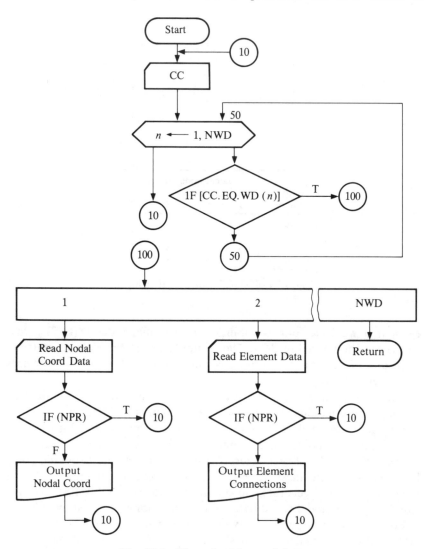

Fig. 15.5   Flow chart for mesh input

analyse, as well as the facilities that are available. The data input scheme included in the program given here is simple and should suffice for most analyses within the capacity of the program. If a user wishes to prepare his own input generation scheme, it can easily be interfaced with subroutine PMESH since each input segment does not interact with any of the others. A flow chart for PMESH is given in Fig. 15.5.

15.2.3 *Material property specification—different element routines.* The above discussion has focused only on the data arrays for nodal coordinates and element connections. It is also necessary to specify the material properties associated with each element, loadings, and the restraints for each node.

Each element has associated material properties, e.g., for linear isotropic elastic materials Young's modulus E and Poisson's ratio v describe the material parameters for an isothermal state. In most situations several elements have the same material properties and it is unnecessary to specify properties for each element individually. In this case an element can be 'keyed' to a material property set by a single number on the element record and the properties can be specified once. For example, if the region shown in Fig. 15.3 is all the same material, only one material property set is required and each element would reference this set.

A more complicated example is that shown in Chapter 1, Fig. 1.4, where elements 1, 2, 4, and 5 might be plane elements while element 3 is a truss element. (In realistic engineering problems several element types may be needed at the same time.) In this case at least two different types of element stiffness formulations must be computed. In the computer program given here facilities exist for using up to four different types of element routines in any analysis. The program has been designed so that all computations associated with any element are contained in one element subroutine called ELMTnn, where nn is between 01 and 04 (see Sec. 15.5.3 for a discussion on organization of ELMTnn). Each element type to be used is specified as part of the material property data. Thus if element type 1, e.g., computed using subroutine ELMT01, is a plane linear elastic three- or four-noded element and element type 4 is a truss element, e.g., computed using subroutine ELMT04, the data given for the example in Fig. 1.4 might be:

(*a*) *Material properties*

Material set number	Element type	Material property data
1	4	E_1, A_1
2	1	E_2, v_2

(b) *Element connections*

Element	Material property set	Connection
1	2	1 3 4
2	2	1 4 2
3	1	2 5
4	2	3 6 7 4
5	2	4 7 8 5

where E is Young's modulus, v is Poisson's ratio, and A is area. Thus elements 1, 2, 4, and 5 have material property set 2 which is associated with element type 1—i.e., it is a plane linear elastic element as Fig. 1.4 shows. In a similar way, element 3 has material property set 1 which is associated with element type 4, the truss element. It will be seen later that the above scheme leads to a simple organization of an element routine which inputs material data sets and computes all necessary arrays for finite element analyses.

More sophisticated schemes could be adopted (e.g., see reference 8); however, for the type and capacity of the program included here the added complexity is not needed.

15.2.4 *Boundary conditions—equation numbers.* The process of specifying the boundary conditions and the procedure for modification for specified displacements is tied to the method adopted to store the global arrays, e.g., stiffness and mass matrices. In the program only those coefficients within a non-zero profile in the global arrays are stored.

The storage of the non-zero profile of the equations leads to considerable savings over the more traditional banded solution storage.† In addition, it is usually more efficient to delete the rows and columns for the equations corresponding to specified boundary displacements. As an example consider the stiffness matrix corresponding to the problem given in Fig. 1.1; storing all terms within the upper profile requires 54 words, whereas if the equations corresponding to the restrained nodes 1 and 6 are deleted only 32 words are required to store the compacted stiffness (see Fig. 15.6). This is a saving of over 40 per cent for the stiffness matrix alone. The effort (as measured by computer time) to solve equations by a profile method is approximately proportional to the sum of the column heights squared. For the example in Fig. 15.6 compacted storage also leads to savings of over 40 per cent in equation solution.

In the frontal solver, the equations associated with specified values of the dependent variable parameters are also not included. In this way the maximum front width is made as small as possible.

† It will be shown in Sec. 15.6 that the profile storage by columns leads to a very efficient direct equation solution method.

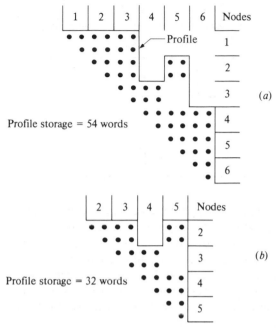

Fig. 15.6   Stiffness matrix. (a) Total stiffness storage. (b) Storage after deletion of boundary condition

To facilitate a compact storage operation in forming the global arrays, a boundary condition array is input for each node. The array is called ID and is dimensioned as shown in Table 15.1. During data input the value of the ID array for any nodal degree of freedom for which either the value is specified *a priori* or for which no unknown exists (i.e., different nodes can have different numbers of associated unknowns) is set to a *non-zero* value; all other degrees of freedom have a *zero* value. Table 15.2 shows the ID values for the example shown in Chapter 1, Fig. 1.1, where it is evident that nodes 1 and 6 are fully restrained.

TABLE 15.2

BOUNDARY RESTRAINT CODE VALUES AFTER
DATA INPUT OF PROBLEM IN FIG. 1.1

Node	Degree of freedom	
	1	2
1	1	1
2	0	0
3	0	0
4	0	0
5	0	0
6	−1	1

TABLE 15.3
COMPACTED EQUATION NUMBERS FOR
PROBLEM IN FIG. 1.1

Node	Degree of freedom	
	1	2
1	0	0
2	1	2
3	3	4
4	5	6
5	7	8
6	0	0

The numbers for the equations associated with the unknowns are constructed from Table 15.2 by replacing each non-zero value with a zero and each zero value by the appropriate equation number. In the program this step is performed in subroutine PROFIL starting with the degrees of freedom associated with node 1, etc. The result for the example leads to values shown in Table 15.3, which contains the boundary restraint information but, in addition assists in the assembly of the equations.

The solution of mixed formulations which have matrices with zero diagonals requires special care in solving for the parameters. For example, in the $\mathbf{q}$, $\boldsymbol{\phi}$ formulation discussed in Sec. 12.2 it is necessary to eliminate all $\bar{\mathbf{q}}_i$ parameters associated with each $\bar{\phi}_j$ when the direct methods described in Sec. 15.6.1 are used. This may be achieved by numbering the ID entries so that the $\bar{\mathbf{q}}_i$ parameters have smaller global equation numbers than the one for the associated $\bar{\phi}_j$ parameter.

The equation number scheme can be further exploited to handle repeating boundaries (see Chapter 8, Sec. 8.18) where nodes on two boundaries are required to have the same displacement but its value is unknown. This is accomplished by setting the equation numbers to the same value (discard the unused one). The equation profile or front must then be adjusted to accommodate the connection (see Sec. 15.2.7).

15.2.5 *Loading.* The non-zero nodal forces or displacements associated with each degree of freedom must be specified. In the program these are both stored in the array F and the distinction between load and displacement is made by comparing the corresponding value of the boundary restraint condition (from the equation number table) for each degree of freedom. For the example of Fig. 1.1, if F(1, 1) was set to 0.01 it would signify that the displacement of the first degree of freedom (i.e., u) is specified to be 0.01 displacement units, whereas F(2, 3) set to 5 would indicate that the force for the second degree of freedom has a value of 5 force units.

In many problems the loading may be distributed and in these cases the loading must first be converted to nodal forces. This can be facilitated in the data input by adding a new macro command DIST which makes a call to subroutine SLDnn to perform the computation of the equivalent nodal loads. Once any distributed load is converted to the generalized forces it is treated in the same way as for the loads input by FORC. A word of caution—the use of the DIST macro must follow specification of the BOUN macro command if the node which has distributed loading applied to it is restrained.

The program also provides facilities to specify nodal temperatures. (The association here is for thermal loading of structural problems; for other classes of problems other interpretations can be inserted by the user.) Temperatures are input for each node in the same way as for forces and coordinates.

Specific instructions for the preparation of data for each of the items discussed above are given in Sec. 15.3.

15.2.6 *Mesh data checking.* Once all data for the geometric, material, and loading are supplied the program is ready to initiate execution of the solution module; however, prior to this step it is usually preferable to perform some checks on the input data. The simplest such check would be a review of the input data (and generated values) as given in the output file from the program. For large problems any checking by this procedure will leave considerable doubt as to the accuracy of the data—it is easy to misenter data such that visual checks reveal nothing! It is advisable to use some automatic plot to scale of the mesh as an alternative check on accuracy. In addition, checks on the value of the jacobians in isoparametric elements, as suggested in Chapter 8, can be used. The program given here provides for these checks. For three-dimensional problems mesh plot routines can also be prepared; however, without capabilities of erasing hidden lines, rotating, slicing, etc., the usual plot is unintelligible for any but the simplest problem. The general topic of automatic plotting is outside the scope of this text and the reader is referred to references 9 and 10 for further information on this important aspect of any practical finite element solution package.

15.2.7 *Front and profile determination.* As discussed above, the global arrays for the coefficient matrix may be stored in either a frontal or a profile mode, depending upon which solution option is used. In both forms the storage mode is by columns above the principal diagonal. In the profile method the diagonal is stored separately, whereas in the frontal method it is retained as part of each column (the difference is related to efficiency in implementing the Gauss and Crout algorithms).

In order to minimize the storage requirements for the variable-band method it is necessary to know beforehand the profile of the equations, and this is determined by first numbering the active equations as described above and then using the element connection array, IX, together with the equation number and boundary condition array, ID, to determine the maximum column height of each equation. Finally the equations are compacted into a vector and the column heights are used to construct the address of the diagonal elements in the storage vector. The programming steps for the profile determination are given in subroutine PROFIL. The total number of equations is determined by the maximum value of the ID array and is called NEQ. The total storage requirement for either the upper (or lower) half of the matrix profile is given by the address for the diagonal of NEQ [i.e., JDIAG(NEQ)]. Thus the storage requirements in common ADATA must be increased by this amount for each profile matrix required, e.g., by NEQ and JDIAG(NEQ) for linear symmetric steady-state problems.

In the frontal method, the current implementation divides the available storage into two parts: one part stores the coefficient terms and right-hand size of the active frontal equations and the other is used as a buffer area to store equations that have already been reduced. Only when the buffer is full will writes to the disk be performed. This minimizes costly input/output operations on the disk and greatly improves efficiency over that of previous implementations (e.g., those in reference 2 or 3). In order to make this division it is necessary to know beforehand how large the front will become during the solution process. This computation is performed by subprogram PREFRT and stored in the variable MAXF.

15.3 User instructions for computer program

The solution of a finite element problem using the program given at the end of this chapter begins with a sketch of a mesh covering the region to be analysed. The user must select a consistent set of units to define the numerical values for data. If boundaries are curved the mesh will only approximate the shape of the region (e.g., see Fig. 15.7). In sketching the mesh the type and order (linear, quadratic, etc.) of elements must be taken into consideration: for triangular elements in two dimensions the mesh is described by a net of triangles, whereas for quadrilateral isoparametric four-noded elements the region can be described by a net of quadrilaterals. The user may wish to use both triangles and quadrilaterals. In this case two element routines may be necessary, one for triangles and one for quadrilaterals. The shape function routine, SHAPE, for quadrilaterals given in this chapter includes the three-noded triangle

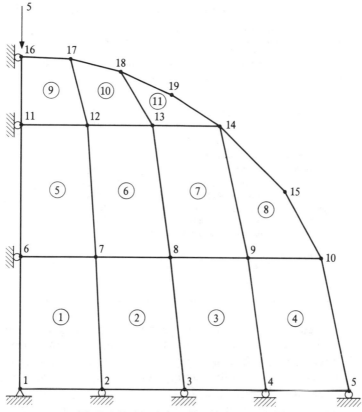

Fig. 15.7   Mesh for circular disk problem

by coalescing the shape functions of two nodes; hence in this case only one element routine need be used.[11]

After a sketch of the mesh has been made the elements and nodes are numbered in consecutive order. The order of numbering nodes is crucial for the profile solution method, whereas the order of numbering the elements is crucial for the frontal method. As a general rule, numbering the elements and nodes along the same directions in the mesh will produce both a minimum front and a minimum profile. Indeed, numbering nodes in the order that a minimum effort front solution would eliminate the unknowns will also produce an efficient profile solution scheme (e.g., see reference 12 where it is shown that the effort in both solution methods is identical). A numbering can usually be improved in some sense by using an automatic renumbering scheme.[13-15] The method in reference 15 produces a near-optimal numbering for the number of terms under the profile—which will give maximum efficiency in a resolve mode. The coding to implement this method is very compact and executes rapidly.

TABLE 15.4
TITLE AND CONTROL INFORMATION FORMATS

TITLE—FORMAT (20A4)

The title also serves as a start of problem record. The first four (4) columns must contain the start word FEAP.

Columns	Description	Variable
1 to 4	Must contain FEAP	TITL(1)
5 to 80	Alphanumeric information for header in output file	TITL(I) I = 2, 20

CONTROL DATA—FORMAT (7I5)

Columns	Description	Variable
1 to 5	Number of nodes	NUMNP
6 to 10	Number of elements	NUMEL
11 to 15	Number of material sets	NUMMAT
16 to 20	Spatial dimension (≤ 3)	NDM
21 to 25	Number of unknowns per node (≤ 6)	NDF
26 to 30	Number of nodes/element	NEN
31 to 35	Added size to element matrices, in excess of NDF*NEN	NAD

Once the sketch and numbering of the mesh is completed the user can proceed to the preparation of data for the program. The first step consists of specifying problem title and control information given in Table 15.4, which is used during subsequent data input and also is used to allocate memory in the program.

In addition to the input data formats, Table 15.4 gives the variable names used in the program. The variables NDF, NEN, and NAD are used to calculate the size of the element arrays, NST. Normally for displacement formulations NDF*NEN is the size of the element array; however, if nodeless variables or mixed methods are used it may be desirable to increase the size of the element array NAD (note that the NAD variables are not assembled as global parameters).

Once the control data is supplied the program expects the data records for the mesh description, e.g., nodal coordinates, element connections, etc. Each problem or problem class may require different types and amounts of data; consequently, the flow of data to the program is controlled by a set of *macro* commands. The available macro commands are given in Table 15.5; others may be added by suitably modifying the data list WD in subroutine PMESH. The macro commands PRINt† and NOPRint allow the user to output and suppress output to the screen or the 'output' file, respectively, of any data which is input subsequently.

† Only capital characters are data. In the program, text may be input in upper or lower case. A free format parser is included in PCFEAP which permits input of text and numerical fields to be separated by a comma (,). This avoids the need to keep count of columns.

TABLE 15.5

DATA INPUT: MACRO CONTROL STATEMENTS

INPUT MACRO CONTROL—FORMAT (A4)
The input of each data segment is controlled by the value assigned to CC. The following values are admissible and each CC record must be immediately followed by the appropriate data (described in Tables 15.6 to 15.14).

CC value	Data to be input
COOR	Coordinate data
ELEM	Element data
MATE	Material data
BOUN	Boundary condition data
FORC	Prescribed nodal force data
TEMP	Temperature data
BLOC	Node and element data
POLA	Convert polar coordinates to cartesian form
ANGL	Angle of sloping two-dimensional boundaries
PRIN	Print subsequent mesh data (default mode)
NOPR	Do not print subsequent mesh data
END	Must be last record in mesh data, terminates mesh input

Except for END and POLA the data segments can be in any order. If the values of ANGL, BOUN, FORC, or TEMP are zero, no input data is required.

Thus, once a mesh has been fully checked and subsequent analyses are desired it is not necessary to reprint all the mesh data.

An analysis will require at least:

(a) coordinate data which follows the macro command COOR and is prepared as described in Table 15.6;

TABLE 15.6

COORDINATE DATA

COORDINATE DATA—FORMAT (2I10, 6F10.0)—must immediately follow a COOR macro.
The coordinate data contains the node number N and the value of the coordinates for the node. Only the values of (XL(I), I = 1,NDM) are used, where NDM is the value input on the control record.

Nodal coordinates can be generated along a straight line described by the values input on two successive records. The value of the node number is computed using the N and NG on the first record to compute the sequence N, N + NG, N + 2NG, etc. NG may be input as a negative number; if it has incorrect sign the sign will be changed. Nodes need not be in order.

Columns	Description	Variable
1 to 10	Node number	N
10 to 20	Generator increment	NG
21 to 30	X1 coordinate	XL(1) → X(1,N)
31 to 40	X2 coordinate	XL(2) → X(2,N)
41 to 50	X3 coordinate	XL(3) → X(3,N)

Note: terminate with blank record(s).

TABLE 15.7
ELEMENT DATA

ELEMENT DATA—FORMAT (16I5)—must immediately follow an ELEM record

The element data contains the element number, material set number (which also selects the element type, see Table 15.8), and the sequence of nodes connected to the element. If there are less than NEN nodes (see Table 15.1 for input of NEN) either leave the appropriate fields blank or zero.

Elements must be in order. If element records are omitted the element data will be generated from the previous element with the same material number and the nodes all incremented by the LX on the previous element. Generation to the maximum element number occurs where a blank record is encountered.

Columns	Description	Variable
1 to 5	Element number	L
6 to 10	Material set number	IX(NEN1,L)
11 to 15	Node 1 number	IX(1,L)
16 to 20	Node 2 number	IX(2,L)
etc.	$\vdots$	$\vdots$
etc.	Node NEN number	IX(NEN,L)
etc.	Generation increment	LX

(b) element data which follows the macro command ELEM and is prepared according to Table 15.7; and

(c) material data which follows the macro command MATE and is prepared according to Table 15.8 and the data required for each particular element (see Sec. 15.8.3).

TABLE 15.8
MATERIAL PROPERTY DATA

MATERIAL DATA SETS—must immediately follow a MATE macro

Each material property set also selects the element type that will be used for the material property data.

RECORD 1). FORMAT (8I10)

Columns	Description	Variable
1 to 10	Property set number	MA
11 to 20	Element type (1 to 4)	IEL
21 to 30	Global DOF number for local DOF 1	IDL (1)
31 to 40	Global DOF number for local DOF 2	IDL (2)
. . .	etc. to NDF	

The IDL(I) may be used to reassign element local variable degree-of-freedom numbers to global degree-of-freedom numbers. The default values are IDL(I) = I. If input values are IDL(1) = 2 and IDL(2) = 1 then element DOF 1 will be assigned to global DOF 2 and element DOF 2 to global DOF 1 (switched). If any IDL is non-zero a zero value of other IDL will not be assembled. Thus, it is easily possible to restrain two-dimensional problems to respond with a one-dimensional solution.

Each material Record 1.) must be followed immediately by the material property data required for the element type IEL being used, e.g., see Sec. 15.8.3.

TABLE 15.9
BOUNDARY RESTRAINT DATA

BOUNDARY CONDITIONS—FORMAT (16I5)—must immediately follow a BOUN macro
For each node that has at least one degree of freedom with a specified displacement, a
boundary condition must be input. The convention used for boundary restraints is
$$= 0 \quad \text{no restraint, force specified}$$
$$\neq 0 \quad \text{restrained, displacement specified}$$
Values of force or displacement input in FORC (Table 15.10):

Columns	Description	Variable
1 to 5	Node number	N
6 to 10	Generation increment	NX
11 to 15	DOF 1 boundary code	$IDL(1) \rightarrow ID(1,N)$
16 to 20	DOF 2 boundary code	$IDL(2) \rightarrow ID(2,N)$
etc.	$\vdots$	$\vdots$
etc.	DOF NDF boundary code	$IDL(NDF) \rightarrow ID(NDF,N)$

Note: when generating boundary condition codes for subsequent nodes IDL is set to zero
if it was input $\geqslant 0$, and is set to -1 if input negative. All degrees of freedom with non-zero
codes are assumed fixed. Terminate with blank record(s).

In addition, most analyses will require specification of nodal boundary
restraint conditions, macro BOUN, and the corresponding nodal force
or displacement value, macro FORC, which are specified according to
Tables 15.9 and 15.10 respectively. The program permits the specification
of sloping boundaries in two dimensions using the macro command
ANGL. The anticlockwise angle, in degrees, that the transformed 1-axis
makes with the $X(1, I)$ direction is input. Subsequently, FORC and
BOUN values are interpreted with respect to the rotated direction. Some
analyses may have auxiliary nodal quantities that specify a loading. For
example, in the analysis of elasticity problems, temperature may provide
loading. The program provides a capability of specifying a temperature

TABLE 15.10
NODAL FORCED BOUNDARY VALUE DATA

FORCE CONDITIONS—FORMAT (2I10, 7F10.0)—must immediately follow a FORC macro
For each node that has a non-zero nodal force or displacement a force record must be
input or generated. Generation is the same as for coordinate data (see Table 15.6). The
value specified is a force if the corresponding restraint code is zero and a displacement if
the corresponding restraint code is non-zero.

Columns	Description	Variable
1 to 10	Node number	N
11 to 20	Generation increment	NG
21 to 30	DOF 1 Force (Displ.)	$XL(1) \rightarrow F(1,N)$
31 to 40	DOF 2 Force (Displ.)	$XL(2) \rightarrow F(2,N)$
etc.	$\vdots$	$\vdots$
etc.	DOF NDF Force (Displ.)	$XL(NDF) \rightarrow F(NDF,N)$

Note: terminate with a blank record(s).

TABLE 15.11
NODAL TEMPERATURE DATA

TEMPERATURE DATA—FORMAT (2I10, F10.0)—must immediately follow a TEMP macro
For each node that has a non-zero temperature the value must be input. Generation of
values can be performed as described for coordinates (see Table 15.6).

Columns	Description	Variable
1 to 10	Node number	N
10 to 20	Generation increment	NG
21 to 30	Nodal temperature	$XL(1) \rightarrow T(N)$

Note: terminate with blank record(s).

TABLE 15.12
BLOCK MESH DATA

BLOCK DATA—must immediately follow a BLOC macro record
Generates a 'block' of nodes and elements. A superelement is described by a four- to
nine-noded quadrilateral whose supernodes are numbered as shown in Fig. 15.25 (page
560). The block data is specified as:

RECORD 1). FORMAT 8I10

Columns	Description	Variable
1 to 10	Number of supernodes	NN
11 to 20	Number of spaces along 1-2 side of superelement	NR
21 to 30	Number of spaces along 1-4 side of superelement	NS
31 to 40	First node number on block at supernode 1	NI
41 to 50	First element in block adjacent to supernode 1	NE
51 to 60	Property set number for block	MA
61 to 70	Node increment between lines	NODINC
71 to 80	Block type: 0 = 4-noded quadrilaterals 8 = 8-noded quadrilaterals 9 = 9-noded quadrilaterals	NTYP

Notes: (1) Nodes increment along the 1-2 side of the superelement
then skip NODINC nodes before starting next line.
(2) If NE is zero no elements are generated.
(3) NR and NS must be even for NTYP of 8 or 9.
(4) Eight-noded quadrilateral elements have a centre node generated but this is
not connected to the element (should be zero with BOUN restraints).

RECORD 2 TO NN + 1.   FORMAT I10,3F10.0

Columns	Description	Variables
1 to 10	Supernode number	L
11 to 20	X1 coordinate of supernode	R
21 to 30	X2 coordinate of supernode	S
31 to 40	X3 coordinate of supernode	T

Note: only NDM supernode coordinates are used.

TABLE 15.13
POLAR COORDINATE DATA

POLAR DATA—FORMAT 3I10,2F10.0—must immediately follow a POLA macro record
Each node that has previously been input in a polar coordinate form r, θ (or r, θ, z), where θ is in degrees measured anticlockwise from the X1 axis, is converted to cartesian components X1, X2 (or X1, X2, X3).

Column	Description	Variable
1 to 10	First node number	NI
11 to 20	Last node number	NE
21 to 30	Increment between nodes	INC
31 to 40	X1 coordinate for centre of polar coordinates	XO
41 to 50	X2 coordinate for centre	YO

Note: coordinates are computed as:

$$X(1,I) = XO + r \cos \theta$$
$$X(2,I) = YO + r \sin \theta$$
$$X(3,I) = X(3,I)$$

where r is $X(1,I)$ and θ is $X(2,I)$ input by COOR or BLOC.

(or corresponding nodal loading) using the TEMP macro followed by data prepared according to Table 15.11.

The end of any mesh data is indicated by use of an END macro. The use of the macro command cards allows the user to specify only those data items actually needed for each analysis. The END macro signifies the end of mesh data input. The use of macro cards also reduces the chance of data input errors due to extraneous blank records. Strict record sequencing is necessary only within each macro segment. Several blank records may exist after actual data without affecting the execution of the program.

Once all necessary mesh data is supplied the user can select to solve the problem or not. If only a check of the mesh is desired, insert either a STOP macro command to stop execution or start a new problem as described in Table 15.1. If a solution to the problem is desired, additional data is required, as discussed in the next section.

TABLE 15.14
ANGLE DATA

ANGLE DATA—FORMAT 2I10,F10.0—must immediately follow an ANGL macro record
The angle data contains the node number N and the value of the angle, in degrees, measured anticlockwise from the X1 axis. The nodal angles may be generated using NG in the same way as for coordinate data.

Column	Description	Variable
1 to 10	Node number	N
11 to 20	Generator increment	NG
21 to 30	Angle in degrees	ANG

TABLE 15.15
DATA INPUT CARD IMAGES FOR DISK PROBLEM

FEAP**QUADRANT OF A CIRCULAR DISK (EXAMPLE PROBLEM)

19	11	1	2	2	4

COOR

1	1	0.		0.	
5	0	5.		0.	
6	1	0.		2.	
10	0	4.5828		2.	
11	1	0.		4.	
14	0	3.0		4.	
15	0	4.0		3.	
16	0	0.		5.	
17	0	0.75		4.9434	
18	0	1.5		4.7697	
19	0	2.25		4.4651	

ELEM

1	1	1	2	7	6	1
5	1	6	7	12	11	1
9	1	11	12	17	16	1

BOUN

1	1	1	−1
5	0	0	1
6	5	−1	0
16	0	1	0

FORC

16	0	0.		−5.

MATE

1	1					
100.		0.3		0.0	2	1	1

END

or in free-format form using the parser and lower case text:

feap * * Quadrant of a circular disk (example problem)
19,11,1,2,2,4
coord
1,1,0., 0.
5,,5.
6,1,,2.
10,,4.5828,2.

etc. to

end
interactive
stop

As an example of the data input required to describe a mesh consider the mesh shown in Fig. 15.7 for the quadrant of a circular disk. The input data card images for this problem are shown in Table 15.15.

15.4 Solution of finite element problems—the macro programming language

At the completion of data input and any checks on the mesh we are prepared to initiate a problem solution. It is at this stage that the particular type of solution mode must be available to the user in the program. In many existing programs only a small number of fixed algorithm solution modes are available to the user. For example, the program may only be able to solve linear steady-state problems, or, in addition, it may be able to solve linear transient problems. In practical engineering problems fixed algorithm programs are often too restrictive and the user must continually modify the program to solve a given problem—often at the expense of another user! For this reason it is desirable to have a program that has modules for variable algorithm capabilities and, if necessary, can be modified without interrupting other users' capabilities. The program that we discuss here is basic and the reader can undoubtedly see many ways to improve and extend the program for other classes of problems. One important extension would be to include a matrix interpretive language so that individual terms or equations can be modified for specific needs.

The macro programming concept described in this section has been used by the authors for more than 10 years and, to date, has not inhibited our research activities by becoming outmoded. Applications are routinely conducted on personal computers, minicomputers, and mainframe computers with the same program. Only the equation solution subprograms are changed in different environments.

15.4.1 *Linear steady-state problems.* The basic aspect of the variable algorithm program is a macro instruction language which can be used to construct modules for specific algorithms as needed. The user only needs to learn the mnemonics of the language to use it. For example, if one wishes to form the global stiffness matrix the program instruction TANG is used (TANG is the mnemonic for a tangent stiffness matrix and for non-linear elements would form and assemble into the global stiffness matrix the element tangent stiffness computed about the current displacement state; for linear elements this is just the linear stiffness matrix). If one wishes to form the right-hand side of the equations modified for specified displacements one uses the program instruction FORM. The resulting equations are solved using the instruction SOLV. Printed

TABLE 15.16
LIST OF MACRO PROGRAMMING COMMANDS

The following table contains a list of the macro instruction commands which may be used to construct solution algorithms. In batch mode the command MACR must be inserted after the mesh data. The other macro commands follow immediately and terminate with an END macro. In interactive mode an INTE command follows the mesh data and users will receive prompts to specify the macro solution algorithm, terminate with a QUIT or EXIT command (EXIT saves restart data). FORMAT—A4,11X,A4,11X,3F15.0

	Columns				
1–4	16–19	31–45	46–60	61–75	Description
BETA		V1	V2		Specify parameters for transient analysis; default (V1 = 0.25, V2 = 0.50)
CHEC					Perform a check of mesh (ISW = 2)†
DATA	T				Read data T macro command (T may be either DT or TOL, only)
DISP	ALL	N1	N2	N3	Output displacement for nodes N1 to N2 at increments of N3; ALL prints all
DT		V1			Set time increment to V1
FORM					Form right side of equations (ISW = 6)
LOOP		N			Loop N times all instructions between a matching NEXT macro
MASS					Form a diagonal mass (ISW = 5)
MESH					Input mesh changes (can not change boundary restraints/ element connections)
NEWF					Set a fixed force pattern at current load state, F_0 (initially, $F_0 = 0$)
NEXT					End of LOOP instruction
PLOT	MESH	V1	V2		Plot mesh, V1 is deformation scale factor, V2 is property set (0 = all)
PLOT	OUTL	V1	V2		Plot mesh outline, V1 and V2 as for MESH
PROP		N1			Input N1 proportional load tables (data follows macro program)
REAC	ALL	N1	N2	N3	Output reactions for nodes N1 to N2 at increments of N3; ALL outputs all values (ISW = 6)
REST					Restart previous problem from restart and read file specified at start of analysis
SOLV					Solve for new displacements (after FORM)
STRE	ALL	N1	N2	N3	Output variables for elements N1 to N2 at increments of N3 (ISW = 4)

† Operations are performed in each element sub-program for specified ISW value.

	Columns				
1–4	16–19	31–45	46–60	61–75	Description
STRE	NODE	N1	N2	N3	Output variables (stress, etc.) for nodes N1 to N2 at increments of N3 (ISW = 8)
STRE	ERRO				Output error and refinement estimates for elements (ISW = 7)
TANG		V1	V2		Compute and factor tangent matrix (ISW = 3). If V1 is non-zero also form residual and solve equations; shift tangent matrix by V2 times mass matrix
TIME					Advance time by DT value
TOL		V1			Set tolerance value to V1

output can be obtained using the instructions DISP for displacements and STRE for element variables such as strains and stresses. An ALL appended to a DISP and STRE indicates all node/elements are to be output. It is possible to output selected patterns of results as shown in Table 15.16. The above instructions are sufficient to solve linear steady-state problems, i.e., the macro instructions

> TANG
> FORM
> SOLV
> DISP,ALL
> STRE,ALL

are precisely the required instruction to solve any linear steady-state problem. The reader will undoubtedly observe at this time that ordering can sometimes be changed without affecting the algorithm. For example, use of the macro instruction

> FORM
> TANG
> SOLV
> STRE,ALL
> DISP,ALL

produces the same algorithm except that element quantities are printed before the nodal displacements.

The variable algorithm program as described by the macro programming language can be extended as necessary. For example, when multiple load problems are analysed the global stiffness matrix is always the same and need only be formed once. The right-hand side changes and the new displacements need to be computed. The procedure to solve two load cases requires changing nodal loads and/or specified displacements.

The macro instruction MESH causes the program to enter the data input module again, and at this stage loads can be changed. Data appears *after* the macro program instructions which terminate with the END statement. Thus the macro program and data for two load cases could be

```
          TANG
          MESH     ⎫
          FORM     ⎪
          SOLV     ⎬ instructions for problem 1
          DISP,ALL ⎪
          STRE,ALL ⎭
          MESH     ⎫
          FORM     ⎪
          SOLV     ⎬ instructions for problem 2
          DISP,ALL ⎪
          STRE,ALL ⎭
          END              end of macro program
          FORC
                loads for  problem 1
          END              end mesh data inputs
          FORC
                loads for  problem 2
          END              end mesh data inputs
```

The reader should notice that the same block of instructions is repeated twice and that if ten load cases were desired, considerable effort is wasted in preparing the macro instruction data cards. To rectify this, looping commands are introduced as the instruction pair

$$\text{LOOP,,n}$$
$$\vdots$$
$$\text{NEXT}$$

which indicates that looping over all instructions between LOOP and NEXT will occur *n* times; hence the macro program for two load cases is now

```
          TANG
          LOOP,,2
          MESH
          SOLV
          DISP,ALL
          STRE,ALL
          NEXT
```

```
          END                    end of macro program
          FORC
                   loads for problem 1
          END
          FORC
                   loads for problem 2
          END
```

It should be noted that in the INTEractive mode execution does not begin until a LOOP is matched by a NEXT statement.

For this program ten load cases are as simple as two (except for the FORC data).

The reader will note that the TANG instruction is executed only once while the SOLV instruction is executed twice. The program will automatically recognize that the execution uses a stiffness matrix for which the triangular decomposition has already been performed and will select a *resolve* mode of operation (see Sec. 15.6).

Many other classes of problems can be solved using the simple macro instruction list given in Table 15.16. We summarize a few algorithms in the subsequent paragraphs.

15.4.2 *Incremental load method.* The macro instruction program to solve a problem in which loads change with time is given next. It is assumed that the problem has a set of loads that vary in time proportionally to one another. The program PCFEAP can change loads at each step as described above; however, for the situation where proportionality occurs it is possible to vary loads by specifying a set of loading functions. The general form of time-dependent loads may assume

$$\mathbf{F}(t) = \mathbf{F}_0 + P(t)\mathbf{F} \tag{15.1}$$

where $\mathbf{F}(t)$ is the current load vector, $\mathbf{F}_0$ is a fixed force vector, $\mathbf{F}$ is the set of nodal forces defined by the FORC mesh command, and $P(t)$ is a proportional loading time function. In PCFEAP the proportional loading may be specified by

$$P(t) = A_1 + A_2 t + A_3[\sin(A_4 t + t_{\min})]^L \qquad t_{\min} \leqslant t \leqslant t_{\max} \tag{15.2}$$

A maximum of 10 proportional load functions may be specified at one time; the function may be reset as often as desired during execution (see Table 15.17).

The NEWF macro command may be used to reset the value for $\mathbf{F}_0$ to the current value and distribution of the $\mathbf{F}(t)$. Accordingly, the distribution of forces may then be reset to describe a new proportional loading state. At the start of execution $\mathbf{F}_0$ is set to zero.

TABLE 15.17
PROPORTIONAL LOADS—FORMAT
2I10,6F10.0

Column	Description	Variable
1 to 10	Not used (load type = 1)	
11 to 20	Exponent in sine function	L
21 to 30	Minimum time of loading	TMIN
31 to 40	Maximum time of loading	TMAX
41 to 50	Parameter 1	A1
51 to 60	Parameter 2	A2
61 to 70	Parameter 3	A3
71 to 80	Parameter 4	A4

Note: a maximum of 10 proportional loading functions may be specified at any one time.

A macro program using the proportional loading concept is

DT,,0.1	$\Delta t = 0.1$
PROP,,1	$\mathbf{F}(t) = P(t)\mathbf{F}$
TANG	compute coefficient matrix (constant)
LOOP,,10	increment loop
TIME	$t = t + \Delta t$
FORM	compute $\mathbf{F}(t) - \mathbf{P}(\mathbf{a})$
SOLV	compute $\Delta\mathbf{a}$, set $\mathbf{a} = \mathbf{a} + \Delta\mathbf{a}$
DISP,,1,12	output $\mathbf{a}$ for nodes 1 to 12
STRE,ALL	output all element stress values
NEXT	end of loop
END	

15.4.3 *The integration of the equations of motion.* The integration of the second-order differential equations of motion for time-dependent structural systems may be treated using PCFEAP. This topic will be considered in considerably more detail in the next volume where general methods based upon finite element and other traditional methods will be developed and analysed. The present version of the program includes the simplest of the beta-*m* methods.[16] The method is the same as the classical Newmark method.[17]

The equations of motion for a linear differential equation are given by

$$\mathbf{M\ddot{a} + Ka = F}$$

where $\mathbf{M}$ is the mass matrix. The elements included in PCFEAP formulate $\mathbf{M}$ as a diagonal (lumped) matrix (e.g., see Appendix 7). Consistent (non-diagonal) masses may be included but either require them to be combined with the computation of the element coefficient matrix or the program to be modified to include this option.[1] While the general treatment and solution of the equations of motion is outside the scope of this volume, we include the macro solution commands that do the solution to indicate how the program may be used in a wider class of problems than originally defined. The appropriate program is

BETA	set solution integration parameters
PROP,,1	use a proportional loading in time
MASS	form the lumped mass matrix
DT,,0.024	set the time step to 0.024
TANG	compute the tangent matrix $(\mathbf{K} + \mathrm{C1}\ \mathbf{M})$
LOOP,time,50	perform 50 time steps of the solution
FORM	form the residual $(\mathbf{F} - \mathbf{P(a)} - \mathbf{M\ddot{a}})$
SOLV	obtain new solution
DISP,,2,24,2	output nodal displacements 2,4,6,,,24
STRE,NODE,2,4	output stresses at nodes 2,3, and 4
NEXT,time	end of time loop
STRE,ALL	output final stress state in all elements
END	

In the above macro program the 'time' notation in LOOP/NEXT statements is not used by the program (e.g., see Table 15.16), but helps make the solution algorithm more readable.

15.4.4 *Non-linear solutions; Newton's method.* The macro programming language may also be used to solve non-linear problems. For example, in Newton's method the set of equations

$$\mathbf{P(a)} = \mathbf{F}$$

may be solved iteratively using

$$\mathbf{K}_T \,\Delta\mathbf{a} = \mathbf{F} - \mathbf{P(a)} = \mathbf{R(a)}$$

where $\mathbf{K}_T$ is some matrix (commonly computed by linearization of $\mathbf{P}$) and iteration continues for a specified number of iterations or until an acceptable error tolerance is reached. In PCFEAP an 'energy' tolerance is used to terminate iteration if convergence is achieved. The 'energy' is computed as

$$E = \mathbf{R} \cdot \Delta\mathbf{a}$$

for the unknown equations only. Several possibilities exist for performing an iteration. One is

LOOP,iteration,10	perform 10 iterations, maximum
TANG,,1	compute a current $\mathbf{K}_T$ matrix, residual, and solve equations
NEXT,iteration	end of iteration loop

An alternative iteration strategy is

TANG	form current $\mathbf{K}_T$ matrix
LOOP,iteration,10	perform 10 iterations with $\mathbf{K}_T$ fixed
FORM	form new residual, $\mathbf{R}$
SOLV	solve equations
NEXT,iteration	end of iteration loop

The reader can observe that the strategies may be mixed; compute a few iterations with the coefficient matrix fixed at some value, change the coefficient matrix, and perform additional iterations.

15.4.5 *Programming for macro instructions.* The macro instruction module for the program PCFEAP is contained in the subroutines PMACR, PMACIO, PMACR1, and PMACR2. The routine PMACR calls the other routines; PMACIO performs inputs and checks on the loops; PMACR1 performs all the operations that involve element calculations; and PMACR2 performs all other macro operations. Currently available macro commands are stored in the character array WD. Thus, to add a macro instruction to the program the array WD must be redimensioned, the list of words stored as data in WD must be modified, and appropriate changes to PCMACR1 or PCMACR2 must be made. The data parameters NW1 and NW2 in PCMACR define the number of commands contained in PCMACR1 and PCMACR2 respectively. The appropriate parameter should be changed when a command is added. If new solution arrays are needed new pointers to the arrays defined through blank common may be defined using the PSETM subprogram (note carefully in subprogram PCONTR how the precision of each variable is set). The logical parameter in the call to PSETM may be used to ensure that memory is allocated only on the first call to PSETM. Before adding any new macro command the programmer is urged to carefully study the program listed in this chapter. In particular the solution option flags stored in common FDATA are used to tell the program which arrays currently exist and should not be changed by any new macro command since this will affect the existing operation of the program.

15.5 Computation of finite element solution modules

In the establishment of the macro instruction for element computations (e.g., TANG and FORM) many of the operations are the same. In this section we discuss aspects that enter into the computation of the element arrays for finite element computation. The first step is to localize all the geometric, material, and displacement data for the element array to be computed. The particular element quantities to be computed include stiffness, mass, internal forces, stresses, strains, etc. We discuss aspects for computing these for the problems of linear elasticity and heat transfer in Sec. 15.5.2. The organization of the element routines for the program is given in Sec. 15.5.3.

15.5.1 *Localization of element data.* When we want to compute an element array, e.g., an element stiffness matrix, S, or a load or internal force vector, **P**, we only need those quantities associated with the element in question—all the other values are superfluous. The nodal and material quantities that are required can be determined from the node and material numbers stored in the IX array for each element. In the program the necessary values are moved from the global arrays to a set of local arrays before the appropriate element routine, i.e., subroutine ELMTnn, is called. This process will be called *localization.* The quantities that are localized are:

(a) nodal coordinates which are stored in the array XL,
(b) nodal displacements and increments which are stored in the array UL,
(c) nodal temperatures which are stored in the array TL, and
(d) equation numbers which are stored in the destination array LD.

The LD array described in step (d) is used to map the element stiffness (or mass) matrix and element load or internal force vector to the global stiffness matrix and load vector respectively. Accordingly, for the following element arrays:

$$\begin{bmatrix} LD(1) & LD(2) & LD(3) & \dots \end{bmatrix}$$
$$\begin{bmatrix} S(1,\,1) & S(1,\,2) & S(1,\,3) & \\ S(2,\,1) & S(2,\,2) & & \\ \vdots & & & \end{bmatrix} \begin{bmatrix} P(1) \\ P(2) \\ \vdots \end{bmatrix}$$

the term S(i, j) would be assembled into the global stiffness (or mass) matrix in the position corresponding to row LD(i) and column LD(j), i.e., the LD array contains the equation numbers of the global matrices. Similarly, P(i) would be assembled into the position corresponding to the LD(i) value. In establishing LD(i) reordering of the degree of freedoms is made in accordance with Table 15.8.

The localization process is the same for every type of finite element and is thus centralized into subroutine PFORM, which organizes all computations associated with elements including looping over the list of elements as described by the IX array. The element properties are stored in the square array S and the vector P. The LD array is a destination vector of element to global equation numbers and is used to map S and P onto the global arrays.

During the localization process the number of nodes actually connected to each element (i.e., NEL which may be less than NEN) is determined by finding the largest non-zero entry in the IX array of that element number. Intermediate zeros are interpreted as no node connected. In this way the program permits mixing of elements with different numbers of connected nodes, e.g., three-noded triangles can be mixed with four-noded quadrilaterals.

Since the current value of the nodal displacements is localized for all element computations, the program can be used to solve non-linear problems. This is, in fact, the only additional information over that for linear problems which is required to construct tangent stiffness matrices, etc., for the solution to non-linear problems as discussed in Chapters 18 and 19 of reference 1 and will be considered further in the next volume.

It should be pointed out before proceeding to the computation of element arrays that the localization step (except for current values of displacements) could be done once and the global data arrays could then be saved on disk or destroyed. This would involve more programming steps and also efficient use of buffering and backing storage to maintain adequate working space in core for subsequent global array determination.

15.5.2 *Element array computations.* The efficient computation of element arrays (in both programmer and computer time) is a crucial aspect of any finite element development. The development of routines to evaluate element stiffness (or tangent stiffness) and load arrays can be accomplished by a combination of appropriate numerical methods. In order to explicate the development, a statement of the essential steps is first given and then some details are shown for the plane stress/strain case.

The essential steps to compute an element stiffness matrix, S, are summarized in Fig. 15.8. The key ingredients are the numerical integration, the use of shape function routines (which are the same for all problems with the same required continuity), and efficient formulation of the matrix products.

Usually Gauss quadrature formulae are utilized to compute all element integrals since they give highest accuracy for effort expended (see Chapter 8). In some instances it is desirable to use other formulae. For example, if one employs a quadrature formula that samples only at the nodes, then evaluation of the mass term leads to a diagonal mass matrix which is often more advantageous for dynamic problems. This concept may be used to obtain the 'nodal' stresses as well (see Chapter 12 and Appendix 7).

Shape function subprograms allow the programmer to develop elements for many problems quickly and reliably. The shape function sub-

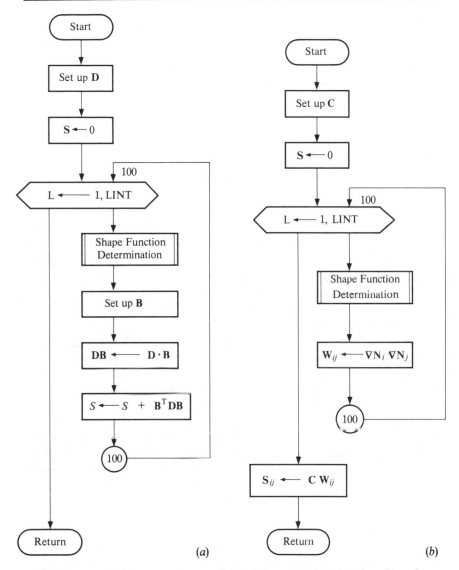

Fig. 15.8   Procedure to compute element stiffness matrix. (a) Using shape function routine and numerical integration. (b) For elements with constant material properties

program evaluates not only the shape function but also its derivatives with respect to the global coordinate frame. As an example, consider the two-dimensional C_0 problem where we need only first derivatives of each shape function N_i. For the four-noded isoparametric quadrilateral we have

$$N_i = \tfrac{1}{4}(1 + \xi_i\,\xi)(1 + \eta_i\,\eta) \tag{15.3}$$

where ξ_i, η_i are the ξ, η coordinates of the nodes.

Using the isoparametric concept we have

$$\begin{aligned} x &= N_i\,x_i \\ y &= N_i\,y_i \end{aligned} \tag{15.4}$$

and derivatives given by

$$\begin{Bmatrix} N_{i,\,\xi} \\ N_{i,\,\eta} \end{Bmatrix} = \begin{bmatrix} x_{,\xi} & y_{,\xi} \\ x_{,\eta} & y_{,\eta} \end{bmatrix} \begin{Bmatrix} N_{i,\,x} \\ N_{i,\,y} \end{Bmatrix} \tag{15.5}$$

$$\begin{Bmatrix} N_{i,\,x} \\ N_{i,\,y} \end{Bmatrix} = \frac{1}{J} \begin{bmatrix} y_{,\eta} & -y_{,\xi} \\ -x_{,\eta} & x_{,\xi} \end{bmatrix} \begin{Bmatrix} N_{i,\,\xi} \\ N_{i,\,\eta} \end{Bmatrix} \tag{15.6}$$

where J is the jacobian determinant and $(\ )_{,x}$ denotes the partial derivative $\partial(\ )/\partial x$, etc. The above relations define steps for the shape function subprogram given in Fig. 15.9 where it is assumed that the nodal coordinates have been transferred to the local coordinate array XL.

This shape function routine can be used for all two-dimensional C_0 problems which use the four-noded element (e.g., two-dimensional plane and axisymmetric elasticity, heat conduction, flow in porous media, fluid flow, etc.). Shape function subprograms can also be used for the generation of mesh data.[7] It is a simple task to extend the shape function routine to higher order elements. As an example the routines SHAPE and SHAP2 (see Sec. 15.8.3) give shape functions for three-noded triangles up to eight-noded serendipity and nine-noded Lagrange quadrilaterals. The elements may even have edges with linear expansion while others have quadratic expansion by simply omitting the mid-side node number for a linear edge. Furthermore, hierarchical interpolations may be constructed by merely avoiding the corrections at the end of subprogram SHAP2.

The generation of the matrix products also deserves special attention since zeros often exist in the **B** and **D** matrices. Several methods can be used to reduce the number of operations. The first is to form explicitly the matrix products. While this at first appears to involve too many hand computations, in fact it is elementary if it is done nodewise. For example, consider the two-dimensional plane linear elasticity problem, where

$$\mathbf{B}_i = \begin{bmatrix} N_{i,\,x} & 0 \\ 0 & N_{i,\,y} \\ N_{i,\,y} & N_{i,\,x} \end{bmatrix} \tag{15.7}$$

```
      SUBROUTINE SHAPEF(SS,TT,XL,XSJ,SHP)                         SHP   1
      IMPLICIT REAL*8 (A-H,O-Z)                                   SHP   2
C....  Shape function routine for 4-node isoparametric quadrilateral SHP  3
C                                                                 SHP   4
C             SS,TT     = Natural coordinates for shape functions SHP   5
C             SHP(1,I)  = X-derivative of I-node shape function    SHP   6
C             SHP(2,I)  = Y-derivative of I-node shape function    SHP   7
C             SHP(3,I)  = Shape function for I-node                SHP   8
C             XS        = Jacobian array                           SHP   9
C             XSJ       = Jacobian determinant                     SHP  10
C             XL(1,I)   = X-coordinate of I-node                   SHP  11
C             XL(2,I)   = Y-coordinate of I-node                   SHP  12
C                                                                 SHP  13
      REAL   XL(2,4),SI(4),TI(4)                                  SHP  14
      REAL*8 SHP(3,4),XS(2,2)                                     SHP  15
      DATA SI/-.5,.5,.5,-.5/, TI/-.5,-.5,.5,.5/                   SHP  16
C                                                                 SHP  17
C....  Compute shape functions and their natural coord. derivatives SHP  18
C                                                                 SHP  19
      DO 100 I = 1,4                                              SHP  20
        SHP(1,I) = SI(I)*(0.5+TI(I)*TT)                           SHP  21
        SHP(2,I) = TI(I)*(0.5+SI(I)*SS)                           SHP  22
        SHP(3,I) = (0.5+SI(I)*SS)*(0.5+TI(I)*TT)                  SHP  23
100   CONTINUE                                                    SHP  24
C                                                                 SHP  25
C....  Compute jacobian transformation from X,Y to SS,TT          SHP  25
C                                                                 SHP  26
      DO 120 I = 1,2                                              SHP  27
      DO 120 J = 1,2                                              SHP  28
        XS(I,J) = 0.0                                             SHP  29
        DO 110 K = 1,4                                            SHP  30
          XS(I,J) = XS(I,J) + XL(I,K)*SHP(J,K)                    SHP  31
110     CONTINUE                                                  SHP  32
120   CONTINUE                                                    SHP  33
C                                                                 SHP  34
C....  Compute jacobian determinant                               SHP  35
C                                                                 SHP  36
      XSJ = XS(1,1)*XS(2,2) - XS(1,2)*XS(2,1)                     SHP  37
C                                                                 SHP  38
C....  Transform to X,Y derivatives                               SHP  39
C                                                                 SHP  40
      DO 130 I = 1,4                                              SHP  41
        TEMP     = ( XS(2,2)*SHP(1,I) - XS(2,1)*SHP(2,I)/XSJ      SHP  42
        SHP(2,I) = (-XS(1,2)*SHP(1,I) + XS(1,1)*SHP(2,I)/XSJ      SHP  43
        SHP(1,I) = TEMP                                           SHP  44
130   CONTINUE                                                    SHP  45
      RETURN                                                      SHP  46
      END                                                         SHP  47
```

Fig. 15.9   Shape function subprogram for four-noded element

and for isotropy

$$\mathbf{D} = \begin{bmatrix} D_{11} & D_{12} & 0 \\ D_{12} & D_{11} & 0 \\ 0 & 0 & D_{33} \end{bmatrix} \tag{15.8}$$

where D_{33} equals $(D_{11} - D_{12})/2$. Thus for typical nodal pairs i and j the element stiffness $\mathbf{k}_{ij}$ is given by

$$\mathbf{k}_{ij} = \mathbf{B}_i^{\mathrm{T}} \mathbf{Q}_j \tag{15.9}$$

where

$$\mathbf{Q}_j = \mathbf{D} \mathbf{B}_j \tag{15.10}$$

Thus

$$\mathbf{Q}_j = \begin{bmatrix} D_{11} N_{j,x} & D_{12} N_{j,y} \\ D_{12} N_{j,x} & D_{11} N_{j,y} \\ D_{33} N_{j,y} & D_{33} N_{j,x} \end{bmatrix} \tag{15.11}$$

and

$$\mathbf{k}_{ij} = \begin{bmatrix} (N_{i,x} Q_{11} + N_{i,y} Q_{31}) & (N_{i,x} Q_{12} + N_{i,y} Q_{32}) \\ (N_{i,y} Q_{21} + N_{i,x} Q_{31}) & (N_{i,y} Q_{22} + N_{i,x} Q_{32}) \end{bmatrix}$$

Accordingly, for each nodal pair it is required to compute 14 multiplications to form the $\mathbf{k}_{ij}$, whereas formal multiplication of $\mathbf{B}^{\mathrm{T}}\mathbf{D}\mathbf{B}$ including all zero operations would give 30 multiplications. Also, when the element matrix is symmetric only one half would be formed during numerical integration (the other half would be formed by reflection). A typical routine for the stiffness computation is given in Fig. 15.10, where it is assumed that the Gauss quadrature points and weights are stored in the arrays SG, TG, and WG.

An extension to anisotropic problems can also be made by replacing the isotropic $\mathbf{D}$ matrix with the appropriate anisotropic one and then recomputing the $\mathbf{Q}_j$ matrix.

The computation of element stiffness matrices for problems that have material properties constant within an element can be made more efficient by noting from Chapter 6 that the internal energy may be written, using index notation, as

$$\frac{1}{2} \bar{u}_a^i D_{abcd} \int_{Ve} N_{,b}^i N_{,d}^j \, dV \bar{u}_c^j \tag{15.12}$$

where a, b, c, d are indices from the elasticity equations and range over the space dimension of the problem, and i, j are nodal indices which in an element range from 1 to NEL. The element stiffness matrix for the

```
C....  Isoparametric element stiffness computation for linear       STF  1
C      elasticity * * plane stress/strain differ only in values     STF  2
C      of elastic parameters * *                                    STF  3
C                                                                    STF  4
C          D(1), D(2), D(3) = Material moduli                        STF  5
C              plane stress: D(1) = E/(1. -NU*NU)                    STF  6
C                            D(2) = NU*D(1)                          STF  7
C              plane strain: D(1) = D*(1.-NU)/((1.+NU)*(1.-2*NU))    STF  8
C                            D(2) = NU*D(1)/(1.-NU)                  STF  9
C                            D(3) = (D(1) - D(2))/2.                 STF 10
C          DV     = Area differential volume * gauss weight          STF 11
C          LINT   = Number of quadrature points                     STF 12
C          NDF    = Number of degrees-of-freedom at a node          STF 13
C          NEL    = Number of nodes on element                      STF 14
C          NST    = Size of S-array                                 STF 15
C          S      = Element stiffness array                         STF 16
C          SG,TG  = Natural coordinates of Gauss points.            STF 17
C          SHP    = Shape functions and their x,y derivatives       STF 18
C          XL     = Nodal coordinates - localized                   STF 19
C          XSJ    = Jacobian determinant                            STF 20
C          WG     = Gauss quadrature weights                        STF 21
C                                                                    STF 22
C....  Compute contribution at each quadrature point.               STF 23
       DO 120 L = 1,LINT                                            STF 24
          CALL SHAPEF(SG(L),GT(L),WL,XSJ,SHP)                       STF 25
          DV = XSJ*WG(L)                                            STF 26
          D11 = D(1)*DV                                             STF 27
          D12 = D(2)*DV                                             STF 28
          D33 = D(3)*DV                                             STF 29
C....  For each J-node compute: DB = D*B                            STF 30
          J1  = 1                                                   STF 31
          DO 110 J = 1,NEL                                          STF 32
             DB11 = D11*SHP(1,J)                                    STF 33
             DB12 = D12*SHP(2,J)                                    STF 34
             DB21 = D12*SHP(1,J)                                    STF 35
             DB22 = D11*SHP(2,J)                                    STF 36
             DB31 = D33*SHP(2,J)                                    STF 37
             DB32 = D33*SHP(1,J)                                    STF 38
C....   For each I-node compute S = Bt*DB                           STF 39
             I1  = 1                                                STF 40
             DO 100 I = 1,J                                         STF 41
                S(I1  ,J1  )=S(I1  ,J1  )+SHP(1,I)*DB11+SHP(2,I)*DB31 STF 42
                S(I1  ,J1+1)=S(I1  ,J1+1)+SHP(1,I)*DB12+SHP(2,I)*DB32 STF 43
                S(I1+1,J1  )=S(I1+1,J1  )+SHP(1,I)*DB31+SHP(2,I)*DB21 STF 44
                S(I1+1,J1+1)=S(I1+1,J1+1)+SHP(1,I)*DB32+SHP(2,I)*DB22 STF 45
                I1 = I1 + NDF                                       STF 46
100          CONTINUE                                              STF 47
             J1 = J1 + NDF                                         STF 48
110       CONTINUE                                                 STF 49
120    CONTINUE                                                    STF 50
C....   Compute lower part by symmetry                             STF 51
       DO 130 I = 2,NST                                            STF 52
       DO 130 J = 1,I                                              STF 53
          S(I,J) = S(J,I)                                          STF 54
130    CONTINUE                                                    STF 55
       RETURN                                                      STF 56
       END                                                         STF 57
```

Fig. 15.10   Element stiffness calculation

nodal pair i, j is thus given by

$$K_{ac}^{ij} = W_{bd}^{ij} D_{abcd} \qquad (15.13)$$

where

$$W_{bd}^{ij} = \int_{V_e} N_{,b}^i N_{,d}^j \, \mathrm{d}V \qquad (15.14)$$

For isotropic materials the elastic constants are given by

$$D_{abcd} = \delta_{ab} \, \delta_{cd} \, \lambda + \mu(\delta_{ac} \, \delta_{bd} + \delta_{ad} \, \delta_{bc}) \qquad (15.15)$$

where λ and μ are the Lamé constants which are related to the usual elastic constants E and v as $\lambda = vE/(1 + v)(1 - 2v)$, $\mu = E/2(1 + v)$.

Thus the stiffness matrix is computed from

$$K_{ac}^{ij} = \lambda W_{ac}^{ij} + \mu(W_{ca}^{ij} + \delta_{ac} \, W_{bb}^{ij}) \qquad (15.16)$$

Using this procedure the steps to compute the element stiffness matrix for plane elasticity are given in Fig. 15.8(b). This procedure for computing stiffness matrices was noted in reference 18 and for plane problems results in about a 25 per cent reduction over the procedure of Fig. 15.8(a). In three dimensions the savings are even greater.

The computation of other element arrays can also be effected using a shape function routine. For example, the computation of the element row sum diagonal mass matrix for transient or eigenvalue computations can be performed. The mass matrix for the two-dimensional plane problem is obtained from

$$\mathbf{M}_{jj}^e = \mathbf{I} \int_{V_e} \rho N_j \, \mathrm{d}V \qquad (15.17)$$

```
C....  Isoparametric element mass matrix for plane problems      MAS  1
C                                                                 MAS  2
C              P  = Diagonal (lumped) mass array (in vector)      MAS  3
C                                                                 MAS  4
C....  Compute matrix at each integration point                  MAS  5
       DO 500 L = 1,LINT                                          MAS  6
C....  Compute shape functions                                   MAS  7
       CALL SHAPEF(SG(L),TG(L),XL,XSJ,SHP)                        MAS  8
C....  For each node I compute contribution                      MAS  9
       DMASS = D(4)*XSJ*WG(L)                                     MAS 10
       DO 510 I = 1,NEL                                           MAS 11
         P(2*I-1) = P(2*I-1) + DMASS*SHP(3,I)                     MAS 12
         P(2*I  ) = P(2*I-1)                                      MAS 13
510    CONTINUE                                                   MAS 14
500    CONTINUE                                                   MAS 15
       RETURN                                                     MAS 16
       END                                                        MAS 17
```

Fig. 15.11   Diagonal (lumped) mass matrix for isoparametric element

where $\mathbf{I}$ is a 2×2 identity matrix and ρ is the mass density (see Appendix 7). A set of statements to compute the mass matrix for the plane problem is shown in Fig. 15.11, where the element lumped, diagonal mass matrix is stored in the array P.

The shape function routine may also be used to compute the element strains, stresses, and internal forces. For the two-dimensional plane problem the strains are computed from

$$\varepsilon = \mathbf{B}_i \mathbf{a}_i \qquad (15.18)$$

and stresses from

$$\sigma = \mathbf{D}\varepsilon = \mathbf{Q}_i \mathbf{a}_i \qquad (15.19)$$

The $\mathbf{B}_i$ matrix is given above for the two-dimensional problem and depends only on the derivatives of the shape functions. It has previously been computed at the Gauss points when the element stiffness matrix was evaluated. If stress output points are also at the Gauss points the $\mathbf{B}$ matrix could be saved on tape or disk and recalled at the time stresses and strains are to be output. (In fact $\mathbf{Q}$, which is the stress recovery matrix, could also be saved.) The program PCFEAP provides capabilities to retain Q for each element. This procedure was adopted in the program given in reference 19. Often, however, the points where the stresses and strains are to be determined do not coincide with the stiffness Gauss points. In these cases the $\mathbf{B}$ matrix must be recomputed. In non-linear problems the computation of strains and stresses must be performed directly; hence, it seems desirable for a computer program to be able to compute strains and stresses as necessary. In addition, in the macro instruction FORM the internal forces given by

$$\mathbf{P}_i = -\int_V \mathbf{B}_i^T \sigma \, dV \qquad (15.20)$$

are computed using current displacements. The programming steps to compute strains, stresses, and internal forces for the two-dimensional problem are given in Fig. 15.12. The local coordinates ξ, η, where stresses and strains are evaluated, are called SG, TG. The user could read these as data or preset them to specify particular output points; on the other hand, they could be specified as Gauss points.

If stresses are computed at Gauss points they can be extrapolated to the nodes and smoothed as suggested in Chapter 12. Shape function subprograms can be used for this purpose also.

The generality of an isoparametric C_0 shape function routine can be exploited to program element routines for other problems. For example, Fig. 15.13 gives the necessary program instructions to compute the 'stiffness' matrix for the problems of the Laplace equation discussed in Chapter 12.

```
C....  Isoparametric Element Stress, Strain, and Internal Forces   STR  1
C                                                                   STR  2
C           P  =  Internal force vector                            STR  3
C           UL =  Nodal displacement vector                        STR  4
C                                                                   STR  5
C....  Compute element stresses, strains and forces                STR  6
C                                                                   STR  7
       DO 440 L = 1,LINT                                           STR  8
C....  Compute element shape functions                            STR  9
       CALL SHAPEF(SG(L),TG(L),XL,XSJ,SHP)                        STR 10
C....  Compute strains and coordinates                            STR 11
       DO 410 I = 1,3                                             STR 12
         EPS(I) = 0.0                                             STR 13
410    CONTINUE                                                   STR 14
       XX = 0.0                                                   STR 15
       YY = 0.0                                                   STR 16
       DO 420 I = 1,NEL                                           STR 17
         XX    = XX + XL(1,I)*SHP(3,I)                            STR 18
         YY    = YY + XL(2,I)*SHP(3,I)                            STR 19
         EPS(1) = EPS(1) + UL(1,I)*SHP(1,I)                       STR 20
         EPS(2) = EPS(2) + UL(2,I)*SHP(2,I)                       STR 21
         EPS(3) = EPS(3) + UL(1,I)*SHP(2,I) + UL(2,I)*SHP(1,I)    STR 22
420    CONTINUE                                                   STR 23
C....  Compute stresses                                           STR 24
       SIG(1) = D(1)*EPS(1) + D(2)*EPS(2)                         STR 25
       SIG(2) = D(2)*EPS(1) + D(1)*EPS(2)                         STR 26
       SIG(3) = D(3)*EPS(3)                                       STR 27
C....  Output stresses and strains to unit ISW                    STR 28
       IF(MCT.GT.0) THEN                                          STR 29
         CALL PRTHED(IOW)                                         STR 30
         WRITE(IOW,2000)                                          STR 31
         MCT = 25                                                 STR 32
       ENDIF                                                      STR 33
       WRITE(IOW,2001) N,MA,XX,YY,SIG,EPS                         STR 34
C....  Compute the internal force vector (stress divergence term) STR 35
       DO 430 I = 1,NEL                                           STR 36
         P(2*I-1) = P(2*I-1) - (SHP(1,I)*SIG(1)+SHP(2,I)*SIG(3))*DV STR 37
         P(2*I  ) = P(2*I  ) - (SHP(2,I)*SIG(2)+SHP(1,I)*SIG(3))*DV STR 38
430    CONTINUE                                                   STR 39
440    CONTINUE                                                   STR 40
       RETURN                                                     STR 41
2000   FORMAT(//'    E L E M E N T    S T R E S S / S T R A I N'//  STR 42
      1 '  Elmt Matl    X-coord    Y-coord',5X,'11-Stress',5X,    STR 43
      2 '22-Stress',5X,'12-Stress'/39X,'11-Strain',5X,'22-Strain', STR 44
      3 '12-Strain')                                              STR 45
2001   FORMAT(2I6,2F11.4,3E14.5/34X,3E14.5)                       STR 46
       END                                                        STR 47
```

Fig. 15.12   Stress, strain, and internal force computation

15.5.3 *Organization of element routines*. The previous discussion has focused on procedures for determining element arrays. The reader will note that the element square matrices are stored in the array S while element vectors are stored in array P. This was intentional since all aspects of computing element arrays for the program are to be consolidated into a single subprogram called the 'element routine'. An element

```
C....   Isoparametric element computation for Laplace operator        HEA   1
C                                                                      HEA   2
C                D(1)  =  K (isotropic material parameter)             HEA   3
C                DV    =  Area weighting                               HEA   4
C                LINT  =  Number of integration points                HEA   5
C                NEL   =  Number of nodes connected to element         HEA   6
C                S     =  Element coefficient matrix                   HEA   7
C                SG,TG =  Integration points in natural coords.        HEA   8
C                SHP   =  Shape function array                         HEA   9
C                XL    =  Nodal coordinate array (localized)           HEA  10
C                XSJ   =  Jacobian determinant                         HEA  11
C                WG    =  Integration weights                          HEA  12
C                                                                      HEA  13
C....   Compute contribution at each integration point                HEA  14
C                                                                      HEA  15
        DO 120 L = 1,LINT                                              HEA  16
          CALL SHAPEF(SG(L),TG(L),XL,XSJ,SHP)                          HEA  17
          DV  = XSJ*WG(L)                                              HEA  18
          D1  = D(1)*DV                                                HEA  19
C                                                                      HEA  20
C....   For each J-node compute: DB = D*B                              HEA  21
C                                                                      HEA  22
          DO 110 J = 1,NEL                                             HEA  23
            DB11 = D1*SHP(1,J)                                         HEA  24
            DB12 = D1*SHP(2,J)                                         HEA  25
C                                                                      HEA  26
C....   For each I-node compute: S = Bt*DB                             HEA  27
C                                                                      HEA  28
            DO 100 I = 1,J                                             HEA  29
              S(I,J) = S(I,J) + SHP(1,I)*DB11 + SHP(2,I)*DB12          HEA  30
100         CONTINUE                                                   HEA  31
110       CONTINUE                                                     HEA  32
120     CONTINUE                                                       HEA  33
C                                                                      HEA  34
C....   Compute symmetric part of S                                    HEA  35
C                                                                      HEA  36
        DO 130 I = 2,NEL                                               HEA  37
        DO 130 J = 1,I                                                 HEA  38
          S(I,J) = S(J,I)                                              HEA  39
130     CONTINUE                                                       HEA  40
C                                                                      HEA  41
        RETURN                                                         HEA  42
        END                                                           HEA  43
```

Fig. 15.13   Coefficient matrix for Laplace operator

routine is called by the subprogram ELMLIB, which is the element library generation routine. As given here, the element library provides space for four element subprograms at any one time, where the names of the element routines are ELMT01, ELMT02, ELMT03, and ELMT04. This can easily be increased by adding more element routine names in ELMLIB. The subroutine ELMLIB is, in turn, called by subroutine PFORM which, as mentioned previously, is the subroutine to loop through all elements, and set up local arrays for coordinates (XL), displacements (UL), and destinations in the global arrays (LD). The subroutine PFORM also assembles element vectors into global vectors

```
      SUBROUTINE ELMTNN(D,UL,XL,IX,TL,S,P,NDF,NDM,NST,ISW)        ELM   1
      IMPLICIT REAL*8 (A-H,O-Z)                                   ELM   2
C                                                                 ELM   3
C....  MOCK ELEMENT ROUTINE                                       ELM   4
C                                                                 ELM   5
      INTEGER*2 IX(NEN,1)                                         ELM   6
      REAL   XL(NDM,1),TL(1),DM                                   ELM   7
      REAL*8 D(18),UL(NDF,1),S(NST,NST),P(NST)                    ELM   8
C                                                                 ELM   9
      COMMON /ADATA/ AA(16000)                                    ELM  10
      COMMON /CDATA/ NUMNP,NUMEL,NUMMAT,NEN,NEQ                   ELM  11
      COMMON /ELDATA/ DM,N,MA,MCT,IEL,NEL                         ELM  12
      COMMON /HDATA/ NH1,NH2                                      ELM  13
      COMMON  H(1)                                                ELM  14
C....  READ AND OUTPUT MATERIAL PROPERTY DATA                    ELM  15
      IF(ISW.EQ.1) THEN                                           ELM  16
c             The array D(18) is used to store up to 18 words of  ELM  17
c             information for each material set                   ELM  18
         NH1 = number of words each element needs in data base.   ELM  19
C....  CHECK ELEMENT FOR ERRORS                                   ELM  20
      ELSEIF(ISW.EQ.2) THEN                                       ELM  21
c             Check element for any negative jacobians, etc.      ELM  22
C....  COMPUTE ELEMENT COEFFICIENT MATRIX - AND RESIDUAL          ELM  23
      ELSEIF(ISW.EQ.3) THEN                                       ELM  24
c             The S(NST,NST) array is used to store matrix, and   ELM  25
c             the P(NST) array is used to store the element residual ELM  26
         H(NH1) = first word in history data base for this element.h ELM  27
C....  OUTPUT ELEMENT QUANTITIES (E.G., STRESSES)                 ELM  28
      ELSEIF(ISW.EQ.4) THEN                                       ELM  29
c             N is the current element number, MCT is a line counter ELM  30
C....  COMPUTE DIAGONAL ELEMENT MASS MATRIX                       ELM  31
      ELSEIF(ISW.EQ.5) THEN                                       ELM  32
c             The P(NST) array is used to store the diagonal matrix ELM  33
C....  COMPUTE A RESIDUAL ONLY                                    ELM  34
      ELSEIF(ISW.EQ.6) THEN                                       ELM  35
c             Compute an element residual in P(NST)              ELM  36
C....  COMPUTE ERROR ESTIMATES                                    ELM  37
      ELSEIF(ISW.EQ.7) THEN                                       ELM  38
c             Compute error quantities for each element accumulate ELM  39
C....  COMPUTE STRESS PROJECTIONS TO NODES                        ELM  40
      ELSEIF(ISW.EQ.8) THEN                                       ELM  41
c             Compute the stress projections in array AA          ELM  42
      ENDIF                                                       ELM  43
      RETURN                                                      ELM  44
      END                                                         ELM  45
```

Fig. 15.14   Mock element routine layout

(loads, residuals, reactions, and mass) and uses subprogram MODIFY to perform appropriate modification for prescribed non-zero displacements. When an element routine is accessed the value of the parameter ISW is specified between 1 and 8. The parameter specifies what action is to be taken in the element routine. Each element routine must provide appropriate transfers for each value of ISW. A mock element routine is shown in Fig. 15.14.

15.6 Solution of simultaneous, linear algebraic equations

In solving problems by the finite element method we are eventually faced with the task of solving a large set of simultaneous, linear algebraic equations. For example, in the analysis of linear steady-state problems the direct assembly of the element stiffness matrices leads to the set of linear algebraic equations. In this section methods are considered which solve the algebraic equation by a direct procedure, where an *a priori* calculation on the number of numerical operations can be made, and by an indirect or iterative method where no *a priori* estimate can be made.

15.6.1 *Direct solution.* Consider first the general problem of direct solution of the set of algebraic equations given by

$$\mathbf{Ka} = \mathbf{r} \tag{15.21}$$

where $\mathbf{K}$ is a square coefficient matrix, $\mathbf{a}$ is a vector of unknowns, and $\mathbf{r}$ is a vector of specified quantities. The reader can associate these with the quantities described previously: namely, the stiffness matrix, the nodal unknowns, and the specified forces or residuals.

In the discussion to follow it is assumed that the coefficient matrix has properties such that interchanges of rows and/or columns is unnecessary in order to solve the equations. This is true in cases where $\mathbf{K}$ is symmetric, positive (or negative) definite.† It may or may not be true when the equations are unsymmetric, or indefinite, conditions which may occur when the finite element formulation is based on some weighted residual methods. In these cases some checks or modifications are necessary to ensure that the equations can be solved.[20, 21, 22]

For the moment consider that the coefficient matrix can be written as the product of a lower triangular matrix with unit diagonals and an upper triangular matrix, i.e.,

$$\mathbf{K} = \mathbf{LU} \tag{15.22}$$

where

$$\mathbf{L} = \begin{bmatrix} 1 & 0 & \dots & 0 \\ L_{21} & 1 & \dots & 0 \\ \vdots & & \ddots & \vdots \\ L_{n1} & L_{n2} & \dots & 1 \end{bmatrix} \tag{15.23}$$

† For mixed methods which lead to Eq. (12.43) the solution is given in terms of a positive definite part for $\bar{\mathbf{q}}$ followed by a negative definite part for $\bar{\phi}$. Thus, interchanges are needed provided ordering of equations is defined as described in Sec. 15.2.4.

and

$$
\mathbf{U} = \begin{bmatrix}
U_{11} & U_{12} & \cdots & U_{1n} \\
0 & U_{22} & \cdots & U_{2n} \\
\vdots & & \ddots & \vdots \\
0 & 0 & \cdots & U_{mn}
\end{bmatrix}
\tag{15.24}
$$

This step is called the triangular decomposition of $\mathbf{K}$. The solution to the equations can now be obtained by solving the pair of equations

$$
\mathbf{Ly} = \mathbf{r}
$$

and   (15.25)

$$
\mathbf{Ua} = \mathbf{y}
$$

where $\mathbf{y}$ is introduced to facilitate the separation, e.g., see references 20 or 21.

The reader can easily observe that the solution to these equations is trivial. In terms of the elements of the equations the solution is

$$
y_1 = r_1
$$

$$
y_i = r_i - \sum_{j=1}^{i-1} L_{ij} y_j \qquad i = 2, 3, \ldots, n
\tag{15.26}
$$

and

$$
a_n = \frac{y_n}{U_{nn}}
$$

$$
a_i = \frac{y_i - \sum_{j=i+1}^{n} U_{ij} a_j}{U_{ii}} \qquad i = n-1, n-2, \ldots, 1
\tag{15.27}
$$

Equation (15.26) is called 'forward elimination' while Eq. (15.27) is called 'back substitution'.

The problem remains to construct the triangular decomposition of the coefficient matrix. This step is accomplished using variations on Gauss elimination. In practice, the operations necessary for the triangular decomposition are performed directly in the coefficient array; however, to make the steps clear, the basic steps are shown in Table 15.18 using separate arrays. The decomposition is performed in the same way that the subprogram DATRI operates; thus, the reader can easily grasp the details of the routine once the steps in Table 15.18 are mastered. Additional details may be found in references 4 and 23.

In DATRI the Crout variation of Gauss elimination is used to successively reduce the original coefficient array to upper triangular form.

TABLE 15.18

TRIANGULAR DECOMPOSITION OF **K**

Active zone

$$
\begin{bmatrix} K_{11} & K_{12} & K_{13} \\ K_{21} & K_{22} & K_{23} \\ K_{31} & K_{32} & K_{33} \end{bmatrix} \qquad \begin{bmatrix} L_{11} = 1 \end{bmatrix} \qquad \begin{bmatrix} U_{11} = K_{11} \end{bmatrix}
$$

Step 1. Active zone. First row and column to principal diagonal.

Reduced zone
Active zone

$$
\begin{bmatrix} & K_{12} & K_{13} \\ K_{21} & K_{22} & K_{23} \\ K_{31} & K_{32} & K_{33} \end{bmatrix} \qquad \begin{bmatrix} 1 & 0 \\ L_{21} = K_{21}/U_{11} & L_{22} = 1 \end{bmatrix} \qquad \begin{bmatrix} U_{11} & U_{12} = K_{12} \\ 0 & U_{22} = K_{22} - L_{21}U_{12} \end{bmatrix}
$$

Step 2. Active zone. Second row and column to principal diagonal. Use first row of **K** to eliminate $L_{21}U_{11}$. The active zone uses only values of **K** from the active zone and values of L and U which have already been computed in steps 1 and 2.

Reduced zone
Active zone

$$
\begin{bmatrix} & K_{13} \\ & K_{23} \\ K_{31} & K_{32} & K_{33} \end{bmatrix} \begin{bmatrix} 1 & 0 & 0 \\ L_{21} & 1 & 0 \\ L_{31} & L_{32} & L_{33} = 1 \end{bmatrix} \begin{bmatrix} U_{11} & U_{12} & U_{13} = K_{13} \\ 0 & U_{22} & U_{23} = K_{23} - L_{21}U_{13} \\ 0 & 0 & U_{33} = K_{33} - L_{31}U_{13} - L_{32}U_{23} \end{bmatrix}
$$

$$ L_{31} = K_{31}/U_{11} $$
$$ L_{32} = (K_{32} - L_{31}U_{12})/U_{22} $$

Step 3. Active zone. Third row and column to principal diagonal. Use first row to eliminate $L_{31}U_{11}$; use second row of reduced terms to eliminate $L_{32}U_{22}$ (reduced coefficient K_{32}). Reduce column 3 to reflect eliminations below diagonal.

The lower triangular portion is not actually set to zero but is used to construct **L**, as shown in Table 15.18. As mentioned above, the upper and lower triangular matrices will replace the original coefficient matrix; consequently, it is not possible to retain the principal diagonal elements of both **L** and **U**. Those of **L** are understood since it is known by definition that they are all unity.

Based upon the organization of Table 15.18, it is convenient to consider the coefficient array to be divided into three parts: part one being the region that is fully reduced, part two the region which is currently being reduced (called the active zone), and part three the region which contains the original unreduced coefficients. These regions are shown in Fig. 15.15 where the *j*th column above the diagonal and the *j*th row below the diagonal constitute the active zone. The algorithm for the triangular decomposition of an $n \times n$ square matrix can be deduced from

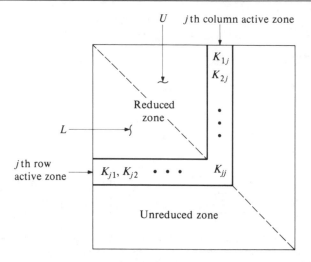

Fig. 15.15

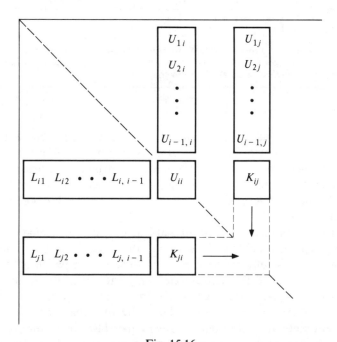

Fig. 15.16

Table 15.18 and Fig. 15.16 as follows:

$$U_{11} = K_{11}$$

$$L_{11} = 1 \tag{15.28a}$$

For each active zone j from 2 to n,

$$L_{j1} = \frac{K_{j1}}{U_{11}}$$

$$U_{1j} = K_{1j} \tag{15.28b}$$

Then

$$L_{ji} = \frac{K_{ji} - \sum\limits_{m=1}^{i-1} L_{jm} U_{mi}}{U_{ii}} \tag{15.28c}$$

$$U_{ij} = K_{ij} - \sum\limits_{m=1}^{i-1} L_{im} U_{mi} \qquad i = 1, 2, \ldots, j-1$$

and finally

$$L_{jj} = 1$$

$$U_{jj} - K_{jj} - \sum\limits_{m=1}^{j-1} L_{jm} U_{mj} \tag{15.28d}$$

The ordering of the reduction process and the terms used are shown in Fig. 15.16. The results from Table 15.18 and Eqs. (15.28) can be verified by the reader using the matrix given in the example shown in Table 15.19.

Once the triangular decomposition of a matrix is computed, several solutions for different right-hand sides r can be computed using Eqs (15.26) and (15.27). This process is often called resolution since it is not necessary to recompute L and U. For very large size coefficient matrices the decomposition process is very costly while a resolution is relatively cheap; consequently, a resolution capability is necessary in any finite element solution system.

The above discussion considered the general case of equation solving (without interchanges). In coefficient matrices resulting from a finite element problem some special properties are usually present. Often the stiffness matrix is symmetric ($K_{ij} = K_{ji}$) and it is easy to verify in this case that

$$U_{ij} = L_{ji} U_{ii} \tag{15.29}$$

For this problem class it is not necessary to store the entire coefficient array. It is sufficient to store only those coefficients above (or below) the

TABLE 15.19

EXAMPLE: TRIANGULAR DECOMPOSITION OF 3×3 MATRIX

$$
\mathbf{K} \qquad\qquad \mathbf{L} \qquad\qquad \mathbf{U}
$$

$$
\begin{bmatrix} 4 & 2 & 1 \\ 2 & 4 & 2 \\ 1 & 2 & 4 \end{bmatrix} \qquad \begin{bmatrix} 1 & & \\ & & \\ & & \end{bmatrix} \qquad \begin{bmatrix} 4 & & \\ & & \\ & & \end{bmatrix}
$$

Step 1. $L_{11} = 1$, $U_{11} = 4$

$$
\begin{bmatrix} & 2 & 1 \\ 2 & 4 & 2 \\ 1 & 2 & 4 \end{bmatrix} \qquad \begin{bmatrix} 1 & & \\ 0.5 & 1 & \\ & & \end{bmatrix} \qquad \begin{bmatrix} 4 & 2 & \\ & 3 & \\ & & \end{bmatrix}
$$

Step 2. $L_{21} = \frac{2}{4} = 0.5$, $U_{12} = 2$, $L_{22} = 1$, $U_{22} = 4 - 0.5 \times 2 = 3$

$$
\begin{bmatrix} & & 1 \\ & & 2 \\ 1 & 2 & 4 \end{bmatrix} \qquad \begin{bmatrix} 1 & & \\ 0.5 & 1 & \\ 0.25 & 0.5 & 1 \end{bmatrix} \qquad \begin{bmatrix} 4 & 2 & 1 \\ & 3 & 1.5 \\ & & 3 \end{bmatrix}
$$

Step 3. $L_{31} = \frac{1}{4} = 0.25$, $U_{13} = 1$, $L_{32} = \dfrac{2 - 0.25 \times 2}{3} = \dfrac{1.5}{3} = 0.5$

$U_{23} = 2 - 0.5 \times 1 = 1.5$, $L_{33} = 1$, $U_{33} = 4 - 0.25 \times 1 - 0.5 \times 1.5 = 3$

$$
\begin{bmatrix} 1 & & \\ 0.5 & 1 & \\ 0.25 & 0.5 & 1 \end{bmatrix} \begin{bmatrix} 4 & 2 & 1 \\ & 3 & 1.5 \\ & & 3 \end{bmatrix} = \begin{bmatrix} 4 & 2 & 1 \\ 2 & 4 & 2 \\ 1 & 2 & 4 \end{bmatrix}
$$

Step 4. Check

---

principal diagonal and use Eq. (15.29) to construct the missing part. This reduces by almost half the required storage for the coefficient array. Still larger savings in storage can be achieved if only the terms within a non-zero *band* are stored. In finite element formulations the maximum 'bandwidth' of non-zero coefficients can usually be made small compared with the number of unknowns—often 10 to 20 per cent, which reduces the storage from $n(n + 1)/2$ to $(0.1$ to $0.2)n^2$ for symmetric problems. A typical storage method for symmetric banded equations is shown in Fig. 15.17. A discussion on band solutions is given by Meyer[24,25] together with an extensive bibliography.

It is possible to reduce the required storage and computational effort still further by storing the necessary parts of the upper triangular portion of the stiffness matrix by columns and the lower triangular portion by rows, as shown in Fig. 15.18. This was noted for symmetric matrices in references 26 to 30. Now it is necessary to store and compute only within the non-zero *profile* of the equations. This method of storage has definite

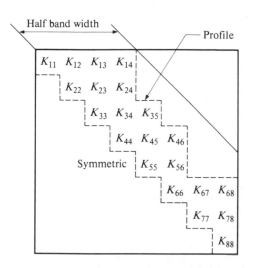

Half band width — Profile

$$
\begin{matrix}
K_{11} & K_{12} & K_{13} & K_{14} \\
 & K_{22} & K_{23} & K_{24} \\
 & & K_{33} & K_{34} & K_{35} \\
 & & & K_{44} & K_{45} & K_{46} \\
\text{Symmetric} & & & & K_{55} & K_{56} \\
 & & & & & K_{66} & K_{67} & K_{68} \\
 & & & & & & K_{77} & K_{78} \\
 & & & & & & & K_{88}
\end{matrix}
$$

Half band width — Banded storage of array

$$
\begin{matrix}
K_{11} & K_{12} & K_{13} & K_{14} \\
K_{22} & K_{23} & K_{24} \\
K_{33} & K_{34} & K_{35} \\
K_{44} & K_{45} & K_{46} \\
K_{55} & K_{56} \\
K_{66} & K & K_{68} \\
K_{77} & K_{78} \\
K_{88}
\end{matrix}
$$

Fig. 15.17

I	AD_i
1	K_{11}
2	K_{22}
3	K_{33}
4	K_{44}
5	K_{55}
6	K_{66}
7	K_{77}
8	K_{88}

Diagonals

I	AU_i	AL_i
1	K_{12}	K_{21}
2	K_{13}	K_{31}
3	K_{23}	K_{32}
4	K_{14}	K_{41}
5	K_{24}	K_{42}
6	K_{34}	K_{43}
7	K_{35}	K_{53}
8	K_{45}	K_{54}
9	K_{46}	K_{64}
10	K_{56}	K_{65}
11	K_{67}	K_{76}
12	K_{18}	K_{81}
•	•	•
18	K_{78}	K_{87}

Storage of arrays

J	JD_j
1	0
2	1
3	3
4	6
5	8
6	10
7	11
8	18

Profile

$$
\begin{matrix}
K_{11} & K_{12} & K_{13} & K_{14} & & & & K_{18} \\
 & K_{22} & K_{23} & K_{24} & & & & K_{28} \\
 & & K_{33} & K_{34} & K_{35} & & & K_{38} \\
 & & & K_{44} & K_{45} & K_{46} & & K_{48} \\
\text{Symmetric} & & & & K_{55} & K_{56} & & K_{58} \\
 & & & & & K_{66} & K_{67} & K_{68} \\
 & & & & & & K_{77} & K_{78} \\
 & & & & & & & K_{88}
\end{matrix}
$$

Fig. 15.18

485

TABLE 15.20

VARIABLES USED IN EQUATION SOLUTION SUBPROGRAMS
DATRI AND DASOL

AD(I)	Diagonals of coefficient matrix, replaced by reciprocal diagonals of U upon return from DATRI
AU(I)	Coefficients above diagonal, replaced by factor AD · U upon return from DATRI
AL(I)	Coefficients below diagonal, replaced by factor L upon return from DATRI (AU = AL if symmetric)
B(I)	Right side of equations on call to DASOL, returned as solution of equations
JP(I)	Pointer to last element in each row/column of AL/AU respectively
NEQ	Number of equations
ENERGY	Energy of active equations
FLG	Flag, if true equations treated as unsymmetric and separate storage must be provided for AU and AL

advantages over a banded storage. Firstly, it requires less storage; secondly, the storage requirements are not severely affected by a few very long columns, as shown in Fig. 15.18, and lastly, it is very easy to use vector dot product routines to effect the triangular decomposition and forward reduction.[20] This last fact is extremely important to modern machines that are vector oriented.

A profile equation-solving subprogram is included for use in the finite element solution package. These are called DATRI and DASOL for symmetric and unsymmetric equations. The profile of both cases must be symmetric. The columns above the principal diagonal or the rows below the diagonal are stored in single subscript arrays, as shown in Fig. 15.18. A pointer array is used to locate the diagonal elements. Table 15.20 defines the labels used in the solution subprograms DASOL and DATRI.

The subprogram DASBLY is used to assemble the parts of the coefficient matrix and force/residual vector. The array LD is used to locate the active equations and position in each column or row.

The frontal solution method may be used to increase the number of equations beyond that which may be solved 'in-core' using the profile solution system. The frontal method was first described by Irons[2] and subsequently adopted for the program described by Hinton and Owen.[3] In the frontal method the active equations are described by the sequence of elements processed. At any one time a small set of active equations will exist. For example, Fig. 15.19 shows a mesh in which elements up to 'E' have been processed. The active equations at the time element E is

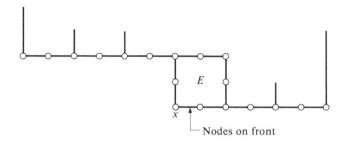

Nodes on front

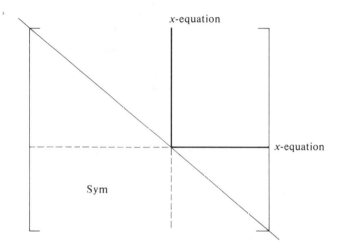

Fig. 15.19 Front stiffness matrix

introduced are indicated in the figure. Once the coefficients for E are introduced the terms associated with node 'x' and its immediate neighbors are complete (i.e., no other elements will affect the terms in the equations associated with x). Accordingly, the equations associated with x may be eliminated by a standard Gauss solution procedure[2,3]. The current front stiffness, coefficient matrix is also shown in Fig. 15.19 together with an equation associated with x. The Gauss solution step must be performed for the x equation and affects all the other terms in the active front coefficient matrix. Any existing symmetry is preserved during the solution step. The Gauss elimination step for the x equation may be written as

$$\mathbf{A}_{x+1} = \mathbf{D}_x \mathbf{L}_x \mathbf{A}_x \qquad (15.30)$$

where

$$\mathbf{L}_x = \begin{bmatrix} 1 & & & & -L_{1x} & & \\ & 1 & & & -L_{2x} & & \\ & & 1 & & -L_{3x} & & \\ & & & \ddots & \vdots & & \\ & & & & 1 & & \leftarrow x \text{ row} \\ & & & & \vdots & \ddots & \\ & & & & -L_{nx} & & 1 \end{bmatrix}$$
$\qquad$ (15.31a)

$\qquad\qquad\qquad\qquad\qquad\qquad$ ↖——x column

is a sparse matrix of the size of the active frontal coefficient matrix where

$$L_{ix} = \frac{A_{ix}}{A_{xx}}$$

and

$$\mathbf{D}_x = \begin{bmatrix} 1 & & & & & & \\ & 1 & & & & & \\ & & 1 & & & & \\ & & & \ddots & & & \\ & & & & 0 & & \leftarrow x \text{ row} \\ & & & & & \ddots & \\ & & & & & & 1 \end{bmatrix}$$
$\qquad$ (15.31b)

$\qquad\qquad\qquad\qquad\qquad\qquad$ ↖——x column

is a diagonal unit matrix except for the x row which is zero. The operation by $\mathbf{D}_x$ and $\mathbf{L}_x$ will give an $\mathbf{A}_{x+1}$ matrix which has a zero x row and column. The x columns of $\mathbf{L}_x$ and the diagonal A_{xx} are retained for the solution step. The frontal triangular decomposition is performed by subprogram PFRTD, the forward solution by PFTRFW, and the back-substitution by PFRTBK. The columns of L are retained in a buffer which is monitored and when space is exhausted is written to disk by routine PBUFF. Only symmetric $\mathbf{A}$ are considered and the columns of the upper triangular part of the array are stored in a single array.

Once a front solution step is performed the equation associated with x is no longer needed and may be removed from the active set of equations. This would create a zero column and row in the x position. New equations could be inserted in this location; however, eventually many zero columns and rows will exist as the number of active equations varies during the solution process. The front array would then be larger and result in additional processing on 'zero' equations when processing each x equation. To avoid this, the solution system provided in PCFEAP will first relocate the x equation into a buffer area. During the elimination step the elements in the rectangular block of terms to the right of the x column are moved one column to the left and the terms in the triangle below the x row are moved up one row and one column to the left. This is accomplished simply by one additional pointer which describes the location to store the reduced term. With this procedure the active frontal equations remain compact and the number of equations may vary throughout the solution procedure (and becomes zero at the end of solution). While the moves described above are desirable for the triangular decomposition step of the solution, in a resolve it is necessary to reverse the step to obtain the solution of each equation. This requires additional integer operations over that required for the profile solution method.

In reference 12 it is shown that solutions using a profile method operating on the equations stored in the order that a frontal program performs eliminations requires exactly the same number of operations as using a frontal method. Differences exist, however, in detail of programming (e.g., the moves described above) and thus the two solution systems do not require the same computation times. The main advantage of the frontal system is the ease of assembly of terms into the coefficient matrix of the active equations. In an out-of-core variable-band solution system an element matrix may have terms that belong to different blocks of storage. Thus, it is necessary to either recompute the element array for each block or to store it in backing storage—either method adds cost to the solution. Accordingly, the frontal system is a good choice for a solution system when it is not possible to keep the entire coefficient matrix in the computer memory at one time. The size of the problem is limited by the maximum size of the active front coefficient matrix which may be retained in the computer memory (in PCFEAP this is about 120 for the 8000 words provided in common ADATA). Once this limit is exceeded an out-of-core variable-band solver will always be more efficient than the frontal system.

15.6.2 *Iterative solution.* For very large problems the number of non-zero terms within the profile of the coefficient matrix is small compared to the number of zero elements. This is particularly true in three-dimensional problems where column heights for problems may be several thousand terms whereas less than a hundred are non-zero before constructing the triangular factors (which will fill in the zero elements). Accordingly, for this class of problems iterative solution methods will generally be more efficient than a direct solution. This is an area of research which is currently receiving considerable attention (e.g., see references 31 and 32). To give an indication of potential advantages of an iterative method we include a simple discussion of the Gauss–Seidel method.

To carry out the Gauss–Seidel iterations we first write an additive decomposition of the coefficient matrix[21]

$$\mathbf{K} = \mathbf{L} + \mathbf{U} \tag{15.32}$$

where $\mathbf{L}$ is lower triangular with

$$L_{ij} = K_{ij} \qquad i = 2, 3, \ldots, n; \quad j = 1, 2, \ldots, i \tag{15.33}$$

$\mathbf{U}$ is upper triangular with

$$U_{ij} = K_{ij} \qquad i = 1, 2, \ldots, n - 1; \quad j = i + 1, i + 2, \ldots, n \tag{15.34}$$

and all other elements of the $\mathbf{L}$ and $\mathbf{U}$ arrays are zero.

The basic Gauss–Seidel iteration scheme is given by the algorithm

$$\mathbf{a}^0 = \mathbf{v}$$
$$\mathbf{L}\mathbf{a}^{n+1} = \mathbf{r} - \mathbf{U}\mathbf{a}^n \tag{15.35}$$

where $\mathbf{v}$ is a starting vector and a superscript refers to iteration number. If the coefficient matrix is symmetric and positive definite the Gauss–Seidel method is known to converge (e.g., see reference 21); however, the rate of convergence may be unacceptably slow. In these cases the computational effort can usually be significantly reduced by using an overrelaxation factor. To facilitate the use of overrelaxation the term $\mathbf{L}\mathbf{a}^n$ is subtracted from both sides of Eq. (15.35), giving

$$\mathbf{L}\,\Delta\mathbf{a} = \mathbf{r} - \mathbf{K}\mathbf{a}^n \tag{15.36}$$

and the solution is advanced using

$$\mathbf{a}^{n+1} = \mathbf{a}^n + \omega\,\Delta\mathbf{a} \tag{15.37}$$

where ω is a problem-dependent overrelaxation factor with a value between 0 and 2. The above process is called 'successive overrelaxation' or SOR. The main advantages of iteration are the reduced central memory storage demands and the elimination of the triangular decomposition which is the most costly part of a direct solution. Only multiplications involving the non-zero terms in A are required; consequently, the cost per iteration will be very small compared to a solution step or triangular decomposition in a direct solution. The disadvantages of iterative methods are: the lack of knowledge on how many iterations are necessary to achieve an acceptable solution; the optimal value of ω to use in the iteration; and the method often fails when applied to indefinite or unsymmetric equations.

Many different iteration methods exist which may be used to solve the equations. Current research that utilizes the finite element structure shows some promise; however, at the present time no procedure that may be used to solve general classes of finite element problems yet exists. Consequently, most finite element solution programs today still use direct solution methods to solve the algebraic equations.

15.6.3 *Computation of energy.* When solving finite element problems that result from minimum (or maximum) principles it is often desirable to compute the minimum (or maximum) value of the functional. In the discrete problem this is equivalent to computing the energy given by

$$-2\Pi(\mathbf{a}) = \mathbf{a}^T\mathbf{K}\mathbf{a} = \mathbf{a}^T\mathbf{r} \tag{15.38}$$

where $\mathbf{K}$ is symmetric, since we now consider only minimum (or maximum) principles. Often it is not convenient to have the right-hand side and the solution in the central memory at the same time. In this case it is possible to compute the value while solving the equations. Using the triangular decomposition and symmetry conditions for K_{ij}, Eq. (15.38) can be written as

$$\sum_{m=1}^{n} \sum_{i=1}^{n} \sum_{j=1}^{n} a_i U_{mi} U_{mm}^{-1} U_{mj} a_j = \sum_{i=1}^{n} a_i r_i \tag{15.39}$$

which becomes

$$\sum_{m=1}^{n} y_m^2 U_{mm}^{-1} = \sum_{i=1}^{n} a_i r_i \tag{15.40}$$

when Eq. (15.25) is used. Thus it is possible to compute the value of the energy during the forward reduction of the right-hand side without having both the solution and right-hand side at the same time.[33]

The value of the computed discrete energy can be used to assess the rate of convergence in energy since the energy of the error is equal to the error of the energy, e.g., see reference 34, where it is shown that

$$\Pi(a - a^h) = \Pi(a) - \Pi(a^h) \tag{15.41}$$

with Π being the energy. The rate of convergence in energy can then be assessed by plotting a curve of log $[\Pi(a) - \Pi(a^h)]$ versus log h, where h is a measure of mesh size and a, a^h are the exact and approximate solutions. In addition to the energy, the program PCFEAP also computes the error estimators described in Chapter 13. The version included reports errors based upon energy estimates as well as those based upon the square errors in stresses (i.e., without use of the material property matrix). Finally, the estimates on needed refinement to produce a 5 per cent error limit are given. The estimates on refinement for both the energy and stress formulation are given. Based upon problem solutions both of the methods indicate where extensive refinement is required; however, the advantage of the stress method is that it may also be extended to non-linear applications.

15.7 Extensions and modification to the computer program

The previous sections describe the program listed in the next section. The capabilities of the program, while quite significant, can still be improved. Improvements could include increased capacity to handle large problems, increased power to the macro programming language, and, finally, postprocessors to prepare additional graphic output of solution features.

In performing finite element analyses on many engineering problems the capacity of the program discussed here will be inadequate. The inadequacy appears first in the number of unknowns that can be treated—primarily the size of the stiffness matrix is limiting the capacity. The capacity can be extended by blocking the stiffness array as shown in Fig. 15.20.[28] It would then only be necessary to have two of these blocks in core at any one time instead of the entire set of equations. When large computers are used this single change would extend the capacity of the

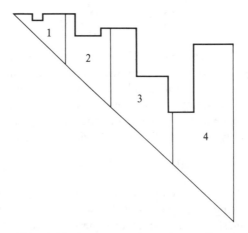

Fig. 15.20 A block storage scheme for profile solution

program to handle thousands of unknowns. Efficiency is enhanced when blocking the equations by writing mesh data onto backing core during the equation-solving steps so that maximum core area is available to the global arrays. In addition, the assembly of the global equations will have to be modified since the whole set of equations is not available at any one time. The element arrays will now have to be saved on backing store and the equations assembled block by block (e.g., see reference 19).

If it is desired to increase the capacity still further, a radical modification for the storage of the data arrays must be made. The data arrays must also be blocked, stored on backing store, and retrieved as needed. In this case it will be necessary to write special software packages to efficiently handle the extensive amounts of I/O and it will be most advantageous in this case to perform the localization step only once.

In addition to extending the capacity of the program to deal with larger numbers of unknowns, it will be necessary to add macro instructions which extend the problem types that can be treated. In Sec. 15.4 we included but a few possible commands and the program we provide cannot, for example, solve general non-linear problems. While it is possible to formulate a general iterative strategy using the commands

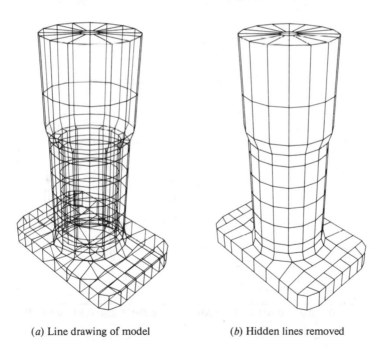

(a) Line drawing of model   (b) Hidden lines removed

Fig. 15.21   Three-dimensional mesh plots with hidden line capability (courtesy of Prof. N. N. Christiansen, Brigham Young University, Provo, Utah)

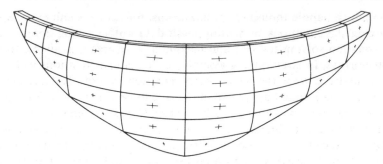

Fig. 15.22 Stress intensity measures by vectors—gives magnitude and direction

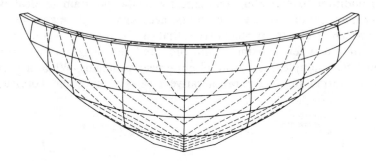

Fig. 15.23 Contour lines for stress levels

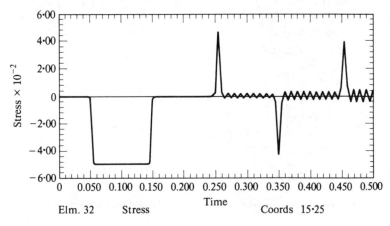

Fig. 15.24 Time history plots—quickly show anomalies such as growing oscillations

included, effective solution of many non-linear problems require the use of 'line-search' and 'arc-length' algorithms. This is a general topic which will be addressed further in the next volume. In addition, the inclusion of procedures to compute the eigenpairs of general problems will provide features to estimate time steps for transient solutions, for computing buckling loads of structures, or for general solutions by mode super-position methods for linear problems. Indeed, general applications often require special macro commands and in larger versions of the program a library feature is maintained to permit users to add commands without modification of the PMACR subprograms.

Finally, we have alluded to the problem of checking meshes and stated that a viable solution package needs a graphics package. While this is true for data checking it is also true for interpretation of output. For large analyses, especially those that are time dependent, it is not possible to interpret very much of a printed output. In these cases a graphics package is certainly a necessity. The graphics package should be capable of plotting deformed grids (magnifying the displacements if necessary), producing stress and strain plots, and producing time history plots.[9]

A few examples of plots are shown in Figs 15.21 to 15.24 and suggest some of the kinds of plotting facilities that should be available.

General plot packages can also be added to the program using the macro language to control the creation of the plot data files.

15.8 Listing of finite element computer program

We include here the complete Fortran listing for the computer program discussed in this chapter. The section is divided into five parts: the first part consists of the routines for the control and data input modules, the second the routines for the macro solution and output modules, the third the equation solution systems, the fourth graphics routines and the fifth contains the element routines for linear elasticity based upon both dis-placement and mixed models plus a simple truss structural element. The routines in each part are separated into groups which are to be placed in a single program file. This organization is used to simplify the creation of the program using a micro- or minicomputer system. The files are named and described briefly in Table 15.21.

15.8.1 *Program installation and execution.* The program has been exten-sively tested using two computing environments. The first is in an IBM or equivalent personal computer system using a Fortran 77 compiler[35] and the Graphics Development Toolkit.[5] It is recommended that the program be installed using the frontal solution system. In linking the files together all routines except PCMAC2.FOR, element, and plot files must

TABLE 15.21

PCFEAP FILES FOR SOURCE PROGRAM

File Name	Description
PCFEAP.FOR	Main program, file assignment, installation
PCDEPT.FOR	Installation or compiler-dependent routines
PCMESH.FOR	Mesh definition routines
PCMAC1.FOR	Macro program routines—part 1
PCMAC.2FOR	Macro program routines—part 2
PCMAC3.FOR	Macro program routines—part 3
PASOLV.FOR	Variable-band solution routines
PFSOLV.FOR	Frontal solution routines
PCPLOT.FOR	Plot routines for mesh and outline
PCELM1.FOR	Element routine for displacement model; plane and axisymmetric linear elasticity
PCELM2.FOR	Element routine mixed model, **B**-bar; plane and axisymmetric linear elasticity
PCELM3.FOR	Element routine for plane linear elasticity; Pian–Sumihara formulation
PCELM4.FOR	Element routine for elastic truss

The program PCFEAP and on-line documentation is available on diskette and magnetic tape. Further information may be obtained by writing to: Professor Robert L. Taylor, Department of Civil Engineering, University of California, Berkeley, CA 94720, USA.

be in the root level—overlays are used for PCMAC2.FOR, elements, and plots to minimize program size. The program requires at least 384 Kbytes of main memory and two floppy disk drives. A hard disk will significantly improve performance.

If the executable program file is called PCFEAP.EXE (or PCFEAP) then the program is run by issuing the command 'pcfeap' at the terminal. When the first analysis is performed the program will ask where disk output files are to be stored. When using the program with only two floppy disk drives, the program disk should be loaded into one disk drive (a:) and the output disk files should be saved on a scratch disk in the second drive (b:), whereas when a hard disk is available the program and output disk files may be saved on the same drive (c:).

In the second environment a virtual memory minicomputer has been used and it is recommended that the variable-band solution system be used. In the minicomputer version the graphics software was replaced by other software routines. Also the arrays in blank common and labelled common ADATA were increased significantly. Both versions used the mixed precisions to define integer, real, and double precision quantities. Installation and execution of the program are performed in the same way as for the micro version; however, the disk parameter will merely be used as part of the file name in the current directory of the computer.

After installation a small disk file named FEAP.NAM will be written which retains the installation parameters and the names of the files used during the last analysis.

The analysis files consist of:

1. Input data file. This file stores the input data which describes the mesh and solution algorithm (e.g., MACR and the macro program or INTE).
2. Output file. This file retains the output created during the mesh generation (if NOPR is used very little information is saved), and in addition any output requested during macro execution.
3. Restart read file. This file may be used to restart a previously run problem only if an appropriate restart file was saved.
4. Restart write file. This file may be used to save a restart file with the current solution state for control parameters, displacements, and history variables.

The output file will retain information that is displayed on the screen during an interactive mode of execution except prompts for the macro commands.

When using PCFEAP, the user will be asked to specify each of the four files. After the first execution of a problem the names of the currently used files are retained as defaults for the next analysis. Whenever a new problem is performed the user is given the option to rename the files. It is suggested that the input file name begin with an 'I' (e.g., IEXAMPL1); the program will then change the default names to begin with an 'O', 'R', 'S' for the output, read restart, and save restart files (e.g., OEXAMPL1, REXAMPL1, SEXAMPL1). If files are to be redirected to another drive, the full path must be specified (e.g., C:IEXAMPL1, C:OEXAMPL1, etc.). File names are limited to 12 characters including any disk redirections.

15.8.2 *Control and data input modules.* The control and data input subprograms are contained in the files PCFEAP.FOR, PCDEPT.FOR, and PCMESH.FOR.

(*a*) File PCFEAP.FOR contains the following set of subprograms:

Name	*Type*	*Description*
PCFEAP	PROGRAM	Main program
PDEFIL	SUBROUTINE	Deletes scratch disk files
FILNAM	SUBROUTINE	Sets up input, output, and restart files
PCONTR	SUBROUTINE	Controls overall problem solution
PLTSTR	SUBROUTINE	Computes nodal stress projections
PRTHED	SUBROUTINE	Prints title header in output file

```
c-------------------------------------------------------------------+
c                                                                   |
c......      F E A P    - - A Finite Element Analysis Program for    |
c                              Mini and Mainframe Computers          |
c                                                                   |
c......      p c F E A P   - - A Finite Element Analysis Program for |
c                              Personal Computers                    |
c                                                                   |
c.... A (PC)  Finite Element Analysis Program for solution of general|
c.... problem classes using the finite element method.   Problem size|
c.... is controlled by the dimension of blank common and value of max|
c.... set below.                                                     |
c                                                                   |
c.... Programmed by:                                                 |
c                R. L. Taylor                                        |
c                Department of Civil Engineering                     |
c                University of California                            |
c                Berkeley, California 94720                          |
c                                                                   |
c.... Mini Version 1.22 of FEAP - January 1987                       |
c                Use Skyline Solution system in file: pasolv.for     |
c                                                                   |
c         Uses: Fortran 77 Compiler System                          |
c                                                                   |
c.... PC Version 1.22 of pcFEAP - January 1987                       |
c                Use Frontal Solution system in file: pfsolv.for     |
c                                                                   |
c         Uses: Micro Soft Fortran Version 3.3x                      |
c                IBM Graphics Toolkit - need forvdi.lib              |
c                                                                   |
c.... (C) Copyright - University of California - 1985,1986,1987      |
c                                                                   |
c-------------------------------------------------------------------+
$NOFLOATCALLS
      program pcfeap                                         pcf  1
      character versn*12,tfile*12                            pcf  2
      integer*2 m                                            pcf  3
      common m(32000)                                        pcf  4
      common /cdata/   numnp,numel,nummat,nen,neq            pcf  5
      common /iofild/  iodr,iodw,ipd,ipr,ipi                 pcf  6
      common /psize/   maxm,ne                               pcf  7
      common /temfl1/  tfile(6)                              pcf  8
      common /vdata/   versn(3)                              pcf  9
c.... set version data                                       pcf 10
      versn(1) = 'IBM PC / AT '                              pcf 11
      versn(2) = ' -- 1.22 -- '                              pcf 12
      versn(3) = '  05/11/87 '                               pcf 13
c.... reserve memory size; set default input/output units    pcf 14
      maxm = 32000                                           pcf 15
      ne   = 1                                               pcf 16
      iodr = 1                                               pcf 17
      iodw = 2                                               pcf 18
c.... set precision values: ipd = double; ipr = real; ipi = integer  pcf 19
      ipd  = 4                                               pcf 20
      ipr  = 2                                               pcf 21
      ipi  = 1                                               pcf 22
c.... clear the screen to start                              pcf 23
      call pclear                                            pcf 24
c.... open files; erase scratch files if they exist; start execution pcf 25
      call filnam                                            pcf 26
```

```
      call pdefil(tfile,1,4)                                      pcf 27
      call pcontr                                                 pcf 28
c.... close input and output files; destroy temporary disk files pcf 29
      close(iodr)                                                 pcf 30
      close(iodw)                                                 pcf 31
      call pdefil(tfile,1,4)                                      pcf 32
      stop                                                        pcf 33
      end                                                         pcf 34
c
      subroutine pdefil(tfile,n1,n2)                              pde  1
c.... destroy temporary files                                    pde  2
      logical lfil                                                pde  3
      character*12 tfile(n2)                                      pde  4
      do 100 n = n1,n2                                            pde  5
        inquire(file=tfile(n),exist=lfil)                        pde  6
        if(lfil) then                                             pde  7
          open (4,file=tfile(n))                                 pde  8
          close(4,status='delete')                               pde  9
        endif                                                     pde 10
100   continue                                                    pde 11
      return                                                      pde 12
      end                                                         pde 13
c
      subroutine filnam                                           fil  1
c.... set filenames for execution                                fil  2
      logical lfil,linp,lout,lres,lsav                            fil  3
      character*1   y,disknm,wd(2)*6,tfile*12                     fil  4
      character*12 finp,fout,fres,fsav,pinp,pout,pres,psav,versn  fil  5
      common /iofild/ iodr,iodw,ipd,ipr,ipi                       fil  6
      common /temfl1/ tfile(6)                                    fil  7
      common /vdata/ versn(3)                                     fil  8
      data wd/'new   ','exists'/                                  fil  9
c.... output version data to screen                              fil 10
      write(*,2000) versn                                         fil 11
c.... look to see if any problem has been run                    fil 12
      inquire(file='FEAP.NAM',exist=lfil)                         fil 13
      if(lfil) then                                               fil 14
         open(3,file='FEAP.NAM',status='old')                     fil 15
         read(3,1000) pinp,pout,pres,psav,disknm                  fil 16
         finp = pinp                                              fil 17
         fout = pout                                              fil 18
         fres = pres                                              fil 19
         fsav = psav                                              fil 20
         go to 200                                                fil 21
      else                                                        fil 22
c.... default installation parameters                            fil 23
         pinp = 'NONE'                                            fil 24
         disknm = 'c'                                             fil 25
c.... set scratch disk name                                      fil 26
         write(*,2007) disknm                                     fil 27
         read(*,1001) pdisknm                                     fil 28
         if(pdisknm.ne.' ') disknm = pdisknm                      fil 29
      endif                                                       fil 30
c.... name file for input data                                   fil 31
100   write(*,2000) versn                                         fil 32
      assign 1 to ix                                              fil 33
1     write(*,2001) pinp                                          fil 34
      read (*,1000,err=900) finp                                  fil 35
      if(finp.eq.' ') finp = pinp                                 fil 36
```

```
c.... check if the input files exists                               fil 37
      inquire(file=finp,exist=linp)                                 fil 38
      if(.not.linp) then                                            fil 39
         write(*,3000)                                              fil 40
         go to 1                                                    fil 41
      else                                                          fil 42
         pout = finp                                                fil 43
         pres = finp                                                fil 44
         psav = finp                                                fil 45
         call pdisk('o',pout)                                       fil 46
         call pdisk('r',pres)                                       fil 47
         call pdisk('s',psav)                                       fil 48
      endif                                                         fil 49
      pinp = finp                                                   fil 50
c.... name file for output data                                     fil 51
      assign 2 to ix                                                fil 52
2     write(*,2002) pout                                            fil 53
      read (*,1000,err=900) fout                                    fil 54
      if(fout.eq.' ') fout = pout                                   fil 55
      pout = fout                                                   fil 56
c.... name file for restart read data                               fil 57
      assign 3 to ix                                                fil 58
3     write(*,2003) pres                                            fil 59
      read (*,1000,err=900) fres                                    fil 60
      if(fres.eq.' ') fres = pres                                   fil 61
      pres = fres                                                   fil 62
c.... name file for restart save data                               fil 63
      assign 4 to ix                                                fil 64
4     write(*,2004) psav                                            fil 65
      read (*,1000,err=900) fsav                                    fil 66
      if(fsav.eq.' ') fsav = psav                                   fil 67
      psav = fsav                                                   fil 68
c.... check file status and input if necessary                     fil 69
200   inquire(file=finp,exist=linp)                                 fil 70
      if(.not.linp) go to 100                                       fil 71
      inquire(file=fout,exist=lout)                                 fil 72
      iop = 1                                                       fil 73
      if(lout) iop = 2                                              fil 74
      inquire(file=fres,exist=lres)                                 fil 75
      irs = 1                                                       fil 76
      if(lres) irs = 2                                              fil 77
      inquire(file=fsav,exist=lsav)                                 fil 78
      isv = 1                                                       fil 79
      if(lsav) isv = 2                                              fil 80
      write(*,2005) finp,wd(2),fout,wd(iop),fres,wd(irs),fsav,wd(isv) fil 81
      read(*,1001) y                                                fil 82
      if(y.ne.'Y' .and. y.ne.'y') go to 100                        fil 83
c.... save a copy of the current filenames                         fil 84
      if(.not.lfil) open(3,file='FEAP.NAM',status='new')           fil 85
      rewind 3                                                      fil 86
      write(3,1000) finp,fout,fres,fsav,disknm                     fil 87
      close(3)                                                      fil 88
c.... erase the output file if it exists                            fil 89
      if(lout) then                                                 fil 90
        open(3,file=fout)                                           fil 91
        close(3,status='delete')                                   fil 92
      endif                                                         fil 93
      write(*,2006)                                                 fil 94
c.... open the files for input and output                           fil 95
      open(unit=iodr,file=finp,status='old')                       fil 96
```

```
      open(unit=iodw,file=fout,status='new')                      fil  97
c.... set the scratch disk names and locations                    fil  98
      tfile(1) = ' :FRNT.TEM'                                      fil  99
      tfile(2) = ' :HIST.TEM'                                      fil 100
      tfile(3) = ' :MESH.TEM'                                      fil 101
      tfile(4) = ' :STRE.TEM'                                      fil 102
      tfile(5) = fres                                              fil 103
      tfile(6) = fsav                                              fil 104
      do 300 i = 1,4                                               fil 105
        call pdisk(disknm,tfile(i))                                fil 106
300   continue                                                    fil 107
      return                                                      fil 108
c.... error trap                                                  fil 109
900   write(*,3001)                                               fil 110
      go to ix                                                    fil 111
c.... format statements                                          fil 112
1000  format(4a12,a1)                                             fil 113
1001  format(a1)                                                  fil 114
2000  format(/////6x,                                             fil 115
     1'F I N I T E    E L E M E N T    A N A L Y S I S    P R O G R A M'//fil 116
     2 13x,'VERSION: ',3a12)                                      fil 117
2001  format(/13x,'I n p u t    F i l e n a m e s'//              fil 118
     1      15x,'Input data    (default: ',a12,') :',$)           fil 119
2002  format(15x,'Output data   (default: ',a12,') :',$)          fil 120
2003  format(15x,'Restart read  (default: ',a12,') :',$)          fil 121
2004  format(15x,'Restart save  (default: ',a12,') :',$)          fil 122
2005  format( /13x,'Files are set as follows :'//                 fil 123
     1   32x,'Filename',9x,'Status'/                              fil 124
     2   15x,'Input    (read ) : ',a12,5x,a6/                     fil 125
     3   15x,'Output   (write) : ',a12,5x,a6/                     fil 126
     4   15x,'Restart  (read ) : ',a12,5x,a6/                     fil 127
     5   15x,'Restart  (write) : ',a12,5x,a6//                    fil 128
     6   13x,'Caution, existing write files will be overwritten.'//   fil 129
     7   13x,'Are filenames correct? ( y or n) : ',$)             fil 130
2006  format(/12x,'R U N N I N G    P C F E A P    P R O B L E M    N O W')fil 131
2007  format(/13x,'I N S T A L L A T I O N    P A R A M E T E R S'//   fil 132
     1      15x,'Disk Name For Scratch Files:'/                   fil 133
     2      17x,'             (default = ',a1,9x,') :',$)         fil 134
3000  format(' *** ERROR - - Specified input file does not exist') fil 135
3001  format(' *** ERROR on read *** reinput')                    fil 136
      end                                                         fil 137
c
      subroutine pcontr                                           pco   1
c.... control program for feap                                    pco   2
      logical prt,tfl,pcomp                                       pco   3
      integer*2 ia,m                                              pco   4
      real*8 d                                                    pco   5
      character*4 head,titl(20),versn*12,yyy*80                   pco   6
      common /bdata/ head(20)                                     pco   7
      common /cdata/ numnp,numel,nummat,nen,neq                   pco   8
      common /iofile/ ior,iow                                     pco   9
      common /iofild/ iodr,iodw,ipd,ipr,ipi                       pco  10
      common /mdata/ nn,n0,n1,n2,n3,n4,n5,n6,n7,n8,n9,n10,n11,n12,n13  pco 11
      common /mdat2/ n11a,n11b,n11c,ia(2,11)                      pco  12
      common /sdata/ ndf,ndm,nen1,nst                             pco  13
      common /vdata/ versn(3)                                     pco  14
      common d(1),r(1),m(1)                                       pco  15
c.... set parameters for rotation dof                             pco  16
      do 4 i = 1,4                                                pco  17
        ia(1,i) = 1                                               pco  18
```

```
          ia(2,i) = 2                                               pco 19
4         continue                                                  pco 20
c.... set files back to default values                             pco 21
1         ior = iodr                                                pco 22
          iow = iodw                                                pco 23
c.... read a card and compare first 4 columns with macro list      pco 24
          read(ior,1000,err=600,end=700) (titl(i),i=1,20)          pco 25
          if(pcomp(titl(1),'feap')) go to 100                      pco 26
          if(pcomp(titl(1),'inte')) go to 200                      pco 27
          if(pcomp(titl(1),'macr')) go to 300                      pco 28
          if(pcomp(titl(1),'stop')) return                         pco 29
          go to 1                                                   pco 30
c.... read and print control information                           pco 31
100       do 101 i = 1,20                                           pco 32
101       head(i) = titl(i)                                         pco 33
          call pintio(yyy,10)                                       pco 34
          read(yyy,1001,err=600) numnp,numel,nummat,ndm,ndf,nen,nad pco 35
          write(iow,2000) head,versn,numnp,numel,nummat,ndm,ndf,nen,nad pco 36
c.... set pointers for allocation of data arrays                   pco 37
          nen1 = nen + 4                                            pco 38
          nst  = nen*ndf + nad                                      pco 39
          call psetm(nn, numnp*max(ndm,ndf,2),ipd,tfl)             pco 40
          call psetm(nn, 4*nen*ndf,        ipd,tfl)                pco 41
          call psetm(n0, nen*ndm,          ipr,tfl)                pco 42
          call psetm(n1, nen,              ipr,tfl)                pco 43
          call psetm(n2, nst,              ipi,tfl)                pco 44
          call psetm(n3, nst,              ipd,tfl)                pco 45
          call psetm(n4, nst*nst,          ipd,tfl)                pco 46
          call psetm(n5, nummat*9,         ipi,tfl)                pco 47
          call psetm(n6, nummat*18,        ipd,tfl)                pco 48
          call psetm(n7, ndf*numnp,        ipi,tfl)                pco 49
          call psetm(n8, ndm*numnp,        ipr,tfl)                pco 50
          call psetm(n9, nen1*numel,       ipi,tfl)                pco 51
          call psetm(n10,numnp*ndf,        ipr,tfl)                pco 52
          call psetm(n11,numnp,            ipr,tfl)                pco 53
          call psetm(n11a,nen,             ipr,tfl)                pco 54
          call psetm(n11b,numnp,           ipr,tfl)                pco 55
          call psetm(n11c,numel,           ipi,tfl)                pco 56
          call psetm(n12,ndf*numnp,        ipi,tfl)                pco 57
c.... call mesh input subroutine to read and print all mesh data   pco 58
          iii = 0                                                   pco 59
          prt = .true.                                              pco 60
          call pmesh(m(n2),m(n5),d(n6),m(n7),r(n8),m(n9),r(n10),r(n11),ndf, pco 61
1         ndm,nen1,iii,prt)                                         pco 62
          go to 1                                                   pco 63
c.... set files for interactive macro execution                    pco 64
200       ior = -iodr                                               pco 65
c.... compute profile                                              pco 66
300       call profil(m(n12),m(n11c),m(n7),m(n9),ndf,nen1)         pco 67
          call psetm(n13,numnp*ndf,    ipr,tfl)                    pco 68
          call psetm(n14,3*numnp*ndf, ipd,tfl)                     pco 69
c.... set up stress history addresses                              pco 70
          call sethis(m(n5),m(n9),m(n11c),4,nen,nen1,numel)        pco 71
c.... zero the initial force and solution vectors                  pco 72
          call pconsr(r(n13),  ndf*numnp,0.0)                      pco 73
          call pconsd(d(n14),3*ndf*numnp,0.0d0)                    pco 74
c.... macro module for establishing solution algorithm             pco 75
          call pmacr(d(nn),r(n0),r(n1),m(n2),d(n3),d(n4),m(n5),d(n6),m(n7), pco 76
1             r(n8),m(n9),r(n10),r(n11),m(n12),r(n13),d(n14),d(1), pco 77
2             ndf,ndm,nen1,nst,prt)                                pco 78
```

```
          go to 1                                                    pco 79
600       call perror('PCONTR',yyy)                                  pco 80
          return                                                     pco 81
700       call pend('pcontr')                                        pco 82
          return                                                     pco 83
c....  input/output formats                                          pco 84
1000      format(20a4)                                               pco 85
1001      format(8i10)                                               pco 86
2000      format(1x,20a4//5x,'VERSION :',3a12//                      pco 87
      x   5x,'Number of nodal points       =',i6/                    pco 88
      1   5x,'Number of elements           =',i6/                    pco 89
      2   5x,'Number of material sets      =',i6/                    pco 90
      3   5x,'Dimension of coordinate space=',i6/                    pco 91
      4   5x,'Degree of freedoms/node      =',i6/                    pco 92
      5   5x,'Nodes per element (maximum)  =',i6/                    pco 93
      6   5x,'Extra d.o.f. to element      =',i6)                    pco 94
          end                                                        pco 95
c
          subroutine pltstr(dt,st,numnp)                             plt  1
c....  stress projections computed by dividing by 'lumped' weightings plt  2
          real dt(numnp),st(numnp,1)                                 plt  3
          real*8 sig(6),eerror,elproj,ecproj,efem,enerr,ebar         plt  4
          common /errind/eerror,elproj,ecproj,efem,enerr,ebar        plt  5
          elproj = 0.0                                               plt  6
          do 100 ii = 1,numnp                                        plt  7
            if(dt(ii).ne.0.0) then                                   plt  8
              styld    = st(ii,5)                                    plt  9
              do 200 j = 1,4                                         plt 10
                sig(j)   = st(ii,j)/dt(ii)                           plt 11
                elproj   = elproj + sig(j)*st(ii,j)                  plt 12
                st(ii,j) = sig(j)                                    plt 13
200           continue                                              plt 14
c....  compute the principal stress values                           plt 15
              call pstres(sig,sig(4),sig(5),sig(6))                  plt 16
              if(st(ii,5).ne.0.0) then                               plt 17
                dt(ii) = st(ii,5)/dt(ii)                             plt 18
              else                                                   plt 19
                dt(ii) = sig(6)                                      plt 20
              endif                                                  plt 21
              st(ii,5) = sig(4)                                      plt 22
              st(ii,6) = sig(5)                                      plt 23
              st(ii,7) = (sig(4)-sig(5))/2.0                         plt 24
            endif                                                    plt 25
100       continue                                                   plt 26
          return                                                     plt 27
          end                                                        plt 28
c
          subroutine prthed(iow)                                     prt  1
c....  output a header to printed outputs                             prt  2
          character*4 head                                           prt  3
          common/bdata/ head(20)                                     prt  4
          write(iow,2000) head                                       prt  5
          return                                                     prt  6
2000      format(1x,20a4//1x)                                        prt  7
          end                                                        prt  8
```

(*b*) File **PCDEPT.FOR** contains the following set of subprograms (note that routines in this file may need to be changed for various computer systems):

Name	Type	Description
PCTIME	SUBROUTINE	Set clock or elapsed time for outputs
PCOMP	FUNCTION	Compare ASCII character data
PINTIO	SUBROUTINE	Reads input data from screen or file
PCLEAR	SUBROUTINE	DOS clear screen routine

```
$NOFLOATCALLS
      INTERFACE TO SUBROUTINE TIME                                     tim  1
c.... interface to dos time function                                  tim  2
      CHARACTER*10 STR [NEAR,REFERENCE]                                tim  3
      INTEGER*2 N [VALUE]                                              tim  4
      END                                                             tim  5
c
      subroutine pctime(etime)                                        pct  1
      character*10 tarry [near,reference]                             pct  2
      character*10 etime                                             pct  3
c.... DOS time call routine: replace by appropriate system call       pct  4
      call time (10,tarry)                                           pct  5
      etime = tarry                                                  pct  6
      return                                                         pct  7
      end                                                           pct  8
c
      logical function pcomp(a,b)                                     fun  1
c.... determine match between alphanumeric data: inc = ascii parameter fun  2
      character*1 a(4),b(4)                                          fun  3
      data inc/32/                                                   fun  4
      pcomp = .false.                                                fun  5
      do 100 i = 1,4                                                 fun  6
        ia = ichar(a(i))                                             fun  7
        ib = ichar(b(i))                                             fun  8
        if(ia.ne.ib .and. ia+inc.ne.ib .and. ia.ne.ib+inc ) return   fun  9
100   continue                                                       fun 10
      pcomp = .true.                                                 fun 11
      return                                                         fun 12
      end                                                           fun 13
c
      subroutine pintio(y,n0)                                         pin  1
c.... input control from current active unit - into a character array  pin  2
c.... for free format processing by 'acheck' into field widths of 'n0' pin  3
      character*80 x,y                                               pin  4
      common /iofile/ ior,iow                                        pin  5
      if(ior.gt.0) read(ior,'(a)',err=100,end=100) x                 pin  6
      if(ior.lt.0) read(*,'(a)',err=100,end=100) x                   pin  7
      call acheck(x,y,n0,80)                                         pin  8
      return                                                         pin  9
100   call perror('PINTIO',x)                                        pin 10
      stop                                                           pin 11
      end                                                           pin 12
c
      subroutine pclear                                              pcl  1
c.... clear PC screen and home cursor on monitor                       pcl  2
      write(*,2000) char(27)                                         pcl  3
      return                                                         pcl  4
2000  format(1x,a1,'[2J')                                           pcl  5
      end                                                           pcl  6
```

(c) File PCMESH.FOR contains the following set of subprograms:

Name	Type	Description
PMESH	SUBROUTINE	Control program for mesh data inputs
PMATIN	SUBROUTINE	Input material property sets
BLKGEN	SUBROUTINE	Input block of nodes and elements
GENVEC	SUBROUTINE	Generate real nodal vectors/arrays
PBCIN	SUBROUTINE	Input boundary restraint codes
PELIN	SUBROUTINE	Input element connection lists
POLAR	SUBROUTINE	Convert polar to cartesian coordinates
SBLK	SUBROUTINE	Generate nodes and elements for BLKGEN
SETHIS	SUBROUTINE	Write initial history data to disk

```
$NOFLOATCALLS
      subroutine pmesh(idl,ie,d,id,x,ix,f,t,ndf,ndm,nen1,iii,prt)    pme   1
c.... data input routine for mesh description                        pme   2
      logical prt,error,pcomp                                        pme   3
      integer*2 ie(9,1),id(ndf,1),ix(nen1,1),idl(6),ia               pme   4
      real    x(ndm,1),f(ndf,1),t(1)                                 pme   5
      real*8 d(18,1),dm                                              pme   6
      character*4 head,wd(12),cc,sc,y*1,yyy*80                       pme   7
      character*12 cds,tmp,fds,an                                    pme   8
      common /bdata/ head(20)                                        pme   9
      common /cdata/ numnp,numel,nummat,nen,neq                      pme  10
      common /eldata/ dq,n,ma,mct,iel,nel                            pme  11
      common /iofile/ ior,iow                                        pme  12
      common /mdat2/ n11a,n11b,n11c,ia(2,11)                         pme  13
      common dm(1),rm(1)                                             pme  14
      data wd/'coor','elem','mate','boun','forc','temp','prin','nopr', pme  15
     1      'bloc','pola','angl','end '/                             pme  16
      data an/'  angles     '/, cds/' coordinates'/, tmp/' temperature'/ pme  17
      data fds/' force/displ'/, list/12/                             pme  18
c.... initialize arrays                                              pme  19
      error = .false.                                                pme  20
      if(iii.ge.0) then                                             pme  21
        prt = .true.                                                 pme  22
        call pconsr(rm(n11b),numnp,0.0)                              pme  23
        call pconsr( f,numnp*ndf,0.0)                                pme  24
        call pconsi(id,numnp*ndf,0)                                  pme  25
        if(iii.eq.0) call pconsr( t,numnp,0.0)                       pme  26
        if(iii.eq.0) call pconsr( x,numnp*ndm,-999.0)                pme  27
      endif                                                          pme  28
102   if(ior.lt.0) write(*,2000)                                     pme  29
      call pintio(yyy,10)                                            pme  30
      read(yyy,1000,err=901,end=900) cc,sc                           pme  31
      if((ior.lt.0).and.pcomp(cc,'help')) then                       pme  32
        call phelp(wd,list,'MESH ',0)                                pme  33
        go to 102                                                    pme  34
      endif                                                          pme  35
20    do 30 i = 1,list                                               pme  36
30    if(pcomp(cc,wd(i))) go to 40                                   pme  37
      go to 102                                                      pme  38
c.... nodal coordinate data input                                    pme  39
40    if(i.eq.1) call genvec(ndm,x,cds,prt,error,.true.)             pme  40
c.... element data input                                             pme  41
      if(i.eq.2) call pelin(idl,ix,nen1,nen,numnp,numel,error,prt)   pme  42
```

```
c.... material data input                                              pme 43
        if(i.eq.3) call pmatin(d,x,ix,idl,ie,numnp,numel,nummat,ndm,ndf, pme 44
     1                    prt)                                          pme 45
c.... read in the restraint conditions for each node                   pme 46
        if(i.eq.4) call pbcin(iii,idl,id,numnp,ndf,prt)                 pme 47
c.... force/displ data input                                           pme 48
        if(i.eq.5) call genvec(ndf,f,fds,prt,error,.false.)            pme 49
c.... temperature data input                                           pme 50
        if(i.eq.6) call genvec(1,t,tmp,prt,error,.false.)             pme 51
c.... set print flag                                                   pme 52
        if(i.eq.7.or.i.eq.8) prt = i.eq.7                              pme 53
c.... generate block of nodes and 4-node elements                      pme 54
        if(i.eq.9) then                                                pme 55
          if(iii.lt.0) write(iow,3000)                                 pme 56
          call blkgen(ndm,ndf,nen,nen1,x,ix,prt)                       pme 57
        endif                                                          pme 58
c.... convert polar/cylindrical to cartesian coordinates               pme 59
        if(i.eq.10) call polar(x,ndm,prt)                              pme 60
c.... set boundary angles                                              pme 61
        if(i.eq.11) call genvec(1,rm(n11b),an,prt,error,.false.)       pme 62
c.... mesh complete, return                                            pme 63
        if(i.eq.12) then                                               pme 64
          if(error) stop                                               pme 65
          return                                                       pme 66
        endif                                                          pme 67
        go to 102                                                      pme 68
c.... end of file encountered                                          pme 69
900     call pend('pmesh ')
        stop
901     call perror('PMESH ',yyy)
        go to 102
1000    format(a4,6x,a4)
2000    format('     Mesh 1 > ',$)
3000    format(' **WARNING** element connections necessary to use '
     1        ,'block in macro program')
4000    format(' ** Current Problem Values **'/i6,' nodes,',i5,' elmts,',
     1 i3,' matls,',i2,' dims,',i2,' dof/node,',i3,' nodes/elmt')
        end
c
        subroutine pmatin(d,x,ix,idl,ie,numnp,numel,nummat,ndm,ndf,prt) pma  1
        logical prt                                                    pma  2
        character yyy*80                                               pma  3
        integer*2 ix(1),idl(1),ie(9,1)                                 pma  4
        real    x(ndm,1)                                               pma  5
        real*8 d(18,1)                                                 pma  6
        common /adata/  ad(16000)                                      pma  7
        common /eldata/ dq,n,ma,mct,iel,nel                            pma  8
        common /hdata/  nh1,nh2                                        pma  9
        common /iofile/ ior,iow                                        pma 10
c.... material data input                                              pma 11
        if(prt) then                                                   pma 12
          call prthed(iow)                                             pma 13
          write(iow,2004)                                              pma 14
          if(ior.lt.0) write(*,2004)                                   pma 15
        endif                                                          pma 16
        do 306 n = 1,nummat                                            pma 17
301       if(ior.lt.0) write(*,3000)                                   pma 18
          call pintio(yyy,10)                                          pma 19
          read(yyy,1002,err=311) ma,iel,(idl(i),i=1,ndf)              pma 20
          if(ma.le.0) return                                           pma 21
```

```
c.... set all zero inputs                                               pma 22
      do 302 i = 1,ndf                                                  pma 23
302   ie(i,ma) = idl(i)                                                 pma 24
      do 303 i = 1,ndf                                                  pma 25
      if(idl(i).ne.0) go to 305                                        pma 26
303   continue                                                         pma 27
      do 304 i = 1,ndf                                                  pma 28
304   ie(i,ma) = i                                                      pma 29
305   ie(7,ma) = iel                                                    pma 30
      mct = 0                                                           pma 31
      nh1 = 0                                                           pma 32
      if(prt) then                                                     pma 33
        write(iow,2003) ma,iel,(i,ie(i,ma),i=1,ndf)                    pma 34
        if(ior.lt.0) write(*,2003) ma,iel,(i,ie(i,ma),i=1,ndf)        pma 35
      endif                                                            pma 36
      call elmlib(d(1,ma),ad,x,ix,ad,ad,ad,ndf,ndm,ndf,iel,1)         pma 37
      if(nh1.eq.0 .and. mct.ne.0) nh1 = mct                           pma 38
      ie(8,ma) = nh1                                                   pma 39
      go to 306                                                        pma 40
311   call perror('PMESH ',yyy)                                        pma 41
      go to 301                                                        pma 42
306   continue                                                         pma 43
      return                                                           pma 44
c.... formats                                                          pma 45
1002  format(8i10)                                                     pma 46
2003  format(/5x,'Material Set',i3,' for Element Type',i2,5x,//        pma 47
     1   10x,'degree of freedom assignments    local    global' /      pma 48
     2   42x, 'number',4x,'number'/(36x,2i10))                        pma 49
2004  format(' M a t e r i a l   P r o p e r t i e s')                 pma 50
3000  format(' Input: matl. no., elmt type'/3x,'>',$)                 pma 51
      end                                                              pma 52
c
      subroutine blkgen(ndm,ndf,nel,nel1,x,ix,prt)                     blk  1
c.... generate a block of nodes/elements                              blk  2
      logical prt                                                      blk  3
      character yyy*80                                                 blk  4
      integer*2 ix(nel1,1),ixl(9)                                      blk  5
      real    x(ndm,1),xl(3,9)                                        blk  6
      real*8 shp(3,9),dr,ds                                           blk  7
      common /cdata/ numnp,numel,nummat,nen,neq                        blk  8
      common /iofile/ ior,iow                                          blk  9
100   if(ior.lt.0) write(*,5000)                                       blk 10
      call pintio(yyy,5)                                               blk 11
      read(yyy,1000,err=101) nn,nr,ns,ni,ne,ma,nodinc,ntyp            blk 12
      go to 102                                                        blk 13
101   call perror('BLKGEN',yyy)                                        blk 14
      go to 100                                                        blk 15
102   nodinc = max(nodinc,0)                                           blk 16
      nr = max(nr,1)                                                   blk 17
      ns = max(ns,1)                                                   blk 18
      ni = max(ni,1)                                                   blk 19
      ma = max(ma,1)                                                   blk 20
      if(prt) then                                                    blk 21
        call prthed(iow)                                              blk 22
        write(iow,2000) nr,ns,ni,ne,ma,nodinc,ntyp                    blk 23
        if(ne.eq.0) write(iow,2005)                                   blk 24
        write(iow,2002) (i,i=1,ndm)                                   blk 25
        if(ior.lt.0) then                                             blk 26
          write(*,2000) nr,ns,ni,ne,ma,nodinc,ntyp                    blk 27
          if(ne.eq.0) write(*,2005)                                   blk 28
```

```
              write(*,2002) (i,i=1,ndm)                              blk 29
           endif                                                     blk 30
        endif                                                        blk 31
        do 10 n = 1,9                                                blk 32
        do 10 j = 1,ndm                                              blk 33
        xl(j,n) = 0.0                                                blk 34
10      ixl(n) = 0                                                   blk 35
        nm = 0                                                       blk 36
        do 20 n = 1,nn                                               blk 37
200     if(ior.lt.0) write(*,5001)                                  blk 38
        call pintio(yyy,10)                                          blk 39
        read(yyy,1001,err=201) l,r,s,t                              blk 40
        go to 202                                                    blk 41
201     call perror('BLKGEN',yyy)                                    blk 42
        go to 200                                                    blk 43
202     if(l.eq.0) l = n                                             blk 44
        nm = max(nm,l)                                               blk 45
        ixl(l) = l                                                   blk 46
        xl(1,l) = r                                                  blk 47
        xl(2,l) = s                                                  blk 48
        xl(3,l) = t                                                  blk 49
20      if(prt) write(iow,2001) l,(xl(i,l),i=1,ndm)                  blk 50
        if(prt.and.ior.lt.0) write(*,2001) l,(xl(i,l),i=1,ndm)       blk 51
        dr = 2.d0/nr                                                 blk 52
        ds = 2.d0/ns                                                 blk 53
        if (ntyp.eq.0) then                                          blk 54
           nf = ne + nr*ns - 1                                       blk 55
        elseif (ntyp.gt.7) then                                      blk 56
           nf = ne + (nr*ns)/4 - 1                                   blk 57
        else                                                         blk 58
           nf = ne + 2*nr*ns - 1                                     blk 59
        endif                                                        blk 60
        if(nf.gt.numel.and.ne.gt.0) go to 401                        blk 61
        nr = nr + 1                                                  blk 62
        ns = ns + 1                                                  blk 63
        if(ndm.eq.1) ns = 1                                          blk 64
        ng = nr*ns + ni -1                                           blk 65
        if(ng.gt.numnp) go to 400                                    blk 66
c.... form block                                                     blk 67
        call sblk(nr,ns,xl,ixl,shp,x,ix,dr,ds,ni,ne,ndm,            blk 68
     1        nel1,nodinc,ntyp,nm,ma,prt)                            blk 69
c.... print lists if wanted                                         blk 70
        if(.not.prt) return                                         blk 71
c.... print element lists                                           blk 72
        if(ne.le.0) return                                          blk 73
        do 502 n = ne,nf,50                                         blk 74
           call prthed(iow)                                         blk 75
           write(iow,2003) (i,i=1,nel)                              blk 76
           if(ior.lt.0) write(*,2003) (i,i=1,nel)                  blk 77
           j = min(nf,n+49)                                        blk 78
           do 501 i = n,j                                          blk 79
              write(iow,2004) i,ma,(ix(k,i),k=1,nel)              blk 80
              if(ior.lt.0) write(*,2004) i,ma,(ix(k,i),k=1,nel)   blk 81
501        continue                                                 blk 82
502     continue                                                    blk 83
        return                                                      blk 84
400     write(iow,2006) ng,numnp                                    blk 85
        if(ior.lt.0) write(*,2006) ng,numnp                         blk 86
        return                                                      blk 87
```

```
401     write(iow,2007) nf,numel                                        blk 88
        if(ior.lt.0) write(*,2007) nf,numel                             blk 89
        return                                                          blk 90
1000    format(8i5)                                                     blk 91
1001    format(i10,3f10.0)                                              blk 92
2000    format('   N o d e   G e n e r a t i o n s'//                   blk 93
     19x,'number of r-increments:',i5/9x,'number of s-increments:',i5/  blk 94
     29x,'first node number      :',i5/9x,'first element number   :',i5/ blk 95
     39x,'element material type :',i5/9x,'node line increment    :',i5/ blk 96
     4  9x,'block type (0-9)       :',i5/1x)                            blk 97
2001    format(i9,1p3e12.3)                                             blk 98
2002    format(5x,'node',3(i6,' coord'))                               blk 99
2003    format('   E l e m e n t   C o n n e c t i o n s'//            blk100
     1  '   element',7x,'matl',9(i5,' node'))                           blk101
2004    format(11i10)                                                   blk102
2005    format('   **WARNING** No elements are generated ')            blk103
2006    format('   **ERROR** insufficient storage for nodes'/          blk104
     1   10x,'final node =',i5,5x,'numnp =',i5)                         blk105
2007    format('   **ERROR** insufficient storage for elements'/       blk106
     1   10x,'final element =',i5,5x,'numel =',i5)                      blk107
5000    format(' Input: nn,nr,ns,ni,ne,ma,nodinc,ntyp'/3x,'>',$)       blk108
5001    format(' Input: node, x-1, x-2, x-3'/3x,'>',$)                 blk109
        end                                                             blk110
c
        subroutine genvec(ndm,x,cd,prt,err,prtz)                        gen  1
c.... generate real data arrays by linear interpolation                gen  2
        logical prt,err,prtz                                           gen  3
        character cd*12,yyy*80                                         gen  4
        real x(ndm,1),xl(6)                                           gen  5
        common /cdata/ numnp,numel,nummat,nen,neq                      gen  6
        common /iofile/ ior,iow                                       gen  7
        mct = 0                                                        gen  8
        n = 0                                                         gen  9
        ng = 0                                                        gen 10
102     l = n                                                         gen 11
        lg = ng                                                       gen 12
100     if(ior.lt.0) write(*,5000) cd                                 gen 13
        call pintio(yyy,10)                                          gen 14
        read(yyy,1000,err=11) n,ng,(xl(i),i=1,ndm)                    gen 15
        go to 12                                                      gen 16
11      call perror('GENVEC',yyy)                                     gen 17
        go to 100                                                     gen 18
12      if(n.gt.numnp) write(iow,3001) n,cd                          gen 19
        if(ior.lt.0.and.n.gt.numnp) write(*,3001) n,cd               gen 20
        if(n.le.0.or.n.gt.numnp) go to 109                           gen 21
        do 103 i = 1,ndm                                             gen 22
103     x(i,n) = xl(i)                                               gen 23
        if(lg) 104,102,104                                           gen 24
104     lg = sign(lg,n-l)                                            gen 25
        li =(abs(n-l+lg)-1)/abs(lg)                                  gen 26
        do 105 i = 1,ndm                                             gen 27
105     xl(i) = (x(i,n)-x(i,l))/li                                   gen 28
106     l = l + lg                                                   gen 29
        if((n-l)*lg.le.0) go to 102                                  gen 30
        if(l.le.0.or.l.gt.numnp) go to 108                           gen 31
        do 107 i = 1,ndm                                             gen 32
107     x(i,l) = x(i,l-lg) + xl(i)                                   gen 33
        go to 106                                                    gen 34
108     write(iow,3000) l,cd                                         gen 35
```

```
         if(ior.lt.0) write(*,3000) I,cd                              gen 36
         err = .true.                                                 gen 37
         go to 102                                                    gen 38
109      if(.not.prt) return                                          gen 39
         do 113 j = 1,numnp                                           gen 40
         if(prtz) go to 111                                           gen 41
         do 110 I = 1,ndm                                             gen 42
         if(x(I,j).ne.0.0) go to 111                                  gen 43
110      continue                                                     gen 44
         go to 113                                                    gen 45
111      mct = mct - 1                                                gen 46
         if(mct.gt.0) go to 112                                       gen 47
         mct = 50                                                     gen 48
         write(iow,2000) cd,(I,cd,I=1,ndm)                            gen 49
         if(ior.lt.0) write(*,2000) cd,(I,cd,I=1,ndm)                 gen 50
112      if(x(1,j).eq.-999.0) write(iow,2001) j                       gen 51
         if(x(1,j).ne.-999.0) write(iow,2002) j,(x(I,j),I=1,ndm)      gen 52
         if(ior.lt.0) then                                            gen 53
           if(x(1,j).eq.-999.0) write(*,2001) j                       gen 54
           if(x(1,j).ne.-999.0) write(*,2002) j,(x(I,j),I=1,ndm)      gen 55
         endif                                                        gen 56
113      continue                                                     gen 57
         return                                                       gen 58
1000     format(2i10,6f10.0)                                          gen 59
2000     format('    N o d a l: ',a12//6x,'node',9(i7,a6))            gen 60
2001     format(i10,' has not been input or generated')              gen 61
2002     format(i10,9f13.4)                                           gen 62
3000     format(' **ERROR** attempt to generate node',i5,' in ',a12) gen 63
3001     format(' **ERROR** attempt to input node',i5,', terminate'   gen 64
     1    ,' input of nodes in ',a12)                                 gen 65
5000     format(' Input ',a12,' values: node, inc, value(i),i=1,nval'/ gen 66
     1        3x,'>',$)                                               gen 67
         end                                                          gen 68
c
         subroutine pbcin(iii,idl,id,numnp,ndf,prt)                   pbc  1
         logical prt                                                  pbc  2
         character yyy*80                                             pbc  3
         integer*2 idl(1),id(ndf,1)                                   pbc  4
         common /iofile/ ior,iow                                      pbc  5
c.... read in the restraint conditions for each node                 pbc  6
         iii = 1                                                      pbc  7
         n = 0                                                        pbc  8
         ng = 0                                                       pbc  9
400      I = n                                                        pbc 10
         Ig = ng                                                      pbc 11
401      if(ior.lt.0) write(*,5000)                                   pbc 12
         call pintio(yyy,10)                                          pbc 13
         read(yyy,1000,err=412) n,ng,(idl(i),i=1,ndf)                 pbc 14
         if(n.gt.0.and.n.le.numnp) then                               pbc 15
           do 402 i = 1,ndf                                           pbc 16
             id(i,n) = idl(i)                                         pbc 17
             if(I.ne.0.and.idl(i).eq.0.and.id(i,I).lt.0) id(i,n) = -1 pbc 18
402        continue                                                   pbc 19
           Ig = sign(Ig,n-I)                                          pbc 20
403        I = I + Ig                                                 pbc 21
           if((n-I)*Ig.le.0) go to 400                                pbc 22
           do 404 i = 1,ndf                                           pbc 23
             if(id(i,I-Ig).lt.0) id(i,I) = -1                         pbc 24
404        continue                                                   pbc 25
```

```
          go to 403                                              pbc 26
        endif                                                    pbc 27
c.... output nodes with nonzero codes                            pbc 28
      if(prt) then                                               pbc 29
        call prthed(iow)                                         pbc 30
        write(iow,2000) (i,i=1,ndf)                              pbc 31
        do 407 n = 1,numnp                                       pbc 32
          do 406 l = 1,ndf                                       pbc 33
            if(id(l,n).ne.0) then                                pbc 34
              write(iow,2001) n,(id(i,n),i=1,ndf)                pbc 35
              go to 407                                          pbc 36
            endif                                                pbc 37
406         continue                                             pbc 38
407       continue                                               pbc 39
        endif                                                    pbc 40
        return                                                   pbc 41
412     call perror('PBCIN ',yyy)                                pbc 42
        go to 401                                                pbc 43
c.... formats                                                    pbc 44
1000    format(8i10)                                             pbc 45
2000    format('      N o d a l   B. C.'//6x,'node',8(i3,' b.c.')/1x)  pbc 46
2001    format(i10,8i8)                                          pbc 47
5000    format(' Input: node, inc, b.c. codes(i),i=1,ndf'/3x,'>',$)    pbc 48
        end                                                      pbc 49
c
      subroutine pelin(idl,ix,nen1,nen,numnp,numel,error,prt)    pel  1
      logical error,prt                                          pel  2
      character yyy*80                                           pel  3
      integer*2 idl(1),ix(nen1,1)                                pel  4
      common /iofile/ ior,iow                                    pel  5
c.... element data input                                         pel  6
      l = 0                                                      pel  7
      do 210 i = 1,numel,50                                      pel  8
        if(prt) call prthed(iow)                                 pel  9
        if(prt) write(iow,2001) (k,k=1,nen)                      pel 10
        if(ior.lt.0.and.prt) write(  *,2001) (k,k=1,nen)         pel 11
        j = min(numel,i+49)                                      pel 12
        do 209 n = i,j                                           pel 13
          if(l-n) 200,202,204                                    pel 14
100       call perror('PELIN ',yyy)                              pel 15
200       if(ior.lt.0) write(*,5000)                             pel 16
          call pintio(yyy,5)                                     pel 17
          read(yyy,1001,err=100) l,lk,(idl(k),k=1,nen),lx        pel 18
          if(l.eq.0) l = numel+1                                 pel 19
          if(lx.eq.0) lx=1                                       pel 20
          if(l-n) 201,202,204                                    pel 21
201       write(iow,3001) l,n                                    pel 22
          if(ior.lt.0) write(*,3001) l,n                         pel 23
          error = .true.                                         pel 24
          go to 209                                              pel 25
202       nx = lx                                                pel 26
          do 203 k = 1,nen                                       pel 27
            if(idl(k).gt.numnp.or.idl(k).lt.0) go to 208         pel 28
            ix(k,l) = idl(k)                                     pel 29
203       continue                                               pel 30
          ix(nen1,l) = lk                                        pel 31
          go to 206                                              pel 32
204       ix(nen1,n) = ix(nen1,n-1)                              pel 33
          do 205 k = 1,nen                                       pel 34
```

```
        ix(k,n) = ix(k,n-1) + nx                                      pel 35
        if(ix(k,n-1).eq.0) ix(k,n) = 0                                pel 36
        if(ix(k,n).gt.numnp.or.ix(k,n).lt.0) go to 208                pel 37
205     continue                                                      pel 38
206     if(prt .and. ior.lt.0) then                                   pel 39
           write(  *,2002) n,ix(nen1,n),(ix(k,n),k=1,nen)             pel 40
        endif                                                         pel 41
        if(prt) write(iow,2002) n,ix(nen1,n),(ix(k,n),k=1,nen)        pel 42
        go to 209                                                     pel 43
208     write(iow,3002) n                                             pel 44
        error = .true.                                                pel 45
209     continue                                                      pel 46
210     continue                                                      pel 47
        return                                                        pel 48
c.... formats                                                         pel 49
1001    format(16i5)                                                  pel 50
2001    format('       E l e m e n t s'//3x,'elmt  matl',8(i3,' node')/  pel 51
       1       (13x,8(i3,' node')))                                   pel 52
2002    format(2i6,8i8/(13x,8i8))                                     pel 53
3001    format(' **ERROR** element',i5,' appears after element',i5)   pel 54
3002    format(' **ERROR** element',i5,' has illegal nodes')          pel 55
5000    format(' Input: elm, mat, ix(i),i=1,nen, inc'/3x,'>',$)       pel 56
        end                                                           pel 57
c
        subroutine polar(x,ndm,prt)                                   pol  1
c.... convert polar to cartesian coordinates                          pol  2
        logical prt                                                   pol  3
        real*8 th                                                     pol  4
        real   x(ndm,1)                                               pol  5
        character yyy*80                                              pol  6
        common /cdata/ numnp,numel,nummat,nen,neq                     pol  7
        common /iofile/ ior,iow                                       pol  8
        if(ndm.eq.1) return                                           pol  9
        mct = 0                                                       pol 10
        th = atan(1.0d0)/45.0                                         pol 11
100     if(ior.lt.0) write(*,5000)                                    pol 12
        call pintio(yyy,10)                                           pol 13
        read(yyy,1000,err=101) ni,ne,inc,x0,y0                        pol 14
        go to 102                                                     pol 15
101     call perror('POLAR ',yyy)                                     pol 16
        go to 100                                                     pol 17
102     if(ni.le.0) return                                            pol 18
        if(ni.gt.numnp.or.ne.gt.numnp) go to 300                      pol 19
        inc = sign(max(abs(inc),1),ne-ni)                             pol 20
        if(ne.eq.0) ne = ni                                           pol 21
        n = ni                                                        pol 22
200     r = x(1,n)                                                    pol 23
        x(1,n) = x0 + r*cos(x(2,n)*th)                                pol 24
        x(2,n) = y0 + r*sin(x(2,n)*th)                                pol 25
        if(mct.gt.0) go to 250                                        pol 26
        if(prt) call prthed(iow)                                      pol 27
        if(prt) write(iow,2000) x0,y0,(i,i=1,ndm)                     pol 28
        if(ior.lt.0.and.prt) write(*,2000) x0,y0,(i,i=1,ndm)          pol 29
        mct = 50                                                      pol 30
250     if(prt) write(iow,2001) n,(x(i,n),i=1,ndm)                    pol 31
        if(ior.lt.0.and.prt) write(*,2001) n,(x(i,n),i=1,ndm)         pol 32
        mct = mct - 1                                                 pol 33
        n = n + inc                                                   pol 34
        if((ne-n)*inc.ge.0) go to 200                                 pol 35
        if(mod(ne-ni,inc).eq.0) go to 100                             pol 36
```

```
      ni = ne                                                pol 37
      n = ne                                                 pol 38
      go to 200                                              pol 39
c.... error                                                  pol 40
300   write(iow,3000) ni,ne                                  pol 41
      if(ior.lt.0) write(*,3000) ni,ne                       pol 42
      stop                                                   pol 43
c.... formats                                                pol 44
1000  format(3i10,2f10.0)                                    pol 45
2000  format('        P o l a r   t o   C a r t e s i a n   C o o r d s.'/ pol 46
     1 8x,'Center: x0 = ',e12.4,' y0 = ',e12.4/6x,'node',6(i7,'-coord'))) pol 47
2001  format(i10,6f13.4)                                     pol 48
3000  format(' **ERROR** attempt to convert nodes ni= ',i6,' - ne= ',i6) pol 49
5000  format(' Input: 1-node, 2-node, inc, x1(cent), x2(cent)'/3x,'>',$) pol 50
      end                                                    pol 51
c
      subroutine sblk(nr,ns,xl,ixl,shp,x,ix,dr,ds,ni,ne,ndm,nel1,  sbl  1
     1   nodinc,ntyp,nm,ma,prt)                              sbl  2
      logical prt,ityp                                       sbl  3
      real    xl(3,1),x(ndm,1)                               sbl  4
      real*8 shp(3,1),r,s,dr,ds,xsj                          sbl  5
      integer*2 ixl(1),ix(nel1,1)                            sbl  6
      common /cdata/ numnp,numel,nummat,nen,neq              sbl  7
      common /iofile/ ior,iow                                sbl  8
      n = ni                                                 sbl  9
      mct = 0                                                sbl 10
      s = -1.0                                               sbl 11
      do 200 j = 1,ns                                        sbl 12
      r = -1.0                                               sbl 13
      do 100 i = 1,nr                                        sbl 14
      call shape(r,s,xl,shp,xsj,3,nm,ixl,.true.)             sbl 15
      do 55 l = 1,ndm                                        sbl 16
        x(l,n) = 0.0                                         sbl 17
        do 50 k = 1,9                                        sbl 18
          m = ixl(k)                                         sbl 19
          if(m.gt.0) x(l,n) = x(l,n) + shp(3,m)*xl(l,m)      sbl 20
50      continue                                             sbl 21
55    continue                                               sbl 22
      if(prt) then                                           sbl 23
        mct = mct + 1                                        sbl 24
        if(mod(mct,50).eq.1) then                            sbl 25
          call prthed(iow)                                   sbl 26
          write(iow,2000) (k,k=1,ndm)                        sbl 27
          if(ior.lt.0) write(*,2000) (k,k=1,ndm)             sbl 28
        endif                                                sbl 29
        write(iow,2001) n,(x(k,n),k=1,ndm)                   sbl 30
        if(ior.lt.0) write(*,2001) n,(x(k,n),k=1,ndm)        sbl 31
      endif                                                  sbl 32
      n = n + 1                                              sbl 33
100   r = r + dr                                             sbl 34
      n = n + nodinc                                         sbl 35
200   s = s + ds                                             sbl 36
      if(ne.le.0) return                                     sbl 37
      me = ne - 1                                            sbl 38
      n = ni                                                 sbl 39
      inc = 1                                                sbl 40
      if(ntyp.ge.8) inc = 2                                  sbl 41
      do 400 j = 1,ns-1,inc                                  sbl 42
      do 300 i = 1,nr-1,inc                                  sbl 43
      n = n + 1                                              sbl 44
```

```
          me = me + 1                                         sbl 45
          ix(nel1,me) = ma                                    sbl 46
          if(ntyp.eq.0) then                                  sbl 47
             ix(1,me)     = n - 1                              sbl 48
             ix(2,me)     = n                                  sbl 49
             if(ndm.ne.1) then                                sbl 50
                ix(3,me)     = n + nr + nodinc                 sbl 51
                ix(4,me)     = n + nr - 1 + nodinc             sbl 52
             endif                                             sbl 53
          elseif(ntyp.ge.8) then                              sbl 54
             ix(1,me) = n-1                                    sbl 55
             ix(5,me) = n                                      sbl 56
             ix(2,me) = n+1                                    sbl 57
             ix(8,me) = nr+nodinc + n-1                        sbl 58
             if(ntyp.gt.8) ix(9,me) = nr+nodinc + n            sbl 59
             ix(6,me) = nr+nodinc + n+1                        sbl 60
             ix(4,me) = 2*(nr+nodinc) + n-1                    sbl 61
             ix(7,me) = 2*(nr+nodinc) + n                      sbl 62
             ix(3,me) = 2*(nr+nodinc) + n+1                    sbl 63
             n = n+1                                           sbl 64
          endif                                               sbl 65
300       continue                                            sbl 66
400       n = n + (inc-1) * nr + nodinc + 1                   sbl 67
          return                                              sbl 68
2000      format(' N o d a l   C o o r d i n a t e s'//6x,'node',  sbl 69
     1        3(i7,' coord'))                                  sbl 70
2001      format(i10,3f13.4)                                  sbl 71
          end                                                 sbl 72
c
          subroutine sethis(ie,ix,idl,ipd,nen,nen1,numel)     set  1
c....     set up history addresses in ix array                set  2
          logical hfl,hout                                    set  3
          character*12 tfile                                  set  4
          integer*2 ie(9,1),ix(nen1,1),idl(1)                 set  5
          double precision dm                                 set  6
          common /hdatb/ nhi,nhf,ihbuff,irec,jrec,nrec,hfl,hout   set  7
          common /psize/ maxm,ne                              set  8
          common /temfl1/ tfile(6)                            set  9
          common /temfl2/ itrec(4),nw1,nw2                    set 10
          common dm(1)                                        set 11
          nh0 = (ne+ipd-1)/ipd                                set 12
          nhi = nh0                                           set 13
          ihbuff = 8001 - nh0                                 set 14
          hfl  = .true.                                       set 15
          hout = .false.                                      set 16
          irec = 0                                            set 17
          nrec = 1                                            set 18
c....     determine buffer size needed for history terms      set 19
          ihmin = 0                                           set 20
          ihdm  = 0                                           set 21
          do 50  nu = 1,numel                                 set 22
            n     = idl(nu)                                   set 23
            ihmin = ihmin + ie(8,ix(nen1,n))                  set 24
            ihdm  = max(ihdm,ie(8,ix(nen1,n)))                set 25
50        continue                                            set 26
c....     set the buffer length and record length to minimum possible   set 27
          ihbuff = min(ihbuff,ihmin)                          set 28
          nhf  = nhi + ihbuff - 1                             set 29
          ihbuff = ihbuff*8                                   set 30
c....     set history area                                    set 31
```

```
      do 100 nu = 1,numel                                             set 32
      n     = idl(nu)                                                 set 33
      nhinc = ie(8,ix(nen1,n))                                        set 34
      if(hfl .and. nhinc.ne.0) then                                  set 35
        itrec(2) = ihbuff                                            set 36
        open(3,file=tfile(2),access='direct', status='new',          set 37
     1         form='unformatted',recl=itrec(2))                     set 38
        close(3)                                                      set 39
        hfl = .false.                                                set 40
        call pconsd(dm(nhi),ihbuff/8,0.0d0)                          set 41
      endif                                                           set 42
      if(nh0+nhinc .gt. 8000) then                                   set 43
        call phstio(3,nrec,dm(nhi),8000-nhi+1,2,tfile(2),itrec(2))   set 44
        nrec = nrec + 1                                               set 45
        nh0 = nhi                                                     set 46
      endif                                                           set 47
      ix(nen+1,n) = nh0                                               set 48
      ix(nen+2,n) = nrec                                              set 49
100   nh0 = nh0 + nhinc                                               set 50
c.... check for errors and finish initialization                     set 51
      if(nrec.gt.numel) then                                         set 52
        write(*,2000)                                                 set 53
        call pdefil(tfile,2,2)                                        set 54
        stop                                                          set 55
      elseif(nh0.gt.nhi) then                                        set 56
        call phstio(3,nrec,dm(nhi),nhf-nhi+1,2,tfile(2),itrec(2))    set 57
      else                                                            set 58
        nrec = nrec - 1                                               set 59
      endif                                                           set 60
      ne    = (nhf+1)*ipd + 1                                         set 61
      close(3)                                                        set 62
      return                                                          set 63
2000  format(' ** ERROR ** insufficient storage for history terms')  set 64
      end                                                             set 65
```

15.8.3 *Macro solution and output modules.* The solution and output for each problem is controlled by the subprograms contained in files PCMAC1.FOR, PCMAC2.FOR, and PCMAC3.FOR

(*a*) File PCMAC1.FOR contains the following set of subprograms:

Name	*Type*	*Description*
PMACR	SUBROUTINE	Controls macro solution sequence
PINITC	SUBROUTINE	Sets initial values to macro parameters
PMACIO	SUBROUTINE	Input and compile of macro commands
PMACR1	SUBROUTINE	Controls FE macro command executions
PMACR2	SUBROUTINE	Controls other macro commands
FORMFE	SUBROUTINE	Sets up call to FE array computations
PHSTIO	SUBROUTINE	Input/output of data to disk

```
$NOFLOATCALLS
c
      subroutine pmacr (ul,xl,tl,ld,p,s,ie,d,id,x,ix,f,t,jd,f0,b,      pma   1
     1   dr,ndf,ndm,nen1,nst,prt)                                       pma   2
c.... macro instruction subprograms                                    pma   3
      logical fl,prt,hfl,hout                                           pma   4
      integer*2 ld(1),ie(1),id(1),ix(1),jd(1),jct,lvs,lve,im            pma   5
      real    ct(3,100),xl(1),tl(1),x(1),f(1),f0(1),t(1)               pma   6
      real*8 d(1),ul(1),p(1),s(1),b(1),dr(1),dm,aengy,rnmax,prop       pma   7
      character*4 wd(20),lct,tary*10,tfile*12                          pma   8
      common dm(1),rm(1),im(1)                                         pma   9
      common /cdata/ numnp,numel,nummat,nen,neq                       pma  10
      common /fdata/ fl(11)                                            pma  11
      common /hdatb/ nhi,nhf,ihbuff,irec,jrec,nrec,hfl,hout            pma  12
      common /iofile/ ior,iow                                          pma  13
      common /ldata/ l,lv,lvs(9),lve(9),jct(100)                      pma  14
      common /ldatb/ lct(100)                                          pma  15
      common /ndata/ nv,nw                                             pma  16
      common /rdata/ aengy,rnmax,tol,shift                             pma  17
      common /tdata/ ttim,dt,c1,c2,c3,c4,c5,c6                        pma  18
      common /temfl1/ tfile(6)                                         pma  19
      common /temfl2/ itrec(4),nw1,nw2                                pma  20
      common /prlod/ prop,a(6,10),iexp(10),ik(10),npld               pma  21
      data wd/'stre','tang','form','mass','reac','chec',              pma  22
     1          'disp','solv','mesh','rest',                          pma  23
     2          'tol ','dt ','loop','next','prop','data','time',      pma  24
     2          'beta','newf',        'plot'/                         pma  25
c.... nmi = no. macro commands in 'pmacri'; nlp = loop number          pma  26
      data nm1,nm2/10,9/                                               pma  27
      nlp = nm1 + 3                                                    pma  28
c.... set initial values of parameters                                 pma  29
      call pinitc(aengy,rnmax,shift,tol,dt,prop,ttim,npld)            pma  30
      nw1  = nm1                                                       pma  31
      nw2  = nm2 + nw1                                                 pma  32
      nneq = ndf*numnp                                                 pma  33
c.... input the macro commands                                         pma  34
100   call pmacio (jct,lct,ct,wd,nw2+1,nlp,ll)                        pma  35
      if(ll.le.0) go to 300                                            pma  36
c.... execute macro instruction program                                pma  37
      lv = 0                                                           pma  38
      l = 1                                                            pma  39
```

```
200      j = jct(l)                                                    pma 40
         i = l - 1                                                     pma 41
         call pctime (tary)                                            pma 42
         if(l.ne.1.and.l.ne.ll) then                                  pma 43
           write(iow,2001) i,wd(j),lct(l),(ct(k,l),k=1,3),tary         pma 44
           if(ior.lt.0) write(*,2001) i,wd(j),lct(l),(ct(k,l),k=1,3),tary   pma 45
         endif                                                         pma 46
         if(j.le.nw1) call pmacr1(id,ie,ix,ld,d,s,p,x,f,f0,t,jd,b,dr,  pma 47
     1                     lct,ct,ndf,ndm,nen1,nst,nneq,prt,j)         pma 48
         if(j.ge.nw1+1.and.j.le.nw2)                                   pma 49
     1 call pmacr2(id,ix,f,f0,b,dr,lct,ct,ndf,nneq,j-nw1)              pma 50
c....  plot macro call                                                 pma 51
         if(j.eq.nw2+1) then                                           pma 52
           call pplotf(x,ix,b,lct(l),ct(1,l),ndf,ndm,nen1,nneq)        pma 53
         endif                                                         pma 54
         l = l + 1                                                     pma 55
         if(l.le.ll) go to 200                                         pma 56
         if (ior.lt.0) go to 100                                       pma 57
300      call pctime(tary)                                             pma 58
         write(iow,2000) tary                                          pma 59
         if(ior.lt.0) write(*,2000) tary                               pma 60
         if(.not.fl(4)) close(4,status='delete')                       pma 61
c....  save restart information                                        pma 62
         if(ll.lt.-1.or.fl(7)) close(3,status='delete')                pma 63
         if(ll.lt.-1.or.fl(7)) return                                  pma 64
         open (7,file=tfile(6),form='unformatted',status='new')        pma 65
         rewind 7                                                      pma 66
         write(7) numnp,numel,nummat,ndm,ndf,nhi,nhf,nrec              pma 67
         write(7) ttim,(b(i),i=1,3*nneq)                               pma 68
         if(fl(9)) write(7) (dm(i),i=nv,nv+2*neq)                      pma 69
         if(nrec.gt.0) then                                            pma 70
           do 400 j = 1,nrec                                           pma 71
             call phstio(3,j,dm(nhi),nhf-nhi+1,1,tfile(2),itrec(2))    pma 72
             call phstio(7,j,dm(nhi),nhf-nhi+1,22,tfile(6),0)          pma 73
400        continue                                                    pma 74
           call pdefil(tfile,2,2)                                      pma 75
         endif                                                         pma 76
         close(7)                                                      pma 77
         return                                                        pma 78
c....  formats                                                         pma 79
2000  format(' *End of macro execution*'/40x,'time=',a10)              pma 80
2001  format(' *Macro ',i3,' *',2(a4,1x),                              pma 81
     1    'v1=',g10.3,' v2=',g10.3,' v3=',g10.3/40x,'time=',a10)       pma 82
      end                                                              pma 83
c
      subroutine pinitc(engy,rnmx,shift,tol,dt,prop,ttim,npld)         pin  1
      logical fl                                                       pin  2
      real*8 engy,rnmx,prop                                           pin  3
      common /fdata/ fl(11)                                            pin  4
      data zero,one,tolc/0.0,1.0,1.e-12/                              pin  5
c....  set initial values of parameters                               pin  6
      npld  = 0                                                        pin  7
      engy  = zero                                                     pin  8
      rnmx  = zero                                                     pin  9
      shift = zero                                                     pin 10
      tol   = tolc                                                     pin 11
      dt    = zero                                                     pin 12
      prop  = one                                                      pin 13
      ttim  = zero                                                     pin 14
      do 50 i = 1,7                                                    pin 15
      fl(i+4) = .false.                                                pin 16
```

```
50      fl(i)    = .true.                                             pin 17
        return                                                        pin 18
        end                                                           pin 19
c
        subroutine pmacio (jct,lct,ct,wd,nwd,nlp,ll)                  pma  1
c.... macro instruction input subprogram                             pma  2
        logical       pcomp                                          pma  3
        integer*2     jct(1),js                                      pma  4
        real          ct(3,1)                                        pma  5
        character*4   wd(nwd),lct(1),clab1,clab2,ljs,tary*10,yyy*80  pma  6
        common /iofile/ ior,iow                                      pma  7
c.... initiate the read of macro statements                          pma  8
        if(ior.gt.0) call prthed(iow)                                pma  9
        if(ior.gt.0) write(iow,2001)                                 pma 10
c.... read macro cards                                               pma 11
        ll = 1                                                       pma 12
        jct(1) = nlp                                                 pma 13
        ct(1,1) = 1.0                                                pma 14
100     if(ior.lt.0) then                                           pma 15
         call pctime(tary)                                           pma 16
         write(*,2002) tary,ll                                       pma 17
        endif                                                        pma 18
        ll = ll + 1                                                  pma 19
        call pintio(yyy,15)                                          pma 20
        read(yyy,1000,err=401) clab1,clab2,(ct(i,ll),i=1,3)          pma 21
        if(ior.lt.0.and.pcomp(clab1,'help')) then                   pma 22
          call phelp(wd,nwd,'MACRO',1)                               pma 23
          ll = ll - 1                                                pma 24
          go to 100                                                  pma 25
        endif                                                        pma 26
        if(ior.gt.0.and.pcomp(clab1,'end ')) go to 150              pma 27
        if(ior.lt.0) then                                           pma 28
          if(pcomp(clab1,'exit')) ll = -1                           pma 29
          if(pcomp(clab1,'q   ').or.pcomp(clab1,'quit')) ll = -2    pma 30
          if(ll.lt.0) return                                        pma 31
        endif                                                        pma 32
c.... set execution flag                                             pma 33
        lct(ll) = clab2                                             pma 34
        do 110 j = 1,nwd                                            pma 35
110     if(pcomp(clab1,wd(j))) go to 130                            pma 36
        call perror('PMACIO',yyy)                                   pma 37
        ll = ll - 1                                                 pma 38
        go to 100                                                   pma 39
130     jct(ll) = j                                                 pma 40
        ll = ll + 1                                                 pma 41
150     jct(ll)= nlp+1                                              pma 42
c.... set loop markers                                              pma 43
        j = 0                                                       pma 44
        do 200 l = 2,ll-1                                           pma 45
          if(jct(l).eq.nlp)   j = j + 1                             pma 46
          if(j.gt.8) go to 400                                      pma 47
          if(jct(l).eq.nlp+1) j = j - 1                             pma 48
          if(j.lt.0) go to 400                                      pma 49
200     continue                                                   pma 50
        if(j.ne.0.and.ior.gt.0) go to 400                           pma 51
        if(j.ne.0.and.ior.lt.0) ll = ll - 1                         pma 52
        if(j.ne.0) go to 100                                        pma 53
        do 230 l = 1,ll-1                                           pma 54
          if(jct(l).ne.nlp) go to 230                               pma 55
          j = 1                                                     pma 56
```

```
         do 210 i = I+1,II                                            pma 57
            if(jct(i).eq.nlp) j = j + 1                               pma 58
            if(jct(i).eq.nlp+1) j = j - 1                             pma 59
            if(j.eq.0) go to 220                                      pma 60
210      continue                                                     pma 61
         go to 400                                                    pma 62
220      ct(2,i) = I                                                  pma 63
         ct(2,I) = i                                                  pma 64
230      continue                                                     pma 65
         return                                                       pma 66
c.... error messages                                                  pma 67
400      write(iow,4000)                                              pma 68
         if(ior.gt.0) stop                                            pma 69
         if(ior.lt.0) write(*,4000)                                   pma 70
         go to 100                                                    pma 71
401      call perror('PMACIO',yyy)                                    pma 72
         go to 100                                                    pma 73
1000     format(2(a4,11x),3f15.0)                                     pma 74
2000     format(7x,a4,1x,a4,1x,3g12.5)                                pma 75
2001     format(' M a c r o    I n s t r u c t i o n s'// 2x,         pma 76
        1'macro statement',2x,'variable 1',2x,'variable 2',2x,'variable 3')pma 77
2002     format(' Input a macro instruction, enter "exit" for restarts',  pma 78
        1    ', "quit" to quit.'/3x,'Time = ',a10,' Macro',i3,'> ',$) pma 79
4000     format(' error in pmacio ** wrong loop/next order, or > 8 loops') pma 80
         end                                                          pma 81
c
         subroutine pmacr1(id,ie,ix,ld,d,s,p,x,f,f0,t,jd,b,dr,         pma  1
        1                  lct,ct,ndf,ndm,nen1,nst,nneq,prt,j)         pma  2
c.... macro instruction subprograms                                   pma  3
         logical fa,tr,fl,pcomp,prt,sflg,tflg,hfl,hout                 pma  4
         character lct(1)*4,tfile*12,tt*10,yyy*80                      pma  5
         integer*2 id(1),ie(1),ix(nen1,1),jd(1),ld(1),lvs,lve,jct,     pma  6
        1           ia,im                                             pma  7
         real    ct(3,1),x(1),f(1),f0(1),t(1)                         pma  8
         real*8  b(1),d(1),dr(1),s(1),p(1),dm,aa,aengy,rnorm,          pma  9
        1 rnmax,dot,prop,eerror,elproj,ecproj,efem,enerr,ebar          pma 10
         common /adata/ aa(1),ap(15998)                               pma 11
         common /cdata/ numnp,numel,nummat,nen,neq                    pma 12
         common /errind/ eerror,elproj,ecproj,efem,enerr,ebar         pma 13
         common /fdata/ fl(11)                                        pma 14
         common /frdata/ maxf                                         pma 15
         common /hdatb/ nhi,nhf,ihbuff,irec,jrec,nrec,hfl,hout        pma 16
         common /iofile/ ior,iow                                      pma 17
         common /iofild/ iodr,iodw,ipd,ipr,ipi                        pma 18
         common /ldata/ l,lv,lvs(9),lve(9),jct(100)                   pma 19
         common /mdat2/ n11a,n11b,n11c,ia(2,11)                       pma 20
         common /ndata/ nv,nw                                         pma 21
         common /prlod/ prop,a(6,10),iexp(10),ik(10),npld             pma 22
         common /rdata/ aengy,rnmax,tol,shift                         pma 23
         common /tdata/ ttim,dt,c1,c2,c3,c4,c5,c6                     pma 24
         common /temfl1/ tfile(6)                                     pma 25
         common /temfl2/ itrec(4),nw1,nw2                             pma 26
         common dm(1),rm(1),im(1)                                     pma 27
         data zero,one/0.0,1.0/,fa,tr/.false.,.true./                 pma 28
c.... transfer to correct process                                     pma 29
         n1 = 1                                                       pma 30
         n3 = 1                                                       pma 31
         go to (1,2,3,4,5,6,5,2,9,10), j                              pma 32
c.... print stress values                                             pma 33
1        n1 = ct(1,l)                                                 pma 34
```

```
      n2 = ct(2,l)                                            pma 35
      n3 = ct(3,l)                                            pma 36
      n3 = max(n3,1)                                          pma 37
      n4 = numnp - 1                                          pma 38
      if (pcomp(lct(l),'node')) then                          pma 39
        n1 = max(1,min(numnp,n1))                             pma 40
        n2 = max(n1,min(numnp,n2))                            pma 41
        enerr = 0.0                                           pma 42
        if(.not.fl(11)) then                                  pma 43
          call pconsr(aa,8*numnp,0.0)                         pma 44
          call formfe(b,dr,fa,fa,fa,fa,8,1,numel,1)           pma 45
          call pltstr(aa,ap(n4),numnp)                        pma 46
        endif                                                 pma 47
        call prtstr(aa,ap(n4),numnp,n1,n2,n3)                 pma 48
        fl(11) = tr                                           pma 49
      elseif (pcomp(lct(l),'erro')) then                      pma 50
        n1 = max(n1,1)                                        pma 51
        n2 = 8*numnp                                          pma 52
        call pconsr(ap(n2-1),n2,0.0)                          pma 53
        enerr = 0.0                                           pma 54
        do 110 i = 1,n1                                       pma 55
          call pconsr(aa,n2,0.0)                              pma 56
          call formfe(b,dr,fa,fa,fa,fa,8,1,numel,1)           pma 57
          call pltstr(aa,ap(n4),numnp)                        pma 58
          call addvec(ap(n2-1),ap(n4),n2-numnp)               pma 59
110     continue                                              pma 60
        fl(11) = tr                                           pma 61
        eerror = 0.0                                          pma 62
        eproj  = 0.0                                          pma 63
        efem   = 0.0                                          pma 64
        ebar   = 0.05*sqrt(enerr/numel)                       pma 65
        ietyp  = 1                                            pma 66
        call formfe(b,dr,fa,fa,fa,fa,7,1,numel,1)             pma 67
        call prterr                                           pma 68
      else                                                    pma 69
        if(pcomp(lct(l),'all ')) then                         pma 70
          n2 = numel                                          pma 71
        else                                                  pma 72
          n1 = max(1,min(numel,n1))                           pma 73
          n2 = max(n1,min(numel,n2))                          pma 74
        endif                                                 pma 75
        call formfe(b,dr,fa,fa,fa,fa,4,n1,n2,n3)              pma 76
      endif                                                   pma 77
      return                                                  pma 78
c.... form tangent stiffness                                  pma 79
2     shft = c1                                               pma 80
      sflg = fl(9)                                            pma 81
      if(j.eq.2) then                                         pma 82
        if(ct(1,l).ne.zero) then                              pma 83
          fl(8) = tr                                          pma 84
          fl(7) = fa                                          pma 85
          call pload(id,f,f0,dr,nneq,prop,dm(nl),dm(nw))      pma 86
        endif                                                 pma 87
        shift= 0.                                             pma 88
        tflg = tr                                             pma 89
        if(.not.fl(9).and.ct(2,l).ne.zero) then               pma 90
          if(fl(2)) then                                      pma 91
            sflg = tr                                         pma 92
            shift= ct(2,l)                                    pma 93
            shft = -shift                                     pma 94
```

```
          if(ior.lt.0) write(*,2006) shift                       pma 95
          write(iow,2006) shift                                  pma 96
        else                                                     pma 97
          if(ior.lt.0) write(*,2007)                             pma 98
          write(iow,2007)                                        pma 99
          if(ior.gt.0) stop                                      pma100
          return                                                 pma101
        endif                                                    pma102
      endif                                                      pma103
    else                                                         pma104
      if(.not.fl(8)) return                                      pma105
      fl(7) = .false.                                            pma106
      tflg  = .false.                                            pma107
    endif                                                        pma108
c.... call the solve routine to assemble and solve the tangent matrix   pma109
    na = maxf + 1                                                pma110
    nal= (maxf*(maxf+1))/2 + na                                  pma111
    call psolve(b,aa(na),aa,dr,aa(nal),dm(nl),s,ld,jd,im(nl1c),nst,1,  pma112
   1         tflg,fl(8),sflg,shft,4,rnorm,aengy,1)               pma113
    call pctime(tt)                                              pma114
    if(fl(8)) then                                               pma115
      fl(8) = fa                                                 pma116
      write(iow,2001) rnorm,tt                                   pma117
      if(ior.lt.0) write(*,2001) rnorm,tt                        pma118
      if (rnmax.eq.0.0d0) rnmax = abs(aengy)                     pma119
      write(iow,2004) rnmax,aengy,tol                            pma120
      if(ior.lt.0) write(*,2004) rnmax,aengy,tol                 pma121
      if(abs(aengy).le.tol*rnmax) then                           pma122
        ct(1,lve(lv)) = ct(1,lvs(lv))                            pma123
        l = lve(lv) - 1                                          pma124
      endif                                                      pma125
      call update(id,f0,f,b,dm(nv),dm(nw),dr,nneq,neq,fl(9),prop,2)   pma126
    else                                                         pma127
      write(iow,2002) tt                                         pma128
      if(ior.lt.0) write(*,2002) tt                              pma129
    endif                                                        pma130
    return                                                       pma131
c.... form out of balance force for time step/iteration          pma132
3   if(fl(8)) return                                             pma133
    call pload(id,f,f0,dr,nneq,prop,dm(nl),dm(nw))               pma134
    call formfe(b,dr,fa,tr,fa,fa,6,1,numel,1)                    pma135
    rnorm = sqrt(dot(dr,dr,neq))                                 pma136
    write(iow,2003) rnorm                                        pma137
    if(ior.lt.0) write(*,2003) rnorm                             pma138
    fl(8) = tr                                                   pma139
    return                                                       pma140
c.... form a lumped mass approximation                           pma141
4   if(fl(5)) call psetm(nl,neq,ipd,fl(5))                       pma142
    call pconsd(dm(nl),neq,0.0d0)                                pma143
    fl(2) = tr                                                   pma144
    call formfe(b,dm(nl),fa,tr,fa,fa,5,1,numel,1)                pma145
    return                                                       pma146
c.... compute reactions and print                                pma147
5   if(pcomp(lct(l),'all ')) then                                pma148
      n2 = numnp                                                 pma149
    else                                                         pma150
      n1 = ct(1,l)                                               pma151
      n2 = ct(2,l)                                               pma152
      n3 = ct(3,l)                                               pma153
      n1 = max(1,min(numnp,n1))                                  pma154
```

```
          n2 = max(n1,min(numnp,n2))                                    pma155
          n3 = max(1,n3)                                                pma156
        endif                                                           pma157
        if(j.eq.5) then                                                 pma158
          call pconsd(dr,nneq,0.0d0)                                    pma159
          call formfe(b,dr,fa,tr,fa,tr,6,1,numel,1)                     pma160
          call prtrea(dr,ndf,numnp,n1,n2,n3)                            pma161
        else                                                            pma162
          call prtdis(x,b,ttim,prop,ndm,ndf,n1,n2,n3)                   pma163
        endif                                                           pma164
        return                                                          pma165
c....  check mesh for input errors                                      pma166
6       call formfe(b,dr,fa,fa,fa,fa,2,1,numel,1)                       pma167
        return                                                          pma168
c....  modify mesh data (cannot change profile of stiffness/mass)       pma169
9       i = -1                                                          pma170
        call pmesh(ld,ie,d,id,x,ix,f,t,ndf,ndm,nen1,i,prt)              pma171
        if (i.gt.0) go to 400                                           pma172
        return                                                          pma173
c....  restart previously run problem                                   pma174
10      open (7,file=tfile(5),form='unformatted',status='old')          pma175
        read(7) nnpo,nnlo,nnmo,ndmo,ndfo,nhio,nhfo,nrco                 pma176
        if((nnpo.eq.numnp).and.(nnlo.eq.numel).and.(nnmo.eq.nummat)     pma177
     1     .and.(ndmo.eq.ndm).and.(ndfo.eq.ndf).and.(nrco.eq.nrec)      pma178
     2     .and.(nhfo-nhio.eq.nhf-nhi) ) then                           pma179
          read(7) ttim,(b(i),i=1,3*nneq)                                pma180
          if(fl(9)) read(7) (dm(i),i=nv,nv+2*neq)                       pma181
          if(nrec.gt.0) then                                            pma182
            do 101 j = 1,nrec                                           pma183
              call phstio(7,j,dm(nhi),nhf-nhi+1,11,tfile(5),0)          pma184
              call phstio(3,j,dm(nhi),nhf-nhi+1,2,tfile(2),itrec(2))    pma185
101         continue                                                    pma186
          endif                                                         pma187
          close(7)                                                      pma188
        else                                                            pma189
          if(ior.gt.0) write(iow,3001)                                  pma190
          if(ior.lt.0) write( *,3001)                                   pma191
        endif                                                           pma192
        return                                                          pma193
c....  error diagnostics                                                pma194
400     write(iow,4000)                                                 pma195
        if(ior.gt.0) stop                                               pma196
        if(ior.lt.0) write( *,4000)                                     pma197
        return                                                          pma198
c....  formats                                                          pma199
2001    format('   residual norm = ',e15.7,6x,'time=',a10)              pma200
2002    format(40x,'time=',a10)                                         pma201
2003    format('   residual norm = ',e15.7)                             pma202
2004    format('   energy convergence test'/'   maximum   =',e24.15/    pma203
     1  '   current   =',e24.15/'   tolerance =',e24.15)                pma204
2006    format('   shift of',e12.5,' applied with mass')                pma205
2007    format('   shift requested but no mass matrix exists.')         pma206
3001    format(' ** ERROR ** Data on restart file incompatible with',   pma207
     1         ' current problem.')                                     pma208
4000    format(' ** ERROR ** attempt to change profile during mesh')    pma209
        end                                                             pma210
c
        subroutine pmacr2(id,ix,f,f0,b,dr,lct,ct,ndf,nneq,j)            pma   1
c....  macro instruction subprograms                                    pma   2
        logical fl,pcomp,hfl,hout                                       pma   3
```

```
      integer*2 id(1),ix(1),lvs,lve,jct,im                     pma   4
      real      f0(1),f(1),ct(3,1),xtl(2)                      pma   5
      real*8    b(1),dr(1),dm,aengy,rnmax,prop                 pma   6
      character*4 lct(1),ctl(2),yyy*80                         pma   7
      common /cdata/ numnp,numel,nummat,nen,neq                pma   8
      common /fdata/ fl(11)                                    pma   9
      common /hdatb/ nhi,nhf,ihbuff,irec,jrec,nrec,hfl,hout     pma  10
      common /iofile/ ior,iow                                  pma  11
      common /iofild/ iodr,iodw,ipd,ipr,ipi                    pma  12
      common /ldata/ l,lv,lvs(9),lve(9),jct(100)               pma  13
      common /ndata/ nv,nw                                     pma  14
      common /prlod/ prop,a(6,10),iexp(10),ik(10),npld         pma  15
      common /rdata/ aengy,rnmax,tol,shift                     pma  16
      common /tdata/ ttim,dt,c1,c2,c3,c4,c5,c6                 pma  17
      common dm(1),rm(1),im(1)                                 pma  18
c.... transfer to correct process                             pma  19
      go to (1,2,3,4,5,6,7,9), j                               pma  20
c.... set solution tolerance                                  pma  21
1     tol = ct(1,l)                                            pma  22
      return                                                   pma  23
c.... set time increment                                      pma  24
2     dt = ct(1,l)                                             pma  25
      if(fl(9)) call setci(ior)                                pma  26
      return                                                   pma  27
c.... set loop start indicators                               pma  28
3     lv = lv + 1                                              pma  29
      lvs(lv) = l                                              pma  30
      lve(lv) = ct(2,l)                                        pma  31
      ct(1,lve(lv)) = 1.                                       pma  32
      return                                                   pma  33
c.... loop terminator control                                 pma  34
4     n = ct(2,l)                                              pma  35
      ct(1,l) = ct(1,l) + 1.0                                  pma  36
      if(ct(1,l).gt.ct(1,n)) lv = lv - 1                       pma  37
      if(ct(1,l).le.ct(1,n)) l = n                             pma  38
      return                                                   pma  39
c.... input proportional load table                           pma  40
5     npld = ct(1,l)                                           pma  41
      prop = propld (ttim,npld)                                pma  42
      return                                                   pma  43
c.... data command                                            pma  44
6     if(ior.lt.0) write(*,3000) lct(l)                        pma  45
      call pintio(yyy,10)                                      pma  46
      read(yyy,1000,err=61) (ctl(i),i=1,2),(xtl(i),i=1,2)      pma  47
      if(.not.pcomp(lct(l),ctl(1))) go to 402                  pma  48
      if(pcomp(ctl(1),'tol ')) tol = xtl(1)                    pma  49
      if(pcomp(ctl(1),'dt  ')) dt  = xtl(1)                    pma  50
      return                                                   pma  51
61    call perror('PMACR2',yyy)                                pma  52
      go to 6                                                  pma  53
c.... increment time                                          pma  54
7     ttim = ttim + dt                                         pma  55
      if(npld.gt.0) prop = propld(ttim,0)                      pma  56
      write(iow,2002) ttim,prop                                pma  57
      if(ior.lt.0) write( *,2002) ttim,prop                    pma  58
      aengy = 0.0                                              pma  59
      rnmax = 0.0                                              pma  60
c.... update history on the disk                              pma  61
      if(.not.hfl) then                                        pma  62
        hout = .true.                                          pma  63
```

```
         call formfe(b,dr,.false.,.false.,.false.,.false.,6,1,numel,1)    pma 64
         hout = .false.                                                     pma 65
      endif                                                                 pma 66
c.... update dynamic vectors for time step                                 pma 67
      if(fl(9)) then                                                        pma 68
         call setci(ior)                                                    pma 69
         call update(id,f0,f,b,dm(nv),dm(nw),dr,nneq,neq,fl(9),prop,1)      pma 70
      endif                                                                 pma 71
c.... zero displacement increment for next time step                       pma 72
      call pconsd(b(nneq+1),nneq+nneq,0.0d0)                                pma 73
      fl(10) = .true.                                                       pma 74
      return                                                                pma 75
c.... input integration parameters and initialize vectors                  pma 76
8     call param(ct(1,l))                                                   pma 77
      if(fl(9)) return                                                      pma 78
      call psetm(nv,neq*2,ipd,fl(9))                                        pma 79
      nw = nv + neq                                                         pma 80
      fl(9) = .true.                                                        pma 81
      call pconsd(dm(nv),neq*2,0.0d0)                                       pma 82
      return                                                                pma 83
c.... update the current force vector f0                                    pma 84
9     call saxpb(f,f0,prop,nneq, f0)                                        pma 85
      return                                                                pma 86
c.... error diagnostics                                                     pma 87
402   if(ior.gt.0) write(iow,4002)                                         pma 88
      if(ior.gt.0)   stop                                                   pma 89
      if(ior.lt.0) write( *,4002)                                          pma 90
      return                                                                pma 91
c.... formats                                                               pma 92
1000  format(a4,6x,a4,6x,2f10.0)                                            pma 93
2002  format('   computing solution for time',e12.5/                       pma 94
     1       '   proportional load value is ',e12.5)                       pma 95
3000  format(' Input ',a4,' macro >',)                                     pma 96
4002  format(' **ERROR** macro label mismatch on data command')            pma 97
      end                                                                   pma 98
c
      subroutine formfe(u,b,af,bf,cf,df,is,ne1,ne2,ne3)                     for  1
c.... form finite element arrays as required                               for  2
      logical af,bf,cf,df,afl,bfl,cfl,dfl,hfl,hout                         for  3
      integer*2 m,ia                                                        for  4
      real*8 u(1),b(1),dm                                                   for  5
      common /mdata/nn,n0,n1,n2,n3,n4,n5,n6,n7,n8,n9,n10,n11,n12,n13        for  6
      common /mdat2/ n11a,n11b,n11c,ia(2,11)                                for  7
      common /sdata/ ndf,ndm,nen1,nst                                       for  8
      common /xdata/ isw,nn1,nn2,nn3,afl,bfl,cfl,dfl                       for  9
      common dm(1),rm(1),m(1)                                               for 10
c.... form appropriate f.e. array                                          for 11
      afl = af                                                              for 12
      bfl = bf                                                              for 13
      cfl = cf                                                              for 14
      dfl = df                                                              for 15
      isw = is                                                              for 16
      nn1 = ne1                                                             for 17
      nn2 = ne2                                                             for 18
      nn3 = ne3                                                             for 19
      call pform(dm(nn),rm(n0),rm(n1),m(n2),dm(n3),dm(n4),m(n5),dm(n6),    for 20
     1 m(n7),rm(n8),m(n9),rm(n10),rm(n11),m(n11c),rm(n13),u,b,ndf,         for 21
     2 ndm,nen1,nst)                                                        for 22
      return                                                                for 23
      end                                                                   for 24
```

```
c
      subroutine phstio(iu,irec,hh,nh,isw,tfile,itrec)          phs  1
c.... i/o control                                               phs  2
      character*12 tfile                                        phs  3
      real*8 hh(nh)                                             phs  4
c....    direct access read/write                               phs  5
      if(isw.lt.10) open(iu,file=tfile,access='direct',recl=itrec)  phs  6
      if(isw.eq.1) read (iu,rec=irec) hh                        phs  7
      if(isw.eq.2) write(iu,rec=irec) hh                        phs  8
      if(isw.lt.10) close(iu)                                   phs  9
c....    sequential access read/write                           phs 10
      if(isw.eq.11) read (iu) hh                                phs 11
      if(isw.eq.22) write(iu) hh                                phs 12
      return                                                    phs 13
      end                                                       phs 14
```

(*b*) File PCMAC2.FOR contains the following set of subprograms (routines in this file may be in an overlay):

Name	Type	Description
ADDVEC	SUBROUTINE	Accumulate a vector
JUST	SUBROUTINE	Justify data for parser
MODIFY	SUBROUTINE	Modify load vector for displacement b.c.
PANGL	SUBROUTINE	Set up local angles for rotation of DOF
PFORM	SUBROUTINE	Form FE arrays
PLOAD	SUBROUTINE	Set up load and inertia vector residual
PROPLD	FUNCTION	Input or set proportional load value
PRTDIS	SUBROUTINE	Output nodal displacements
PRTERR	SUBROUTINE	Output error estimate values
PRTREA	SUBROUTINE	Output nodal reactions and sums
PRTSTR	SUBROUTINE	Output nodal stresses
PTRANS	SUBROUTINE	Transform arrays for sloping boundary
SETCI	SUBROUTINE	Set transient integration parameters
UPDATE	SUBROUTINE	Update solution vectors

```
$NOFLOATCALLS
      subroutine addvec(a,b,nn)                                     add   1
      real a(1),b(1)                                                add   2
      do 100 n = 1,nn                                               add   3
        a(n) = a(n) + b(n)                                          add   4
100   continue                                                     add   5
      return                                                       add   6
      end                                                          add   7
c
      subroutine just(y,k,n0)                                      jus   1
c.... complete the parser alignment of data                        jus   2
      character*1 y(k),yi,ze,ni,mi,pl,dt,sp                         jus   3
      data ze,ni,mi,pl,dt,sp/'0','9','-','+','.',' '/              jus   4
      n1 = n0 - 1                                                   jus   5
      n2 = n1 - 1                                                   jus   6
      do 140 i = 1,k,n0                                             jus   7
        do 100 j = i,i+n1                                          jus   8
          if(y(j).ne.sp) go to 115                                 jus   9
100     continue                                                   jus  10
        y(i+n1) = ze                                               jus  11
115     if(y(i+n1).ne.sp) go to 140                                jus  12
        yi = y(i)                                                  jus  13
        if((yi.ge.ze.and.yi.le.ni).or.(yi.eq.mi).or.(yi.eq.pl)     jus  14
     1   .or.(yi.eq.dt)) then                                      jus  15
          do 110 j = i+n2,i,-1                                     jus  16
            if(y(j).ne.sp) go to 120                               jus  17
110       continue                                                 jus  18
120       kl = n1 + i - j                                          jus  19
          do 130 l = j,i,-1                                        jus  20
            y(l+kl) = y(l)                                         jus  21
            y(l) = sp                                              jus  22
130       continue                                                 jus  23
        endif                                                      jus  24
```

```
140   continue                                                    jus 25
      return                                                      jus 26
      end                                                         jus 27
c
      subroutine modify(b,s,dul,nst)                              mod 1
c.... modify for non-zero displacement boundary conditions        mod 2
      real*8 b(1),s(nst,1),dul(1)                                 mod 3
      do 100 i = 1,nst                                            mod 4
      do 100 j = 1,nst                                            mod 5
100   b(i) = b(i) - s(i,j)*dul(j)                                 mod 6
      return                                                      mod 7
      end                                                         mod 8
c
      subroutine pangl(ix,nen,angl,angg,nrot)                     pan 1
c.... set up table of rotation angles                             pan 2
      integer*2 ix(nen)                                           pan 3
      real      angl(nen),angg(1)                                 pan 4
      nrot = 0                                                    pan 5
      do 100 n = 1,nen                                            pan 6
        angl(n) = 0.0                                             pan 7
        ii = abs(ix(n))                                           pan 8
        if (ii.gt.0) then                                         pan 9
          angl(n) = angg(ii)                                      pan 10
          if (angg(ii).ne.0.0) nrot = nrot + 1                    pan 11
        endif                                                     pan 12
100   continue                                                    pan 13
      return                                                      pan 14
      end                                                         pan 15
c
      subroutine param(ct)                                        par 1
c.... set appropriate time integration parameters                 par 2
      common /iofile/ ior,iow                                     par 3
      common /tbeta/ beta,gamm                                    par 4
      real ct(1)                                                  par 5
      beta = ct(1)                                                par 6
      gamm = ct(2)                                                par 7
      if(beta.eq.0.0) beta = 0.25                                 par 8
      if(gamm.eq.0.0) gamm = 0.50                                 par 9
      write(iow,2000) beta,gamm                                   par 10
      if(ior.lt.0) write(*,2000) beta,gamm                        par 11
      return                                                      par 12
2000  format(' Newmark Method Parameters'/                        par 13
     1       ' beta = ',f9.4,' ;   gamma = ',f9.4)                par 14
      end                                                         par 15
c
      subroutine pform(ul,xl,tl,ld,p,s,ie,d,id,x,ix,f,t,idl,      pfo 1
     1                 f0,u,b,ndf,ndm,nen1,nst)                   pfo 2
c.... compute element arrays and assemble global arrays           pfo 3
      integer*2 ld(ndf,1),ie(9,1),id(ndf,1),ix(nen1,1),idl(1),ia,im  pfo 4
      real      xl(ndm,1),x(ndm,1),f(ndf,1),f0(ndf,1),tl(1),t(1)  pfo 5
      real*8    d(18,1),p(1),s(nst,1),b(1),ul(ndf,1),u(ndf,1),dm, pfo 6
     1          dun,un,prop                                       pfo 7
      logical afl,bfl,cfl,dfl,efl,hfl,hout                        pfo 8
      character*12 tfile                                          pfo 9
      common /cdata/ numnp,numel,nummat,nen,neq                   pfo 10
      common /eldata/ dq,n,ma,mct,iel,nel                         pfo 11
      common /mdat2/ n11a,n11b,n11c,ia(2,11)                      pfo 12
      common /hdata/ nh1,nh2                                      pfo 13
      common /hdatb/ nhi,nhf,ihbuff,irec,jrec,nrec,hfl,hout       pfo 14
      common /prlod/ prop,ap(6,10),iexp(10),ik(10),npld           pfo 15
```

```
      common /temfl1/ tfile(6)                                   pfo 16
      common /temfl2/ itrec(4),nw1,nw2                           pfo 17
      common /xdata/ isw,nn1,nn2,nn3,afl,bfl,cfl,dfl             pfo 18
      common dm(1),rm(1),im(1)                                   pfo 19
c.... set up local arrays before calling element library        pfo 20
      iel = 0                                                    pfo 21
      efl = .false.                                              pfo 22
      if(.not.dfl.and.isw.eq.6) efl = .true.                    pfo 23
      if(bfl.and.isw.eq.3)      efl = .true.                    pfo 24
      if(isw.ne.3.or.nn1.eq.1)  irec = 0                        pfo 25
      ne2 = nen + nen                                            pfo 26
      ne3 = nen + ne2                                            pfo 27
      numnp2 = numnp + numnp                                     pfo 28
      do 110 nu = 1,numel                                        pfo 29
       n = idl(nu)                                               pfo 30
       if( (n.ge.nn1 .and. n.le.nn2) .and. (mod(n-nn1,nn3).eq.0) ) then pfo 31
c.... set up history terms                                       pfo 32
         ma  = ix(nen1,n)                                        pfo 33
         nh1 = ix(nen+1,n)                                       pfo 34
         nh2 = nh1                                               pfo 35
         if(.not.hfl) then                                       pfo 36
           jrec= ix(nen+2,n)                                     pfo 37
           if(jrec.ne.irec) then                                pfo 38
             if(hout .and. irec.ne.0) then                      pfo 39
               call phstio(3,irec,dm(nhi),nhf-nhi+1,2,tfile(2),itrec(2))pfo 40
             endif                                               pfo 41
             call phstio(3,jrec,dm(nhi),nhf-nhi+1,1,tfile(2),itrec(2))  pfo 42
             irec = jrec                                         pfo 43
           endif                                                 pfo 44
         endif                                                   pfo 45
         call pconsd(ul,4*nen*ndf,0.0d0)                        pfo 46
         call pconsr(xl,nen*ndm,0.0)                            pfo 47
         call pconsr(tl,nen,0.0)                                pfo 48
         call pconsr(rm(n11a),nen,0.0)                          pfo 49
         call pconsi(ld,nst,0)                                   pfo 50
         un = 0.0                                                pfo 51
         dun= 0.0                                                pfo 52
         call pangl(ix(1,n),nen,rm(n11a),rm(n11b),nrot)         pfo 53
         do 108 i = 1,nen                                        pfo 54
           ixi= ix(i,n)                                          pfo 55
           ii = abs(ixi)                                         pfo 56
           if(ii.ne.0) go to 105                                 pfo 57
           do 104 j = 1,ndf                                      pfo 58
104        ld(j,i) = 0                                           pfo 59
           go to 108                                             pfo 60
105        iid = ii*ndf - ndf                                    pfo 61
           nel = i                                               pfo 62
           tl(i) = t(ii)                                         pfo 63
           do 106 j = 1,ndm                                      pfo 64
106        xl(j,i) = x(j,ii)                                     pfo 65
           do 107 j = 1,ndf                                      pfo 66
             jj = ie(j,ma)                                       pfo 67
             if(jj.le.0) go to 107                               pfo 68
             k = id(jj,ii)                                       pfo 69
             ul(j,i) = u(jj,ii)                                  pfo 70
             ul(j,i+nen) = u(jj,ii+numnp)                        pfo 71
             ul(j,i+ne2) = u(jj,ii+numnp2)                       pfo 72
             if(k.le.0) ul(j,i+ne3) = f0(jj,ii) + f(jj,ii)*prop  pfo 73
1                                      - u(jj,ii)                pfo 74
```

```
                dun = max(dun,abs(ul(j,i+ne3)))                   pfo 75
                un  = max( un,abs(ul(j,i)))                       pfo 76
                if(dfl) then                                      pfo 77
                  ld(j,i) = iid + jj                              pfo 78
                else                                              pfo 79
                  ld(j,i) = k                                     pfo 80
                  if(ixi.lt.0) ld(j,i) = -k                       pfo 81
                endif                                             pfo 82
107         continue                                             pfo 83
108       continue                                               pfo 84
c.... form element array                                         pfo 85
          if(ie(7,ma).ne.iel) mct = 0                            pfo 86
          iel = ie(7,ma)                                         pfo 87
          isx = isw                                              pfo 88
          if(efl .and. dun.gt.0.0000001d0*un .and. .not.afl) isx = 3   pfo 89
          if(nrot.gt.0)                                          pfo 90
     1    call ptrans(ia(1,iel),rm(n11a),ul,p,s,nel,nen,ndf,nst,1)    pfo 91
          call elmlib(d(1,ma),ul,xl,ix(1,n),tl,s,p,ndf,ndm,nst,iel,isx)   pfo 92
          if(nrot.gt.0)                                          pfo 93
     1    call ptrans(ia(1,iel),rm(n11a),ul,p,s,nel,nen,ndf,nst,2)    pfo 94
c.... modify for non-zero displacement boundary conditions       pfo 95
          if(efl) call modify(p,s,ul(1,ne3+1),nst)               pfo 96
c.... assemble a vector if needed                                pfo 97
          if(bfl) then                                           pfo 98
            do 109 i = 1,nst                                     pfo 99
              j = abs(ld(i,1))                                   pfo100
              if(j.ne.0) b(j) = b(j) + p(i)                      pfo101
109         continue                                             pfo102
          endif                                                  pfo103
        endif                                                    pfo104
110   continue                                                   pfo105
c.... put the last history state on the disk                     pfo106
      if(hout) call phstiu(3,jrec,dm(nhi),nhf-nhi+1,2,tfile(2),itrec(2))   pfo107
      return                                                     pfo108
      end                                                        pfo109
c
      subroutine pload(id,f,f0,b,nn,p,xm,ac)                     plo  1
c.... form load vector in compact form                           plo  2
      logical fl,pfr                                             plo  3
      integer*2 id(1)                                            plo  4
      real f(1),f0(1)                                            plo  5
      real*8 b(1),p,xm(1),ac(1)                                  plo  6
      common /fdata/ fl(11),pfr                                  plo  7
      fl(11) = .false.                                           plo  8
      call pconsd(b,nn,0.0d0)                                    plo  9
      do 100 n = 1,nn                                            plo 10
      j = id(n)                                                  plo 11
      if(j.gt.0) then                                            plo 12
        b(j) = f(n)*p + f0(n) + b(j)                             plo 13
        if(fl(9)) b(j) = b(j) - xm(j)*ac(j)                      plo 14
      endif                                                      plo 15
100   continue                                                   plo 16
      return                                                     plo 17
      end                                                        plo 18
c
      function propld(t,j)                                       pro  1
c.... proportional load table (j load cards, maximum 10)         pro  2
      real*8 prop                                                pro  3
      character*80 yyy                                           pro  4
```

```
      common /iofile/ ior,iow                                       pro   5
      common /prlod/ prop,a(6,10),iexp(10),ik(10),npld              pro   6
      if(j.le.0) go to 200                                          pro   7
c.... input table of proportional loads                            pro   8
      write(iow,2000)                                               pro   9
      if(ior.lt.0) then                                             pro  10
        write( *,2000)                                              pro  11
        write( *,2003)                                              pro  12
      endif                                                         pro  13
      do 100 i=1,j                                                  pro  14
101   call pintio(yyy,10)                                           pro  15
      read(yyy,1000,err=102) ik(i),iexp(i),(a(m,i),m=1,6)           pro  16
      go to 103                                                     pro  17
c.... error message                                                pro  18
102   call perror('PROPLD',yyy)                                     pro  19
      go to 101                                                     pro  20
c.... set a default ramp table if a type "0" input                 pro  21
103   if(ik(i).eq.0) then                                          pro  22
        a(2,i) = 1.e+6                                              pro  23
        a(4,i) = 1.                                                 pro  24
      endif                                                         pro  25
      write(iow,2001) i,ik(i),(a(m,i),m=1,6),iexp(i)                pro  26
      if(ior.lt.0) write(*,2001) i,ik(i),(a(m,i),m=1,6),iexp(i)     pro  27
100   continue                                                      pro  28
      nprop = j                                                     pro  29
c.... compute value at time t                                      pro  30
200   propld = 0.0                                                  pro  31
      do 220 i = 1,nprop                                            pro  32
        tmin = a(1,i)                                               pro  33
        tmax = a(2,i)                                               pro  34
        if(t.lt.tmin.or.t.gt.tmax) go to 220                        pro  35
        l = max(iexp(i),1)                                          pro  36
        propld = a(3,i)+a(4,i)*t+ a(5,i)*(sin(a(6,i)*t+tmin))**l    pro  37
     1         + propld                                             pro  38
220   continue                                                      pro  39
      return                                                        pro  40
c.... formats                                                       pro  41
1000  format(2i10,6f10.0)                                           pro  42
2000  format(30x,'P r o p o r t i o n a l   L o a d   T a b l e')   pro  43
2001  format(/,' number     type        tmin',10x,'tmax',/,i3,i10,7x,g10.4, pro  44
     1 4x,g10.4,/6x,'a(1)',10x,'a(2)',10x,'a(3)',10x,'a(4)',10x,    pro  45
     2 'exp',/4x,4(g10.4,4x),i5/)                                   pro  46
2003  format(' Input: type, exponent, tmin, tmax, a(i),i=1,4'/' >',$) pro  47
      end                                                           pro  48
c
      subroutine prtdis(x,b,ttim,prop,ndm,ndf,n1,n2,n3)             prt   1
c.... output nodal displacement values                             prt   2
      real       x(ndm,1)                                           prt   3
      real*8     b(ndf,1),prop                                      prt   4
      character*4 cd*6,di*6                                         prt   5
      common /iofile/ ior,iow                                       prt   6
      data cd/' coord'/,di/' displ'/                                prt   7
      kount = 0                                                     prt   8
      do 100 n = n1,n2,n3                                           prt   9
      kount = kount - 1                                             prt  10
      if(kount.le.0) then                                           prt  11
      call prthed(iow)                                              prt  12
      write(iow,2000) ttim,prop,(i,cd,i=1,ndm),(i,di,i=1,ndf)       prt  13
      if(ior.lt.0) write(*,2000) ttim,prop,(i,cd,i=1,ndm),(i,di,i=1,ndf)prt 14
```

```
      kount = 48                                                     prt 15
      endif                                                          prt 16
      if(x(1,n).ne. -999.) then                                      prt 17
        write(iow,2001) n,(x(i,n),i=1,ndm),(b(i,n),i=1,ndf)          prt 18
        if(ior.lt.0) write(*,2001) n,(x(i,n),i=1,ndm),(b(i,n),i=1,ndf) prt 19
      else                                                           prt 20
        write(iow,2002) n                                            prt 21
        if(ior.lt.0) write(*,2002) n                                 prt 22
      endif                                                          prt 23
100   continue                                                       prt 24
      return                                                         prt 25
2000  format(' N o d a l   D i s p l a c e m e n t s',5x,            prt 26
     1 'time',e18.5/31x,'prop. Id. (eigenvalue)',e13.5//            prt 27
     2 ' node',9(i7,a6))                                            prt 28
2001  format(i6,1p9e13.6)                                            prt 29
2002  format(i6,' not input.')                                       prt 30
      end                                                            prt 31
c
      subroutine prterr                                              prt  1
      implicit real*8 (a-h,o-z)                                      prt  2
      common /iofile/ ior,iow                                        prt  3
      common /errind/ eerror,elproj,ecproj,efem,enerr,ebar           prt  4
c.... output the error indicator values                             prt  5
      elrind = 0.0                                                   prt  6
      ecrind = 0.0                                                   prt  7
      if(elproj.ne.0.0d0) elrind = sqrt((efem-elproj)/elproj)        prt  8
      if(ecproj.ne.0.0d0) ecrind = sqrt((efem-ecproj)/ecproj)        prt  9
      if(ecproj.ne.0.0d0) eerror = sqrt(eerror/ecproj)               prt 10
      write(iow,2000) efem,elproj,ecproj,elrind,ecrind,eerror        prt 11
      if(ior.lt.0)write(*,2000) efem,elproj,ecproj,elrind,ecrind,eerror prt 12
2000  format(/'  Finite Element Stress Measure      =',e15.8/        prt 13
     1       '  L u m p e d   Projected Stress Measure =',e15.8/     prt 14
     2       '  Consistent   Projected Stress Measure =',e15.8/      prt 15
     3       '  L u m p e d   Error Indicator      =',e15.8/         prt 16
     4       '  Consistent   Error Indicator      =',e15.8/          prt 17
     5       '  Direct Error Indicator      =',e15.8/)               prt 18
      return                                                         prt 19
      end                                                            prt 20
c
      subroutine prtrea(r,ndf,numnp,n1,n2,n3)                        prt  1
c.... print nodal reactions                                         prt  2
      real*8     rr(6),r(ndf,1),rsum(6),asum(6),psum(6)              prt  3
      common /iofile/ ior,iow                                        prt  4
      call pconsd(rsum,ndf,0.0d0)                                    prt  5
      call pconsd(psum,ndf,0.0d0)                                    prt  6
      call pconsd(asum,ndf,0.0d0)                                    prt  7
      do 75 i = 1,numnp                                              prt  8
      do 75 k = 1,ndf                                                prt  9
      rsum(k) = rsum(k) - r(k,i)                                     prt 10
75    asum(k) = asum(k) + abs(r(k,i))                                prt 11
      kount = 0                                                      prt 12
      do 100 n = n1,n2,n3                                            prt 13
      kount = kount - 1                                              prt 14
      if(kount.le.0) then                                            prt 15
        call prthed(iow)                                             prt 16
        write(iow,2000) (k,k=1,ndf)                                  prt 17
        if(ior.lt.0) write(*,2000) (k,k=1,ndf)                       prt 18
        kount = 50                                                   prt 19
      endif                                                          prt 20
```

```
      do 80 k = 1,ndf                                              prt 21
      rr(k) = -r(k,n)                                              prt 22
80    psum(k) = psum(k) + rr(k)                                    prt 23
      write(iow,2001) n, (rr(k),k=1,ndf)                           prt 24
      if(ior.lt.0) write(*,2001) n, (rr(k),k=1,ndf)                prt 25
100   continue                                                     prt 26
c.... print statics check                                         prt 27
      write(iow,2002) (rsum(k),k=1,ndf)                            prt 28
      write(iow,2003) (psum(k),k=1,ndf)                            prt 29
      write(iow,2004) (asum(k),k=1,ndf)                            prt 30
      if(ior.lt.0) then                                            prt 31
        write(*,2002) (rsum(k),k=1,ndf)                            prt 32
        write(*,2003) (psum(k),k=1,ndf)                            prt 33
        write(*,2004) (asum(k),k=1,ndf)                            prt 34
      endif                                                        prt 35
      return                                                       prt 36
2000  format(' N o d a l    R e a c t i o n s'//6x,'node',6(i9,' dof'))prt 37
2001  format(i10,6e13.4)                                           prt 38
2002  format(/7x,'sum',6e13.4)                                     prt 39
2003  format( 3x,'prt sum',6e13.4)                                 prt 40
2004  format( 3x,'abs sum',6e13.4)                                 prt 41
      end                                                          prt 42
c
      subroutine prtstr(dt,ds,numnp,n1,n2,n3)                      prt  1
c.... output projected nodal stress values                        prt  2
      real dt(numnp),ds(numnp,7)                                   prt  3
      common /iofile/ ior,iow                                      prt  4
      kount = 0                                                    prt  5
      do 200 n = n1,n2,n3                                          prt  6
      kount = kount - 1                                            prt  7
      if(kount.le.0) then                                          prt  8
        call prthed(iow)                                           prt  9
        write(iow,2000)                                            prt 10
        if(ior.lt.0) write(*,2000)                                 prt 11
        kount = 17                                                 prt 12
      endif                                                        prt 13
      write(iow,2001) n, (ds(n,i),i=1,7),dt(n)                     prt 14
      if(ior.lt.0) write(*,2001) n, (ds(n,i),i=1,7),dt(n)          prt 15
200   continue                                                     prt 16
      return                                                       prt 17
2000  format(' N o d a l    S t r e s s e s'//' node',4x,'11-stress',4x,prt 18
     1 '12-stress',4x,'22-stress',4x,'33-stress'/10x,'1-stress',5x, prt 19
     2 '2-stress',4x,'max-shear',8x,'angle')                      prt 20
2001  format(i5,4e13.5/5x,4e13.5/1x)                               prt 21
      end                                                          prt 22
c
      subroutine ptrans(ia,angl,ul,p,s,nel,nen,ndf,nst,isw)        ptr  1
c.... subroutine to make two-dimesional rotations                 ptr  2
      integer*2 ia(2)                                              ptr  3
      real      angl(1)                                            ptr  4
      real*8    ul(ndf,nen,1),p(ndf,1),s(nst,nst),ang,cs,sn,tm     ptr  5
c.... recover dof to be rotated                                   ptr  6
      ij1 = ia(1)                                                  ptr  7
      ij2 = ia(2)                                                  ptr  8
      go to (1,2), isw                                             ptr  9
c.... transform the displacement quantities to element coordinates ptr 10
1     i1 = 0                                                       ptr 11
      do 110 i = 1,nel                                             ptr 12
        if(angl(i).ne.0.0) then                                   ptr 13
          ang = angl(i)*3.1415926d0/180.                          ptr 14
```

```
              cs  = cos(ang)                                         ptr 15
              sn  = sin(ang)                                         ptr 16
              do 100 j = 1,4                                         ptr 17
                tm          = cs*ul(ij1,i,j) - sn*ul(ij2,i,j)        ptr 18
                ul(ij2,i,j) = sn*ul(ij1,i,j) + cs*ul(ij2,i,j)        ptr 19
                ul(ij1,i,j) = tm                                     ptr 20
100           continue                                              ptr 21
            endif                                                   ptr 22
110     i1 = i1 + ndf                                               ptr 23
        return                                                      ptr 24
c.... transform the element arrays to global coordinates            ptr 25
2       i1 = 0                                                      ptr 26
        do 220 i = 1,nel                                            ptr 27
          if(angl(i).ne.0.0) then                                   ptr 28
            ang = angl(i)*3.1415926d0/180.                          ptr 29
            cs  = cos(ang)                                          ptr 30
            sn  = sin(ang)                                          ptr 31
c.... transform load vector                                         ptr 32
                tm      = cs*p(ij1,i) + sn*p(ij2,i)                 ptr 33
                p(ij2,i) =-sn*p(ij1,i) + cs*p(ij2,i)               ptr 34
                p(ij1,i) = tm                                       ptr 35
c.... postmultiply s by the transformation                          ptr 36
              do 210 j = 1,nst                                      ptr 37
                tm          = s(j,i1+ij1)*cs + s(j,i1+ij2)*sn       ptr 38
                s(j,i1+ij2)=-s(j,i1+ij1)*sn + s(j,i1+ij2)*cs        ptr 39
                s(j,i1+ij1)= tm                                     ptr 40
210           continue                                              ptr 41
c.... premultiply s by the transformation                           ptr 42
              do 215 j = 1,nst                                      ptr 43
                tm          = cs*s(i1+ij1,j) + sn*s(i1+ij2,j)       ptr 44
                s(i1+ij2,j)=-sn*s(i1+ij1,j) + cs*s(i1+ij2,j)        ptr 45
                s(i1+ij1,j)= tm                                     ptr 46
215           continue                                              ptr 47
            endif                                                   ptr 48
220     i1 = i1 + ndf                                               ptr 49
        return                                                      ptr 50
        end                                                         ptr 51
c
        subroutine setci(ior)                                       set  1
c.... compute integration constants 'c1' to 'c5' for current 'dt'   set  2
        common /tbeta/ beta,gamm                                    set  3
        common /tdata/ ttim,dt,c1,c2,c3,c4,c5,c6                    set  4
        if(dt.le.0.0 .or. beta.le.0.0) then                        set  5
          write(*,2000) dt,beta                                     set  6
          if(ior.gt.0) stop                                         set  7
          return                                                    set  8
        endif                                                       set  9
c.... compute integration constants 'c1' to 'c5' for current 'dt'   set 10
        c1 = 1.d0/(beta*dt*dt)                                      set 11
        c2 = gamm/(dt*beta)                                         set 12
        c3 = 1.d0 - 1.d0/(beta+beta)                                set 13
        c4 = 1.d0 - gamm/beta                                       set 14
        c5 = (1.d0 - gamm/(beta+beta))*dt                           set 15
        c6 = dt*c1                                                  set 16
        return                                                      set 17
2000    format(' ** ERROR ** A dynamic solution parameter is zero'/  set 18
     1          '             dt = ',1pe10.3,' , beta = ',0p1f9.4/   set 19
     2          '             Reinput dt or beta as nonzero number.')  set 20
        end                                                         set 21
```

```
c
      subroutine update(id,f0,f,u,v,a,du,nneq,neq,fdyn,prop,isw)      upd  1
c.... update the displacements (and velocities and accelerations)     upd  2
      logical fdyn                                                    upd  3
      integer*2 id(1)                                                 upd  4
      real      f0(1),f(1)                                            upd  5
      real*8    u(nneq,1),v(1),a(1),du(1),ur1,ur2,prop,dot            upd  6
      common /iofile/ ior,iow                                         upd  7
      common /tdata/ ttim,dt,c1,c2,c3,c4,c5,c6                        upd  8
c.... update solution vectors to begin a step                        upd  9
      if(isw.eq.1) then                                               upd 10
        ur1 = sqrt(dot(v,v,neq))                                      upd 11
        ur2 = sqrt(dot(a,a,neq))                                      upd 12
        write(iow,2000) ur1,ur2                                       upd 13
        if(ior.lt.0) write(*,2000) ur1,ur2                            upd 14
        do 100 n = 1,neq                                              upd 15
          ur2  = - c6*v(n) + c3*a(n)                                  upd 16
          v(n) =   c4*v(n) + c5*a(n)                                  upd 17
          a(n) =   ur2                                                upd 18
100     continue                                                      upd 19
      elseif(isw.eq.2) then                                           upd 20
c.... update displacement and its increments within the time step     upd 21
        do 200 n = 1,nneq                                             upd 22
          j = id(n)                                                   upd 23
          if (j.gt.0) then                                            upd 24
c.... for the active degrees-of-freedom compute values from solution  upd 25
            u(n,1) = du(j) + u(n,1)                                   upd 26
            u(n,2) = du(j) + u(n,2)                                   upd 27
            u(n,3) = du(j)                                            upd 28
          else                                                        upd 29
c.... for the fixed degrees-of-freedom compute values from forced inputsupd 30
            ur1    = f0(n) + f(n)*prop                                upd 31
            u(n,3) = ur1 - u(n,1)                                     upd 32
            u(n,2) = ur1 - u(n,1) + u(n,2)                            upd 33
            u(n,1) = ur1                                              upd 34
          endif                                                       upd 35
200     continue                                                      upd 36
c.... for time dependent solutions update the rate terms              upd 37
        if(fdyn) then                                                 upd 38
          do 210 n = 1,neq                                            upd 39
            v(n) = v(n) + c2*du(n)                                    upd 40
            a(n) = a(n) + c1*du(n)                                    upd 41
210       continue                                                    upd 42
        endif                                                         upd 43
      endif                                                           upd 44
      return                                                          upd 45
2000  format('  N o r m s   f o r   D y n a m i c s'/                 upd 46
     1   10x,'Velocity:',e13.5,' Acceleration:',e13.5)                upd 47
      end                                                             upd 48
```

(c) File PCMAC3.FOR contains the following set of subprograms (routines in this file may *not* be in an overlay):

Name	*Type*	*Description*
ACHECK	SUBROUTINE	Parser for alphanumeric data
CKISOP	SUBROUTINE	Check isoparametric element for errors
DOT	FUNCTION	Vector dot product (DOUBLE PRECISION)
ELMLIB	SUBROUTINE	Element library
PCONSD	SUBROUTINE	Set double precision vector to constant
PCONSI	SUBROUTINE	Set integer vector to constant
PCONSR	SUBROUTINE	Set real vector to constant
PDISK	SUBROUTINE	Add disk character to filename (c:, etc.)
PEND	SUBROUTINE	Print error on read of end file
PERROR	SUBROUTINE	Print for read error
PGAUSS	SUBROUTINE	Gauss points/weights for two-dimensional problems
PHELP	SUBROUTINE	Help information prints
PSETM	SUBROUTINE	Data management with byte alignments
PSTRES	SUBROUTINE	Principal stress determination for two dimensions
SAXPB	SUBROUTINE	Scalar times vector plus vector
SHAP2	SUBROUTINE	Quadratic shape function determination
SHAPE	SUBROUTINE	Linear shape function determination

```
$NOFLOATCALLS
        subroutine acheck(x,y,n0,nl)                          ach  1
c.... data parser                                            ach  2
        character*1 x(nl),y(nl)                               ach  3
        do 100 ii = nl,1,-1                                   ach  4
          if(x(ii).ne.' ') go to 110                          ach  5
100     continue                                             ach  6
110     do 150 i = 1,nl                                       ach  7
          y(i) = ' '                                          ach  8
150     continue                                             ach  9
        k = 0                                                ach 10
        il= 0                                                ach 11
        do 200 i = 1,ii                                       ach 12
        if(x(i).eq.',') then                                  ach 13
          k   = k + n0                                        ach 14
          if(k.gt.nl-n0) go to 210                            ach 15
          il = k - i                                          ach 16
        else                                                 ach 17
          y(i+il) = x(i)                                      ach 18
        endif                                                ach 19
200     continue                                             ach 20
        k   = k + n0                                          ach 21
210     call just(y,k,n0)                                     ach 22
        do 220 i = n0,nl,n0                                   ach 23
          if(y(i).eq.' ') y(i) = '0'                          ach 24
```

```
220   continue                                                    ach 25
      return                                                      ach 26
      end                                                         ach 27
c
      subroutine ckisop(ix,xl,shp,ndm)                            cki  1
c.... check isoparametric elements                                cki  2
      integer*2 xn(9),yn(9),ic(18),ix(1)                          cki  3
      real    xl(ndm,1)                                           cki  4
      real*8 shp(3,1),ss,tt,xsj                                   cki  5
      common /eldata/ dm,n,ma,mct,iel,nel                         cki  6
      common /iofile/ ior,iow                                     cki  7
      data xn/-1,1,1,-1,0,1,0,1,0/, yn/-1,-1,1,1,-1,0,1,0,0/      cki  8
c.... check the element for input errors                          cki  9
      ineg = 0                                                    cki 10
      do 100 l = 1,nel                                            cki 11
        if(xl(1,l).eq. -999.0 .and. ix(l).ne.0) then              cki 12
          ic(ineg+1) = l                                          cki 13
          ic(ineg+2) = abs(ix(l))                                 cki 14
          ineg = ineg + 2                                         cki 15
        endif                                                     cki 16
100   continue                                                    cki 17
      if(ineg.gt.0) then                                          cki 18
        write(iow,2000) n, (ic(i),i=1,ineg)                       cki 19
        if(ior.lt.0) write(*,2000) n, (ic(i),i=1,ineg)            cki 20
      else                                                        cki 21
        do 110 l = 1,nel                                          cki 22
          ss = xn(l)                                              cki 23
          tt = yn(l)                                              cki 24
          call   shape (ss,tt,xl,shp,xsj,ndm,nel,ix,.false.)      cki 25
          if(xsj.le.0.0d0) then                                   cki 26
            ic(ineg+1) = l                                        cki 27
            ic(ineg+2) = abs(ix(l))                               cki 28
            ineg = ineg + 2                                       cki 29
          endif                                                   cki 30
110     continue                                                  cki 31
        if(ineg.gt.0) then                                        cki 32
          write(iow,2001) n, (ic(i),i=1,ineg)                     cki 33
          if(ior.lt.0) write(*,2001) n, (ic(i),i=1,ineg)          cki 34
        endif                                                     cki 35
      endif                                                       cki 36
      return                                                      cki 37
2000  format(' >Element',i4,' coordinates not input for nodes:'/  cki 38
     1        ('            Local =',i3,' Global =',i4))           cki 39
2001  format(' >Element',i4,' has negative jacobian at nodes:'/   cki 40
     1        ('            Local =',i3,' Global =',i4))           cki 41
      end                                                         cki 42
c
      double precision function dot (a,b,n)
      real*8 a(1),b(1)
c.... dot product function
      dot = 0.0d0
      do 10 k=1,n
        dot = dot + a(k)*b(k)
10    continue
      return
      end
c
      subroutine elmlib(d,u,x,ix,t,s,p,i,j,k,iel,isw)             elm  1
c.... element library                                            elm  2
      real*8 d(1),p(1),s(1),u(1)                                 elm  3
```

```
      real    x(1),t(1)                                       elm  4
      integer*2 ix(1)                                         elm  5
      common /iofile/ ior,iow                                 elm  6
c.... total elmts loaded ----+                                elm  7
c                            V                                elm  8
      if(iel.ge.1.and.iel.le.4) then                          elm  9
        if(isw.ge.3.and.isw.le.6) then                        elm 10
          call pconsd(p,k,0.0d0)                              elm 11
          call pconsd(s,k*k,0.0d0)                            elm 12
        endif                                                 elm 13
        if(iel.eq.1) then                                     elm 14
          call elmt01(d,u,x,ix,t,s,p,i,j,k,isw)               elm 15
        elseif(iel.eq.2) then                                 elm 16
          call elmt02(d,u,x,ix,t,s,p,i,j,k,isw)               elm 17
        elseif(iel.eq.3) then                                 elm 18
          call elmt03(d,u,x,ix,t,s,p,i,j,k,isw)               elm 19
        elseif(iel.eq.4) then                                 elm 20
          call elmt04(d,u,x,ix,t,s,p,i,j,k,isw)               elm 21
        endif                                                 elm 22
      else                                                    elm 23
400   write(iow,4000) iel                                     elm 24
      if(ior.lt.0) write(*,4000) iel                          elm 25
      stop                                                    elm 26
      endif                                                   elm 27
      return                                                  elm 28
4000  format(' **ERROR** Element type number',i3,' found.')   elm 29
      end                                                     elm 30
c
      subroutine pconsd(v,nn,cc)                              pco  1
c.... zero real*8 array                                       pco  2
      real*8 v(nn),cc                                         pco  3
      do 100 n = 1,nn                                         pco  4
100   v(n) = cc                                               pco  5
      return                                                  pco  6
      end                                                     pco  7
c
      subroutine pconsi(iv,nn,ic)                             pco  1
c.... zero integer*2 array                                    pco  2
      integer*2 iv(nn),ic                                     pco  3
      do 100 n = 1,nn                                         pco  4
100   iv(n) = ic                                              pco  5
      return                                                  pco  6
      end                                                     pco  7
c
      subroutine pconsr(v,nn,cr)                              pco  1
c.... zero real array                                         pco  2
      real v(nn)                                              pco  3
      do 100 n = 1,nn                                         pco  4
100   v(n) = cr                                               pco  5
      return                                                  pco  6
      end                                                     pco  7
c
      subroutine pdisk(disk,files)                            pdi  1
c.... set disk name character                                pdi  2
      character*1 disk,files(1)                               pdi  3
      files(1) = disk                                         pdi  4
      return                                                  pdi  5
      end                                                     pdi  6
c
      subroutine pend(subnam)                                 pen  1
```

```
      character*6 subnam                                          pen  2
      common /iofile/ ior,iow                                     pen  3
      write(iow,4000) subnam                                      pen  4
      if(ior.lt.0) write(*,4000) subnam                           pen  5
      stop                                                        pen  6
4000  format(' ** ERROR in ',a6,' ** end of file encountered')    pen  7
      end                                                         pen  8
c
      subroutine perror(subnam,yy)                                per  1
      character*80 yy,subnam*6                                    per  2
      common /iofile/ ior,iow                                     per  3
      write(iow,4000) subnam,yy                                   per  4
      if(ior.gt.0) stop                                           per  5
      write(*,4000) subnam                                        per  6
      return                                                      per  7
4000  format(' ** ERROR in ',a6,' ** reinput last record:'/1x,a80) per 8
      end                                                         per  9
c
      subroutine pgauss(l,lint,r,z,w)                             pga  1
c.... gauss points and weights for two dimensions                pga  2
      integer*2 lr(9),lz(9),lw(9)                                 pga  3
      real*8 r(1),z(1),w(1),g,h                                   pga  4
      common /eldata/ dm,n,ma,mct,iel,nel                         pga  5
      data lr/-1,1,1,-1,0,1,0,-1,0/,lz/-1,-1,1,1,-1,0,1,0,0/      pga  6
      data lw/4*25,4*40,64/                                       pga  7
      lint = l*l                                                  pga  8
      go to (1,2,3),l                                             pga  9
c.... 1x1 integration                                            pga 10
1     r(1) = 0.                                                   pga 11
      z(1) = 0.                                                   pga 12
      w(1) = 4.                                                   pga 13
      return                                                      pga 14
c.... 2x2 integration                                            pga 15
2     g = 1.0/sqrt(3.d0)                                          pga 16
      do 21 i = 1,4                                               pga 17
      r(i) = g*lr(i)                                              pga 18
      z(i) = g*lz(i)                                              pga 19
21    w(i) = 1.                                                   pga 20
      return                                                      pga 21
c.... 3x3 integration                                            pga 22
3     g = sqrt(0.60d0)                                            pga 23
      h = 1.0/81.0d0                                              pga 24
      do 31 i = 1,9                                               pga 25
      r(i) = g*lr(i)                                              pga 26
      z(i) = g*lz(i)                                              pga 27
31    w(i) = h*lw(i)                                              pga 28
      return                                                      pga 29
      end                                                         pga 30
c
      subroutine phelp(wd,nwd,wrd,isw)                            phe  1
      character wd(nwd)*4,wrd*5                                   phe  2
      common /iofile/ ior,iow                                     phe  3
c.... help file for macro command list                           phe  4
      if(ior.gt.0) return                                         phe  5
      if(isw.eq.1) write(*,2000)                                  phe  6
      if(isw.ne.1) write(*,2001) wrd                              phe  7
      write(*,2002) wd                                            phe  8
      write(*,2003) wrd                                           phe  9
      return                                                      phe 10
2000  format(//' The following macro commands are available'//    phe 11
```

```
      1              ' use of loop must terminate with a matching next'//     phe 12
      2              ' multiple loop-next pairs may occur up to depth of 8')   phe 13
2001  format(//' The following ',a5,'commands are available:')                phe 14
2002  format(/8(3x,a4))                                                       phe 15
2003  format(/' Terminate ',a5,' execution with an "end" command.')           phe 16
      end                                                                     phe 17
c
      subroutine psetm(na,nl,np,afl)                                          pse  1
      logical afl                                                             pse  2
      common /iofile/ ior,iow                                                 pse  3
      common /psize/ maxm,ne                                                  pse  4
c.... set data management pointers for arrays                                 pse  5
      na = ne                                                                 pse  6
      ne = na + nl*np + mod(4 - mod(nl*np,4),4)                               pse  7
      na  = (na + np - 1)/np - max(0,6-(np-1)*4)                              pse  8
      afl = .false.                                                           pse  9
      amx = maxm                                                              pse 10
      amx = ne/amx                                                            pse 11
      if(amx.gt.0.90) write(*,1001) ne,maxm,amx                              pse 12
      if(ne.le.maxm) return                                                   pse 13
      write(iow,1000) ne,maxm                                                 pse 14
      if(ior.lt.0) write(*,1000) ne,maxm                                     pse 15
      stop                                                                    pse 16
1000  format(2x,'**ERROR** insufficient storage in blank common'/            pse 17
     1   10x,'required  =',i6/10x,'available =',i6/)                          pse 18
1001  format('   **Memory warning** used =',i6,' avail =',i6,' % =',f6.3)     pse 19
      end                                                                     pse 20
c
      subroutine pstres(sig,p1,p2,p3)                                         pst  1
c.... principal stresses (2 dimensions): sig = tau-xx,tau-xy,tau-yy           pst  2
      real*8 sig(3),p1,p2,p3,xi1,xi2,rho                                      pst  3
      xi1 = (sig(1) + sig(3))/2.                                              pst  4
      xi2 = (sig(1) - sig(3))/2.                                              pst  5
      rho = sqrt(xi2*xi2 + sig(2)*sig(2))                                     pst  6
      p1 = xi1 + rho                                                          pst  7
      p2 = xi1 - rho                                                          pst  8
      p3 = 45.0                                                               pst  9
      if(xi2.ne.0.0d0) p3 = 22.5*atan2(sig(2),xi2)/atan(1.0d0)               pst 10
      return                                                                  pst 11
      end                                                                     pst 12
c
      subroutine saxpb (a,b,x,n,c)                                            sax  1
      real*8 a(1),b(1),c(1),x                                                 sax  2
c... vector times scalar added to second vector                              sax  3
      do 10 k=1,n                                                             sax  4
        c(k) = a(k)*x +b(k)                                                   sax  5
   10 continue                                                                sax  6
      return                                                                  sax  7
      end                                                                     sax  8
c
      subroutine shap2(s,t,shp,ix,nel)                                        sha  1
c.... add quadratic functions as necessary                                   sha  2
      real*8    shp(3,1),s,t,s2,t2                                            sha  3
      integer*2 ix(1)                                                         sha  4
      s2 = (1.-s*s)/2.                                                        sha  5
      t2 = (1.-t*t)/2.                                                        sha  6
      do 100 i=5,9                                                            sha  7
      do 100 j = 1,3                                                          sha  8
100   shp(j,i) = 0.0                                                          sha  9
c.... midside nodes (serendipity)                                            sha 10
```

```
            if(ix(5).eq.0) go to 101                              sha 11
            shp(1,5) = -s*(1.-t)                                  sha 12
            shp(2,5) = -s2                                        sha 13
            shp(3,5) = s2*(1.-t)                                  sha 14
101         if(nel.lt.6) go to 107                                sha 15
            if(ix(6).eq.0) go to 102                              sha 16
            shp(1,6) = t2                                         sha 17
            shp(2,6) = -t*(1.+s)                                  sha 18
            shp(3,6) = t2*(1.+s)                                  sha 19
102         if(nel.lt.7) go to 107                                sha 20
            if(ix(7).eq.0) go to 103                              sha 21
            shp(1,7) = -s*(1.+t)                                  sha 22
            shp(2,7) = s2                                         sha 23
            shp(3,7) = s2*(1.+t)                                  sha 24
103         if(nel.lt.8) go to 107                                sha 25
            if(ix(8).eq.0) go to 104                              sha 26
            shp(1,8) = -t2                                        sha 27
            shp(2,8) = -t*(1.-s)                                  sha 28
            shp(3,8) = t2*(1.-s)                                  sha 29
c....   interior node (lagrangian)                               sha 30
104         if(nel.lt.9) go to 107                                sha 31
            if(ix(9).eq.0) go to 107                              sha 32
            shp(1,9) = -4.*s*t2                                   sha 33
            shp(2,9) = -4.*t*s2                                   sha 34
            shp(3,9) = 4.*s2*t2                                   sha 35
c....   correct edge nodes for interior node (lagrangian)        sha 36
            do 106 j= 1,3                                         sha 37
            do 105 i = 1,4                                        sha 38
105         shp(j,i) = shp(j,i) - 0.25*shp(j,9)                  sha 39
            do 106 i = 5,8                                        sha 40
106         if(ix(i).ne.0) shp(j,i) = shp(j,i) - .5*shp(j,9)     sha 41
c....   correct corner nodes for presense of midside nodes       sha 42
107         do 108 i = 1,4                                        sha 43
            k = mod(i+2,4) + 5                                   sha 44
            l = i + 4                                            sha 45
            do 108 j = 1,3                                       sha 46
108         shp(j,i) = shp(j,i) - 0.5*(shp(j,k)+shp(j,l))        sha 47
            return                                               sha 48
            end                                                  sha 49
c
            subroutine shape(ss,tt,xl,shp,xsj,ndm,nel,ix,flg)    sha  1
c....   shape function routine for two dimensional elements      sha  2
            logical flg                                          sha  3
            real    xl(ndm,1),s(4),t(4)                          sha  4
            real*8 shp(3,1),xs(2,2),sx(2,2),ss,tt,xsj,tp         sha  5
            integer*2 ix(1)                                      sha  6
            data s/-.5,.5,.5,-.5/, t/-.5,-.5,.5,.5/              sha  7
c....   form 4-node quadrilateral shape functions               sha  8
            do 100 i = 1,4                                        sha  9
              shp(3,i) = (0.5+s(i)*ss)*(0.5+t(i)*tt)             sha 10
              shp(1,i) = s(i)*(0.5+t(i)*tt)                      sha 11
              shp(2,i) = t(i)*(0.5+s(i)*ss)                      sha 12
100         continue                                             sha 13
            if(nel.eq.3) then                                    sha 14
c....   form triangle by adding third and fourth together        sha 15
              do 110 i = 1,3                                     sha 16
110           shp(i,3) = shp(i,3)+shp(i,4)                       sha 17
            endif                                                sha 18
c....   add quadratic terms if necessary                         sha 19
            if(nel.gt.4) call shap2(ss,tt,shp,ix,nel)            sha 20
```

```
c.... construct jacobian and its inverse                           sha 21
      do 125 i = 1,2                                               sha 22
      do 125 j = 1,2                                               sha 23
        xs(i,j) = 0.0                                              sha 24
        do 120 k = 1,nel                                          sha 25
          xs(i,j) = xs(i,j) + xl(i,k)*shp(j,k)                    sha 26
120     continue                                                  sha 27
125   continue                                                    sha 28
      xsj = xs(1,1)*xs(2,2)-xs(1,2)*xs(2,1)                       sha 29
      if(flg) return                                              sha 30
      if(xsj.le.0.0d0) xsj = 1.0                                  sha 31
      sx(1,1) = xs(2,2)/xsj                                       sha 32
      sx(2,2) = xs(1,1)/xsj                                       sha 33
      sx(1,2) =-xs(1,2)/xsj                                       sha 34
      sx(2,1) =-xs(2,1)/xsj                                       sha 35
c.... form global derivatives                                     sha 36
      do 130 i = 1,nel                                            sha 37
        tp        = shp(1,i)*sx(1,1)+shp(2,i)*sx(2,1)            sha 38
        shp(2,i)  = shp(1,i)*sx(1,2)+shp(2,i)*sx(2,2)            sha 39
        shp(1,i) = tp                                             sha 40
130   continue                                                    sha 41
      return                                                      sha 42
      end                                                         sha 43
```

15.8.4 *Equation solution modules.* The type of equation solution for each problem is controlled by the program system selected. One of the following solution systems must be chosen from files PASOLV.FOR or PFSOLV.FOR.

(*a*) File PASOLV.FOR contains the set of subprograms to solve the equations by the variable-band, profile method. The file consists of the following subprograms:

Name	Type	Description
PSOLVE	SUBROUTINE	Controls equation solution
DASBLY	SUBROUTINE	Assemble variable-band arrays
DASOL	SUBROUTINE	Solve equations
DATRI	SUBROUTINE	Triangular decomposition
DREDU	SUBROUTINE	Computes reduced diagonal in decomposition
PROFIL	SUBROUTINE	Computes equation numbers and profile

```
$NOFLOATCALLS
c
      subroutine psolve(u,a,b,dr,m,xm,s,ld,ig,idl,nst,nrs,afac,solv,     pso  1
     1                  dyn,c1,ipd,rnorm,aengy,ifl)                      pso  2
c...  active column assembly and solution of equations                  pso  3
      logical afac,solv,dyn,fl,fa                                       pso  4
      character*12 tfile                                                pso  5
      real*8 u(1),a(1),b(1),dr(1),xm(1),s(nst,1),aengy,rnorm,dot        pso  6
      integer*2 m(1),ld(1),ig(1),idl(1)                                 pso  7
      common /cdata/  numnp,numel,nummat,nen,neq                        pso  8
      common /fdata/  fl(11)                                            pso  9
      common /iofile/ ior,iow                                           pso 10
      common /temfl1/ tfile(6)                                          pso 11
      common /temfl2/ itrec(4),nw1,nw2                                  pso 12
      data fa/.false./                                                  pso 13
c.... form and assemble the matrix                                      pso 14
      if(afac) then                                                     pso 15
        if(fl(4)) then                                                  pso 16
          ibuf = ig(neq)+neq                                            pso 17
          if(ibuf.gt.8000) stop 'profile too large'                    pso 18
          itrec(1) = ibuf*8                                             pso 19
          open (4,file=tfile(1),status='new',access='direct',          pso 20
     1          form='unformatted',recl=itrec(1))                       pso 21
          close(4)                                                      pso 22
          fl(4) = fa                                                    pso 23
        endif                                                           pso 24
        call pconsd(a,ibuf,0.0d0)                                       pso 25
c.... modify tangent form lumped mass effects                           pso 26
        if(dyn) then                                                    pso 27
          do 310 n = 1,neq                                              pso 28
            a(n)  = c1*xm(n)                                            pso 29
310       continue                                                      pso 30
        endif                                                           pso 31
        do 320 n = 1,numel                                             pso 32
c...  compute and assemble element arrays                               pso 33
          ne = n                                                        pso 34
```

```
            call formfe(u,dr,.true.,solv,fa,fa,3,ne,ne,1)              pso 35
            if(ior.lt.0 .and. mod(n,20).eq.0) write(*,2000) n          pso 36
            call dasbly(s,s,ld,ig,nst,fa,afac,fa,dr,a(neq+1),a(neq+1),a)  pso 37
320         continue                                                   pso 38
            rnorm = sqrt(dot(dr,dr,neq))                               pso 39
            call datri(a(neq+1),a(neq+1),a,ig,neq,.false.)            pso 40
            call phstio(4,1,a,ibuf,2,tfile(1),itrec(1))               pso 41
          endif                                                        pso 42
          if(solv) then                                                pso 43
            if(.not.afac) call phstio(4,1,a,ibuf,1,tfile(1),itrec(1))  pso 44
            do 330 n = 1,nrs                                           pso 45
              ne = (n-1)*neq + 1                                       pso 46
              call dasol(a(neq+1),a(neq+1),a,dr(ne),ig,neq, aengy)    pso 47
330         continue                                                   pso 48
          endif                                                        pso 49
          return                                                       pso 50
2000      format(5x,'**',i4,' elements completed.')                    pso 51
2001      format(i4,' Elmts completed.')                               pso 52
          end                                                          pso 53
c
      subroutine dasbly(s,p,ld,jp,ns,alfl,aufl,bfl,  b,al,au,ad)       das  1
      implicit real*8 (a-h,o-z)                                        das  2
c.... assemble the symmetric or unsymmetric arrays for 'dasol'        das  3
      logical alfl,aufl,bfl                                            das  4
      integer*2 ld(ns),jp(1)                                           das  5
      real*8 al(1),au(1),ad(1),b(1),s(ns,ns),p(ns)                     das  6
c.... loop through the rows to perform the assembly                    das  7
      do 200 i = 1,ns                                                  das  8
        ii = ld(i)                                                     das  9
        if(ii.gt.0) then                                               das 10
          if(aufl) then                                                das 11
c.... loop through the columns to perform the assembly                 das 12
            do 100 j = 1,ns                                            das 13
              if(ld(j).eq.ii) then                                     das 14
                ad(ii) = ad(ii) + s(i,j)                               das 15
              elseif(ld(j).gt.ii) then                                 das 16
                jc = ld(j)                                             das 17
                jj = ii + jp(jc) - jc + 1                              das 18
                au(jj) = au(jj) + s(i,j)                               das 19
                if(alfl) al(jj) = al(jj) + s(j,i)                      das 20
              endif                                                    das 21
100         continue                                                   das 22
          endif                                                        das 23
          if(bfl) b(ii)  = b(ii)  + p(i)                               das 24
        endif                                                          das 25
200   continue                                                         das 26
      return                                                           das 27
      end                                                              das 28
c
      subroutine dasol(al,au,ad,b,jp,neq, energy)                      das  1
      implicit real*8 (a-h,o-z)                                        das  2
c.... solution of symmetric equations stored in profile form          das  3
c.... coefficient matrix must be decomposed into its triangular        das  4
c.... factors using datri before using dasol.                          das  5
      integer*2 jp(1)                                                  das  6
      real*8 al(1),au(1),ad(1),b(1)                                    das  7
      common /iofile/ ior,iow                                          das  8
      data zero/0.0d0/                                                 das  9
c.... find the first nonzero entry in the right hand side              das 10
      do 100 is = 1,neq                                                das 11
        if(b(is).ne.zero) go to 200                                    das 12
```

```
100     continue                                                     das 13
        write(iow,2000)                                              das 14
        if(ior.lt.0) write(*,2000)                                   das 15
        return                                                       das 16
200     if(is.lt.neq) then                                           das 17
c....   reduce the right hand side                                   das 18
        do 300 j = is+1,neq                                          das 19
          jr = jp(j-1)                                               das 20
          jh = jp(j) - jr                                            das 21
          if(jh.gt.0) then                                           das 22
            b(j) = b(j) - dot(al(jr+1),b(j-jh),jh)                   das 23
          endif                                                      das 24
300     continue                                                     das 25
        endif                                                        das 26
c....   multiply by inverse of diagonal elements                     das 27
        energy = zero                                                das 28
        do 400 j = is,neq                                            das 29
          bd = b(j)                                                  das 30
          b(j) = b(j)*ad(j)                                          das 31
          energy = energy + bd*b(j)                                  das 32
400     continue                                                     das 33
c....   backsubstitution                                             das 34
        if(neq.gt.1) then                                            das 35
        do 500 j = neq,2,-1                                          das 36
          jr = jp(j-1)                                               das 37
          jh = jp(j) - jr                                            das 38
          if(jh.gt.0) then                                           das 39
            call saxpb(au(jr+1),b(j-jh),-b(j),jh, b(j-jh))           das 40
          endif                                                      das 41
500     continue                                                     das 42
        endif                                                        das 43
        return                                                       das 44
2000    format(' ***DASOL WARNING 1*** Zero right-hand-side vector') das 45
        end                                                          das 46
        subroutine datest(au,jh,daval)                               dat  1
        implicit real*8 (a-h,o-z)                                    dat  2
        real*8 au(jh)                                                dat  3
c....   test for rank                                                dat  4
        daval = 0.0d0                                                dat  5
        do 100 j = 1,jh                                              dat  6
          daval = daval + abs(au(j))                                 dat  7
100     continue                                                     dat  8
        return                                                       dat  9
        end                                                          dat 10
c
        subroutine datri(al,au,ad,jp,neq,flg)                        dat  1
        implicit real*8 (a-h,o-z)                                    dat  2
c....   triangular decomposition of a matrix stored in profile form  dat  3
        logical flg                                                  dat  4
        integer*2 jp(1)                                              dat  5
        real*8 al(1),au(1),ad(1)                                     dat  6
        common /iofile/ ior,iow                                      dat  7
c....   n.b.  tol should be set to approximate half-word precision.  dat  8
        data zero,one/0.0d0,1.0d0/, tol/0.5d-07/                     dat  9
c....   set initial values for conditioning check                    dat 10
        dimx = zero                                                  dat 11
        dimn = zero                                                  dat 12
        do 50 j = 1,neq                                              dat 13
        dimn = max(dimn,abs(ad(j)))                                  dat 14
50      continue                                                     dat 15
```

```
      dfig = zero                                                     dat 16
c.... loop through the columns to perform the triangular decomposition  dat 17
      jd = 1                                                          dat 18
      do 200 j = 1,neq                                                dat 19
        jr = jd + 1                                                   dat 20
        jd = jp(j)                                                    dat 21
        jh = jd - jr                                                  dat 22
        if(jh.gt.0) then                                              dat 23
          is = j - jh                                                 dat 24
          ie = j - 1                                                  dat 25
c.... if diagonal is zero compute a norm for singularity test          dat 26
          if(ad(j).eq.zero) call datest(au(jr),jh,daval)              dat 27
          do 100 i = is,ie                                            dat 28
            jr = jr + 1                                               dat 29
            id = jp(i)                                                dat 30
            ih = min(id-jp(i-1),i-is+1)                               dat 31
            if(ih.gt.0) then                                          dat 32
              jrh = jr - ih                                           dat 33
              idh = id - ih + 1                                       dat 34
              au(jr) = au(jr) - dot(au(jrh),al(idh),ih)               dat 35
              if(flg) al(jr) = al(jr) - dot(al(jrh),au(idh),ih)       dat 36
            endif                                                     dat 37
100         continue                                                  dat 38
        endif                                                         dat 39
c.... reduce the diagonal                                              dat 40
        if(jh.ge.0) then                                              dat 41
          dd = ad(j)                                                  dat 42
          jr = jd - jh                                                dat 43
          jrh = j - jh - 1                                            dat 44
          call dredu(al(jr),au(jr),ad(jrh),jh+1,flg  ,ad(j))          dat 45
c.... check for possible errors and print warnings                     dat 46
          if(abs(ad(j)).lt.tol*abs(dd))  write(iow,2000) j            dat 47
          if(dd.lt.zero.and.ad(j).gt.zero) write(iow,2001) j          dat 48
          if(dd.gt.zero.and.ad(j).lt.zero) write(iow,2001) j          dat 49
          if(ad(j) .eq.  zero)             write(iow,2002) j          dat 50
          if(dd.eq.zero.and. jh.gt.0) then                            dat 51
            if(abs(ad(j)).lt.tol*daval)    write(iow,2003) j          dat 52
          endif                                                       dat 53
          if(ior.lt.0) then                                           dat 54
            if(abs(ad(j)).lt.tol*abs(dd))  write(*,2000) j            dat 55
            if(dd.lt.zero.and.ad(j).gt.zero) write(*,2001) j          dat 56
            if(dd.gt.zero.and.ad(j).lt.zero) write(*,2001) j          dat 57
            if(ad(j) .eq.  zero)             write(*,2002) j          dat 58
            if(dd.eq.zero.and. jh.gt.0) then                          dat 59
              if(abs(ad(j)).lt.tol*daval)    write(*,2003) j          dat 60
            endif                                                     dat 61
          endif                                                       dat 62
        endif                                                         dat 63
c.... store reciprocal of diagonal, compute condition checks           dat 64
        if(ad(j).ne.zero) then                                        dat 65
          dimx  = max(dimx,abs(ad(j)))                                dat 66
          dimn  = min(dimn,abs(ad(j)))                                dat 67
          dfig  = max(dfig,abs(dd/ad(j)))                             dat 68
          ad(j) = one/ad(j)                                           dat 69
        endif                                                         dat 70
200   continue                                                        dat 71
c.... print conditioning information                                   dat 72
      dd = zero                                                       dat 73
      if(dimn.ne.zero) dd = dimx/dimn                                 dat 74
      ifig = dlog10(dfig) + 0.6                                       dat 75
```

```
      write(iow,2004) dimx,dimn,dd,ifig                              dat 76
      if(ior.lt.0) write(*,2004) dimx,dimn,dd,ifig                   dat 77
      return                                                         dat 78
c.... formats                                                        dat 79
2000  format(' ***DATRI WARNING 1*** Loss of at least 7 digits in', dat 80
     1 ' reducing diagonal of equation',i5)                         dat 81
2001  format(' ***DATRI WARNING 2*** Sign of diagonal changed when',dat 82
     1 ' reducing equation',i5)                                     dat 83
2002  format(' ***DATRI WARNING 3*** Reduced diagonal is zero for', dat 84
     1 ' equation',i5)                                              dat 85
2003  format(' ***DATRI WARNING 4*** Rank failure for zero unreduced',dat 86
     1 ' diagonal in equation',i5)                                  dat 87
2004  format(' Condition check: D-max',e11.4,'; D-min',e11.4,       dat 88
     1 '; Ratio',e11.4/' Maximum no. diagonal digits lost:',i3)     dat 89
2005  format('Cond ck: Dmax',1p1e9.2,'; Dmin',1p1e9.2,'; Ratio',1p1e9.2)dat 90
      end                                                            dat 91
c
      subroutine dredu(al,au,ad,jh,flg  ,dj)                         dre  1
      implicit real*8 (a-h,o-z)                                      dre  2
c.... reduce diagonal element in triangular decomposition           dre  3
      logical flg                                                    dre  4
      real*8 al(jh),au(jh),ad(jh)                                    dre  5
      do 100 j = 1,jh                                                dre  6
        ud = au(j)*ad(j)                                             dre  7
        dj = dj - al(j)*ud                                           dre  8
        au(j) = ud                                                   dre  9
100   continue                                                       dre 10
c.... finish computation of column of al for unsymmetric matrices    dre 11
      if(flg) then                                                   dre 12
        do 200 j = 1,jh                                              dre 13
          al(j) = al(j)*ad(j)                                        dre 14
200     continue                                                     dre 15
      endif                                                          dre 16
      return                                                         dre 17
      end                                                            dre 18
c
      subroutine profil (jd,idl,id,ix,ndf,nen1)                      pro  1
      implicit real*8 (a-h,o-z)                                      pro  2
c.... compute profile of global arrays                              pro  3
      integer*2 jd(1),idl(1),id(ndf,1),ix(nen1,1)                    pro  4
      common /cdata/ numnp,numel,nummat,nen,neq                      pro  5
      common /frdata/ maxf                                           pro  6
      common /iofile/ ior,iow                                        pro  7
c.... set up the equation numbers                                   pro  8
      neq = 0                                                        pro  9
      nneq = ndf*numnp                                               pro 10
      do 10 n = 1,nneq                                               pro 11
        j = id(n,1)                                                  pro 12
        if(j.eq.0) then                                             pro 13
          neq = neq + 1                                              pro 14
          id(n,1) = neq                                              pro 15
        else                                                         pro 16
          id(n,1) = 0                                                pro 17
        endif                                                        pro 18
10    continue                                                       pro 19
c.... compute column heights                                        pro 20
      call pconsi(jd,neq,0)                                          pro 21
      do 50 n = 1,numel                                              pro 22
        mm  = 0                                                      pro 23
        nad = 0                                                      pro 24
```

```
      do 30 i = 1,nen                                         pro 25
        ii = iabs(ix(i,n))                                    pro 26
        if(ii.gt.0) then                                      pro 27
          do 20 j = 1,ndf                                     pro 28
            jj = id(j,ii)                                     pro 29
            if(jj.gt.0) then                                  pro 30
              if(mm.eq.0) mm = jj                             pro 31
              mm = min(mm,jj)                                 pro 32
              nad = nad + 1                                   pro 33
              idl(nad) = jj                                   pro 34
            endif                                             pro 35
20        continue                                            pro 36
        endif                                                 pro 37
30      continue                                              pro 38
        if(nad.gt.0) then                                     pro 39
          do 40 i = 1,nad                                     pro 40
            ii = idl(i)                                       pro 41
            jj = jd(ii)                                       pro 42
            jd(ii) = max(jj,ii-mm)                            pro 43
40        continue                                            pro 44
        endif                                                 pro 45
50    continue                                                pro 46
c.... compute diagonal pointers for profile                  pro 47
      nad = 0                                                 pro 48
      jd(1) = 0                                               pro 49
      if(neq.gt.1) then                                       pro 50
        do 60 n = 2,neq                                       pro 51
          jd(n) = jd(n) + jd(n-1)                             pro 52
60      continue                                              pro 53
        nad - jd(noq)                                         pro 54
      endif                                                   pro 55
c.... set element search order to sequential                 pro 56
      do 70 n = 1,numel                                       pro 57
        idl(n) = n                                            pro 58
70    continue                                                pro 59
c.... equation summary                                       pro 60
      maxf = 0                                                pro 61
      mm   = 0                                                pro 62
      if(neq.gt.0) mm = (nad+neq)/neq                         pro 63
      write(iow,2001) neq,numnp,mm,numel,nad,nummat           pro 64
      if(ior.lt.0) write(*,2001) neq,numnp,mm,numel,nad,nummat pro 65
      return                                                  pro 66
      end                                                     pro 67
```

(b) File PFSOLV.FOR contains the set of subprograms to solve the equation by the frontal method. The file consists of the following subprograms:

Name	Type	Description
PSOLVE	SUBROUTINE	Controls equation solution
PBUFF	SUBROUTINE	Disk input/output of front buffer array
PFRTAS	SUBROUTINE	Assemble front arrays
PFRTBK	SUBROUTINE	Backsubstitution solution
PFRTB	SUBROUTINE	Backsubstitution macro
PFRTD	SUBROUTINE	Triangular decomposition
PFRTFW	SUBROUTINE	Forward solution
PFRTF	SUBROUTINE	Forward solution macro
PREFRT	SUBROUTINE	Computes frontal equation order
PROFIL	SUBROUTINE	Computes equation numbers and front size

```
$NOFLOATCALLS
      subroutine psolve(u,a,b,dr,m,xm,s,ld,ig,idl,nst,nrs,afac,solv,    pso  1
     1                       dyn,c1,ipd,rnorm,aengy,ifl)                pso  2
c...  frontal assembly and solution of equations                       pso  3
      logical afac,solv,dyn,fl,fa                                       pso  4
      character*12 tfile                                                pso  5
      real*8 u(1),a(1),b(1),dr(1),xm(1),s(1),dimx,dimn,r,rnorm,aengy    pso  6
      integer*2 m(1),ld(1),ig(1),idl(1)                                pso  7
      common /cdata/  numnp,numel,nummat,nen,neq                       pso  8
      common /fdata/  fl(11)                                           pso  9
      common /frdata/ maxf                                             pso 10
      common /iofile/ ior,iow                                          pso 11
      common /nfrta/  dimx,dimn,nv,npl                                 pso 12
      common /temfl1/ tfile(6)                                         pso 13
      common /temfl2/ itrec(4),nw1,nw2                                 pso 14
      data fa/.false./                                                 pso 15
c...  set control data and zero                                        pso 16
      if(.not.afac) go to 400                                          pso 17
      if(fl(4)) then                                                   pso 18
        nal= 1 + (maxf*(maxf+3))/2                                     pso 19
        if(nal+maxf.gt.8000) stop 'front too large'                   pso 20
        ibuf = (min(8000 - nal,(maxf+2)*neq))*4                       pso 21
        itrec(1) = ibuf*2 + 4                                          pso 22
        open (4,file=tfile(1),status='new',access='direct',           pso 23
     1         form='unformatted',recl=itrec(1))                       pso 24
        close(4)                                                       pso 25
        fl(4) = fa                                                     pso 26
      endif                                                            pso 27
      call pconsd(a,maxf*(maxf+1)/2,0.0d0)                            pso 28
      call pconsd(b,maxf,0.0d0)                                        pso 29
      ig(1)=1                                                          pso 30
      ig(maxf+1)=0                                                     pso 31
      do 100 n=2,maxf                                                  pso 32
       ig(n)=ig(n-1)+n                                                 pso 33
       ig(maxf+n)=0                                                    pso 34
100   continue                                                         pso 35
```

```
c....  begin loop through elements to perform front solution          pso 36
       np  = 0                                                        pso 37
       nv  = 0                                                        pso 38
       nfrt= 0                                                        pso 39
       dimx  = 0.0                                                    pso 40
       dimn  = 0.0                                                    pso 41
       rnorm = 0.0                                                    pso 42
c...   frontal elimination program                                    pso 43
       do 320 n = 1,numel                                             pso 44
c...   pick next element                                              pso 45
       ne = idl(n)                                                    pso 46
c...   compute element arrays                                         pso 47
         call formfe(u,dr,.true.,solv,fa,fa,3,ne,ne,1)                pso 48
         if(ior.lt.0 .and. mod(n,20).eq.0) write(*,2000) n,nfrt       pso 49
c....  assemble element and determine eliminations                    pso 50
         call pfrtas(a,s,ld,ig,ig(maxf+1),ie,maxf,nfrt,nst)           pso 51
         if(ie.ne.0) then                                             pso 52
c...   eliminate equations                                            pso 53
           do 310 i=1,ie                                              pso 54
             k=ld(i)                                                  pso 55
             jj=-ig(maxf+k)                                           pso 56
             if(solv) then                                            pso 57
               rnorm = rnorm + dr(jj)**2                              pso 58
               r=b(k)+dr(jj)                                          pso 59
               dr(jj) = r                                             pso 60
             endif                                                    pso 61
             if(dyn) then                                             pso 62
               ii = ig(k)                                             pso 63
               a(ii) = a(ii) + c1**m(jj)                              pso 64
             endif                                                    pso 65
             if(np+8+ntrt*ipd.gt.ibuf) then                           pso 66
               call pbuff(m,ibuf,np,nv,2,ifl)                         pso 67
             endif                                                    pso 68
             m(np+1) = nfrt                                           pso 69
             m(np+2) = k                                              pso 70
             m(np+3) = jj                                             pso 71
             call pfrtd(a,b,r,ig,ig(maxf+1),solv,m(np+5),nfrt,k)      pso 72
c.........align last word of buffer for backsubstitution reads        pso 73
             np = np + 8 + nfrt*ipd                                   pso 74
             m(np) = nfrt                                             pso 75
             nfrt = nfrt - 1                                          pso 76
310          continue                                                 pso 77
         endif                                                        pso 78
c...   end of forward elimination and triangularization               pso 79
320    continue
       rnorm = sqrt(rnorm)
       rxm   = 0.0
       if(dimn.ne.0.0d0) rxm = dimx/dimn
       if(ior.lt.0) write(*,2001) dimx,dimn,rxm
       write(iow,2001) dimx,dimn,rxm
c....  clear buffer one last time
       if(np.gt.0) call pbuff(m,ibuf,np,nv,2,ifl)
       go to 500
c....  forward and back substitutions for solution
400    if(solv) call pfrtfw(b,dr,m,ipd,ibuf,maxf,nv,neq,nrs,ifl)
500    if(solv) call pfrtbk(b,dr,m,ipd,ibuf,maxf,nv,neq,nrs,aengy,ifl)
       return
2000   format(5x,'**',i4,' elements completed.  Current front width is',
     1            i4)
```

```
2001  format(5x,'Condition check: D-max',1p1e11.4,'; D-min',1p1e11.4,
     1         '; Ratio',1p1e11.4)
2002  format(i4,' Elmts completed.  Current front width is',i4)
2003  format('Cond ck: Dmax',1p1e9.2,'; Dmin',1p1e9.2,'; Ratio',1p1e9.2)
      end
c
      subroutine pbuff(m,ibuf,ilast,nv,is,ifl)                          pbu  1
c.... tape input/output routine for frontal program                    pbu  2
      integer*2 m(ibuf)                                                 pbu  3
      real*8    dimx,dimn                                               pbu  4
      character*12 tfile                                                pbu  5
      common /nfrta/ dimx,dimn,nvp,npl                                  pbu  6
      common /temfl1/ tfile(6)                                          pbu  7
      common /temfl2/ itrec(4),nw1,nw2                                  pbu  8
      open(4,file=tfile(ifl),access='direct',recl=itrec(ifl))          pbu  9
c.... read record 'nv' from the file                                   pbu 10
      if(is.eq.1) then                                                  pbu 11
        read(4,rec=nv,end=901,err=902)  ilast,m                        pbu 12
c.... write record 'nv' from the file                                  pbu 13
      elseif(is.eq.2) then                                             pbu 14
        nv = nv + 1                                                     pbu 15
        if(nv.eq.1) npl = ilast                                        pbu 16
        write(4,rec=nv,err=901) ilast,m                               pbu 17
        ilast = 0                                                       pbu 18
      endif                                                             pbu 19
      close(4)                                                          pbu 20
      return                                                            pbu 21
c.... error messages                                                   pbu 22
901   write(*,4000)                                                    pbu 23
      stop                                                              pbu 24
902   call pend('PBUFF ')                                              pbu 25
      stop                                                              pbu 26
4000  format(' ** ERROR IN PBUFF ** records do not match problem?')    pbu 27
      end                                                               pbu 28
c
      subroutine pfrtas(a,s,ld,ig,lg,ie,maxf,nfrt,nst)                  pfr  1
c.... assembly and elimination determination for frontal program       pfr  2
      real*8 a(1),s(nst,1)                                             pfr  3
      integer*2 ld(1),ig(1),lg(1)                                      pfr  4
c... convert ld to front order                                         pfr  5
      do 203 j=1,nst                                                   pfr  6
        ii=abs(ld(j))                                                   pfr  7
        i = 0                                                           pfr  8
c... check if ii is already in list                                   pfr  9
        if(ii.ne.0) then                                               pfr 10
          if(nfrt.ne.0) then                                           pfr 11
            do 200 i=1,nfrt                                             pfr 12
              if(ii.eq.abs(lg(i))) go to 202                           pfr 13
200         continue                                                   pfr 14
          endif                                                         pfr 15
c... assign ii to next available entry and increase front width        pfr 16
          nfrt = nfrt + 1                                               pfr 17
          i = nfrt                                                      pfr 18
c... replace destination value by new value                           pfr 19
202       lg(i)=ld(j)                                                   pfr 20
c... set ld for assembly                                               pfr 21
          ld(j)=i                                                       pfr 22
        endif                                                           pfr 23
203   continue                                                          pfr 24
c... assemble element into front                                       pfr 25
```

```
          do 205 j = 1,nst                                              pfr 26
            k = abs(ld(j))                                              pfr 27
            if(k.gt.0) then                                             pfr 28
              l = ig(k) - k                                             pfr 29
              do 204 i = 1,nst                                          pfr 30
                m = abs(ld(i))                                          pfr 31
                if(m.gt.0 .and. m.le.k) a(m+l) = a(m+l) + s(i,j)        pfr 32
204           continue                                                  pfr 33
            endif                                                       pfr 34
205       continue                                                      pfr 35
c...  set up equations to be eliminated                                 pfr 36
          ie=0                                                          pfr 37
          do 206 i = nfrt,1,-1                                          pfr 38
            if(lg(i).lt.0) then                                         pfr 39
              ie = ie + 1                                               pfr 40
              ld(ie) = i                                                pfr 41
            endif                                                       pfr 42
206       continue                                                      pfr 43
          return                                                        pfr 44
          end                                                           pfr 45
c
          subroutine pfrtbk(b,dr,m,ipd,ibuf,maxf,nv,neq,nev,aengy,ifl)  pfr  1
c....  backsubstitution for frontal solution                            pfr  2
          real*8 b(maxf,1),dr(neq,1),aengy,dimx,dimn                    pfr  3
          integer*2 m(1)                                                pfr  4
          common /nfrta/ dimx,dimn,nvp,npl                              pfr  5
          call pconsd(b,maxf*ncv,0.0d0)                                 pfr  6
          aengy = 0.0                                                   pfr  7
c....  recover block                                                    pfr  8
          np = npl                                                      pfr  9
          do 503 n = nv,1,-1                                            pfr 10
            if(nv.gt.1) call pbuff(m,ibuf,np,n,1,ifl)                   pfr 11
501         nfrt = m(np)                                                pfr 12
            np = np - 8 - nfrt*ipd                                      pfr 13
            k  = m(np+2)                                                pfr 14
            jj = m(np+3)                                                pfr 15
            do 502 i = 1,nev                                            pfr 16
              call pfrtb(b(1,i),dr(1,i),nfrt,k,jj,m(np+5),aengy)        pfr 17
502         continue                                                    pfr 18
            if(np.gt.0) go to 501                                       pfr 19
503       continue                                                      pfr 20
          return                                                        pfr 21
          end                                                           pfr 22
c
          subroutine pfrtb(b,dr,nfrt,k,jj,eq,aengy)                     pfr  1
c....  backsubstitution macro for frontal program                       pfr  2
          real*8 b(1),dr(1),eq(1),dot,aengy                             pfr  3
c...  expand b array                                                    pfr  4
          kk = nfrt                                                     pfr  5
500       if(kk.le.k) go to 501                                         pfr  6
          b(kk) = b(kk-1)                                               pfr  7
          kk = kk - 1                                                   pfr  8
          go to 500                                                     pfr  9
501       b(k) = 0.0                                                    pfr 10
c...  extract pivot and solve, also compute energy                      pfr 11
          aengy = aengy + dr(jj)**2/eq(k)                               pfr 12
          dr(jj)=(dr(jj)-dot(eq,b,nfrt))/eq(k)                          pfr 13
          b(k) = dr(jj)                                                 pfr 14
          return                                                        pfr 15
          end                                                           pfr 16
```

```
c
      subroutine pfrtd(a,b,r,ig,lg,solv,eq,nfrt,k)            pfr   1
c.... triangular decomposition for frontal program           pfr   2
      logical solv                                           pfr   3
      real*8 a(1),b(1),eq(1),dimx,dimn,pivot,term,r          pfr   4
      integer*2 ig(1),lg(1)                                  pfr   5
      common /nfrta/ dimx,dimn,nv,npl                        pfr   6
c...  extract equation                                       pfr   7
      do 301 j=1,nfrt                                        pfr   8
        l = max(j,k)                                         pfr   9
        l=ig(l)-l+min(j,k)                                   pfr  10
        eq(j)=a(l)                                           pfr  11
301   continue                                               pfr  12
      pivot=eq(k)                                            pfr  13
      if(pivot.eq.0.0d0) then                                pfr  14
        call pconsd(eq,nfrt,0.0d0)                           pfr  15
        write(*,2000)                                        pfr  16
        pivot = 1.0                                          pfr  17
        eq(k) = 1.0                                          pfr  18
      else                                                   pfr  19
        dimx = max(dimx,abs(pivot))                          pfr  20
        if(dimn.eq.0.0d0) dimn = abs(pivot)                  pfr  21
        dimn = min(dimn,abs(pivot))                          pfr  22
      endif                                                  pfr  23
      km = k - 1                                             pfr  24
      kp = k + 1                                             pfr  25
      kk = 1                                                 pfr  26
      if(km.gt.0) then                                       pfr  27
        do 302 jj = 1,km                                     pfr  28
          if(eq(jj).ne.0.0d0) then                           pfr  29
            term = -eq(jj)/pivot                             pfr  30
            if(solv) b(jj) = b(jj) + term*r                  pfr  31
            call saxpb(eq,a(kk),term,jj,a(kk))               pfr  32
          endif                                              pfr  33
          kk = ig(jj) + 1                                    pfr  34
302     continue                                             pfr  35
      endif                                                  pfr  36
      kk = ig(k) + 1                                         pfr  37
      if(kp.le.nfrt) then                                    pfr  38
        do 303 jj = kp,nfrt                                  pfr  39
          lg(jj-1) = lg(jj)                                  pfr  40
          kl = kk - jj                                       pfr  41
          term = -eq(jj)/pivot                               pfr  42
          if(solv) b(jj-1) = b(jj) + term*r                  pfr  43
          call saxpb(eq(kp),a(kk+k),term,jj-k,a(kl+k))       pfr  44
          if(km.gt.0) then                                   pfr  45
            kl = kl + 1                                       pfr  46
            call saxpb(eq,a(kk),term,km, a(kl))              pfr  47
          endif                                              pfr  48
          kk = ig(jj) + 1                                    pfr  49
303     continue                                             pfr  50
      endif                                                  pfr  51
      call pconsd(a(kk-nfrt),nfrt,0.0d0)                     pfr  52
      b(nfrt) = 0.0                                          pfr  53
      lg(nfrt) = 0                                           pfr  54
      return                                                 pfr  55
2000  format(' WARNING -- Zero pivot, check boundary codes.'/pfr  56
     1        '            Pivot set to 1.0 and solution continued.')  pfr  57
      end                                                    pfr  58
```

```
c                                                                    pfr  1
      subroutine pfrtfw(b,dr,m,ipd,ibuf,maxf,nv,neq,nev,ifl)         pfr  1
c.... forward solution for frontal resolutions                       pfr  2
      real*8 b(maxf,1),dr(neq,1)                                     pfr  3
      integer*2 m(1)                                                 pfr  4
      call pconsd(b,maxf*nev,0.0d0)                                  pfr  5
      do 403 n = 1,nv                                                pfr  6
        call pbuff(m,ibuf,ilast,n,1,ifl)                             pfr  7
        np = 1                                                       pfr  8
401     nfrt = m(np)                                                 pfr  9
        k    = m(np+1)                                               pfr 10
        jj   = m(np+2)                                               pfr 11
        do 402 i = 1,nev                                             pfr 12
          call pfrtf(b(1,i),dr(1,i),nfrt,k,jj,m(np+4))               pfr 13
402     continue                                                     pfr 14
c.........align last word of buffer for backsubstitution reads       pfr 15
        np = np + 8 + m(np)*ipd                                      pfr 16
        if(np.lt.ilast) go to 401                                    pfr 17
403   continue                                                       pfr 18
      return                                                         pfr 19
      end                                                            pfr 20
c
      subroutine pfrtf(b,dr,nfrt,k,jj,eq)                            pfr  1
c.... forward elimination macro for front program                    pfr  2
      real*8 b(1),dr(1),eq(1),r                                      pfr  3
      dr(jj) = dr(jj) + b(k)                                         pfr  4
      r = dr(jj)/eq(k)                                               pfr  5
      km = k - 1                                                     pfr  6
      kp = k + 1                                                     pfr  7
      if(km.gt.0) then                                               pfr  8
        do 402 ii = 1,km                                             pfr  9
          b(ii) = b(ii) - eq(ii)*r                                   pfr 10
402     continue                                                     pfr 11
      endif                                                          pfr 12
      if(kp.le.nfrt) then                                            pfr 13
        do 404 ii = kp,nfrt                                          pfr 14
          b(ii-1) = b(ii) - eq(ii)*r                                 pfr 15
404     continue                                                     pfr 16
      endif                                                          pfr 17
      b(nfrt) = 0.0                                                  pfr 18
      return                                                         pfr 19
      end                                                            pfr 20
c
      subroutine prefrt(il,idl,ix,maxf,ndf,nen,nen1,numel,numnp)     pre  1
      integer*2 il(1),idl(1),ix(nen1,1)                              pre  2
c... prefrontal routine to flag last occurance of nodes              pre  3
c... preset check array                                              pre  4
      do 100 n=1,numnp                                               pre  5
        il(n)=0                                                      pre  6
100   continue                                                       pre  7
c... set last occurance of nodes                                     pre  8
      do 102 nu=numel,1,-1                                           pre  9
        n = idl(nu)                                                  pre 10
        do 101 i=nen,1,-1                                            pre 11
          ii=abs(ix(i,n))                                            pre 12
          if((ii.ne.0).and.(il(ii).eq.0)) then                      pre 13
            il(ii)=n                                                 pre 14
            ii=-ii                                                   pre 15
          endif                                                      pre 16
```

```
          ix(i,n)=ii                                                    pre 17
101       continue                                                      pre 18
102    continue                                                         pre 19
c...   get estimate to maximum frontwith                                pre 20
       maxf=0                                                           pre 21
       nowf=0                                                           pre 22
       do 107 nu=1,numel                                                pre 23
         n = idl(nu)                                                    pre 24
         do 105 i=1,nen                                                 pre 25
           ii=ix(i,n)                                                   pre 26
           if(ii.ne.0) then                                            pre 27
             jj=abs(ii)                                                 pre 28
             if(il(jj).ne.0) nowf=nowf+ndf                             pre 29
             maxf=max(maxf,nowf)                                        pre 30
             il(jj)=0                                                   pre 31
           endif                                                        pre 32
105       continue                                                      pre 33
         do 106 i = 1,nen                                               pre 34
           if(ix(i,n).lt.0) nowf = max(0,nowf-ndf)                     pre 35
106       continue                                                      pre 36
107    continue                                                         pre 37
       write(*,3000) maxf                                               pre 38
3000   format(' ** Estimate of maximum front width is ',i5)            pre 39
       return                                                           pre 40
       end                                                              pre 41
c
       subroutine profil (jd,idl,id,ix,ndf,nen1)                        pro  1
c....  compute front profile of global arrays                           pro  2
       integer*2 jd(1),idl(1),id(1),ix(nen1,1)                         pro  3
       common /cdata/ numnp,numel,nummat,nen,neq                       pro  4
       common /iofile/ ior,iow                                          pro  5
       common /frdata/ maxf                                             pro  6
c....  set up the equation numbers                                      pro  7
       neq = 0                                                          pro  8
       nneq = ndf*numnp                                                 pro  9
       do 10 n = 1,nneq                                                 pro 10
         j = id(n)                                                      pro 11
         if(j.eq.0) then                                                pro 12
           neq = neq + 1                                                pro 13
           id(n) = neq                                                  pro 14
         else                                                           pro 15
           id(n) = 0                                                    pro 16
         endif                                                          pro 17
10     continue                                                         pro 18
       do 11 n = 1,numel                                                pro 19
         idl(n) = n                                                     pro 20
11     continue                                                         pro 21
c....  compute front width                                              pro 22
       call prefrt(jd,idl,ix,maxf,ndf,nen,nen1,numel,numnp)            pro 23
       if(maxf.gt.120) then                                            pro 24
         write(*,2001) maxf                                             pro 25
         if(ior.lt.0) write(iow,2001) maxf                            pro 26
       endif                                                            pro 27
       return                                                           pro 28
2001   format(' ** ERROR ** front requires too much storage ')         pro 29
       end                                                              pro 30
```

15.8.5 *Plot module.* The plot of meshes for two-dimensional problems is controlled by the subprograms contained in file PCPLOT.FOR.

(*a*) File PCPLOT.FOR contains the following set of subprograms:

Name	Type	Description
PPLOTF	SUBROUTINE	Controls plot solution sequence
DPLOT	SUBROUTINE	Draws lines on screen
PDEVCL	SUBROUTINE	Close plot device and returns to macro
PDEVOP	SUBROUTINE	Open plot device, draw box around region
PLOTL	SUBROUTINE	Scale plot point, calls DPLOT
FRAME	SUBROUTINE	Determines plot region, scales to screen
PDEFM	SUBROUTINE	Compute coordinates of deformed plot
PLINE	SUBROUTINE	Draw mesh or outline

```
$NOFLOATCALLS
        subroutine pplotf(x,ix,b,lci,ct,ndf,ndm,nen1,nneq)          ppl    1
c.... plot control subroutine for feap                             ppl    2
        logical pcomp,oflg                                         ppl    3
        character lci*4                                            ppl    4
        integer*2 ixy,devnam,status,vslcol,il,ix(1),coli           ppl    5
        real    x(ndm,1),ct(2)                                     ppl    6
        real*8  b(1),prop                                          ppl    7
        common dr(1)                                               ppl    8
        common /adata/ il(32000)                                   ppl    9
        common /cdata/ numnp,numel,nummat,nen,neq                  ppl   10
        common /pdata2/ ixy(4),devnam                              ppl   11
c.... open kernel system and plot mesh or outline of parts         ppl   12
        call pdevop(devnam)                                        ppl   13
        call frame(x,ndm,numnp)                                    ppl   14
c.... plot mesh or outline of parts                                ppl   15
        oflg = .not.pcomp(lci,'outl')                              ppl   16
        c     = 0.0                                                ppl   17
        ic    = 1                                                  ppl   18
        if(ct(1).ne.0.0) ic = 2                                    ppl   19
        do 100 i = ic,1,-1                                         ppl   20
          coli    = 8 - 2*i                                        ppl   21
          status = vslcol(devnam,coli)                             ppl   22
          call pdefm(x,b,c,ndm,ndf,numnp, dr)                      ppl   23
          call pline(dr,ix,il,numnp,numel,ndm,nen1,nen,ct(2),oflg) ppl   24
          c       = ct(1)                                          ppl   25
100     continue                                                   ppl   26
c.... close plot                                                   ppl   27
        call pdevcl(devnam)                                        ppl   28
        return                                                     ppl   29
        end                                                        ppl   30
c
        subroutine dplot(x,y,ipen)                                 dpl    1
        integer*2 ixy,devnam,status,vpline                         dpl    2
        common /pdata2/ ixy(4),devnam                              dpl    3
c.... pen command motions   (ipen = 2, pendown)                    dpl    4
        ixy(3) = 22000*x                                           dpl    5
```

```
      ixy(4) = 22000*y                                          dpl  6
      if(ipen.eq.2) then                                        dpl  7
        status = vpline(devnam,2,ixy)                           dpl  8
      endif                                                     dpl  9
      ixy(1) = ixy(3)                                           dpl 10
      ixy(2) = ixy(4)                                           dpl 11
      return                                                    dpl 12
      end                                                       dpl 13
c
      subroutine pdevcl(devnam)                                 pde  1
      integer*2 devnam,status,vclrwk,vencur,vclswk,vrqstr,ixy(2) pde  2
      character*1 xxx                                           pde  3
c.... close the plotting device                                pde  4
      status = vrqstr(devnam,1,0,ixy,xxx)                       pde  5
      status = vclrwk(devnam)                                   pde  6
      status = vencur(devnam)                                   pde  7
      status = vclswk(devnam)                                   pde  8
      return                                                    pde  9
      end                                                       pde 10
c
      subroutine pdevop(devnam)                                 pde  1
c.... open graphics workstation                                pde  2
      integer*2 devnam,wkin(19),wkout(66),status,vopnwk,vclrwk,vslcol pde  3
      data wkin/1,1,1,4,1,1,1,1,1,1,1,68,73,83,80,76,65,89,32/  pde  4
c.... open kernel system                                       pde  5
      status = vopnwk(wkin,devnam,wkout)                        pde  6
      if(status.lt.0) then                                      pde  7
        write(*,2000)                                           pde  8
        return                                                  pde  9
      endif                                                     pde 10
      status = vclrwk(devnam)                                   pde 11
      status = vslcol(devnam,2)                                 pde 12
      call dplot(0.0000,0.0000,1)                               pde 13
      call dplot(1.4545,0.0000,2)                               pde 14
      call dplot(1.4545,1.0090,2)                               pde 15
      call dplot(0.0000,1.0090,2)                               pde 16
      call dplot(0.0000,0.0000,2)                               pde 17
      return                                                    pde 18
2000  format('  Graphics Device Driver not installed correctly.') pde 19
      end                                                       pde 20
c
      subroutine plotl(x1,x2,x3,ipen)                           plo  1
c.... line drawing command                                     plo  2
      common /pdata1/ scale,dx(2),sx(2)                         plo  3
c.... compute the normal coordinates                           plo  4
      s1 = max(0.0,min(1.45,scale*(x1 + x1 - sx(1)) + 0.5))     plo  5
      s2 = max(0.0,min(1.00,scale*(x2 + x2 - sx(2)) + 0.5))     plo  6
      call dplot(s1,s2,ipen)                                    plo  7
      return                                                    plo  8
      end                                                       plo  9
c
      subroutine frame(x,ndm,numnp)                             fra  1
c.... compute scaling for plot area                            fra  2
      logical iflg                                              fra  3
      real      x(ndm,1),xmn(2),xmx(2),xmin(3),xmax(3)          fra  4
      common /pdata1/ scale,dx(2),sx(2)                         fra  5
c.... determine window coordinates                             fra  6
      if(ndm.eq.1) then                                         fra  7
        dx(2) = 0.                                              fra  8
        sx(2) = 0.0                                             fra  9
      endif                                                     fra 10
```

```
      ii = min(ndm,3)                                              fra 11
      ij = min(ndm,2)                                              fra 12
c.... find the minimum and maximum coordinate of input nodes      fra 13
      iflg = .true.                                                fra 14
      do 104 n = 1,numnp                                           fra 15
        if(x(1,n).ne. -999.) then                                 fra 16
          if(iflg) then                                           fra 17
            do 100 i = 1,ii                                        fra 18
              xmin(i) = x(i,n)                                     fra 19
              xmax(i) = x(i,n)                                     fra 20
100         continue                                              fra 21
            iflg = .false.                                        fra 22
          else                                                    fra 23
            do 102 i = 1,ii                                       fra 24
              xmin(i) = min(xmin(i),x(i,n))                       fra 25
              xmax(i) = max(xmax(i),x(i,n))                       fra 26
102         continue                                              fra 27
          endif                                                   fra 28
        endif                                                     fra 29
104   continue                                                    fra 30
      scale  = max(xmax(1)-xmin(1),xmax(2)-xmin(2))               fra 31
c.... plot region determination                                   fra 32
      do 110 i = 1,ij                                             fra 33
        xmn(i) = min(xmin(i),xmax(i))                             fra 34
        xmx(i) = max(xmin(i),xmax(i))                             fra 35
        dx(i) = xmx(i) - xmn(i)                                   fra 36
        sx(i) = xmx(i) + xmn(i)                                   fra 37
110   continue                                                    fra 38
c.... rescale window                                              fra 39
      if(dx(1).gt.dx(2)) then                                     fra 40
        xmn(2) = (sx(2) - dx(1))/2.0                              fra 41
        xmx(2) = (sx(2) + dx(1))/2.0                              fra 42
      else                                                        fra 43
        xmn(1) = (sx(1) - dx(2))/2.0                              fra 44
        xmx(1) = (sx(1) + dx(2))/2.0                              fra 45
      endif                                                       fra 46
      do 112 i = 1,ij                                             fra 47
        xmin(i) = max(xmin(i),xmn(i)) - scale/100.                fra 48
        xmax(i) = min(xmax(i),xmx(i)) + scale/100.                fra 49
112   continue                                                    fra 50
c.... default values                                              fra 51
      scale  = max(xmax(1)-xmin(1),xmax(2)-xmin(2))               fra 52
c.... reset values for deformed plotting                          fra 53
      do 114 i = 1,ij                                             fra 54
        xcen = xmax(i)+xmin(i)                                    fra 55
        xmax(i) = (xcen + 1.1*scale)/2.                           fra 56
        xmin(i) = (xcen - 1.1*scale)/2.                           fra 57
114   continue                                                    fra 58
      scale = 0.4/scale                                           fra 59
      return                                                      fra 60
      end                                                         fra 61
c
      subroutine pdefm(x,b,c,ndm,ndf,numnp, dr)                   pde  1
c.... compute the deformed position of two-dimensional meshes     pde  2
      real  x(ndm,1),dr(ndm,1)                                    pde  3
      real*8 b(ndf,1)                                             pde  4
      call pconsr(dr,ndm*numnp,0.0)                               pde  5
      do 120 n = 1,numnp                                          pde  6
        if(x(1,n).ne. -999.) then                                pde  7
          do 110 i = 1,ndm                                        pde  8
            dr(i,n) = x(i,n) + c*b(i,n)                           pde  9
```

```
110        continue                                                  pde 10
        endif                                                        pde 11
120     continue                                                     pde 12
        return                                                       pde 13
        end                                                          pde 14
c
        subroutine pline(x,ix,ic,numnp,numel,ndm,nen1,nen,ct,isw)    pli  1
c.... plot mesh or outline                                           pli  2
        logical ifl,iend,isw                                         pli  3
        integer*2 ix(nen1,1),iplt(8),ic(numnp,1)                     pli  4
        real      x(ndm,1)                                           pli  5
        data iplt/5,2,6,3,7,4,8,1/                                   pli  6
c.... initialize connection array                                    pli  7
        call pconsi(ic,numnp*4,0)                                    pli  8
c.... loop through elements to set up list                           pli  9
        do 206 n = 1,numel                                           pli 10
          jj = abs(ct)                                               pli 11
          if(jj.eq.0) go to 197                                      pli 12
          ii = abs(ix(nen1,n))                                       pli 13
          if(ii.ne.jj) go to 206                                     pli 14
197       i = 1                                                      pli 15
          ii = abs(ix(i,n))                                          pli 16
          do 205 ij = 1,8                                            pli 17
            j = iplt(ij)                                             pli 18
            if(j.le.nen.and.ix(j,n).ne.0) then                      pli 19
              jj = abs(ix(j,n))                                      pli 20
              if(jj.ne.ii) then                                     pli 21
                n1 = min(ii,jj)                                      pli 22
                n2 = max(ii,jj)                                      pli 23
                do 203 k = 1,4                                       pli 24
                  if(ic(n1,k).eq.0) then                            pli 25
                    ic(n1,k) = n2                                    pli 26
                    go to 204                                        pli 27
                  elseif(ic(n1,k).eq.n2) then                       pli 28
                    ic(n1,k) = -n2                                   pli 29
                    go to 204                                        pli 30
                  endif                                              pli 31
203             continue                                            pli 32
              endif                                                 pli 33
204           ii = jj                                               pli 34
            endif                                                    pli 35
205       continue                                                  pli 36
206     continue                                                     pli 37
c.... change signs to permit mesh plot                               pli 38
        if(isw) then                                                 pli 39
          do 250 n = 1,numnp                                         pli 40
          do 250 i = 1,4                                             pli 41
            ic(n,i) = abs(ic(n,i))                                   pli 42
250       continue                                                  pli 43
        endif                                                        pli 44
c.... plot outline of part with continuous lines                     pli 45
        x3 = 0.                                                      pli 46
        do 304 ni = 1,numnp                                          pli 47
          iend = .true.                                              pli 48
          do 303 n = 1,numnp                                         pli 49
            ifl = .true.                                             pli 50
            n1 = n                                                   pli 51
300         do 301 i = 1,4                                           pli 52
              if(ic(n1,i)) 301,303,302                              pli 53
301         continue                                                pli 54
            go to 303                                                pli 55
```

```
302       iend = .false.                                          pli 56
          if(ndm.ge.3)  x3 = x(3,n1)                              pli 57
          if(ifl) call plotl(x(1,n1),x(2,n1),x3,1)                pli 58
          ifl = .false.                                           pli 59
          n2 = ic(n1,i)                                           pli 60
          ic(n1,i) = -n2                                          pli 61
          if(ndm.ge.3)  x3 = x(3,n2)                              pli 62
          call plotl(x(1,n2),x(2,n2),x3,2)                        pli 63
          n1 = n2                                                 pli 64
          go to 300                                               pli 65
303     continue                                                 pli 66
          if(iend) go to 305                                      pli 67
304   continue                                                   pli 68
305   return                                                     pli 69
      end                                                        pli 70
```

15.8.6 *Element modules.* The element modules for linear elastic elements are contained in files PCELM1.FOR, PCELM2.FOR, PCELM3.FOR, and PCELM4.FOR.

(*a*) File PCELM1.FOR contains an element module for two-dimensional, isotropic linear elasticity. The control information must be input so that

$$NDM = 2$$
$$NDF \geqslant 2$$
$$NEN \geqslant 3$$

Each of the three elasticity elements has two degrees of freedom at each node (i.e., u, v); thus the user must set NDF > 2 as indicated above. The elements will work if NDF > 2 but produce zero displacements at the added degrees of freedom. Additional degrees of freedom should only be used if the element is used with another element that requires them (e.g., a typical two-dimensional frame element requires 3 degrees of freedom per node; the plane stress elements may be used as shear panels).

The element, ELMTO1, has features to analyse plane stress, plane strain, and axisymmetric geometries with specified nodal force, temperature, or body force loading. The element is based upon the irreducible, displacement method of analysis and uses the three- to nine-noded isoparametric shape function subprograms contained in file PCMAC3.FOR. Local node numbering for this element is shown in Fig. 15.25. If the input value of the global node number is zero at any local node the shape function is not formed. Thus it is possible to have elements that have linear shape functions on one side while other sides

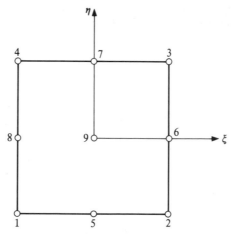

Fig. 15.25   Local node number sequence for a four- to nine-noded element

have quadratic variations. The three-node element is generated either by repeating the global node number for adjacent corner nodes or by specifying only three corner nodes. Quadratic order triangles may be formed by assigning the same node number to an entire edge. It is suggested that both mid-side and centre (i.e., local node 9) be included for the quadratic order triangle. The element computes both element and nodal stress values. The nodal stress projections are used to compute and report error and refinement estimates.

The material property set parameters which follow the data described in Table 15.8 are input as follows:

Property Record 1.) FORMAT—3F10.0,3I10

Column	Description
1 to 10	E, Young's modulus
11 to 20	v, Poisson's ratio
21 to 30	ρ, mass density
31 to 40	L, number of quadrature pts/dir for array computation
41 to 50	K, number of pts/dir for element stress outputs
51 to 60	I, problem type: $I = 1$ for plane stress
	$I = 2$ for plane strain
	$I = 3$ for axisymmetric

Property Record 2.) FORMAT—5F10.0

Column	Description
1 to 10	Thickness (plane stress only)
11 to 20	1-body force (per unit volume)
21 to 30	2-body force (per unit volume)
31 to 40	α, coefficient of thermal expansion
41 to 50	T_0, stress-free temperature

File PCELM1.FOR consists of the following subprograms:

Name	Type	Description
ELMTO1	SUBROUTINE	Inputs parameters, computes FE arrays
STREO1	SUBROUTINE	Computes stresses and strains at point
STCNO1	SUBROUTINE	Computes nodal stress integrals
STERO1	SUBROUTINE	Computes and outputs error estimates

```
$NOFLOATCALLS
      subroutine elmt01(d,ul,xl,ix,tl,s,p,ndf,ndm,nst,isw)        elm  1
c                                                                 elm  2
c.... plane linear elastic element routine                        elm  3
c                                                                 elm  4
      implicit real*8 (a-h,o-z)                                   elm  5
      integer*2 ix(1)                                             elm  6
      real    xl(ndm,1),tl(1),aa,dm                               elm  7
```

```
      real*8 d(1),ul(ndf,1),s(nst,1),p(1),eps(4),sigr(6),shp(3,9),    elm  8
     1       sg(16),tg(16),wg(16),ang                                 elm  9
      character wd(3)*12,yyy*80                                       elm 10
      common /adata/ aa(16000)                                        elm 11
      common /cdata/ numnp,numel,nummat,nen,neq                       elm 12
      common /eldata/ dm,n,ma,mct,iel,nel                             elm 13
      common /iofile/ ior,iow                                         elm 14
      data wd/'Plane Stress','Plane Strain','Axisymmetric'/           elm 15
c.... go to correct array processor                                  elm 16
      l    = d(5)                                                     elm 17
      k    = d(6)                                                     elm 18
      ityp = d(15)                                                    elm 19
      go to(1,2,3,3,3,3,7,8), isw                                     elm 20
c.... input material properties                                      elm 21
1     if(ior.lt.0) write(*,5000)                                      elm 22
      call pintio(yyy,10)                                             elm 23
      read(yyy,1000,err=110) e,xnu,d(4),l,k,ityp                      elm 24
11    if(ior.lt.0) write(*,5001)                                      elm 25
      call pintio(yyy,10)                                             elm 26
      read(yyy,1001,err=111) d(14),d(11),d(12),alp,t0                 elm 27
c.... set material parameter type and flags                          elm 28
      ityp = max(1,min(ityp,3))                                       elm 29
      j = min(ityp,2)                                                 elm 30
      d(1) = e*(1.+(1-j)*xnu)/(1.+xnu)/(1.-j*xnu)                     elm 31
      d(2) = xnu*d(1)/(1.+(1-j)*xnu)                                  elm 32
      d(3) = e/2./(1.+xnu)                                            elm 33
      d(13)= d(2)*(j-1)                                               elm 34
      if(d(14).le.0.0d0 .or. ityp.ge.2) d(14) = 1.0                   elm 35
      d(15) = ityp                                                    elm 36
      d(16) = e                                                       elm 37
      d(17) = xnu                                                     elm 38
      d(18) = -xnu/e                                                  elm 39
      l = min(4,max(1,l))                                             elm 40
      k = min(4,max(1,k))                                             elm 41
      d(5) = l                                                        elm 42
      d(6) = k                                                        elm 43
      d(9) = t0                                                       elm 44
      d(10)= e*alp/(1.-j*xnu)                                         elm 45
      lint = 0                                                        elm 46
      write(iow,2000) wd(ityp),d(16),d(17),d(4),l,k,d(14),d(11),d(12), elm 47
     1          alp,t0                                                elm 48
      if(ior.lt.0) write(*,2000) wd(ityp),d(16),d(17),d(4),l,k,       elm 49
     1          d(14),d(11),d(12),alp,t0                              elm 50
      d(4) = d(4)*d(14)                                               elm 51
      return                                                         elm 52
c.... read error messages                                           elm 53
110   call perror('PCELM1',yyy)                                      elm 54
      go to 1                                                        elm 55
111   call perror('PCELM1',yyy)                                      elm 56
      go to 11                                                       elm 57
2     call ckisop(ix,xl,shp,ndm)                                     elm 58
      return                                                         elm 59
c.... stiffness/residual computation                                elm 60
3     if(isw.eq.4) l = k                                             elm 61
      if(l*l.ne.lint) call pgauss(l,lint,sg,tg,wg)                   elm 62
c.... compute integrals of shape functions                          elm 63
      do 340 l = 1,lint                                              elm 64
         call shape(sg(l),tg(l),xl,shp,xsj,ndm,nel,ix,.false.)       elm 65
         call stre01(d,xl,ul,tl,shp,eps,sigr,xx,yy,ndm,ndf,nel,ityp) elm 66
         xsj = xsj*wg(l)*d(14)                                       elm 67
```

```
c.... compute jacobian correction                                       elm 68
          if(ityp.le.2) then                                            elm 69
              dv  = xsj                                                  elm 70
              xsj = 0.0                                                  elm 71
              zz  = 0.0                                                  elm 72
              sigr4 = -d(11)*dv                                          elm 73
          else                                                          elm 74
              dv  = xsj*xx                                              elm 75
              zz  = 1./xx                                               elm 76
              sigr4 = sigr(4)*xsj - d(11)*dv                            elm 77
          endif                                                         elm 78
          j1 = 1                                                        elm 79
c.... compute the mass term                                             elm 80
          if(isw.eq.5) then                                             elm 81
              dv = dv*d(4)                                              elm 82
              do 315 j = 1,nel                                          elm 83
                p(j1  ) = p(j1) + shp(3,j)*dv                           elm 84
                p(j1+1) = p(j1)                                         elm 85
                j1 = j1 + ndf                                           elm 86
315           continue                                                 elm 87
          elseif(isw.eq.4) then                                         elm 88
              call pstres(sigr,sigr(5),sigr(6),ang)                     elm 89
c.... output stresses and strains                                       elm 90
              mct = mct - 2                                             elm 91
              if(mct.le.0) then                                         elm 92
                call prthed(iow)                                        elm 93
                write(iow,2001)                                         elm 94
                if(ior.lt.0) write(*,2001)                              elm 95
                mct = 50                                                elm 96
              endif                                                     elm 97
              write(iow,2002) n,xx,sigr,ma,yy,epo,ang                   elm 98
              if(ior.lt.0) write(*,2002) n,xx,sigr,ma,yy,eps,ang        elm 99
          else                                                          elm100
c.... loop over rows                                                    elm101
              do 330 j = 1,nel                                          elm102
                w11 = shp(1,j)*dv                                       elm103
                w12 = shp(2,j)*dv                                       elm104
                w22 = shp(3,j)*xsj                                      elm105
c.... compute the internal forces                                       elm106
                p(j1  ) = p(j1  ) - (shp(1,j)*sigr(1)+shp(2,j)*sigr(2))*dv  elm107
     1                            - shp(3,j)*sigr4                      elm108
                p(j1+1) = p(j1+1) - (shp(1,j)*sigr(2)+shp(2,j)*sigr(3))*dv  elm109
     1                            + d(12)*shp(3,j)*dv                   elm110
c.... loop over columns (symmetry noted)                                elm111
                if(isw.eq.3) then                                       elm112
                  k1 = j1                                               elm113
                  a11 = d(1)*w11 + d(2)*w22                             elm114
                  a21 = d(2)*w11 + d(1)*w22                             elm115
                  a31 = d(2)*(w11+w22)                                  elm116
                  a41 = d(3)*w12                                        elm117
                  a12 = d(2)*w12                                        elm118
                  a32 = d(1)*w12                                        elm119
                  a42 = d(3)*w11                                        elm120
                  do 320 k = j,nel                                      elm121
                    w11 = shp(1,k)                                      elm122
                    w12 = shp(2,k)                                      elm123
                    w22 = shp(3,k)*zz                                   elm124
                    s(j1  ,k1  ) = s(j1  ,k1  ) + w11*a11+w22*a21+w12*a41  elm125
                    s(j1+1,k1  ) = s(j1+1,k1  ) + (w11 + w22)*a12+w12*a42  elm126
                    s(j1  ,k1+1) = s(j1  ,k1+1) + w12*a31 + w11*a41     elm127
```

```
              s(j1+1,k1+1) = s(j1+1,k1+1) + w12*a32 + w11*a42       elm128
              k1 = k1 + ndf                                         elm129
320         continue                                               elm130
          endif                                                    elm131
          j1 = j1 + ndf                                            elm132
330       continue                                                 elm133
        endif                                                      elm134
340   continue                                                     elm135
c.... make stiffness symmetric and compute a residual             elm136
      if(isw.eq.3) then                                            elm137
        do 360 j = 1,nst                                           elm138
        do 360 k = j,nst                                           elm139
          s(k,j) = s(j,k)                                          elm140
360     continue                                                   elm141
      endif                                                        elm142
      return                                                       elm143
c.... compute the stress errors                                   elm144
7     if(l*l.ne.lint) call pgauss(l,lint,sg,tg,wg)                elm145
      call ster01(ix,d,xl,ul,tl,shp,aa,aa(numnp+1),ndf,ndm,       elm146
     1   numnp,numel,sg,tg,wg,sigr,eps,lint,ityp)                 elm147
      return                                                       elm148
c.... compute the nodal stress values                             elm149
8     if(l*l.ne.lint) call pgauss(l,lint,sg,tg,wg)                elm150
      call stcn01(ix,d,xl,ul,tl,shp,aa,aa(numnp+1),ndf,ndm,nel,   elm151
     1   numnp,sg,tg,sigr,eps,lint,ityp)                          elm152
      return                                                       elm153
c.... formats for input-output                                    elm154
1000  format(3f10.0,3i10)                                          elm155
1001  format(8f10.0)                                               elm156
2000  format(/5x,a12,' Linear Elastic Element'//                  elm157
     1 10x,'Modulus',e18.5/10x,'Poisson ratio',f8.5/10x,'Density',e18.5/elm158
     2 10x,'Gauss pts/dir',i3/10x,'Stress pts',i6/10x,'Thickness',e16.5/elm159
     3 10x,'1-gravity',e16.5/10x,'2-gravity',e16.5/10x,'Alpha',e20.5/   elm160
     4 10x,'Base temp',e16.5/)                                     elm161
2001  format(5x,'Element Stresses'//' elmt  1-coord',2x,'11-stress',2x, elm162
     1 '12-stress',2x,'22-stress',2x,'33-stress',3x,'1-stress',3x,elm163
     2 '2-stress'/' matl  2-coord',2x,'11-strain',2x,'12-strain',2x,elm164
     3 '22-strain',2x,'33-strain',6x,'angle'/39(' -'))             elm165
2002  format(i5,0p1f9.3,1p6e11.3/i5,0p1f9.3,1p4e11.3,0p1f11.2/)    elm166
5000  format(' Input: E, nu, rho, pts/stiff, pts/stre',           elm167
     1 ', type(1=stress,2=strain,3=axism)',/3x,'>',$)              elm168
5001  format(' Input: Thickness, 1-body force, 1-body force, alpha,'elm169
     1        ,' Temp-base'/3x,'>',$)                              elm170
      end                                                          elm171
c
      subroutine stre01(d,xl,ul,tl,shp,eps,sig,xx,yy,ndm,ndf,nel,ityp) str   1
      implicit real*8 (a-h,o-z)                                    str   2
      real    xl(ndm,1),tl(1)                                      str   3
      real*8 d(1),ul(ndf,1),shp(3,1),eps(4),sig(4)                 str   4
c.... compute strains and coordinates                             str   5
      call pconsd(eps,4,0.0d0)                                     str   6
      xx = 0.0                                                     str   7
      yy = 0.0                                                     str   8
      ta = -d(9)                                                   str   9
      do 100 j = 1,nel                                             str  10
        xx = xx + shp(3,j)*xl(1,j)                                 str  11
        yy = yy + shp(3,j)*xl(2,j)                                 str  12
        ta = ta + shp(3,j)*tl(j)                                   str  13
        eps(1) = eps(1) + shp(1,j)*ul(1,j)                         str  14
        eps(2) = eps(2) + shp(1,j)*ul(2,j) + shp(2,j)*ul(1,j)      str  15
        eps(3) = eps(3) + shp(2,j)*ul(2,j)                         str  16
```

```
            if(ityp.eq.3) eps(4) = eps(4) + shp(3,j)*ul(1,j)         str 17
100    continue                                                      str 18
       ta = ta*d(10)                                                 str 19
c.... compute stresses                                               str 20
       if(ityp.gt.2) then                                            str 21
         if(xx.ne.0.0d0) then                                        str 22
           eps(4) = eps(4)/xx                                        str 23
         else                                                        str 24
           eps(4) = eps(1)                                           str 25
         endif                                                       str 26
         sig(4) = d(1)*eps(4) + d(2)*(eps(1) + eps(3)) - ta          str 27
       else                                                          str 28
         sig(4) = d(13)*(eps(1) + eps(3)) - ta                       str 29
       endif                                                         str 30
       sig(1) = d(1)*eps(1) + d(2)*(eps(3) + eps(4)) - ta            str 31
       sig(2) = d(3)*eps(2)                                          str 32
       sig(3) = d(1)*eps(3) + d(2)*(eps(1) + eps(4)) - ta            str 33
       if(ityp.eq.1) eps(4) = d(18)*(sig(1) + sig(3))                str 34
       return                                                        str 35
       end                                                           str 36
c
       subroutine stcn01(ix,d,xl,ul,tl,shp,dt,st,ndf,ndm,nel,numnp,   stc  1
     1                   sg,tg,sig,eps,lint,ityp)                     stc  2
c.... project stresses onto nodes                                    stc  3
       implicit real*8 (a-h,o-z)                                     stc  4
       integer*2 ix(1)                                               stc  5
       real    dt(numnp),st(numnp,1),xl(ndm,1),tl(1)                 stc  6
       real*8 eps(4),sig(4),ul(ndf,1),shp(3,4),d(1),sg(9),tg(9)      stc  7
       common /errind/ eeror,elproj,ecproj,efem,enerr,ebar           stc  8
       gr = (1.+d(17))/d(16)                                         stc  9
       do 130 l = 1,lint                                             stc 10
         call shape(sg(l),tg(l),xl,shp,xsj,ndm,nel,ix,.false.)       stc 11
         call stre01(d,xl,ul,tl,shp,eps,sig,xx,yy,ndm,ndf,nel,ityp)  stc 12
         enerr = enerr+((sig(1)+sig(3)+sig(4))**2)*d(18)*xsj         stc 13
     1       + gr*xsj*sig(2)**2                                      stc 14
         do 110 i = 1,4                                              stc 15
           enerr   = enerr   + gr*sig(i)**2*xsj                      stc 16
110      continue                                                    stc 17
         do 120 ii = 1,nel                                           stc 18
           ll = abs(ix(ii))                                          stc 19
           if(ll.gt.0) then                                          stc 20
             xsji = xsj*shp(3,ii)                                    stc 21
             dt(ll) = dt(ll) + xsji                                  stc 22
             st(ll,1) = st(ll,1) + sig(1)*xsji                       stc 23
             st(ll,2) = st(ll,2) + sig(2)*xsji                       stc 24
             st(ll,3) = st(ll,3) + sig(3)*xsji                       stc 25
             st(ll,4) = st(ll,4) + sig(4)*xsji                       stc 26
           endif                                                     stc 27
120      continue                                                    stc 28
130    continue                                                      stc 29
       return                                                        stc 30
       end                                                           stc 31
c
       subroutine ster01(ix,d,xl,ul,tl,shp,dt,st,ndf,ndm,            ste  1
     1   numnp,numel,sg,tg,wg,sig,eps,lint,ityp)                     ste  2
       implicit real*8 (a-h,o-z)                                     ste  3
       integer*2 ix(1)                                               ste  4
       real    dt(numnp),st(numnp,1),xl(ndm,1),tl(1),dm              ste  5
       real*8 shp(3,4),sig(6),sigp(4),dsig(4),d(1),eps(4),ul(ndf,1), ste  6
     1       sg(16),tg(16),wg(16),deps(4)                            ste  7
       common /iofile/ ior,iow                                       ste  8
```

```
          common /eldata/ dm,n,ma,mct,iel,nel                      ste  9
          common /errind/ eerror,elproj,ecproj,efem,enerr,ebar      ste 10
c.... stress error computations                                     ste 11
          psis = 0.0                                                ste 12
          psi  = 0.0                                                ste 13
          gr   = (1.+d(17))/d(16)                                   ste 14
          do 200 ii = 1,lint                                        ste 15
            call shape(sg(ii),tg(ii),xl,shp,xsj,ndm,nel,ix,.false.) ste 16
            call stre01(d,xl,ul,tl,shp,eps,sig,xx,yy,ndm,ndf,nel,ityp)ste 17
            xsj = xsj*wg(ii)                                        ste 18
            do 100 i = 1,4                                          ste 19
              sigp(i)= 0.0d0                                        ste 20
100         continue                                                ste 21
            do 110 i = 1,nel                                        ste 22
              ll = iabs(ix(i))                                      ste 23
              if(ll.ne.0) then                                      ste 24
                do 105 j = 1,4                                      ste 25
                  sigp(j) = sigp(j) + shp(3,i)*st(ll,j)             ste 26
105             continue                                            ste 27
              endif                                                 ste 28
110         continue                                                ste 29
c.... compute the integral of the stress squares for error indicator usesteste 30
            do 120 i = 1,4                                          ste 31
              dsig(i) = sigp(i)-sig(i)                              ste 32
              efem    = efem   + sig(i)*sig(i)*xsj                  ste 33
              ecproj  = ecproj + sigp(i)*sigp(i)*xsj                ste 34
              psis    = psis   + (dsig(i)**2)*xsj                   ste 35
              psi     = psi    + gr*(dsig(i)**2)*xsj                ste 36
120         continue                                                ste 37
            psi = psi + gr*dsig(2)**2*xsj +                         ste 38
     1          ((dsig(1)+dsig(3)+dsig(4))**2)*d(18)*xsj            ste 39
200       continue                                                  ste 40
          eerror = eerror + psis                                    ste 41
          if(elproj.ne.0.0d0) then                                  ste 42
            psi    = sqrt(abs(psi))/ebar                            ste 43
            psis   = 20.0*sqrt(abs(psis)/elproj*numel)              ste 44
            if(mct.eq.0) then                                       ste 45
              write(iow,2000)                                       ste 46
              if(ior.lt.0) write(*,2000)                            ste 47
              mct = 50                                              ste 48
            endif                                                   ste 49
            mct = mct - 1                                           ste 50
            write(iow,2001) n,psis,psi                              ste 51
            if(ior.lt.0) write(*,2001) n,psis,psi                   ste 52
          endif                                                     ste 53
          return                                                    ste 54
2000      format(' M e s h   R e f i n e m e n t s    f o r   5%',  ste 55
     1          '    E r r o r'//'     elmt   h-sigma    h-energy'/)ste 56
2001      format(i8,1p2e12.4)                                       ste 57
          end                                                       ste 58
```

(b) File PCELM2.FOR contains an element module for two-dimensional, isotropic linear elasticity. The control information must be input so that

$$NDM = 2$$
$$NDF \geqslant 2$$
$$NEN \geqslant 4$$

The element has features to analyse plane strain and axisymmetric geometries with specified nodal force loading. The element is based upon the three-field mixed formulation discussed in Chapter 12 which leads to the modified strain displacement matrix, **B**-bar. Accordingly, the element may be used for materials that are nearly incompressible. While the element uses the isoparametric shape function subprograms contained in file PCMAC3.FOR it is restricted for use with four-noded quadrilateral elements only. Local node numbering for this element is shown in Fig. 15.25. The element computes both element and nodal stress values. No error estimates are included in this element due to limitations in space.

The material property set parameters which follow the data described in Table 15.8 are input as follows:

Property Record 1.) FORMAT—3F10.0

Column	Description
1 to 10	E, Young's modulus
11 to 20	v, Poisson's ratio

Property Record 2.) FORMAT—2I10

Column	Description	
1 to 10	IT, problem type:	IT = 1 for plane stress
		IT = 2 for plane strain
11 to 20	IB, strain matrix:	IB = 0 for **B**-bar
		IB = 1 for **B**, displacement form
21 to 30	ρ, mass density	

File PCELM2.FOR consists of the following subprograms:

Name	Type	Description
ELMTO2	SUBROUTINE	Inputs parameters, computes FE arrays
GVC2	SUBROUTINE	Compute volume integrals of shape functions
BMATO2	SUBROUTINE	Compute strain-displacement matrix
STRNO2	SUBROUTINE	Computes strains at point
STCNO2	SUBROUTINE	Computes nodal stress integrals

MATLO2 SUBROUTINE Input material parameter values
MODLO2 SUBROUTINE Compute stresses and moduli

```
$NOFLOATCALLS
      subroutine elmt02(d,ul,xl,ix,tl,s,p,ndf,ndm,nst,isw)          elm  1
c.... plane/axisymmetric linear element routine -- bbar formulation  elm  2
      implicit real*8 (a-h,o-z)                                      elm  3
      logical flg                                                    elm  4
      integer*2 ix(1)                                                elm  5
      real    xl(ndm,1),aa,dm                                        elm  6
      real*8 d(1),ul(ndf,1),s(nst,1),p(1),eps(4),sig(6),bbar(4,2,4), elm  7
     1    bbd(4,2),sg(4),tg(4),wg(4),shp3(4,4),shp(3,4,4),xsj(4),ang elm  8
      common /adata/ aa(16000)                                       elm  9
      common /cdata/ numnp,numel,nummat,nen,neq                      elm 10
      common /eldata/ dm,n,ma,mct,iel,nel                            elm 11
      common /elcom2/ g(2,4),ad(4,4)                                 elm 12
      common /iofile/ ior,iow                                        elm 13
c.... go to correct array processor                                 elm 14
      go to(1,2,3,3,5,3,7,3), isw                                    elm 15
      return                                                         elm 16
c.... input material properties                                     elm 17
    1 call matl02(d,ityp,ib)                                         elm 18
      d(5) = ityp                                                    elm 19
      d(6) = ib                                                      elm 20
      l    = 2                                                       elm 21
      call pgauss(l,lint,sg,tg,wg)                                   elm 22
      return                                                         elm 23
c.... check for element errors                                      elm 24
    2 call ckisop(ix,xl,shp,ndm)                                     elm 25
      return                                                         elm 26
c                                                                    elm 27
c.... compute tangent stiffness and residual force vector           elm 28
    3 type = d(5)                                                    elm 29
      ib   = d(6)                                                    elm 30
c.... compute volumetric integrals                                  elm 31
      call pconsd(g,8,0.0d0)                                         elm 32
      do 300 l = 1,lint                                              elm 33
        call shape(sg(l),tg(l),xl,shp(1,1,l),xsj(l),ndm,4,ix,.false.) elm 34
        call gvc2(shp(1,1,l),shp3(1,l),xsj(l),wg(l),xl,type,ndm)     elm 35
  300 continue                                                       elm 36
      vol  = xsj(1) + xsj(2) + xsj(3) + xsj(4)                       elm 37
      do 310 i = 1,4                                                 elm 38
        g(1,i) = g(1,i)/vol                                         elm 39
        g(2,i) = g(2,i)/vol                                         elm 40
  310 continue                                                       elm 41
      if(isw.eq.4) go to 4                                           elm 42
      if(isw.eq.8) go to 8                                           elm 43
      flg  = isw .eq. 3                                              elm 44
      do 400 l = 1,lint                                              elm 45
c.... compute stress, strain, and material moduli                   elm 46
        call strn02(shp(1,1,l),xl,ul,type,xr0,xz0,ndm,ndf,eps)      elm 47
        call modl02(d,ul,eps,sig,xsj(l),ndf,ib)                     elm 48
        i1 = 0                                                       elm 49
        do 380 i = 1,4                                               elm 50
c.... compute the internal stress divergence term                   elm 51
          p(i1+1) = p(i1+1) - shp(1,i,l)*sig(1) - shp(2,i,l)*sig(2) elm 52
     1                      - shp3(i,l)*sig(4)                       elm 53
          p(i1+2) = p(i1+2) - shp(2,i,l)*sig(3) - shp(1,i,l)*sig(2) elm 54
```

```
c.... compute stiffness                                                 elm 55
            if(flg) then                                                elm 56
c.... compute b-bar matrix                                              elm 57
              call bmat02(shp3(i,l),shp(1,i,l),g(1,i),bbar(1,1,i),ib)   elm 58
              do 320 ii = 1,2                                           elm 59
              do 320 jj = 1,4                                           elm 60
                bbd(jj,ii) = dot(bbar(1,ii,i),ad(1,jj),4)               elm 61
320           continue                                                  elm 62
              j1 = 0                                                    elm 63
              do 360 j  = 1,i                                           elm 64
                do 340 ii = 1,2                                         elm 65
                do 340 jj = 1,2                                         elm 66
                  s(ii+i1,jj+j1) = s(ii+i1,jj+j1)                       elm 67
      1                          + dot(bbd(1,ii),bbar(1,jj,j),4)        elm 68
340             continue                                                elm 69
360           j1 = j1 + ndf                                             elm 70
            endif                                                       elm 71
380         i1 = i1 + ndf                                               elm 72
400     continue                                                        elm 73
c.... form lower part by symmetry                                       elm 74
        if(flg) then                                                    elm 75
          do 410 i = 1,nst                                              elm 76
          do 410 j = 1,i                                                elm 77
            s(j,i) = s(i,j)                                             elm 78
410       continue                                                      elm 79
        endif                                                           elm 80
        return                                                          elm 81
c                                                                       elm 82
c.... compute stresses at center of element                            elm 83
    4   call shape(0.0d0,0.0d0,xl,shp,xsj,ndm,4,ix,.false.)             elm 84
        call strn02(shp,xl,ul,type,rr,zz,ndm,ndf,eps)                   elm 85
        call modl02(d,ul,eps,sig,1.0d0,ndf,ib)                          elm 86
        call pstres(sig,sig(5),sig(6),ang)                             elm 87
c.... output stresses                                                   elm 88
        mct = mct - 2                                                   elm 89
        if(mct.le.0) then                                               elm 90
          call prthed(iow)                                              elm 91
          write(iow,2001)                                               elm 92
          if(ior.lt.0) write(*,2001)                                    elm 93
          mct = 50                                                      elm 94
        endif                                                           elm 95
        write(iow,2002) n,ma,sig,rr,zz,ang                             elm 96
        if(ior.lt.0) write(*,2002) n,ma,sig,rr,zz,ang                  elm 97
        return                                                         elm 98
c                                                                       elm 99
c.... compute lumped mass matrix                                        elm100
    5   do 530 l = 1,lint                                               elm101
          call shape(sg(l),tg(l),xl,shp,xsj,ndm,4,ix,.false.)          elm102
c.... compute radius and multiply into jacobian for axisymmetry         elm103
          if(d(5).ne.0.0d0) then                                       elm104
            rr = 0.0                                                    elm105
            do 500 j = 1,4                                              elm106
              rr = rr + shp(3,j,1)*xl(1,j)                              elm107
500         continue                                                    elm108
            xsj(1) = xsj(1)*rr                                          elm109
          endif                                                         elm110
          xsj(1) = wg(l)*xsj(1)*d(4)                                    elm111
c.... for each node j compute db = rho*shape*dv                         elm112
          j1 = 1                                                        elm113
          do 520 j = 1,4                                                elm114
```

```
              p(j1  ) = p(j1) + shp(3,j,1)**sj(1)                    elm115
              p(j1+1) = p(j1)                                        elm116
              j1 = j1 + ndf                                          elm117
520        continue                                                 elm118
530     continue                                                    elm119
        return                                                      elm120
     7  return                                                      elm121
c                                                                   elm122
c.... stress computations for nodes                                 elm123
     8  call stcn02(ix,d,xl,ul,shp,aa,aa(numnp+1),ndf,ndm,          elm124
     1     numnp,sg,tg,sig,eps,lint,type,ib)                        elm125
        return                                                      elm126
c                                                                   elm127
c.... formats for input-output                                      elm128
1000    format(2i5)                                                 elm129
1001    format(a5,1x,2i10,f10.0/9f10.0)                             elm130
2001    format(' Element Stresses'//'  elmt  matl  11-stress  12-stress',elm131
     1  '  22-stress  33-stress   1-stress   2-stress'/'  1-coord',elm132
     2  '  2-coord ',42x,'angle')                                  elm133
2002    format(2i6,6e11.3/2f9.3,41x,f8.2/1x)                        elm134
        end                                                         elm135
c
        subroutine gvc2(shp,shp3,xsj,wg,xl,type,ndm)                gvc   1
        implicit real*8 (a-h,o-z)                                   gvc   2
c.... compute the volumetric integrals for the bbar                gvc   3
        real    xl(ndm,4)                                           gvc   4
        real*8 shp(3,4),shp3(4)                                     gvc   5
        common /elcom2/ g(2,4),ad(4,4)                              gvc   6
        if(type.ne.0.0d0) then                                      gvc   7
          rr = 0.0                                                  gvc   8
          do 100 i = 1,4                                            gvc   9
            rr = rr + shp(3,i)*xl(1,i)                              gvc  10
100       continue                                                 gvc  11
          xsj = xsj*rr                                             gvc  12
        else                                                        gvc  13
        endif                                                       gvc  14
        xsj = xsj*wg                                                gvc  15
        do 110 i = 1,4                                              gvc  16
          shp3(i) = 0.0                                            gvc  17
          if(type.ne.0.0d0) shp3(i) = shp(3,i)/rr                 gvc  18
          g(1,i) = g(1,i) + (shp(1,i) + shp3(i))*xsj              gvc  19
          g(2,i) = g(2,i) +  shp(2,i)*xsj                         gvc  20
110     continue                                                   gvc  21
        return                                                      gvc  22
        end                                                         gvc  23
c
        subroutine bmat02(sh3,shp,g,bbar,ib)                        bma   1
        implicit real*8 (a-h,o-z)                                   bma   2
        real*8 shp(3),g(1),bbar(4,2)                                bma   3
c.... bbar matrix for plane and axisymmetric problems              bma   4
        bbar(1,1) = shp(1)                                          bma   5
        bbar(2,1) = 0.0                                             bma   6
        bbar(3,1) = sh3                                             bma   7
        bbar(4,1) = shp(2)                                          bma   8
        bbar(1,2) = 0.0                                             bma   9
        bbar(2,2) = shp(2)                                          bma  10
        bbar(3,2) = 0.0                                             bma  11
        bbar(4,2) = shp(1)                                          bma  12
c.... correct for the b-bar effects                                bma  13
        if(ib.eq.0) then                                           bma  14
          bb1 = (g(1) - shp(1) - sh3)/3.0                          bma  15
          bb2 = (g(2) - shp(2))/3.0                                bma  16
```

```
          do 100 i = 1,3                                            bma 17
          bbar(i,1) = bbar(i,1) + bb1                               bma 18
          bbar(i,2) = bbar(i,2) + bb2                               bma 19
100       continue                                                 bma 20
        endif                                                       bma 21
        return                                                      bma 22
        end                                                         bma 23
c
        subroutine strn02(shp,xl,ul,type,xr0,xz0,ndm,ndf,eps)       str  1
        implicit real*8 (a-h,o-z)                                   str  2
        real    xl(ndm,1)                                           str  3
        real*8 ul(ndf,1),eps(4),shp(3,1)                            str  4
c.... compute strain and incremental tensors for constitutive equations str 5
        xr0    = 0.0                                                str  6
        xz0    = 0.0                                                str  7
        call pconsd(eps,4,0.0d0)                                    str  8
        do 310 k = 1,4                                              str  9
          xr0    = xr0    + shp(3,k)*xl(1,k)                        str 10
          xz0    = xz0    + shp(3,k)*xl(2,k)                        str 11
          eps(1) = eps(1) + ul(1,k)*shp(1,k)                       str 12
          eps(2) = eps(2) + ul(2,k)*shp(2,k)                       str 13
          eps(3) = eps(3) + shp(3,k)*ul(1,k)                       str 14
          eps(4) = eps(4) + (ul(2,k)*shp(1,k) + ul(1,k)*shp(2,k))/2. str 15
310       continue                                                 str 16
        eps(3) = type*eps(3)/xr0                                    str 17
        return                                                      str 18
        end                                                         str 19
c
        subroutine stcn02(ix,d,xl,ul,shp,dt,st,ndf,ndm,             stc  1
     1                    nnp,sg,tg,sig,eps,lint,type,ib)           stc  2
        Implicit real*8 (a-h,o-z)                                   stc  3
        integer*2 ix(1)                                             stc  4
        real    dt(nnp),st(nnp,1),xl(ndm,1)                         stc  5
        real*8 d(1),ul(ndf,1),sig(1),eps(1),shp(3,4),sg(1),tg(1)    stc  6
        do 200 jj = 1,4                                             stc  7
c.... compute stresses at nodes from history terms                 stc  8
          call shape(sg(jj),tg(jj),xl,shp,xsj,ndm,4,ix,.false.)    stc  9
          call strn02(shp,xl,ul,type,rr,zz,ndm,ndf,eps)            stc 10
          call modl02(d,ul,eps,sig,1.0d0,ndf,ib)                   stc 11
          do 100 ii = 1,4                                           stc 12
            ll = abs(ix(ii))                                        stc 13
            if(ll.gt.0) then                                       stc 14
              shpj = shp(3,i)*xsj                                   stc 15
              dt(ll) = dt(ll) + shpj                                stc 16
              do 50 i = 1,4                                         stc 17
50            st(ll,i) = st(ll,i) + sig(i)*shpj                    stc 18
            endif                                                   stc 19
100       continue                                                 stc 20
200     continue                                                    stc 21
        return                                                      stc 22
        end                                                         stc 23
c
        subroutine matl02(d,it,ib)                                  mat  1
        implicit real*8 (a-h,o-z)                                   mat  2
        real*8 d(1)                                                 mat  3
        character*12 wa(2),wb(2),yyy*80                             mat  4
c.... parameter specification for FEAP materials                   mat  5
        common /iofile/ ior,iow                                     mat  6
c                                                                   mat  7
        data wa/' P l a n e ',' A x i s y m'/                       mat  8
        data wb/' S t r a i n',' m e t r i c'/                      mat  9
```

```
1         if(ior.lt.0) write(*,3000)                                           mat 10
          call pintio(yyy,10)                                                  mat 11
          read(yyy,1000,err=101) ee,xnu                                        mat 12
          if(ior.lt.0) write(*,3001)                                           mat 13
          call pintio(yyy,10)                                                  mat 14
          read(yyy,1001,err=101) it,ib,d(4)                                    mat 15
          it = max(0,min(1,it))                                                mat 16
          ib = max(0,min(1,ib))                                                mat 17
          d(1) = ee/(1. - 2.*xnu)/3.0                                          mat 18
          d(2) = ee/(1.+xnu)/2.                                                mat 19
          write(iow,2000) wa(it+1),wb(it+1),ee,xnu,d(4),ib                     mat 20
          if(ior.lt.0) write(*,2000) wa(it+1),wb(it+1),ee,xnu,d(4),ib          mat 21
          return                                                              mat 22
101       call perror('PCELM4',yyy)                                            mat 23
          go to 1                                                            mat 24
c....  formats                                                                mat 25
1000   format(2f10.0)                                                         mat 26
1001   format(2i10,f10.0)                                                     mat 27
2000   format(/' E l a s t i c    M a t e r i a l  '/5x,2a12/                 mat 28
     2    10x,'Young modulus (E)   ',e15.5/                                   mat 29
     3    10x,'Poisson Ratio (nu)  ',e15.5/                                   mat 30
     4    10x,'Density       (rho) ',e15.5/                                   mat 31
     5    10x,'Strain-B  (0=bbar) ',i5/)                                      mat 32
3000   format(' Input: E, nu'/3x,'>',$)                                       mat 33
3001   format(' Input: it(0=plane,1=axisy), ib(0=bbar), rho'/' >',$)          mat 34
          end                                                               mat 35
c
          subroutine modl02(d,ul,eps,sig,xsj,ndf,ib)                           mod  1
          implicit real*8 (a-h,o-z)                                           mod  2
          real*8 d(15),ul(ndf,1),eps(4),sig(4)                                mod  3
          common /elcom2/ g(2,4),ad(4,4)                                      mod  4
c....  constitutive equation                                                  mod  5
          twog = (d(2) + d(2))*xsj                                            mod  6
          elam = d(1)*xsj - twog/3.                                           mod  7
c....  compute trace of incremental strain                                    mod  8
          treps = 0.0                                                         mod  9
          do 100 i = 1,4                                                      mod 10
            treps = treps + g(1,i)*ul(1,i) + g(2,i)*ul(2,i)                    mod 11
100       continue                                                           mod 12
c....  compute the stress                                                     mod 13
          epstr  = eps(1)+eps(2)+eps(3)                                       mod 14
          press = elam*epstr                                                  mod 15
          if(ib.eq.0) press = d(1)*xsj*treps - twog*epstr/3.0                  mod 16
          sig(1) = twog*eps(1) + press                                        mod 17
          sig(2) = twog*eps(4)                                                mod 18
          sig(3) = twog*eps(2) + press                                        mod 19
          sig(4) = twog*eps(3) + press                                        mod 20
c....  set up elastic moduli                                                  mod 21
          do 110 i = 1,3                                                      mod 22
            do 105 j = 1,3                                                    mod 23
              ad(i,j) = elam                                                  mod 24
105         continue                                                         mod 25
            ad(i,4) = 0.0                                                     mod 26
            ad(4,i) = 0.0                                                     mod 27
            ad(i,i) = elam + twog                                            mod 28
110       continue                                                           mod 29
          ad(4,4) = d(2)*xsj                                                  mod 30
c                                                                            mod 31
          return                                                             mod 32
          end                                                               mod 33
```

(c) File PCELM3.FOR contains an element module for two-dimensional, isotropic linear elasticity. The control information must be input so that

$$NDM = 2$$
$$NDF \geqslant 2$$
$$NEN \geqslant 4$$

The element has features to analyse plane stress and plane strain geometries with specified nodal force loading. The element is based upon the two-field mixed formulation discussed in Chapter 13. The original development is due to Pian and Sumihara[36] and the element is among the most accurate four-noded quadrilateral elements developed to date. The element also may be used for materials that are nearly incompressible. Local node numbering for this element is shown in Fig. 15.25. The element computes both element and nodal stress values. No error estimates are included in this element due to limitations in space.

The material property set parameters which follow the data described in Table 15.8 are input as follows:

Property Record 1.) FORMAT—4F10.0,I10

Column	Description
1 to 10	E, Young's modulus
11 to 20	v, Poisson's ratio
21 to 30	ρ, mass density
31 to 40	Thickness (plane stress only)
41 to 50	I, problem type: $I = 1$ for plane stress
	$I = 2$ for plane strain

File PCELM3.FOR consists of the following subprograms:

Name	Type	Description
ELMTO3	SUBROUTINE	Inputs parameters, computes FE arrays
PIANO3	SUBROUTINE	Computes stiffness and residual force
STCNO3	SUBROUTINE	Computes nodal stress integrals

```
$NOFLOATCALLS
      subroutine elmt03(d,ul,xl,ix,tl,s,p,ndf,ndm,nst,isw)     elm  1
c.... Plane stress/strain Pian Sumihara Element               elm  2
      implicit real*8 (a-h,o-z)                                elm  3
      integer*2 ix(1)                                          elm  4
      real    xl(ndm,1),dm,ap                                  elm  5
      real*8 d(1),ul(ndf,1),s(nst,1),p(1),sig(4),p1,p2,p3      elm  6
      character*4 wd(2),yyy*80                                 elm  7
      common /adata/ ap(16000)                                 elm  8
```

```
      common /cdata/   numnp,numel,nummat,nen,neq                elm  9
      common /eldata/ dm,n,ma,mct,iel,nel                        elm 10
      common /elcom3/ xs,xt,xh,ys,yt,yh,xj0,xj1,xj2,a1(3),a2(3),beta(5)  elm 11
      common /iofile/ ior,iow                                    elm 12
      data wd/'ress','rain'/                                     elm 13
c.... go to correct array processor                             elm 14
      go to(1,2,3,3,3,3,7,3), isw                                elm 15
c.... input/output material properties                          elm 16
    1 if(ior.lt.0) write(*,3000)                                 elm 17
      call pintio(yyy,10)                                        elm 18
      read(yyy,1000,err=11) (d(i),i=8,11),is                     elm 19
      is = max(1,min(is,2))                                      elm 20
      d(12)= is                                                  elm 21
      d(2) = d(9)*d(8)/(1.+d(9))/(1.-is * d(9))                  elm 22
      d(4) = d(8)/(1.+d(9))                                      elm 23
      d(3) = d(4)/2.                                             elm 24
      d(1) = d(2) + d(4)                                         elm 25
c.... set parameters for plane stress (is = 1)                  elm 26
      if(is.eq.1) then                                          elm 27
         if(d(11).le.0.0d0) d(11) = 1.0                          elm 28
         d(5) =  1./d(8)                                         elm 29
         d(6) = -d(9)/d(8)                                       elm 30
         d(13)=  0.0                                             elm 31
c.... set parameters for plane strain (is = 2)                  elm 32
      else                                                       elm 33
         d(11)=  1.0                                             elm 34
         d(5) =  (1.-d(9))/d(4)                                  elm 35
         d(6) = -d(9)/d(4)                                       elm 36
         d(13)=  d(9)                                            elm 37
      endif                                                      elm 38
      write(iow,2000) wd(is),(d(i),i=8,11)                       elm 39
      if(ior.lt.0) write(*,2000) wd(is),(d(i),i=8,11)            elm 40
      d(10) = d(10)*d(11)                                        elm 41
      return                                                     elm 42
   11 call perror('PCEL03',yyy)                                  elm 43
      go to 1                                                    elm 44
c.... check mesh                                                 elm 45
    2 call ckisop(ix,xl,ap,ndm)                                  elm 46
      return                                                     elm 47
c.... Compute the Pian-Sumihara arrays for elastic: compute jacobian  elm 48
    3 xs = (-xl(1,1)+xl(1,2)+xl(1,3)-xl(1,4))/4.                 elm 49
      ys = (-xl(2,1)+xl(2,2)+xl(2,3)-xl(2,4))/4.                 elm 50
      xt = (-xl(1,1)-xl(1,2)+xl(1,3)+xl(1,4))/4.                 elm 51
      yt = (-xl(2,1)-xl(2,2)+xl(2,3)+xl(2,4))/4.                 elm 52
      xh = ( xl(1,1)-xl(1,2)+xl(1,3)-xl(1,4))/4.                 elm 53
      yh = ( xl(2,1)-xl(2,2)+xl(2,3)-xl(2,4))/4.                 elm 54
      xj0 = xs*yt - xt*ys                                        elm 55
      xj1 = xs*yh - xh*ys                                        elm 56
      xj2 = xh*yt - xt*yh                                        elm 57
      if(isw.eq.5) go to 5                                       elm 58
      ssa = xj1/xj0/3.                                           elm 59
      tta = xj2/xj0/3.                                           elm 60
c.... form stiffness for elastic part and compute the beta parameters  elm 61
      call pian03(d,ul,s,p,nst,ndf,isw)                          elm 62
      if(isw.eq.4) go to 4                                       elm 63
      if(isw.eq.8) go to 8                                       elm 64
c.... compute symetric part of s                                elm 65
      do 334 i = 1,nst                                           elm 66
        do 332 j = i,nst                                         elm 67
          s(j,i) = s(i,j)                                        elm 68
```

```
332       continue                                                elm 69
334     continue                                                  elm 70
        return                                                    elm 71
c.... compute the stresses                                        elm 72
    4 is = d(12)                                                  elm 73
c.... compute the stresses at the center and the specified points elm 74
        ssg     = -ssa*beta(5)                                    elm 75
        ttg     = -tta*beta(4)                                    elm 76
        sig(1) =  beta(1) + a1(1)*ttg + a2(1)*ssg                 elm 77
        sig(2) =  beta(2) + a1(2)*ttg + a2(2)*ssg                 elm 78
        sig(3) =  beta(3) + a1(3)*ttg + a2(3)*ssg                 elm 79
        sig(4) =  d(13)*(sig(1)+sig(2))                           elm 80
  330 continue                                                    elm 81
        call pstres(sig,p1,p2,p3)                                 elm 82
        xx = (xl(1,1)+xl(1,2)+xl(1,3)+xl(1,4))/4.0                elm 83
        yy = (xl(2,1)+xl(2,2)+xl(2,3)+xl(2,4))/4.0                elm 84
        mct = mct - 1                                             elm 85
        if(mct.gt.0) go to 450                                    elm 86
        call prthed(iow)                                          elm 87
        write(iow,2001) wd(is)                                    elm 88
        if(ior.lt.0) write(*,2001) wd(is)                         elm 89
        mct = 25                                                  elm 90
  450 write(iow,2002) n,ma,xx,yy,p1,p2,p3,(sig(i),i=1,4)          elm 91
        if(ior.lt.0) write(*,2002) n,ma,xx,yy,p1,p2,p3,           elm 92
    1              (sig(i),i=1,4)                                  elm 93
        return                                                    elm 94
c.... compute a lumped mass matrix                                elm 95
    5 p(      1) = (xj0-(xj1+xj2)/3.)*d(10)                       elm 96
        p(  ndf+1) = (xj0+(xj1-xj2)/3.)*d(10)                     elm 97
        p(2*ndf+1) = (xj0+(xj1+xj2)/3.)*d(10)                     elm 98
        p(3*ndf+1) = (xj0-(xj1-xj2)/3.)*d(10)                     elm 99
        do 510 i = 2,nst,ndf                                      elm100
          p(i) = p(i-1)                                           elm101
  510 continue                                                    elm102
        return                                                    elm103
c.... error estimator goes here!                                  elm104
    7 continue                                                    elm105
        return                                                    elm106
c.... compute the nodal stress values                             elm107
    8 call stcn03(ix,d,ssa,tta,ap,ap(numnp+1),numnp)              elm108
        return                                                    elm109
c.... format statements                                           elm110
 1000 format(4f10.0,i10)                                          elm111
 2000 format(5x,'Plane St',a4,' Element'//10x,'modulus      =',e12.5/  elm112
    1  10x,'poisson ratio=', f8.5/10x,'mass density =',e12.5/    elm113
    2  10x,'thickness    =', e12.5)                               elm114
 2001 format(5x,'Plane St',a4,' Stresses'//'  element material', elm115
    1  5x,'1-coord',5x,'2-coord',8x,'sig1',8x,'sig2',3x,'angle'/ elm116
    2  38x,'s-11',8x,'s-12',8x,'s-22',8x,'s-33'/1x)              elm117
 2002 format(2i9,2f12.4,2e12.4,f8.2/30x,4e12.4)                   elm118
 3000 format(' Input: e, nu, rho, th, is (1=pl.stress,2=pl.strain)'/  elm119
    1    3x,'mate>',$)                                            elm120
        end                                                       elm121
c
        subroutine pian03(d,ul,s,p,nst,ndf,isw)                   pia  1
        implicit real*8 (a-h,o-z)                                 pia  2
c.... pian-sumihara stiffness matrix developed explicitly         pia  3
        real*8 ul(ndf,1),s(nst,1),p(1),d(1),rr(5),x1(4),x2(4),y1(4),y2(4) pia  4
    1      ,ax(4),bx(4),cx(4),ay(4),by(4),cy(4)                   pia  5
        common /elcom3/ xs,xt,xh,ys,yt,yh,xj0,xj1,xj2,a1(3),a2(3),beta(5) pia  6
```

```
c..1.) set up stress interpolants for the 4-5 term          pia  7
      r1 = xj1/xj0                                          pia  8
      r2 = xj2/xj0                                          pia  9
      a1(1) = xs*xs                                         pia 10
      a1(2) = ys*ys                                         pia 11
      a1(3) = xs*ys                                         pia 12
      a2(1) = xt*xt                                         pia 13
      a2(2) = yt*yt                                         pia 14
      a2(3) = xt*yt                                         pia 15
c..2.) set up shape function coefficients - jacobian weighted pia 16
      ax(1) = -yt + ys                                      pia 17
      ax(2) =  yt + ys                                      pia 18
      ax(3) = -ax(1)                                        pia 19
      ax(4) = -ax(2)                                        pia 20
      bx(1) = -yh - ys                                      pia 21
      bx(2) = -bx(1)                                        pia 22
      bx(3) =  yh - ys                                      pia 23
      bx(4) = -bx(3)                                        pia 24
      cx(1) =  yt + yh                                      pia 25
      cx(2) = -yt + yh                                      pia 26
      cx(3) = -cx(2)                                        pia 27
      cx(4) = -cx(1)                                        pia 28
      ay(1) =  xt - xs                                      pia 29
      ay(2) = -xt - xs                                      pia 30
      ay(3) = -ay(1)                                        pia 31
      ay(4) = -ay(2)                                        pia 32
      by(1) =  xh + xs                                      pia 33
      by(2) = -by(1)                                        pia 34
      by(3) = -xh + xs                                      pia 35
      by(4) = -by(3)                                        pia 36
      cy(1) = -xt - xh                                      pia 37
      cy(2) =  xt - xh                                      pia 38
      cy(3) = -cy(2)                                        pia 39
      cy(4) = -cy(1)                                        pia 40
c..3.) compute volume and stabilization h-array             pia 41
      vol = 4.*xj0                                          pia 42
      d11 = d(1)/vol                                        pia 43
      d12 = d(2)/vol                                        pia 44
      d33 = d(3)/vol                                        pia 45
      hy  = vol*3.                                          pia 46
      hx  = hy*d(5)                                         pia 47
      h44 = hx*(1.-r2*r2/3.)*(a1(1)+a1(2))**2               pia 48
      h55 = hx*(1.-r1*r1/3.)*(a2(1)+a2(2))**2               pia 49
      h45 =-(r1*r2/3.)*(hx*(xs*xt+ys*yt)**2+d(6)*hy*(ys*xt-xs*yt)**2) pia 50
c..4.) Invert stabilization h-array                         pia 51
      hx  = h44*h55 - h45*h45                               pia 52
      hy  = h55/hx                                          pia 53
      h55 = h44/hx                                          pia 54
      h45 =-h45/hx                                          pia 55
      h44 = hy                                              pia 56
c..5.) Compute the current stress parameters                pia 57
      call pconsd(rr,5,0.0d0)                               pia 58
      do 50 j = 1,4                                         pia 59
        hx = cx(j) - r2*ax(j)                               pia 60
        hy = cy(j) - r2*ay(j)                               pia 61
        x1(j) = a1(1)*hx + a1(3)*hy                         pia 62
        x2(j) = a1(2)*hy + a1(3)*hx                         pia 63
        hx = bx(j) - r1*ax(j)                               pia 64
        hy = by(j) - r1*ay(j)                               pia 65
        y1(j) = a2(1)*hx + a2(3)*hy                         pia 66
```

```
        y2(j) = a2(2)*hy + a2(3)*hx                              pia 67
        rr(1) = rr(1) + ax(j)*ul(1,j)                            pia 68
        rr(2) = rr(2) + ay(j)*ul(2,j)                            pia 69
        rr(3) = rr(3) + ay(j)*ul(1,j) + ax(j)*ul(2,j)            pia 70
c...... (stabilization terms)                                    pia 71
        rr(4) = rr(4) + x1(j)*ul(1,j) + x2(j)*ul(2,j)            pia 72
        rr(5) = rr(5) + y1(j)*ul(1,j) + y2(j)*ul(2,j)            pia 73
50      continue                                                 pia 74
        beta(1) = d11*rr(1) + d12*rr(2)                          pia 75
        beta(2) = d12*rr(1) + d11*rr(2)                          pia 76
        beta(3) = d33*rr(3)                                      pia 77
        beta(4) = (h44*rr(4) + h45*rr(5))*3.                     pia 78
        beta(5) = (h45*rr(4) + h55*rr(5))*3.                     pia 79
c..6.) Form stiffness matrix for 1-pt (constant) terms           pia 80
      if(isw.eq.3) then                                          pia 81
        d11 = d11*d(11)                                          pia 82
        d12 = d12*d(11)                                          pia 83
        d33 = d33*d(11)                                          pia 84
        i1 = 1                                                   pia 85
        do 110 i = 1,2                                           pia 86
          bd11 = ax(i)*d11                                       pia 87
          bd12 = ax(i)*d12                                       pia 88
          bd13 = ay(i)*d33                                       pia 89
          bd21 = ay(i)*d12                                       pia 90
          bd22 = ay(i)*d11                                       pia 91
          bd23 = ax(i)*d33                                       pia 92
          j1   = i1                                              pia 93
          do 100 j = i,2                                         pia 94
            s(i1  ,j1  ) = bd11*ax(j) + bd13*ay(j)               pia 95
            s(i1  ,j1+1) = bd12*ay(j) + bd13*ax(j)               pia 96
            s(i1+1,j1  ) = bd21*ax(j) + bd23*ay(j)               pia 97
            s(i1+1,j1+1) = bd22*ay(j) + bd23*ax(j)               pia 98
            j1 - j1 + ndf                                        pia 99
100         continue                                            pia100
          i1 - i1 + ndf                                         pia101
110     continue                                                pia102
c..7.) Copy other parts from computed terms                     pia103
        i1 = ndf + ndf                                          pia104
        do 120 i = 1,i1                                         pia105
        do 120 j = i,i1                                         pia106
          s(i  ,j+i1) =-s(i,j)                                  pia107
          s(i+i1,j+i1) = s(i,j)                                 pia108
120     continue                                                pia109
        j1 = i1 + ndf                                           pia110
        s(2   ,i1+1) =-s(2,1)                                   pia111
        s(ndf+2,j1+1) =-s(ndf+2,ndf+1)                          pia112
        s(ndf+1,i1+1) =-s(1,ndf+1)                              pia113
        s(ndf+1,i1+2) =-s(2,ndf+1)                              pia114
        s(ndf+2,i1+1) =-s(1,ndf+2)                              pia115
        s(ndf+2,i1+2) =-s(2,ndf+2)                              pia116
c..8.) Add stabilization matrix                                 pia117
        h44 = h44*d(11)                                         pia118
        h45 = h45*d(11)                                         pia119
        h55 = h55*d(11)                                         pia120
        j1 = 1                                                  pia121
        do 210 j = 1,4                                          pia122
          bd11 = h44*x1(j) + h45*y1(j)                          pia123
          bd12 = h44*x2(j) + h45*y2(j)                          pia124
          bd21 = h45*x1(j) + h55*y1(j)                          pia125
          bd22 = h45*x2(j) + h55*y2(j)                          pia126
```

```
            i1 = 1                                                     pia127
            do 200 i = 1,j                                             pia128
              s(i1  ,j1  ) = s(i1  ,j1  ) + x1(i)*bd11 + y1(i)*bd21    pia129
              s(i1  ,j1+1) = s(i1  ,j1+1) + x1(i)*bd12 + y1(i)*bd22    pia130
              s(i1+1,j1  ) = s(i1+1,j1  ) + x2(i)*bd11 + y2(i)*bd21    pia131
              s(i1+1,j1+1) = s(i1+1,j1+1) + x2(i)*bd12 + y2(i)*bd22    pia132
              i1 = i1 + ndf                                            pia133
200         continue                                                  pia134
            j1 = j1 + ndf                                             pia135
210       continue                                                    pia136
        endif                                                         pia137
c..9.) compute the residual force vector                              pia138
        if(mod(isw,3).eq.0) then                                      pia139
c...a.) compute the constant part                                     pia140
          do 300 i = 1,5                                              pia141
            beta(i) = beta(i)*d(11)                                   pia142
300       continue                                                    pia143
          do 310 i = 1,2                                              pia144
            p(2*i-1) = -(ax(i)*beta(1) + ay(i)*beta(3))               pia145
            p(2*i+3) = -p(2*i-1)                                      pia146
            p(2*i  ) = -(ay(i)*beta(2) + ax(i)*beta(3))               pia147
            p(2*i+4) = -p(2*i  )                                      pia148
310       continue                                                    pia149
c...b.) compute the stabilization part                                pia150
          do 320 i = 1,4                                              pia151
            p(2*i-1) = p(2*i-1) - (x1(i)*beta(4) + y1(i)*beta(5))/3.0  pia152
            p(2*i  ) = p(2*i  ) - (x2(i)*beta(4) + y2(i)*beta(5))/3.0  pia153
320       continue                                                    pia154
        endif                                                         pia155
        return                                                        pia156
        end                                                           pia157
c
      subroutine stcn03(ix,d,ssa,tta,dt,st,nnp)                       stc   1
      implicit real*8 (a-h,o-z)                                       stc   2
      integer*2 ix(1)                                                 stc   3
      real    dt(nnp),st(nnp,1),ss(4),tt(4)                           stc   4
      real*8 d(1)                                                     stc   5
      common /elcom3/ xs,xt,xh,ys,yt,yh,xj0,xj1,xj2,a1(3),a2(3),beta(5) stc 6
      data ss/-1.0,1.0,1.0,-1.0/,tt/-1.0,-1.0,1.0,1.0/               stc   7
c.... compute stress projections                                     stc   8
      do 200 jj = 1,4                                                 stc   9
        ll = abs(ix(jj))                                             stc  10
        if(ll.gt.0) then                                            stc  11
c.... compute weighted stresses at nodes                             stc  12
          xsj = xj0 + ss(jj)*xj1 + tt(jj)*xj2                        stc  13
          ssg = (ss(jj) - ssa)*beta(5)                               stc  14
          ttg = (tt(jj) - tta)*beta(4)                               stc  15
          dt(ll)   = dt(ll)    + xsj                                 stc  16
          sig1     = (beta(1) + a1(1)*ttg + a2(1)*ssg)*xsj           stc  17
          sig2     = (beta(2) + a1(2)*ttg + a2(2)*ssg)*xsj           stc  18
          sig3     = (beta(3) + a1(3)*ttg + a2(3)*ssg)*xsj           stc  19
          st(ll,1) = st(ll,1) + sig1                                 stc  20
          st(ll,2) = st(ll,2) + sig3                                 stc  21
          st(ll,3) = st(ll,3) + sig2                                 stc  22
          st(ll,4) = st(ll,4) + d(13)*(sig1+sig2)                    stc  23
        endif                                                        stc  24
200     continue                                                     stc  25
      return                                                         stc  26
      end                                                            stc  27
```

(d) File PCELM4.FOR contains an element module for one-, two-, or three-dimensional linear elastic truss calculations. The control information must be input so that

$$NDM \quad 1, 2, \text{ or } 3$$
$$NDF \quad = NDM$$
$$NEN \quad \geqslant 2$$

The element has features to analyse truss problems with specified nodal force loading. The element is based upon the displacement method and uses linear displacement fields between the ends of the truss element. The element computes only element force values; no error estimates are included in this element.

The material property set parameters that follow the data described in Table 15.8 are input as follows:

Property Record 1.) FORMAT—3F10.0

Column   Description

1 to 10 E, Young's modulus
11 to 20 A, cross-sectional area
21 to 30 ρ, mass density (per unit volume)

File PCELM4.FOR consists of the following subprogram:

Name	*Type*	*Description*
ELMT04	SUBROUTINE	Inputs parameters, computers FE arrays

```
$NOFLOATCALLS
        subroutine elmt04(d,u,x,ix,t,s,p,ndf,ndm,nst,isw)        elm  1
c.... elastic 1,2, or 3d truss element routine                  elm  2
        implicit real*8 (a-h,o-z)                               elm  3
        real    x(ndm,1),xx(3),dm                               elm  4
        real*8  d(1),u(ndf,1),s(nst,1),p(1),db(3),dx(3)         elm  5
        character yyy*80                                        elm  6
        common /eldata/ dm,n,ma,mct,iel,nel                     elm  7
        common /iofile/ ior,iow                                 elm  8
        go to (1,3,3,3,3,3,8,8),isw                             elm  9
c.... Input material properties                                 elm 10
1       if(ior.lt.0) write(*,3000)                              elm 11
        call pintio(yyy,10)                                     elm 12
        read(yyy,1000,err=110) d(1),d(2),d(3)                   elm 13
        d(4) = d(1)*d(2)                                        elm 14
        d(5) = d(3)*d(2)                                        elm 15
        call pconsr(xx,3,0.0)                                   elm 16
        write(iow,2000) d(1),d(2),d(3)                          elm 17
        if(ior.lt.0) write(*,2000) d(1),d(2),d(3)               elm 18
        return                                                  elm 19
110     call perror('PCELM9',yyy)                               elm 20
        go to 1                                                 elm 21
c.... compute element properties                                elm 22
```

```
3      if(ndf.ne.ndm) then                                              elm 23
         write(iow,4001)                                                elm 24
         if(ior.lt.0) write(*,4001)                                     elm 25
         stop                                                           elm 26
       endif                                                            elm 27
       xl = 0.0                                                         elm 28
       eps = 0.0                                                        elm 29
       do 31 i = 1,ndm                                                  elm 30
         dx(i) = x(i,2) - x(i,1)                                        elm 31
         xl = xl + dx(i)**2                                             elm 32
         eps = eps + dx(i)*(u(i,2)-u(i,1))                              elm 33
         xx(i) = (x(i,2) + x(i,1))/2.                                   elm 34
31     continue                                                         elm 35
c....  error check on length                                           elm 36
       if(isw.eq.2) then                                                elm 37
         if(xl.le.0.0d0. or. x(1,1).eq.-999.0 .or. x(1,2).eq.-999.) then elm 38
           write(iow,4000) n,x(1,1),x(1,2)                             elm 39
           if(ior.lt.0) write(*,4000) n,x(1,1),x(1,2)                  elm 40
         endif                                                          elm 41
         return                                                         elm 42
       endif                                                            elm 43
       eps = eps/xl                                                     elm 44
       sig = d(4)*eps                                                   elm 45
c....  compute the element stiffness matrix                            elm 46
       if(isw.eq.3) then                                                elm 47
         xl = xl*sqrt(xl)                                               elm 48
         do 32 i = 1,ndm                                                elm 49
           db(i) = d(4)*dx(i)                                           elm 50
           dx(i) = dx(i)/xl                                             elm 51
32       continue                                                       elm 52
         do 33 i = 1,ndm                                                elm 53
         do 33 j = 1,ndm                                                elm 54
           s(i,j) = db(i)*dx(j)                                         elm 55
           s(i+ndf,j+ndf) = s(i,j)                                      elm 56
           s(i,j+ndf) = -s(i,j)                                         elm 57
           s(i+ndf,j) = -s(j,i)                                         elm 58
33       continue                                                       elm 59
c....  output stress and strain in element                             elm 60
       elseif(isw.eq.4) then                                            elm 61
         mct = mct - 1                                                  elm 62
         if(mct.le.0) then                                              elm 63
           call prthed(iow)                                             elm 64
           write(iow,2001)                                             elm 65
           if(ior.lt.0) write(*,2001)                                  elm 66
           mct = 50                                                     elm 67
         endif                                                          elm 68
         write(iow,2002) n,ma,xx,sig,eps                               elm 69
         if(ior.lt.0) write(*,2002) n,ma,xx,sig,eps                    elm 70
c....  compute element lumped mass matrix                              elm 71
       elseif(isw.eq.5) then                                            elm 72
         xl = d(5)*sqrt(xl)/2.0                                         elm 73
         do 34 i = 1,ndm                                                elm 74
           p(i    ) = xl                                                elm 75
           p(i+ndf) = xl                                                elm 76
34       continue                                                       elm 77
       endif                                                            elm 78
c....  form a residual if needed                                       elm 79
       if(mod(isw,3).eq.0) then                                         elm 80
         sig = sig/sqrt(xl)                                             elm 81
         do 35 i = 1,ndf                                                elm 82
```

```
                p(i) = dx(i)*sig                                          elm 83
                p(i+ndf) = -p(i)                                          elm 84
35          continue                                                     elm 85
         endif                                                           elm 86
8       return                                                           elm 87
c.... formats                                                            elm 88
1000    format(3f10.0)                                                   elm 89
2000    format(5x,'T r u s s    E l e m e n t'//10x,'Modulus =',e13.5/10x,elm 90
     1   'Area   =',e13.5/10x,'Density =',e13.5)                         elm 91
2001    format(5x,'T r u s s    E l e m e n t'//' elem mate',            elm 92
     1  4x,'1-coord',4x,'2-coord',4x,'3-coord',5x,'force',7x,'strain')   elm 93
2002    format(2i5,3f11.4,2e13.5)                                        elm 94
3000    format(' Input: E, A, rho'/3x,'>',$)                             elm 95
4000    format(' ** ERROR ** Truss element ',i4,' has zero length or'/   elm 96
     1    '           undefined points: 1=',e10.3,' , 2=',e10.3)         elm 97
4001    format(' ** ERROR ** Truss element must have ndm = ndf!')        elm 98
        end                                                              elm 99
```

References

1. O. C. ZIENKIEWICZ, *The Finite Element Method*, 3rd ed., McGraw-Hill, London, 1977.
2. B. M. IRONS, 'A frontal solution program', *Int. J. Num. Meth. Eng.*, **2**, 5–32, 1970.
3. E. HINTON and D. R. J. OWEN, *Finite Element Programming*, Academic Press, 1977.
4. R. L. TAYLOR, 'Solution of linear equations by a profile solver', *Eng. Comput.*, **2**, 344–50, 1985.
5. Graphics Development Toolkit, IBM Personal Computer Software, Boca Raton, Florida, 1984.
6. P. HOOD, 'Frontal solution program for unsymmetric matrices', *Int. J. Num. Meth. Eng.*, **10**, 379–400, 1976.
7. O. C. ZIENKIEWICZ and D. V. PHILLIPS, 'An automatic mesh generation scheme for plane and curved surfaces by isoparametric coordinates', *Int. J. Num. Meth. Eng.*, **3**, 519–28, 1971.
8. THOMAS J. R. HUGHES, *The Finite Element Method*, Prentice-Hall, 1987.
9. W. PILKEY, K. SACZALSKI, and H. SCHAEFFER (eds), *Structural Mechanics Computer Programs*, University Press of Virginia, Charlottesville, 1974.
10. H. H. FONG, 'Interactive graphics and commercial finite element codes', Mechanical Engineering ASME 106, June 1984.
11. B. M. IRONS, 'A technique for degenerating brick type isoparametric elements using hierarchical midside nodes', *Int. J. Num. Meth. Eng.*, **8**, 209–11, 1973.
12. R. L. TAYLOR, E. L. WILSON, and S. J. SACKETT, 'Direct solution of equations by frontal and variable band, active column methods', in *Nonlinear Finite Element Analysis in Structural Mechanics* (eds W. Wunderlich, E. Stein, and K.-J. Bathe), Springer-Verlag, 1981.
13. N. E. GIBBS, W. G. POOLE, JR, and P. K. STOCKMEYER, 'An algorithm for reducing the bandwidth and profile of a sparse matrix', *SIAM J. Num. Anal.*, **13**, 236–50, 1976.
14. W.-H. LIU and A. H. SHERMAN, 'Comparative analysis of the Cuthill–McKee and the reversed Cuthill–McKee ordering algorithms for sparse matrices', *SIAM J. Num. Anal.*, **13**, 198–213, 1976.
15. M. HOIT and E. L. WILSON, 'An equation numbering algorithm based on a minimum front criteria', *Comp. Struct.*, **16**, 225–39, 1983.
16. M. G. KATONA and O. C. ZIENKIEWICZ, 'A unified set of single step algorithms, Part 3: The Beta-m method, a generalization of the Newmark scheme', *Int. J. Num. Meth. Eng.*, **21**, 1345–59, 1985.
17. N. M. NEWMARK, 'A method of computation for structural dynamics', *J. Eng. Mech. Div. ASCE*, **85**, 67–94, 1959.
18. A. K. GUPTA and B. MOHRAZ, 'A method of computing numerically integrated stiffness matrices', *Int. J. Num. Meth. Eng.*, **5**, 83–9, 1972.
19. E. L. WILSON, 'SAP—a general structural analysis program for linear systems', *Nudl. Engr. Des.*, **25**, 257–74, 1973.
20. A. RALSTON, *A First Course in Numerical Analysis*, McGraw-Hill, 1965.
21. L. FOX, *An Introduction to Numerical Linear Algebra*, Oxford University Press, 1965.
22. J. H. WILKINSON and C. REINSCH, *Linear Algebra. Handbook for Automatic Computation*, Vol. II, Springer-Verlag, 1971.
23. G. STRANG, *Linear Algebra and Its Applications*, Academic Press, 1976.

24. C. MEYER, 'Solution of equations; state-of-the-art', *J. Struct. Div. ASCE*, **99** (7), 1507–26, 1973.
25. C. MEYER, 'Special problems related to linear equation solvers', *J. Struct. Div. ASCE*, **101** (4), 869–90, 1975.
26. K.-J. BATHE and E. L. WILSON, *Numerical Methods in Finite Element Analysis*, Prentice-Hall, 1976.
27. A. JENNINGS, 'A compact storage scheme for the solution of symmetric simultaneous equations', *Comp. J.*, **9**, 281–5, 1966.
28. D. P. MONDKAR and G. H. POWELL, 'Towards optimal in-core equation solving', *Comp. Struct.*, **4**, 531–48, 1974.
29. C. A. FELIPPA, 'Solution of linear equations with skyline-stored symmetric matrix', *Comp. Struct.*, **5**, 13–30, 1975.
30. E. L. WILSON, 'SAP-80 structural analysis program for small or large computer systems', *Proc. CEPA Fall Conf.*, 1980.
31. T. J. R. HUGHES, I. LEVIT, and J. WINGET, 'Element-by-element implicit algorithms for heat conduction', *J. Eng. Mech. ASCE*, **109** (2), 576–85, 1983.
32. T. J. R. HUGHES, I. LEVIT, and J. WINGET, 'Element-by-element implicit algorithms for problems of structural and solid mechanics', *Comp. Meth. Appl. Mech. Eng.*, **36**, 241–54, 1983.
33. B. M. IRONS, Personal communication, 1970.
34. G. STRANG and G. J. FIX, *An Analysis of the Finite Element Method*, Prentice-Hall, 1973.
35. Fortran 77 Compiler for Personal Computers, MicroSoft Corporation, 1985.
36. T. H. H. PIAN and K. SUMIHARA, 'Rational approach for assumed stress finite elements', *Int. J. Num. Meth. Eng.*, **20**, 1685–95, 1984.

Matrix algebra

The mystique surrounding matrix algebra is perhaps due to the texts on the subject requiring the student to 'swallow too much' in one operation. It will be found that in order to follow the present text and carry out the necessary computation only a limited knowledge of a few basic definitions is required.

Definition of a matrix

The linear relationship between a set of variables x and b:

$$a_{11}x_1 + a_{12}x_2 + a_{13}x_3 + a_{14}x_4 = b_1$$
$$a_{21}x_1 + a_{22}x_2 + a_{23}x_3 + a_{24}x_4 = b_2 \qquad \text{(A1.1)}$$
$$a_{31}x_1 + a_{32}x_2 + a_{33}x_3 + a_{34}x_4 = b_3$$

can be written, in a shorthand way, as

$$[A]\{x\} = \{b\}$$

or

$$\mathbf{Ax} = \mathbf{b}$$

(A1.1a)

where

$$\mathbf{A} \equiv [A] = \begin{bmatrix} a_{11}, a_{12}, a_{13}, a_{14} \\ a_{21}, a_{22}, a_{23}, a_{24} \\ a_{31}, a_{32}, a_{33}, a_{34} \end{bmatrix} \qquad \text{(A1.2)}$$

$$\mathbf{x} \equiv \{x\} = \begin{Bmatrix} x_1 \\ x_2 \\ x_3 \\ x_4 \end{Bmatrix}$$

$$\mathbf{b} \equiv \{b\} = \begin{Bmatrix} b_1 \\ b_2 \\ b_3 \end{Bmatrix}$$

The above notation contains within it both the definition of a matrix and of the process of multiplication. Matrices are *defined* as 'arrays of numbers' of the type shown in (A1.2). The particular form listing a single column of numbers is often referred to as a vector or column matrix. The multiplication of a matrix by a column vector is *defined* by the equivalence of the left sides of Eqs (A1.1) and (A1.1a).

The use of bold characters to define both vectors and matrices will be followed throughout the text—generally lower case letters denoting vectors and capitals matrices.

If another relationship, using the same constants, but a different set of x and b, exists, and is written as

$$a_{11}x'_1 + a_{12}x'_2 + a_{13}x'_3 + a_{14}x'_4 = b'_1$$

$$a_{21}x'_1 + a_{22}x'_2 + a_{23}x'_3 + a_{24}x'_4 = b'_2 \qquad \text{(A1.3)}$$

$$a_{31}x'_1 + a_{32}x'_2 + a_{33}x'_3 + a_{34}x'_4 = b'_3$$

then we could write

$$[A][X] = [B]$$

or $\qquad\qquad\qquad\qquad\qquad\qquad\qquad\qquad\qquad\qquad$ (A1.4)

$$\mathbf{AX = B}$$

in which

$$\mathbf{X} \equiv [X] = \begin{bmatrix} x_1, & x'_1 \\ x_2, & x'_2 \\ x_3, & x'_3 \\ x_4, & x'_4 \end{bmatrix} \qquad \mathbf{B} \equiv [B] = \begin{bmatrix} b_1, & b'_1 \\ b_2, & b'_2 \\ b_3, & b'_3 \end{bmatrix} \qquad \text{(A1.5)}$$

implying both the statements (A1.1) and (A1.3) arranged simultaneously as

$$\begin{bmatrix} a_{11}x_1 + \cdots, & a_{11}x'_1 + \cdots \\ a_{21}x_1 + \cdots, & a_{21}x'_1 + \cdots \\ a_{31}x_1 + \cdots, & a_{31}x'_1 + \cdots \end{bmatrix} = \begin{bmatrix} b_1, & b'_1 \\ b_2, & b'_2 \\ b_3, & b'_3 \end{bmatrix} \qquad \text{(A1.4a)}$$

It is seen, incidentally, that matrices can be equal only if each of the individual terms is equal.

The multiplication of full matrices is defined above, and it is obvious that it has a meaning only if the number of columns in $\mathbf{A}$ is equal to the number of rows in $\mathbf{X}$ for a relation of type (A1.4). One property that distinguishes matrix multiplication is that, in general,

$$\mathbf{AX \neq XA}$$

i.e., multiplication of matrices is not commutative as in ordinary algebra.

Matrix addition or subtraction

If relations of the form from (A1.1) and (A1.3) are added then we have

$$a_{11}(x_1 + x_1') + a_{12}(x_2 + x_2') + a_{13}(x_3 + x_3') + a_{14}(x_4 + x_4')$$
$$= b_1 + b_1'$$

$$a_{21}(x_1 + x_1') + a_{22}(x_2 + x_2') + a_{23}(x_3 + x_3') + a_{24}(x_4 + x_4')$$
$$= b_2 + b_2' \quad \text{(A1.6)}$$

$$a_{31}(x_1 + x_1') + a_{32}(x_2 + x_2') + a_{33}(x_3 + x_3') + a_{34}(x_4 + x_4')$$
$$= b_3 + b_3'$$

which will also follow from

$$\mathbf{Ax} + \mathbf{Ax'} = \mathbf{A(x + x')} = \mathbf{b} + \mathbf{b'} = \mathbf{b'} + \mathbf{b}$$

if we define the addition of matrices by a simple addition of the individual terms of the array. Clearly this can be done only if the size of the matrices is identical, i.e., for example,

$$\begin{bmatrix} a_{11}, a_{12}, a_{13} \\ a_{21}, a_{22}, a_{23} \end{bmatrix} + \begin{bmatrix} b_{11}, b_{12}, b_{13} \\ b_{21}, b_{22}, b_{23} \end{bmatrix} = \begin{bmatrix} a_{11} + b_{11}, a_{12} + b_{12}, a_{13} + b_{13} \\ a_{21} + b_{21}, a_{22} + b_{22}, a_{23} + b_{23} \end{bmatrix}$$

or   (A1.7)

$$\mathbf{A + B = C}$$

implies that every term of $\mathbf{C}$ is equal to the sum of the appropriate terms of $\mathbf{A}$ and $\mathbf{B}$.

Subtraction obviously follows similar rules.

Transpose of matrix

This is simply a definition for reordering of the number of an array in the following manner:

$$\begin{bmatrix} a_{11} & a_{12} & a_{13} \\ a_{21} & a_{22} & a_{23} \end{bmatrix}^{\mathrm{T}} = \begin{bmatrix} a_{11} & a_{21} \\ a_{12} & a_{22} \\ a_{13} & a_{23} \end{bmatrix} \quad \text{(A1.8)}$$

and will be indicated by the symbol T as shown.

Its use is not immediately obvious but will be indicated later and can be treated here as a simple prescribed operation.

Inverse of a matrix

If in the relationship (A1.1a) the matrix $\mathbf{A}$ is 'square', i.e., it represents the coefficients of simultaneous equations of type (A1.1) equal in number to the number of unknowns **x**, then in general it is possible to solve for the unknowns **x** in terms of the known coefficients **b**. This solution can be written as

$$\mathbf{x} = \mathbf{A}^{-1}\mathbf{b} \tag{A1.9}$$

in which the matrix $\mathbf{A}^{-1}$ is known as the 'inverse' of the square matrix **A**. Clearly $\mathbf{A}^{-1}$ is also square and of the same size as **A**.

We could obtain (A1.9) by multiplying both sides of (A1.1a) by $\mathbf{A}^{-1}$ and hence

$$\mathbf{A}\mathbf{A}^{-1} = \mathbf{A}^{-1}\mathbf{A} = \mathbf{I} \tag{A1.10}$$

where **I** is an identity matrix having zero on all 'off-diagonal' positions and unity on each of the diagonal positions.

If the equations are singular and have no solution then clearly an inverse does not exist.

A sum of products

In problems of mechanics we often encounter a number of quantities such as forces that can be listed as a matrix 'vector':

$$\mathbf{f} = \begin{Bmatrix} f_1 \\ f_2 \\ \vdots \\ f_n \end{Bmatrix} \tag{A1.11}$$

These, in turn, are often associated with the same number of displacements given by another vector, say,

$$\mathbf{u} = \begin{Bmatrix} u_1 \\ u_2 \\ \vdots \\ u_n \end{Bmatrix} \tag{A1.12}$$

It is known that the work is represented as a sum of products of force and displacement:

$$W = \sum f_n u_n$$

Clearly the transpose becomes useful here as we can write, by the first rule of matrix multiplication,

$$W = [f_1, f_2, \ldots, f_n] \begin{Bmatrix} a_1 \\ a_2 \\ \vdots \\ a_n \end{Bmatrix} = \mathbf{f}^{\mathrm{T}}\mathbf{a} \equiv \mathbf{a}^{\mathrm{T}}\mathbf{f} \tag{A1.13}$$

Use of this fact is made frequently in this book.

Transpose of a product

An operation that sometimes occurs is that of taking the transpose of a matrix product. It can be left to the reader to prove from previous definitions that

$$(\mathbf{AB})^\mathrm{T} = \mathbf{B}^\mathrm{T}\mathbf{A}^\mathrm{T} \tag{A1.14}$$

Symmetric matrices

In structural problems symmetric matrices are often encountered. If a term of a matrix $\mathbf{A}$ is defined as a_{ij}, then for a symmetric matrix

$$a_{ij} = a_{ji}$$

It can be shown that the inverse of a symmetric matrix is always symmetric.

Partitioning

It is easy to verify that a matrix product

$$\mathbf{AB}$$

in which, for example,

$$\mathbf{A} = \begin{bmatrix} a_{11} & a_{12} & a_{13} & a_{14} & a_{15} \\ a_{21} & a_{22} & a_{23} & a_{24} & a_{25} \\ a_{31} & a_{32} & a_{33} & a_{34} & a_{35} \end{bmatrix}$$

$$\mathbf{B} = \begin{bmatrix} b_{11} & b_{12} \\ b_{21} & b_{22} \\ b_{31} & b_{32} \\ b_{41} & b_{42} \\ b_{51} & b_{52} \end{bmatrix}$$

could be obtained by dividing each matrix into submatrices, indicated by the dotted lines, and applying the rules of matrix multiplication first to each of such submatrices as if it were a scalar number and then carrying out further multiplication in the usual way. Thus, if we write

$$\mathbf{A} = \begin{bmatrix} \mathbf{A}_{11} & \mathbf{A}_{12} \\ \mathbf{A}_{21} & \mathbf{A}_{22} \end{bmatrix} \qquad \mathbf{B} = \begin{bmatrix} \mathbf{B}_1 \\ \mathbf{B}_2 \end{bmatrix}$$

then

$$\mathbf{AB} = \begin{bmatrix} \mathbf{A}_{11}\mathbf{B}_1 + \mathbf{A}_{12}\mathbf{B}_2 \\ \mathbf{A}_{21}\mathbf{B}_1 + \mathbf{A}_{22}\mathbf{B}_2 \end{bmatrix}$$

can be verified as representing the complete product by further multiplication.

The essential feature of partitioning is that the size of subdivisions has to be such as to make the products of type $\mathbf{A}_{11}\mathbf{B}_1$ meaningful, i.e., the number of columns in $\mathbf{A}_{11}$ must be equal to the number of rows in $\mathbf{B}_1$, etc. If the above definition holds, then all further operations can be conducted on partitioned matrices, treating each partition as if it were a scalar.

It should be noted that any matrix can be multiplied by a scalar (number). Here, obviously, the requirements of equality of appropriate rows and columns no longer apply.

If a symmetric matrix is divided into an equal number of submatrices $\mathbf{A}_{ij}$ in rows and columns then

$$\mathbf{A}_{ij} = \mathbf{A}_{ji}^{\mathrm{T}}$$

The Eigenvalue problem

An eigenvalue of a symmetric matrix $\mathbf{A}$ of size $n \times n$ is a scalar λ_i which allows the solution of

$$(\mathbf{A} - \lambda_i \mathbf{I})\boldsymbol{\phi}_i = 0 \quad \text{and} \quad \det|\mathbf{A} - \lambda_i I| = 0 \qquad (A1.15)$$

There are, of course, n such eigenvalues λ_i to each of which corresponds a vector $\boldsymbol{\phi}_i$. Such vectors can be shown to be orthonormal and we write

$$\boldsymbol{\phi}_i \boldsymbol{\phi}_j = 0 \quad i \neq j \quad = 1 \quad i = j$$

The full set of eigenvalues and eigenvectors can be written as

$$\Lambda = \begin{bmatrix} \lambda_1 & & 0 \\ & \ddots & \\ 0 & & \lambda_n \end{bmatrix} \quad \boldsymbol{\Phi} = [\boldsymbol{\phi}_1, \ldots, \boldsymbol{\phi}_n]$$

The matrix $\mathbf{A}$ can be written in its *spectral form* which the reader can verify

$$\mathbf{A} = \boldsymbol{\Phi}\Lambda\boldsymbol{\Phi}^{\mathrm{T}} \qquad (A.1.16)$$

The condition number κ (which is related to equation-solution roundoff) is defined as

$$\kappa = \frac{|\lambda|_{\max}}{|\lambda|_{\min}} \qquad (A1.17)$$

Basic equations of displacement analysis (Chapter 2)

Displacement

(2.1) $$\mathbf{u} \approx \hat{\mathbf{u}} = \sum \mathbf{N}_i \mathbf{a}_i = \mathbf{N}\mathbf{a}$$

Strain

(2.2) and (2.3)
(2.4)
$$\boldsymbol{\varepsilon} = \mathbf{S}\mathbf{u} \approx \sum \mathbf{B}_i \mathbf{a}_i = \mathbf{B}\mathbf{a}$$
$$\mathbf{B}_i = \mathbf{S}\mathbf{N}_i$$
$$\mathbf{B} = \mathbf{S}\mathbf{N}$$

Stress–strain—constitutive relation of linear elasticity

(2.5) $$\boldsymbol{\sigma} = \mathbf{D}(\boldsymbol{\varepsilon} - \boldsymbol{\varepsilon}_0) + \boldsymbol{\sigma}_0$$

Approximate equilibrium equations

(2.23) $$\mathbf{K}\mathbf{a} + \mathbf{f} = \mathbf{r}$$

(2.24) $$\mathbf{K}_{ij} = \int_V \mathbf{B}_i^{\mathsf{T}} \mathbf{D} \mathbf{B}_j \, \mathrm{d}V$$

$$\mathbf{f}_i = - \int_V \mathbf{N}_i^{\mathsf{T}} \mathbf{b} \, \mathrm{d}V - \int_A \mathbf{N}_i^{\mathsf{T}} \mathbf{t} \, \mathrm{d}A$$

$$- \int_V \mathbf{B}_i^{\mathsf{T}} \mathbf{D} \boldsymbol{\varepsilon}_0 \, \mathrm{d}V + \int_V \mathbf{B}_i^{\mathsf{T}} \boldsymbol{\sigma}_0 \, \mathrm{d}V$$

Some integration formulae for a triangle (Fig. 3.1)

Let a triangle be defined in the xy plane by three points (x_i, y_i), (x_j, y_j), (x_m, y_m) with the origin at the coordinates taken at the centroid, i.e.,

$$\frac{x_i + x_j + x_m}{3} = \frac{y_i + y_j + y_m}{3} = 0$$

Then integrating over the triangle area

$$\int x \, dx \, dy = \int y \, dx \, dy = 0$$

$$\int dx \, dy = \frac{1}{2} \begin{vmatrix} 1 & x_i & y_i \\ 1 & x_j & y_j \\ 1 & x_m & y_m \end{vmatrix} = \Delta = \text{area of triangle}$$

$$\int x^2 \, dx \, dy = \frac{\Delta}{12} (x_i^2 + x_j^2 + x_m^2)$$

$$\int y^2 \, dx \, dy = \frac{\Delta}{12} (y_i^2 + y_j^2 + y_m^2)$$

$$\int xy \, dx \, dy = \frac{\Delta}{12} (x_i y_i + x_j y_j + x_m y_m)$$

Some integration formulae for a tetrahedron (Fig. 5.1)

Let a tetrahedron be defined in the coordinate system (x, y, z) by four points (x_i, y_i, z_i), (x_j, y_j, z_j), (x_m, y_m, z_m), (x_p, y_p, z_p) with the origin at the coordinates taken at the centroid, i.e.,

$$\frac{x_i + x_j + x_m + x_p}{4} = \frac{y_i + y_j + y_m + y_p}{4} = \frac{z_i + z_j + z_m + z_p}{4} = 0$$

Then integrating over the tetrahedron volume

$$\int dx \, dy \, dz = \frac{1}{6} \begin{vmatrix} 1 & x_i & y_i & z_i \\ 1 & x_j & y_j & z_j \\ 1 & x_m & y_m & z_m \\ 1 & x_p & y_p & z_p \end{vmatrix} = V = \text{volume of tetrahedron}$$

Provided the order of numering is as indicated on Fig. 6.1 then also:

$$\int x \, dx \, dy \, dz = \int y \, dx \, dy \, dz = \int z \, dx \, dy \, dz = 0$$

$$\int x^2 \, dx \, dy \, dz = \frac{V}{20} (x_i^2 + x_j^2 + x_m^2 + x_p^2)$$

$$\int y^2 \, dx \, dy \, dz = \frac{V}{20} (y_i^2 + y_j^2 + y_m^2 + y_p^2)$$

$$\int z^2 \, dx \, dy \, dz = \frac{V}{20} (z_i^2 + z_j^2 + z_m^2 + z_p^2)$$

$$\int xy \, dx \, dy \, dz = \frac{V}{20} \left(x_i y_i + x_j y_j + x_m y_m + x_p y_p \right)$$

$$\int xz \, dx \, dy \, dz = \frac{V}{20} \left(x_i z_i + x_j z_j + x_m z_m + x_p z_p \right)$$

$$\int yz \, dx \, dy \, dz = \frac{V}{20} \left(y_i z_i + y_j z_j + y_m z_m + y_p z_p \right)$$

Some vector algebra

Some knowledge and understanding of basic vector algebra is needed in dealing with complexities of elements oriented in space such as occur in shells, etc. Some of the operations are summarized here.

Vectors (in the geometric sense) can be described by their components along the directions of the x, y, z axis.

Thus the vector $\mathbf{V}_{01}$ shown in Fig. A5.1 can be written as

$$\mathbf{V}_{01} = \mathbf{i}x_1 + \mathbf{j}y_1 + \mathbf{k}z_1 \tag{A5.1}$$

in which $\mathbf{i}, \mathbf{j}, \mathbf{k}$ are unit vectors in the directions of the axes x, y, z.

Alternatively, the same vector could be written as

$$\mathbf{V}_{01} = \begin{Bmatrix} x_1 \\ y_1 \\ z_1 \end{Bmatrix} \tag{A5.2}$$

(now a 'vector' in the matrix sense) in which the components are distinguished by positions in the column.

Addition and subtraction

Addition and subtraction is defined by addition and subtraction of components. Thus, for example,

$$\mathbf{V}_{02} - \mathbf{V}_{01} = \mathbf{V}_{21} = \mathbf{i}(x_2 - x_1) + \mathbf{j}(y_2 - y_1) + \mathbf{k}(z_2 - z_1) \tag{A5.3}$$

The same result is achieved by the definitions of matrix algebra; thus

$$\mathbf{V}_{02} - \mathbf{V}_{01} = \mathbf{V}_{21} = \begin{Bmatrix} x_2 - x_1 \\ y_2 - y_1 \\ z_2 - z_1 \end{Bmatrix} \tag{A5.4}$$

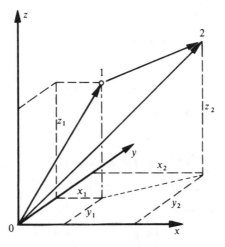

Fig. A5.1 Vector addition

Length of vector

The length of the vector V_{21} is given, purely geometrically, as

$$l_{21} = \sqrt{(x_2 - x_1)^2 + (y_2 - y_1)^2 + (z_2 - z_1)^2} \qquad (A5.5)$$

or in terms of matrix algebra as

$$l_{12} = \sqrt{V_{12}^T V_{12}} \qquad (A5.6)$$

Direction cosines

Direction cosines of a vector are simply, from the definition of the projected component lengths, given as

$$\cos \alpha_x = \lambda_{vx} = \frac{x_2 - x_1}{l_{12}}, \text{ etc.} \qquad (A5.7)$$

where α_x is the angle between the vector and x axis.

'Scalar' products

A scalar product of two vectors is *defined* as the product of the length of one vector by the scalar projection of the other vector on it. Or if γ is the angle between two vectors A and B and their length l_a and l_b respectively

$$A \cdot B = l_a l_b \cos \gamma = B \cdot A \qquad (A5.8)$$

If

$$\mathbf{A} = \mathbf{i}a_x + \mathbf{j}a_y + \mathbf{k}a_z$$

and   (A5.9)

$$\mathbf{B} = \mathbf{i}b_x + \mathbf{j}b_y + \mathbf{k}b_z$$

$$\mathbf{A} \cdot \mathbf{B} = a_x b_x + a_y b_y + a_z b_z \qquad (A5.10)$$

if we note that, by above definition,

$$\mathbf{i} \cdot \mathbf{i} = \mathbf{j} \cdot \mathbf{j} = \mathbf{k} \cdot \mathbf{k} = 1$$

$$\mathbf{i} \cdot \mathbf{j} = \mathbf{j} \cdot \mathbf{k} = \mathbf{k} \cdot \mathbf{i} = 0, \text{ etc.}$$

Using the matrix notation

$$\mathbf{A} = \begin{Bmatrix} a_x \\ a_y \\ a_z \end{Bmatrix} \qquad \mathbf{B} = \begin{Bmatrix} b_x \\ b_y \\ b_z \end{Bmatrix} \qquad (A5.11)$$

$$\mathbf{A} \cdot \mathbf{B} = \mathbf{A}^{\mathrm{T}}\mathbf{B} = \mathbf{B}^{\mathrm{T}}\mathbf{A} \qquad (A5.12)$$

'Vector' or cross product

Another product of the vector is *defined* as a vector oriented normally to the plane given by the two vectors and equal in magnitude to the product of the length of the two vectors, multiplied by the sine of the angle between them. Further, its direction follows the right-hand rule as shown in Fig. A5.2 in which

$$\mathbf{A} \times \mathbf{B} = \mathbf{C} \qquad (A5.13)$$

is shown.

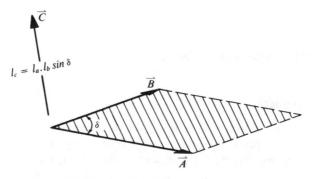

Fig. A5.2 Vector multiplication (cross product)

Thus

$$\mathbf{A} \times \mathbf{B} = -\mathbf{B} \times \mathbf{A} \tag{A5.14}$$

It is worth noting that the magnitude (*or length*) of $\mathbf{C}$ is equal to the area of the parallelogram shown in Fig. A5.2.

Using the definition of Eq. (A5.9) and noting that

$$\mathbf{i} \times \mathbf{i} = \mathbf{j} \times \mathbf{j} = \mathbf{k} \times \mathbf{k} = 0$$
$$\mathbf{i} \times \mathbf{j} = \mathbf{k} \qquad \mathbf{j} \times \mathbf{k} = \mathbf{i} \qquad \mathbf{k} \times \mathbf{i} = \mathbf{j} \tag{A5.15}$$

we have

$$\mathbf{A} \times \mathbf{B} = \det \begin{vmatrix} \mathbf{i} & \mathbf{j} & \mathbf{k} \\ a_x & a_y & a_z \\ b_x & b_y & b_z \end{vmatrix}$$

$$= (a_y b_z - a_z b_y)\mathbf{i} + (a_z b_x - a_x b_z)\mathbf{j} + (a_x b_y - a_y b_x)\mathbf{k} \tag{A5.16}$$

In matrix algebra this does not find a simple counterpart but we can use the above to define the vector $\mathbf{C}$†

$$\mathbf{C} = \mathbf{A} \times \mathbf{B} = \begin{Bmatrix} a_y b_z - a_z b_y \\ a_z b_x - a_x b_z \\ a_x b_y - a_y b_x \end{Bmatrix} \tag{A5.17}$$

The vector product will be found particularly useful when the problem of erecting a normal direction to a surface is considered.

Elements of area and volume

If ξ and η are some curvilinear coordinates then the following vectors in two-dimensional plane

$$d\boldsymbol{\xi} = \begin{Bmatrix} \dfrac{\partial x}{\partial \xi} \\ \dfrac{\partial y}{\partial \xi} \end{Bmatrix} d\xi \qquad d\boldsymbol{\eta} = \begin{Bmatrix} \dfrac{\partial x}{\partial \eta} \\ \dfrac{\partial y}{\partial \eta} \end{Bmatrix} d\eta \tag{A5.18}$$

† If we rewrite $\mathbf{A}$ as a skew symmetric matrix

$$\hat{\mathbf{A}} = \begin{bmatrix} 0, & -a_z, & a_y \\ a_z, & 0, & -a_x \\ -a_y, & a_x, & 0 \end{bmatrix}$$

then the reader can verify that an alternative representation of the vector product in matrix form is (T. Crouch, personal communication)

$$\mathbf{C} = \hat{\mathbf{A}}\mathbf{B}$$

defined from the relationship between the cartesian and curvilinear coordinates, are vectors directed tangentially to the $\xi = $ constant and $\eta = $ constant contours respectively. As the *length* of the vector resulting from a cross product of $d\xi \times d\eta$ is equal to the area of the elementary parallelogram we can write

$$d(\text{area}) = \det \begin{vmatrix} \dfrac{\partial x}{\partial \xi} & \dfrac{\partial x}{\partial \eta} \\ \dfrac{\partial y}{\partial \xi} & \dfrac{\partial y}{\partial \eta} \end{vmatrix} d\xi \; d\eta \qquad (A5.19)$$

by Eq. (A5.17).

Similarly, if we have three curvilinear coordinates ξ, η, ζ in the cartesian space, the 'triple scalar' or box product defines a unit volume

$$d(\text{vol}) = d\xi(d\eta \cdot d\zeta) = \det \begin{vmatrix} \dfrac{\partial x}{\partial \xi} & \dfrac{\partial x}{\partial \eta} & \dfrac{\partial x}{\partial \zeta} \\ \dfrac{\partial y}{\partial \xi} & \dfrac{\partial y}{\partial \eta} & \dfrac{\partial y}{\partial \zeta} \\ \dfrac{\partial z}{\partial \xi} & \dfrac{\partial z}{\partial \eta} & \dfrac{\partial z}{\partial \zeta} \end{vmatrix} d\xi \; d\eta \; d\zeta \quad (A5.20)$$

This follows simply from the geometry. The bracketed product, by definition, forms a vector whose length is equal to the parallelogram area with sides tangent to two of the coordinates. The second scalar multiplication by a length and cosine of the angle between that length and the normal to the parallelogram establishes an elementary volume.

The above equations serve in changing the variables in surface and volume integrals.

Integration by parts in two or three dimensions (Green's theorem)

Consider the integration by parts of the following two-dimensional expression

$$\iint_\Omega \phi \, \frac{\partial \psi}{\partial x} \, \mathrm{d}x \, \mathrm{d}y \tag{A6.1}$$

Integrating first with respect to x and using the well-known relation for integration by parts

$$\int_{x_L}^{x_R} u \, \mathrm{d}v = - \int_{x_L}^{x_R} v \, \mathrm{d}u + (uv)_{x=x_R} - (uv)_{x=x_L} \tag{A6.2}$$

we have, using the symbols of Fig. A6.1,

$$\iint_\Omega \phi \, \frac{\partial \psi}{\partial x} \, \mathrm{d}x \, \mathrm{d}y = - \iint_\Omega \frac{\partial \phi}{\partial x} \, \psi \, \mathrm{d}x \, \mathrm{d}y$$
$$+ \int_{y=y_B}^{y=y_T} [(\phi\psi)_{x=x_R} - (\phi\psi)_{x=x_L}] \, \mathrm{d}y \tag{A6.3}$$

If now we consider a direct segment of the boundary $\mathrm{d}\Gamma$ on the right-hand boundary we note that

$$\mathrm{d}y = \mathrm{d}\Gamma n_x \tag{A6.4}$$

where n_x is the direction cosine between the normal and the x direction. Similarly, on the left-hand section we have

$$\mathrm{d}y = - \mathrm{d}\Gamma n_x \tag{A6.5}$$

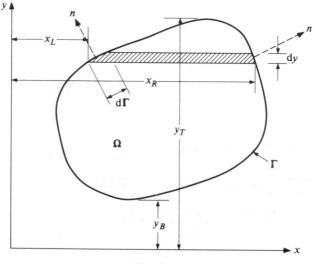

Fig. A6.1

The final term of Eq. (A6.2) can thus be expressed as the integral taken around an anticlockwise direction around a complete closed boundary:

$$\oint_\Gamma \phi \psi n_x \, d\Gamma \qquad (A6.6)$$

If several closed contours are encountered this integration has to be taken around each such contour. The general expression is in all cases

$$\iint_\Omega \phi \frac{\partial \psi}{\partial x} \, dx \, dy \equiv - \iint_\Omega \frac{\partial \phi}{\partial x} \psi \, dx \, dy + \oint_\Gamma \phi \psi n_x \, d\Gamma \qquad (A6.7)$$

Similarly, if differentiation in the y direction arises we can write

$$\iint_\Omega \phi \frac{\partial \psi}{\partial y} \, dx \, dy \equiv - \iint_\Omega \frac{\partial \phi}{\partial y} \psi \, dx \, dy + \oint_\Gamma \phi \psi n_y \, d\Gamma \qquad (A6.8)$$

where n_y is the cosine between the outward normal and the y axis.

In three dimensions by identical procedure we can write

$$\iiint_\Omega \phi \frac{\partial \psi}{\partial x} \, dx \, dy \, dz = - \iiint_\Omega \frac{\partial \phi}{\partial x} \psi \, dx \, dy \, dz + \oint_\Gamma \phi \psi n_x \, d\Gamma \qquad (A6.9)$$

where Γ becomes the element of the surface area and the last integral is taken over the whole surface.

7

Solutions exact at nodes

The finite element solution of ordinary differential equations may be made exact at the interelement nodes by a proper choice of the *weighting* function in the weak form. To be more specific, let us consider the set of ordinary differential equations given by

$$\mathbf{A(u) + f = 0} \tag{A7.1}$$

where **u** is the set of dependent variables which are functions of the single independent variable x and **f** is a vector of specified forcing functions. The weak form of the differential equation is given by

$$\int_L \mathbf{v}^T (\mathbf{A(u) + f}) \, dx = 0 \tag{A7.2}$$

The weak form may be transformed using integration by parts to remove all derivatives from **u** and place them on **v**. The result is given by

$$\int_L (\mathbf{A^*(v)}^T \mathbf{u} + \mathbf{v}^T \mathbf{f}) \, dx + \mathbf{B^*(v)}^T \mathbf{B(u)} |_\Gamma = 0 \tag{A7.3}$$

where $\mathbf{A^*(v)}$ is the *adjoint differential equation* and $\mathbf{B^*(v)}$ and $\mathbf{B(u)}$ are the terms on the boundary, Γ, resulting from integration by parts.

As an example problem, consider the single differential equation

$$\frac{d^2 u}{dx^2} + P \frac{du}{dx} + f = 0 \tag{A7.4}$$

with the associated weak form

$$\int_L v \left(\frac{d^2 u}{dx^2} + P \frac{du}{dx} + f \right) dx = 0 \tag{A7.5}$$

After integration by parts the weak form becomes

$$\int_L \left[\left(\frac{d^2 v}{dx^2} - P \frac{dv}{dx} \right) u + vf \right] dx + \left[v \left(\frac{du}{dx} + Pu \right) - \frac{dv}{dx} u \right]\Bigg|_\Gamma \quad (A7.6)$$

The adjoint differential equation is given by

$$a^*(v) = \frac{d^2 v}{dx^2} - P \frac{dv}{dx} \quad (A7.7)$$

and the boundary terms by

$$\mathbf{B}^*(\mathbf{v}) = \left\{ \begin{array}{c} v \\ -\dfrac{dv}{dx} \end{array} \right\} \quad (A7.8)$$

and

$$\mathbf{B}(\mathbf{u}) = \left\{ \begin{array}{c} \dfrac{du}{dx} + Pu \\ u \end{array} \right\} \quad (A7.9)$$

In the above example two cases are identified:

(a) P zero; in which case the adjoint differential equation is the same as the original equation. This is the *self-adjoint* case.
(b) P not zero; in which case the adjoint differential equation is different from the original equation. This is a *non-self-adjoint* problem.

The finite element solution for these two cases is often quite different. In the first case an equivalent variational theorem exists, whereas, in case (b) no such theorem exists.†

If we now define the domain of the weak form to be a single element L_e *and require the adjoint differential equation to be satisfied in each element* , the weak form simplifies considerably. Indeed, for elements with nodes at the ends only, the weak form will be expresssed in terms of *known* quantities in L_e and unknown values of the dependent variable and its derivatives at the interelement boundaries, Γ_e. If we consider that all values of the forcing function are contained in f, the terms in $\mathbf{B}(\mathbf{u})$ must be continuous for an exact solution between adjacent elements. In addition, the weighting function must satisfy continuity conditions

---

† An integrating factor often may be introduced to make the weak form generate a self-adjoint problem; however, the approximation problem will remain the same. Viz. p. 240.

imposed by the weak form (in the example problem v must be continuous across element boundaries). Addition over all elements e yields a simple requirement that

$$\sum_e \left(\int_{L_e} \mathbf{v}^T \mathbf{f} \, dx + \mathbf{B}^*(\mathbf{v})^T \hat{\mathbf{B}}(\mathbf{u}) \big|_{\Gamma_e} \right) + \mathbf{B}^*(\mathbf{v})^T \mathbf{Q} \big|_{\Gamma} = 0 \qquad (A7.10)$$

which may be used to generate the finite element arrays. After noting continuity and balance requirements of the appropriate quantities, the remaining terms on the boundaries are included in the above by defining

$$\hat{\mathbf{B}}(\mathbf{u}) = \begin{Bmatrix} 0 \\ u \end{Bmatrix} \qquad (A7.11)$$

The boundary of the entire domain is here denoted by Γ and the term $\mathbf{Q}$ is the set of specified *boundary conditions*. The result for the example problem is

$$\sum_e \left(\int_{L_e} vf \, dx - \frac{dv}{dx} u \big|_{\Gamma_e} \right) + vQ \big|_{\Gamma} = 0 \qquad (A7.12)$$

where Q is the flux and is equal to $\dfrac{du}{dx} + Pu$ on the boundary.

In case (a) the solution to the adjoint equation is given by

$$v = Ax + B \qquad (A7.12)$$

which may be written as conventional linear shape functions in each element. Thus, for linear shape functions in each element used as the weighting function the interelement nodal displacements for u will be exact (e.g., see Fig. 9.4) irrespective of the interpolation used for u.

For case (b) the exact solution is

$$v = Ae^{Px} + B \qquad (A7.13)$$

and again u is exact at the interelement nodes.

In both cases, the constants of integration A and B may be expressed in terms of the *nodal parameters* at the left and right ends of each element, v_1 and v_2, respectively. After substitution of the shape functions into the weak form, the final expression will involve values of the dependent variable $u(x)$ at the nodes only. Thus, *without any approximation of the dependent variable*, exact solutions may be computed at each node. If interpolation is used, the result will give exact values at each interelement node, and approximation between nodes. After constructing exact nodal solutions for u, exact solutions for the derivatives may be computed from the weak form for each element. The above process was first given by Tong for self-adjoint differential equations from a variational formulation.[1]

Reference

1. P. TONG, 'Exact solutions of certain problems by the finite element method,' *AIAA J.*, **7**, 179–80, 1969.

8

Matrix diagonalization or lumping

Some of the algorithms discussed in this volume become more efficient if one of the global matrices can be diagonalized (also called 'lumped' by many engineers). For example, the solutions of mixed problems, stress smoothing, or transient problems (to be discussed in Vol. 2) are more efficient if a gobal matrix to be inverted (or equations solved) is diagonal [viz. Chapters 12 and 14 Eqs (12.67) and (14.17)]. Engineers have persisted with purely physical concepts of lumping; however, there is clearly a need for devising a systematic and mathematically acceptable procedure for such lumping exercises.

We shall define the matrix to be considered as

$$\mathbf{A} = \int_{\Omega} \mathbf{N}^{\mathrm{T}} \mathbf{c} \mathbf{N} \, d\Omega \qquad (A8.1)$$

where $\mathbf{c}$ is a matrix with small dimension. Often $\mathbf{c}$ is a diagonal matrix (e.g., in mass or least square problems $\mathbf{c}$ is an identity matrix times some scalar). When $\mathbf{A}$ is computed exactly it has full rank and is not diagonal—this is called the 'consistent' form of $\mathbf{A}$ as it is computed consistently with the other terms in the finite element model. The diagonalized form is defined with respect to 'nodes' of the shape functions, N_i; hence, the matrix will have small diagonal blocks, each with a dimension of $\mathbf{c}$. Only when $\mathbf{c}$ is diagonal can the matrix $\mathbf{A}$ be completely diagonalized. Four basic lines of argument may be followed in constructing a diagonal form.

The first procedure is to use different shape functions to approximate each term in the finite element discretization. For the $\mathbf{A}$ matrix we use shape functions $\bar{N}_i$ for the lumping process. No derivatives exist on the shape functions defining $\mathbf{A}$. Accordingly, for this term the shape functions may be piecewise continuous and still lead to acceptable approx-

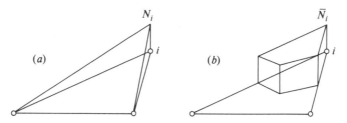

Fig. A8.1   Linear (a) and piecewise constant (b) shape functions for a triangle

imations. If the shape functions used to define **A** are piecewise constants, such that $N_i = 1$ in a certain part of the element surrounding the node i and zero elsewhere, and if such parts are not overlapping, then clearly the matrix of Eq. (A8.1) becomes diagonal as

$$\int_\Omega \bar{\mathbf{N}}_i^T \mathbf{c} \bar{\mathbf{N}}_j \, d\Omega = 0 \qquad i \neq j \qquad (A8.2)$$

Such an approximation with different shape functions is permissible since the usual finite element criteria of integrability and completeness are satisfied. The functions selected need only satisfy the condition

$$\sum_i \bar{N}_i = 1 \qquad \bar{\mathbf{N}}_i = \bar{N}_i \mathbf{I} \qquad (A8.3)$$

In Fig. A8.1 we show the functions N_i and $\bar{N}_i$ for a triangular element.

The second method to diagonalize a matrix is to note that condition (A8.3) is simply a requirement that ensures preservation of the quantity **c** over the element. For structural dynamics applications this is the conservation of mass at the element level. Accordingly, it has been noted that any lumping that preserves the integral of **c** on the element will lead to convergent results, although the rate of convergence may be lower than use of a consistent **A**. Many alternatives have been proposed based upon this method. The earliest procedures performed the diagonalization using physical intuition only. Later alternative algorithms were proposed. One suggestion, often called a 'row sum' method, is to compute the diagonal matrix from

$$\mathbf{A}_{ij} = \begin{cases} \sum_k \int_\Omega \mathbf{N}_i^T \mathbf{c} \mathbf{N}_k \, d\Omega & i = j \\ \mathbf{0} & i \neq j \end{cases} \qquad (A8.4)$$

This algorithm is analogous to replacing the $\mathbf{N}_i$ by the condition (A8.3), which in effect makes the diagonal the sum of all terms in the row. Such an algorithm makes sense only when the degrees of freedom of the problem all have the same physical interpretation. An alternative is to scale the diagonals of the consistent matrix to satisfy the conservation requirement. In this case the diagonal matrix is deduced from

$$
\mathbf{A}_{ij} = \begin{cases} a \int_\Omega \mathbf{N}_i^T \mathbf{c} \mathbf{N}_i \, d\Omega & i = j \\ 0 & i \neq j \end{cases} \tag{A8.5}
$$

where a is selected so that

$$
\sum_i \mathbf{A}_{ii} = \int_\Omega \mathbf{c} \, d\Omega \tag{A8.6}
$$

The third procedure uses numerical integration to obtain diagonalization without apparently introducing additional shape functions. If numerical integration is used to evaluate the $\mathbf{A}$ matrix of Eq. (A8.1) we may write a typical term as a summation (following Chapter 8):

$$
\mathbf{A}_{ij} = \int \mathbf{N}_i^T \mathbf{c} \mathbf{N}_j \, d\Omega = \sum_q W_q (\mathbf{N}_i^T \mathbf{c} \mathbf{N}_j)_q J_q \tag{A8.7}
$$

where q refers to the sampling point at which the integrand is evaluated, J is the jacobian volume transformation, and W_q gives the appropriate quadrature weighting.

If the sampling points for the numerical integration are located at nodes then, as all except one shape function N_i are zero at any node,

$$
\mathbf{A}_{ij} - \mathbf{0} \qquad i \neq j \tag{A8.8}
$$

and the matrix becomes diagonal.

Appropriate weighting values may be deduced by requiring the quadrature to exactly integrate particular polynomials in the natural coordinate system. In general the quadrature should integrate a complete polynomial of the order of the interpolation polynomial. Thus for four-noded quadrilateral elements, linear functions should be exactly integrated. For eight- or nine-noded elements, quadratic functions should be evaluated. Integrating additional terms may lead to diagonal matrices which are more accurate in their performance.

For low order elements, symmetry arguments may be used to lump the matrix. It is, for instance, obvious that in a simple triangle little improvement can be obtained by any other lumping than the simple one in which the total $\mathbf{c}$ is distributed in three equal parts. For an eight-noded two-dimensional isoparametric element no such obvious procedure is available. In Fig. A8.2 we show the case of rectangular elements of four-, eight-, and nine-noded type and lumping by (A8.4), (A8.5), and (A8.7). Is noted that for the eight-noded element some of the lumped quantities are negative when (A8.4) is used. This may have some

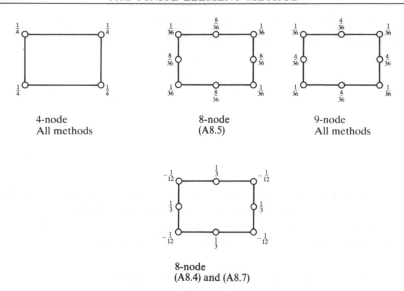

4-node
All methods

8-node
(A8.5)

9-node
All methods

8-node
(A8.4) and (A8.7)

Fig. A8.2   Diagonalization of rectangular elements by various methods

adverse effects in certain algorithms (e.g., time-marching schemes to integrate transient problems) and preclude their use. In Fig. A8.3 we show some lumped matrices for triangular elements computed by quadrature. It should be noted that the cubic element again has negative terms while the quadratic element has zero terms. The zero terms are more difficult to handle as the resulting diagonal **A** matrix no longer has full rank and thus may not be inverted.

Another aspect of lumping is the performance as the element is distorted from its parent shape. For example, as a rectangular element is distorted and approaches a triangular shape it is desirable to have the limit cases behave appropriately. In the case of the four-node rectangular element the lumped matrix for all the lumping procedures gives the same answer. However if the element is distorted by a transformation defined

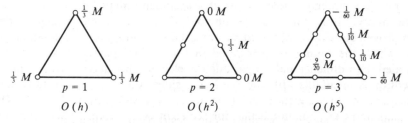

Fig. A8.3   Lumping by numerical integration for triangles

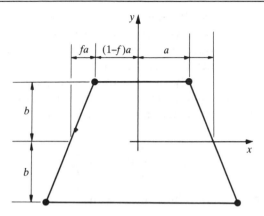

Fig. A8.4 Distorted four-noded element

by one parameter as shown in Fig. A8.4 then the three lumping procedures discussed give different answers. The jacobian transformation is given by

$$J = ab(1 - f) \tag{A8.9}$$

and c is the identity matrix. The form (A8.4) gives

$$\mathbf{A}_{ii} = \begin{cases} ab(1 - f/3) & \text{at top} \\ ab(1 + f/3) & \text{at bottom} \end{cases} \tag{A8.10}$$

the form (A8.5) gives

$$\mathbf{A}_{ii} = \begin{cases} ab(1 - f/2) & \text{at top} \\ ab(1 + f/2) & \text{at bottom} \end{cases} \tag{A8.11}$$

and the quadrature form (A8.7) yields

$$\mathbf{A}_{ii} = \begin{cases} ab(1 - f) & \text{at top} \\ ab(1 + f) & \text{at bottom} \end{cases} \tag{A8.12}$$

The four-noded element has the property that a triangle may be defined by coalescing two nodes and assigning them the same number in the mesh. Thus the quadrilateral becomes a triangle when the parameter f is one. The limit value for the lumped method (A8.4) will give equal lumped terms at the three nodes [method (A8.5) yields a lumped value for the coalesced node which is two-thirds the other nodes and (A8.7) yields zero at this node for the limit case]. The methods (A8.5) and (A8.7) give limit cases which depend on how the nodes are numbered on each triangular element. This lack of invariance is not desirable in applications software; hence for the four-noded quadrilateral, method (A8.4) appears to be superior to (A8.5) or (A8.7). However, we have observed above that use of (A8.4) leads to negative diagonal elements for the eight-noded element; hence it is evident that there is no universal method for diagonalizing the matrix which applies to all elements.

A fourth but not widely used method is available which may be explored for deducing a consistent matrix that is diagonal. This consists of making a mixed representation for the term leading to $\mathbf{A}$. If this term is

$$\mathbf{A}\bar{\mathbf{u}} = \left(\int_{\Omega} \mathbf{N}^{\mathrm{T}} c \mathbf{N} \, d\Omega \right) \bar{\mathbf{u}} \tag{A8.13}$$

then a mixed problem may be deduced with

$$\mathbf{p} = \mathbf{u} \tag{A8.14}$$

with approximations

$$\mathbf{u} = \mathbf{N}\bar{\mathbf{u}} \tag{A8.15}$$

and

$$\mathbf{p} = \mathbf{n}\bar{\mathbf{p}} \tag{A8.16}$$

The weak form of the above is given by

$$\bar{\mathbf{A}}\bar{\mathbf{p}} = \left(\int_{\Omega} \mathbf{N}^{\mathrm{T}} c \mathbf{n} \, d\Omega \right) \bar{\mathbf{p}} = \mathbf{A}\bar{\mathbf{u}} \tag{A8.17}$$

and

$$\mathbf{H}\bar{\mathbf{p}} = \bar{\mathbf{A}}^{\mathrm{T}}\bar{\mathbf{u}} \tag{A8.18}$$

where

$$\mathbf{H} = \int_{\Omega} \mathbf{n}^{\mathrm{T}} c \mathbf{n} \, d\Omega \tag{A8.19}$$

Elimination of $\bar{\mathbf{p}}$ gives the mixed form matrix

$$\mathbf{A} = \bar{\mathbf{A}} \mathbf{H}^{-1} \bar{\mathbf{A}}^{\mathrm{T}} \tag{A8.20}$$

for which diagonal forms may now be sought.

Author index

Numbers in bold type refer to the list of references at the end of each chapter

611

Subject Index

Words in capitals, such as BOUN, refer to the finite element macro programming language, Chapter 15